Concepts of Proof in Mathematics, Philosophy, and Computer Science

Ontos Mathematical Logic

Edited by
Wolfram Pohlers, Thomas Scanlon,
Ernest Schimmerling, Ralf Schindler,
Helmut Schwichtenberg

Volume 6

Concepts of Proof in Mathematics, Philosophy, and Computer Science

Edited by
Dieter Probst and Peter Schuster

DE GRUYTER

ISBN 978-1-5015-1080-9
e-ISBN (PDF) 978-1-5015-0262-0
e-ISBN (EPUB) 978-1-5015-0264-4
ISSN 2198-2341

Library of Congress Cataloging-in-Publication Data
A CIP catalog record for this book has been applied for at the Library of Congress.

Bibliographic information published by the Deutsche Nationalbibliothek
The Deutsche Nationalbibliothek lists this publication in the Deutsche Nationalbibliografie; detailed bibliographic data are available on the Internet at http://dnb.dnb.de.

Printing: CPI books GmbH, Leck
♾ Printed on acid-free paper
Printed in Germany

MIX
Papier aus verantwortungsvollen Quellen
FSC® C083411
www.fsc.org

www.degruyter.com

Preface

The Humboldt-Kolleg "Proof" held in Bern, Switzerland, from 9th to 13th September 2013 was a try to give due consideration to the depth and breadth of the time-honoured concept of proof, by gathering leading scholars from mathematics, informatics and philosophy whose studies are centered at the concept of proof. The present volume is an attempt to represent many of their approaches in print, which can briefly be summarized as follows.

In "Herbrand Confluence for First-Order Proofs with Π_2-Cuts" Bahareh Afshari, Stefan Hetzl and Graham E. Leigh assign to each first-order proof with cuts of complexity at most Π_2 a well-behaved formal grammar of limited size; all normal forms obtained by non-erasing cut reductions result in the same Herbrand expansion.

"Proof-Oriented Categorical Semantics" is Marco Benini's alternative interpretation of the entity of which the first-order proposition-as-types correspondence and the associated type systems are equivalent but different presentations in distinct categories; his framework embraces standard first-order categorical semantics.

In "Logic for Gray-Code Computation" Ulrich Berger, Kenji Miyamoto, Helmut Schwichtenberg and Hideki Tsuiki study real number computation with Gray code from a constructive angle, to extract algorithms from proofs with inductive and coinductive definitions; their case studies are formalized in the proof assistant Minlog.

Douglas S. Bridges in "The Continuum Hypothesis Implies Excluded Middle" shows within the framework of Bishop-style constructive mathematics that the Continuum Hypothesis resembles the Axiom of Choice also inasmuch as it implies the Law of Excluded Middle, and discusses a more explicit Brouwerian counterexample.

"Theories of Proof-Theoretic Strength $\psi(\Gamma_{\Omega+1})$" by Ulrik Buchholtz, Gerhard Jäger and Thomas Strahm is about a range of theories with proof-theoretic ordinal $\psi(\Gamma_{\Omega+1})$; it is not only that this ordinal parallels the one of predicative analysis, Γ_0, but also that some of those theories are parallel to classical theories of strength Γ_0.

The highlight of Thierry Coquand and Henri Lombardi's "Some Remarks about Normal Rings" is a constructive proof that if a commutative ring R is normal, then so is the polynomial ring $R[X]$; this is based on a special technique to replace the use of minimal primes by explicit localizations in a suitable tree.

In "On Sets of Premises" Kosta Došen sheds light from categorial proof theory on the phenomenon that when collecting premises into (multi)sets rather than sequences one might have to face the unwanted consequence that any two deductions with the same premises and the same conclusions would be identified.

Hajime Ishihara and Takako Nemoto in "Non-Deterministic Inductive Definitions and Fullness" prove that a special form of non-deterministic inductive definition is tantamount to the principle of fullness characteristic of constructive Zermelo–Fraenkel set theory as a constructive version of the power set axiom.

In "Cyclic Proofs for Linear Temporal Logic" Ioannis Kokkinis and Thomas Studer establish weakening for annotated sequents by purely syntactical methods, and thus solve an open problem that Brünnler and Lange have brought up in the proof theory of temporal logic.

The main achievement of Roman Kuznets's paper "Craig Interpolation via Hypersequents" is a novel constructive method of proving the Craig interpolation property based on cut-free hypersequent calculi; he tests the method by verifying that property for the modal logic S5.

With "A General View on Normal Form Theorems for Łukasiewicz logic with product" Serafina Lapenta and Ioana Leuştean explore the connection between the Pierce–Birkhoff conjecture and Łukasiewicz logic with product, the models of which reflect an algebraic hierarchy of lattice-ordered structures, from groups to algebras.

In Maria Emilia Maietti and Giuseppe Rosolini's "Relating Quotient Completions via Categorical Logic" the authors show that the elementary quotient completion of an elementary existential doctrine coincides with an exact completion when a choice rule holds in the starting existential elementary doctrine.

Roman Murawski's "Some Historical, Philosophical and Methodological Remarks on Proof in Mathematics" is about task and meaning of proof in mathematics, including discussions of the role of informal proofs in mathematical research practice, of the concept of formal proof, and of the distinction between provability and truth.

In "Cut Elimination in Sequent Calculi with Implicit Contraction, with a conjecture on the origin of Gentzen's altitude line construction" Sara Negri and Jan von Plato settle the issue that standard cut elimination fails if the principal formula of a rule occurs in a premiss, and adapt this to sequent calculi with multisets and contraction.

While one objective of Hilbert's Programme, finitistic consistency proofs, has been dashed by Gödel's second incompleteness theorem, in "Hilbert's Programme and Ordinal Analysis" Wolfram Pohlers argues that ordinal analysis not only helps to the other objective—the elimination of ideal elements—but is actually based on it.

In "Aristotle's Deductive Logic: a Proof-Theoretical Study" Jan von Plato shows that derivations based on Aristotle's rules of proof can so be transformed that the method of indirect proof is invoked at most once as a last step, which is the only way in which—as the author claims—Aristotle used indirect proof.

Michael Rathjen's "Remarks on Barr's Theorem: Proofs in Geometric Theories" gives a constructive proof of Gentzen's Hauptsatz for $\mathcal{L}_{\infty\omega}$–logic which entails in a simple way the so-called Barr theorem that in geometric $\mathcal{L}_{\infty\omega}$–logic classical and intuitionistic provability coincide; also the Axiom of Choice is put into context.

Acknowledgment: First and foremost the editors of this volume would like to express their gratitude to the Alexander-von-Humboldt Foundation for offering the patronage of the Humboldt-Kolleg "Proof" and for generously giving financial support. Open-handed financial assistance further came from the following institutions: Burgergemeinde Bern, Deutsche Vereinigung für Mathematische Logik und für Grundlagenforschung der Exakten Wissenschaften, the Logic and Theory Group at the Institute of Computer Science of the University of Bern, Swiss National Science Foundation, Swiss Academy of Science, and Swiss Society for Logic and Philosophy of Science.

When co-editing this volume the second editor received individual funding by the Alexander-von-Humboldt Foundation, in the form of a Further Research Fellowship which he spent at the Munich Center for Mathematical Philosophy upon kind invitation by Hannes Leitgeb; and by the John Templeton Foundation within the project "Abstract Mathematics for Actual Computation: Hilbert's Program in the 21st Century".

Maik Bierwirth, Project Editor at De Gruyter, provided considerate guidance throughout the present book project. Special thanks are own to Olena Gainulina, Production Editor at De Gruyter, for her advice and patience during the phase of typesetting the final version of this volume; and to Peppo Brambilla, system administrator of the Institut für Informatik, Universität Bern, for his technical support. Laura Crosilla and Daniel Wessel have given helpful advice about certain aspects of the introduction.

Last but not least, the editors are indebted to the authors and to the anonymous peer reviewers, without whose respective contributions this volume would hardly have emerged.

Bern D.P.
Verona P.S.

June 2016

Contents

Dieter Probst and Peter Schuster

Introduction

In a first attempt to explain the concept of proof in mathematics, one could simply say that a proof is a way to convince someone of the truth of a claim. Here "someone" may mean oneself, a colleague, the audience of a seminar or conference talk, the students present at a lecture, or even the entire community of mathematicians.

In parallel to the different context of law one could also say that the truth of a claim consists in that the claim corresponds to the actual facts—which, however, is anything but uncontroversial from certain epistemic perspectives. But to tell what is true is not unproblematic even in law, where the objects of discourse largely belong to the real world and are fairly concrete: just think of the only witness in a case being lying throughout, or of scientific experts called by the court who fail to come to a common assessment.

Mathematics, on the other hand, is to a large extent about an ideal world populated by more-or-less abstract entities such as numbers, points, lines, and more recently by even more puzzling objects such as sets, functions, categories etc. The task of definining truth in mathematics thus is of an entirely different character.

In fact, numerous serious foundational and philosophical issues are about and around. These include the time-honoured problem whether universals exist, or just particulars; the question whether mathematical knowledge is possible at all, e.g. as Kant's synthetic judgements a priori; and the various issues that have arisen from and since what became known as *Grundlagenkrise der Mathematik* (foundational crisis of mathematics).

A way to avoid many of these problems, and still have a reasonable concept of proof, is to move from absolute truth to relative certainty. This also is the working attitude of most mathematicians no matter whether on Sundays they believe that absolute truth is possible.

In a nutshell, relative certainty means that a proof is a successful demonstration that a mathematical theorem necessarily follows by logical reasoning from axioms which are considered evident for the given context and/or agreed upon by the community. It is this very concept of proof that sets mathematics apart from other fields, distinguishing it as the prototype of a deductive discipline especially vis-à-vis the inductive sciences.

Dieter Probst: Institut für Informatik, Universität Bern, Neubrückstrasse 10, 3012 Bern, Switzerland, e-mail: probst@inf.unibe.ch
Peter Schuster: Dipartimento di Informatica , Università di Verona, Strada Le Grazie 15, 37134 Verona, Italy, e-mail: peter.schuster@univr.it

Euclid allegedly was the first to employ the method of proof rigorously, to link definitions and theorems the way that is now standard mathematical practice. However, the idea of a proof was arguably known to the Babylonians, and certainly to the Pythagoreans as they are remembered for establishing the necessity of proofs in mathematics. Yet Euclid's geometry is generally considered to be the paradigm of mathematical certainty by rigorous proof, which has been aspired ever since even for the whole of mathematics.

The quest for certainty became more urgent with the introduction of abstract terms into mathematics, especially from the second half of the 19th century. Real pressure emerged with the paradoxes resulting from this development, the prime example being Russell's discovery that naive set theory is inconsistent. This was particularly relevant to Hilbert, by many considered the last mathematician to be conversant with the entire subject, who indeed held the formalist view that especially with the use of abstract terms mathematics is no more, no less, than a symbolic language in the spirit of Leibniz's *characteristica universalis* intended to serve as an alphabet of human thought.

Slightly later Brouwer advocated—to some extent paralleled by Weyl and preceded by Kronecker—the so-called intuitionist view of mathematics as a constructive mental activity. For proof praxis this means that logical reasoning and set formation cannot be used as easily with infinite or abstract entities as it can with most finite or concrete objects. It took about half a century until Bishop with his constructive analysis proved that doing mathematics with intuitionistic logic need not face the limitations which Hilbert saw in Brouwer's proposal.

Encouraged by the earlier successful axiomatization of geometry, already in 1900 Hilbert asked—with the second item of his famous list of 23 open problems—for a proof that the axioms of arithmetic would never lead to a contradiction: the axioms by which, following Dedekind, he had characterized the real numbers. He made explicit that, apart from entailing the consistency (i.e., non-contradictoriness) of geometry, any consistency proof for arithmetic would constitute a proof of mathematical existence of the real numbers or the continuum.

By a consistency proof for all mathematics one could also do away, at once and in principle, with the constraints perceivedly imposed by Kronecker and Brouwer in particular. Hilbert in fact aimed at a consistent axiomatization of the customary unlimited way of doing mathematics, an axiomatization capable of deriving all mathematical truth including of course its own consistency. Hilbert requested, moreover, that the proof of consistency be done by finite methods only, to justify once and for all the use of ideal objects and transfinite methods.

By Gödel's celebrated incompleteness theorems all hope was dashed to complete Hilbert's programme as a whole. Its initiation, however, was the dawn of a novel direction of mathematics, soon under the identifiers *Metamathematik* and

Beweistheorie (proof theory), which features mathematical reasoning and the concept of proof as main objects of study.

Early but essential achievements were made by Gentzen. With natural deduction and sequent calculus he made precise formal notions of proof flexible enough to accommodate and relate classical and intuitionistic reasoning as propagandized by Hilbert and Brouwer, respectively. Gentzen showed that consistency of arithmetic can be proved at least from outside the system, and his *Hauptsatz* (main theorem) is without doubt the most versatile tool of proof theory not only to establish consistency for wider areas of mathematics.

Gentzen's main tool is what is known as cut elimination: a way of getting rid of unnecessary detours in order to convert formal proofs into normal form as is needed for in-depth proof-theoretical studies. As a late impact hardly to be expected in Gentzen's time, his achievements have proved useful in theoretical computer science: for the development of provably correct software, e.g. by extracting programs directly from formal proofs.

So concerns that initially were mainly foundational—if not partly ideological—ignited a thorough exploration of the concept of formal proof, and prompted adjacent disciplines to discuss and apply this concept from their own perspectives. Despite all this, the prima facie rather philosophical question of what should be considered as a proof has become more pressing with the rapid progress of mathematics in combination with the racily increasing potential of computing technology—and perhaps also just with the quest for formal proofs.

The concept of proof shared by the community is wider than the notion of formal proof. For instance, the famous Four-Colour Theorem is widely accepted although it was proved using a computer to carry out an exhaustive case checking unfeasible for a human agent. The working mathematician anyway takes a pragmatic stance, rarely provides a formalized proof and usually does not even care much which formal system would allow to formalize it.

It needs to be said that the task of formalization is anything but easy, for which Wiles's distinguished but highly complex proof of Fermat's Last Theorem is a good example. These days only relatively few mathematicians are able to comprehend this proof as a whole, and there is demand for a more down-to-earth proof, e.g. in Peano Arithmetic, which would be eligible for formalisation or even for computer-based mechanical verification.

The idea of proof which the formal mathematical counterpart tries to model has evolved with time; it has transgressed the boundaries of logic and permeated most parts of mathematics and even computer science. Proof has become a truly interdisciplinary topic.

Bahareh Afshari, Stefan Hetzl, and Graham E. Leigh

Herbrand Confluence for First-Order Proofs with Π_2-Cuts

Abstract: To each first-order proof π with cuts of at most complexity Π_2, we assign a formal grammar in which the number of production rules is bounded by the size of π and the set of generated terms is finite and forms a Herbrand expansion for the end-sequent of π. Using these grammars we prove that all (possibly infinitely many) normal forms of π obtained by non-erasing cut reductions have the same Herbrand expansion.

Keywords: Cut elimination, First-order logic, Herbrand's theorem, Formal language theory

Mathematics Subject Classification 2010: 03F05, 68Q42

1 Introduction

In classical first-order logic a proof can be considered as being composed of two layers: on the one hand the terms by which quantifiers are instantiated, and on the other hand, the propositional structure. This separation is most clearly illustrated by Herbrand's theorem [Herbrand, 1930, Buss, 1995]: a formula is valid if and only if there is a finite expansion (of existential quantifiers to disjunctions and universal quantifiers to conjunctions of instances) which is a propositional tautology. Such Herbrand expansions can be transformed to and obtained from cut-free sequent calculus proofs in a quite straightforward way.

Standard cut reduction is, however, not confluent, i.e. it permits the computation of many essentially different cut-free proofs. It was shown in [Baaz and Hetzl, 2011] (for pure first-order logic) and in [Hetzl, 2012b] (for arithmetical theories) that the number of different Herbrand expansions obtainable from a single proof with cut grows at least as fast as the size of the cut-free proofs. Still, it is not clear whether these results can be strengthened to obtain even more normal forms. In particular, it is an open question whether in general cut-elimination can produce infinitely many different Herbrand expansions.

In [Hetzl and Straßburger, 2012, 2013] an upper bound for the obtainable normal forms has been provided for proofs with Π_1-cuts in the following strong sense:

Bahareh Afshari, Stefan Hetzl, and Graham E. Leigh: Institute of Discrete Mathematics and Geometry, Vienna University of Technology, Vienna, Austria,
email: bahareh.afshari@tuwien.ac.at, stefan.hetzl@tuwien.ac.at, graham.leigh@tuwien.ac.at

a proof π with Π_1-cuts induces a finite set $[\![\pi]\!]$ such that every cut-free proof π' obtained from π via standard cut elimination has a Herbrand expansion, $\mathcal{H}(\pi')$, which is contained in $[\![\pi]\!]$. Moreover, if π' is obtained from π by *non-erasing* reductions (reductions that do not eliminate sub-proofs) then we even have $\mathcal{H}(\pi') = [\![\pi]\!]$. Consequently all normal forms of the non-erasing reduction (of which there infinitely many) have the same Herbrand expansion. This property of classical logic has been called *Herbrand-confluence* in [Hetzl and Straßburger, 2012, 2013] and provides a general way of defining the computational content of a classical proof.

The present paper extends Herbrand-confluence to proofs with Π_2-cuts. To a simple Π_2-proof[1] π we associate a recursively defined tree grammar $\mathcal{G}_\pi$ whose set of production rules is bounded by the size of π and generates a finite language $\mathcal{L}(\mathcal{G}_\pi)$ satisfying the following confluence result.

Theorem 1.1. *Let $\pi_0, \pi_1, \ldots, \pi_k$ be a sequence of simple Π_2-proofs such that π_{i+1} is obtained from π_i by a standard reduction rule (see Figures 1 and 2) other than weakening reduction. Then $\mathcal{L}(\mathcal{G}_{\pi_0}) = Ł(\mathcal{G}_{\pi_k})$. In particular, if the proof π_k contains only quantifier-free cuts $\mathcal{L}(\mathcal{G}_{\pi_0}) = \mathcal{H}(\pi_k)$.*

Theorem 1.1 can be seen as a refinement of [Afshari et al., 2015, Theorem 2]. Therein each simple Π_2-proof π is associated an acyclic context-free tree grammar $\mathcal{F}_\pi$ such that for $\pi_0, \pi_1, \ldots, \pi_k$ being a reduction sequence (possibly allowing reduction of weakening), $\mathcal{L}(\mathcal{F}_{\pi_0}) \supseteq Ł(\mathcal{F}_{\pi_k})$. For simple proofs, the grammars defined here and in [Afshari et al., 2015] can be shown to have the same language. There are, however, a number of technical differences between the two grammars motivated by the combinatorial nature of proving Herbrand confluence. Most notably, $\mathcal{G}_\pi$ may be cyclic (but permit only 'well-founded' derivations).

The grammar $\mathcal{G}_\pi$ can be considered as a directed graph whose nodes are quantifier occurrences and whose edges describe the information flow between them. In this sense it is also similar to the graphical formalisms of [Heijltjes, 2010, McKinley, 2013]. Other related structures are proof nets, which capture information flow on the propositional level and have been extensively studied starting with [Girard, 1987], as well as the logical flow graphs used by Buss [Buss, 1991] in the solution of the k-provability problem and further investigated by Carbone (see e.g. [Carbone and Semmes, 2000]).

1 See Definition 2.1.

2 First-order logic

We work with a Tait-style sequent calculus for first-order logic with explicit weakening and contraction rules. Terms and formulæ of first-order logic are defined as usual using the connectives $\wedge$, $\vee$ and quantifiers $\forall$, $\exists$, as well as a selection of predicate and function symbols. We assume two sets of variable symbols, *free* variables, denoted α, β, etc., and *bound* variables, x, y, z, with a formula only able to contain the latter sort in bound contexts.

Upper-case Roman letters, A, B, etc. denote formulæ and upper-case Greek letters Γ, Δ, etc. will range over *sequents*, finite unordered collections of formulæ with possible repetition. We abbreviate by Γ, Δ the disjoint union of Γ and Δ; and Γ, A is shorthand for $\Gamma, \{A\}$. We write $\bar{A}$ to denote the dual of the formula A obtained by de Morgan laws, and $A[x/t]$ for the formula obtained from A by replacing x with the term t if this will not induce any variable capture, and A otherwise.

Table 1. Axioms and rules of sequent-calculus

$A, \bar{A}$ (for A an atomic formula)

$$\frac{\Gamma, A, B}{\Gamma, A \vee B}\vee \qquad \frac{\Gamma, A \quad \Delta, B}{\Gamma, \Delta, A \wedge B}\wedge \qquad \frac{\Gamma, A[x/\alpha]}{\Gamma, \forall x A}\forall \qquad \frac{\Gamma, A[y/s]}{\Gamma, \exists y A}\exists$$

$$\frac{\Gamma}{\Gamma, \Delta}\mathrm{w} \qquad \frac{\Gamma, \Delta, \Delta^*}{\Gamma, \Delta}\mathrm{c} \qquad \frac{\Gamma, A \quad \Delta, \bar{A}}{\Gamma, \Delta}\mathrm{cut}$$

A *proof* is a finite binary tree of sequents obtained from the axioms and rules laid out in Table 1. In the ($\forall$) rule, α is called the *eigenvariable* and must not appear in $\Gamma, \forall x A$. In the ($\exists$) rule the term s is assumed to be free for y. In the contraction rule (c), Δ^* denotes a distinct copy of Δ. In each inference rule, those formulæ which are explicitly mentioned in the premise are said to be *principal* in the rules applied, for example A and B are principal in ($\wedge$) rule, every formula from Δ^* is principal in (c), and there are no principal formulæ in the weakening rule (w).

We assume all proofs are *regular*, namely all quantifiers' eigenvariables are distinct and different from any free variables. $\mathrm{EV}(\pi)$ denotes the set of eigenvariables in π and $\pi[\alpha/t]$ is the result of replacing throughout the proof π each occurrence of the variable symbol α by the term t. We write $\pi \vdash \Gamma$ to express that π is a proof with Γ being the sequent appearing at the root of π. A *position* in a proof π is a finite binary sequence pointing to a node in the proof-tree π. $\mathrm{Pos}(\pi)$ denotes the set of all positions in π. For $p \in \mathrm{Pos}(\pi)$, $\pi|p$ denotes the subproof of π at position

p with the convention that $\pi|\langle\rangle = \pi$ and $\pi|ip = \pi'|p$, where π' is the immediate left (or only) subproof of π if $i = 0$ and the immediate right subproof otherwise. The *size* of π, denoted $|\pi|$, is the total number of inference rules and axioms in π.

In this paper we primarily consider the following class of first-order proofs.

Definition 2.1 (Simple formulæ and simple Π_2-proofs). *We call a formula* simple *if it is a prenex Π_2 or prenex Σ_2 formula with at most one universal and one existential quantifier. A* simple Π_2-proof *is a proof in which each sequent is a finite multiset of simple formulæ and every universally quantified formula appearing above a cut is principal in the inference directly after its introduction (which is either a cut or existential introduction).*

Lemma 2.2. *If $\pi \vdash \Gamma$ in which all cut formulæ in π and all formulæ in Γ are simple then there is a simple Π_2-proof $\pi' \vdash \Gamma$ such that $|\pi'| \leq |\pi|$.*

Proof: Apply inversion to all principal occurrences of universally quantified formulæ in π that appear above some cut to 'shift' the quantifier introduction rule $(\forall)$ downwards in the proof resulting in a simple Π_2-proof $\pi' \vdash \Gamma$. This operation will not introduce any new inference rules to π so $|\pi'| \leq |\pi|$. □

2.1 Cut reduction

The standard cut reduction rules are given in Figures 1 and 2.

Axiom:

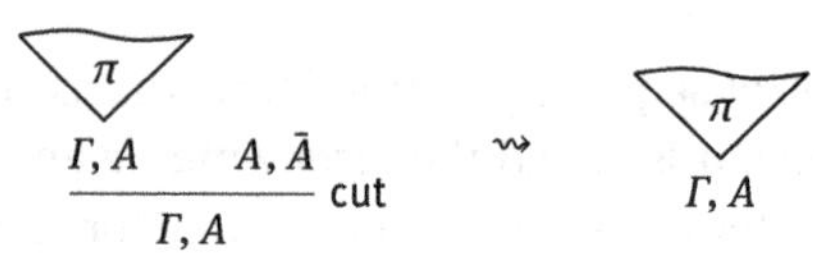

Boolean:

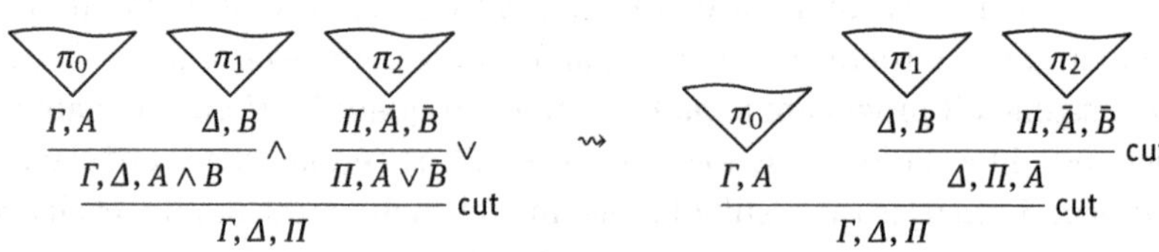

Fig. 1. One-step cut reduction and permutation rules I.

Quantifier:

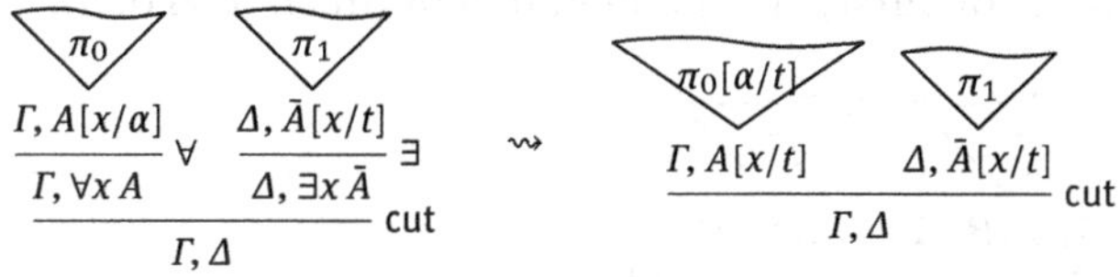

Weakening:

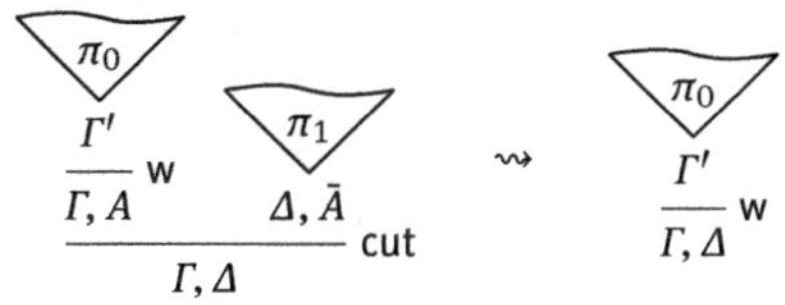

Contraction:

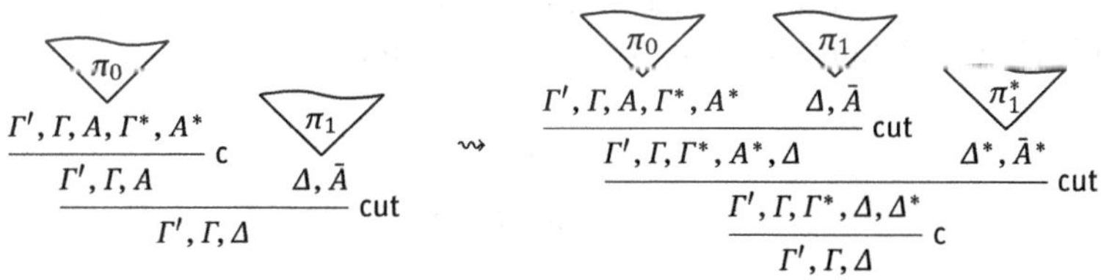

Unary inf.:

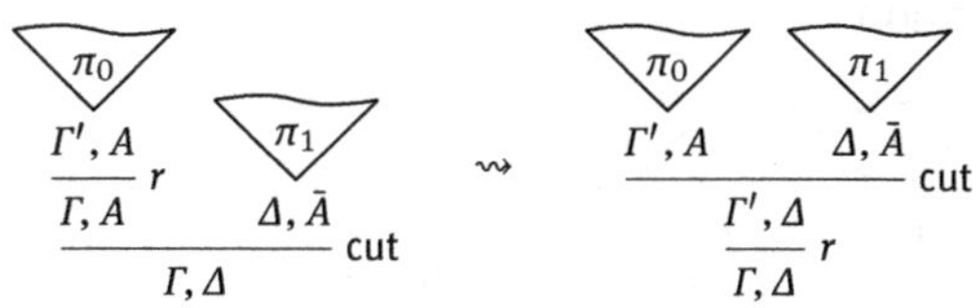

Binary inf.:

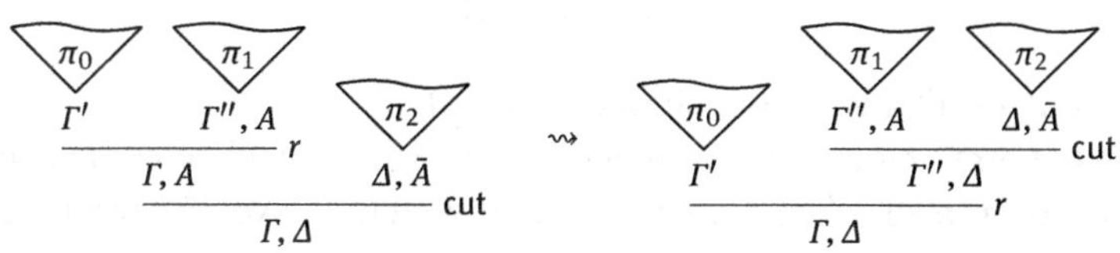

Fig. 2. One-step cut reduction and permutation rules II.

There is one reduction step that may not preserve simple Π_2-proofs, namely the particular instance of binary rule permutation in which (r) is (cut):

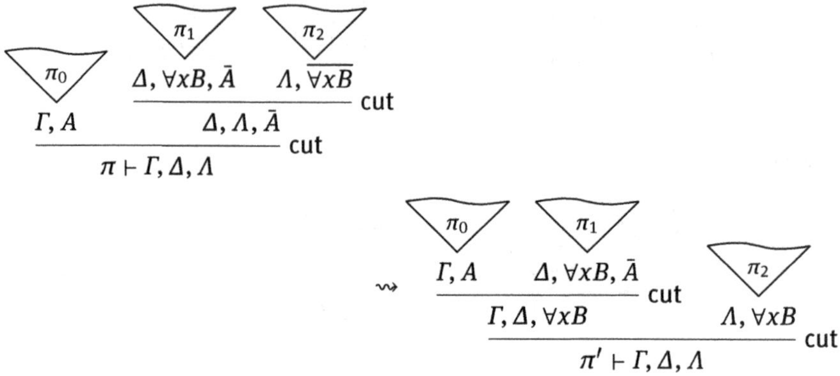

In the right-hand proof, $\forall xB$ is principal in the cut but is not immediately preceded by the rule ($\forall$) introducing it, so is not a simple proof. If the left-hand proof is a simple Π_2-proof then it follows that π_1 has the form

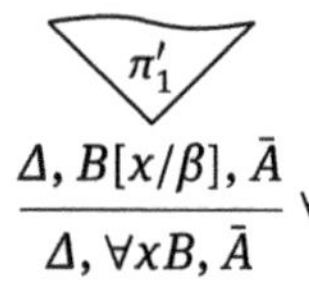

whence we see that applying unary rule permutation to the upper cut in π' we may obtain the derivation

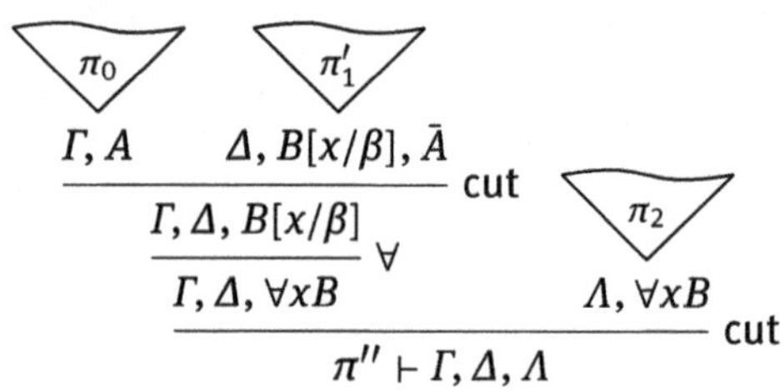

We have $\pi \rightsquigarrow \pi' \rightsquigarrow \pi''$ and π'' is a simple Π_2-proof.

In order to permit reduction strategies of this form it is convenient to consider the reduction from π to π'' as a single reduction step, so we add the following additional rule to the definition of $\rightsquigarrow$:

$$\dfrac{\overset{\pi_0}{\Gamma, A} \qquad \dfrac{\dfrac{\overset{\pi_1}{\Delta, B[x/\beta], \bar{A}}}{\Delta, \forall x B, \bar{A}}\,\forall \qquad \overset{\pi_2}{\Lambda, \overline{\forall x B}}}{\Delta, \Lambda, \bar{A}}\,\text{cut}}{\pi \vdash \Gamma, \Delta, \Lambda}\,\text{cut}$$

$$\rightsquigarrow \quad \dfrac{\dfrac{\dfrac{\overset{\pi_0}{\Gamma, A} \qquad \overset{\pi_1}{\Delta, B[x/\beta], \bar{A}}}{\Gamma, \Delta, B[x/\beta]}\,\text{cut}}{\Gamma, \Delta, \forall x B}\,\forall \qquad \overset{\pi_2}{\Lambda, \forall x B}}{\pi'' \vdash \Gamma, \Delta, \Lambda}\,\text{cut} \tag{1}$$

Definition 2.3. *For proofs π, π' we write $\pi \rightsquigarrow \pi'$ if π' is the result of applying one of the rules of Figures 1 or 2, or* (1) *above to (a sub-proof of) π. Notice that in contraction reduction the left sub-proof is duplicated and care is taken to rename the eigenvariables (expressed by annotating the proof/sequent/formula in question by an asterisk) to maintain regularity.*

2.2 Herbrand's theorem

Suppose Γ is a set of simple formulæ and $\pi \vdash \Gamma$ is a proof in which all cuts are on non-quantified formulæ. For each $F \in \Gamma$ let the *Herbrand set* for F, denoted $\mathcal{H}(\pi, F)$, be the set of terms that occur in π as witnesses to the existential quantifier in F (if there is one). The *Herbrand set for π* is the set $\mathcal{H}(\pi) = \{(F, t) \mid F \in \Gamma \wedge t \in \mathcal{H}(\pi, F)\}$.

Given a set X of terms and simple formula F, let F^X denote the prenex Π_1 formula given by

$$(\forall x F)^X = \forall x F^X \qquad\qquad (\exists x F)^X = \bigvee_{t \in X} F[x/t]$$

and $F^X = F$ if F is quantifier-free.

Theorem 2.4 (Herbrand's theorem for simple formulæ). *Let Γ be a finite set of simple Σ_1 formulæ. $\bigvee \Gamma$ is valid iff there exist finite sets $\{X_F \subseteq \text{Terms} \mid F \in \Gamma\}$ such that the formula $\bigvee_{F \in \Gamma} F^{X_F}$ is valid.*

Proof: The right-to-left direction is immediate. For the left-to-right direction, suppose $\bigvee \Gamma$ is valid formula of first-order logic. Fix a cut-free proof $\pi \vdash \Gamma$ and for each $F \in \Gamma$, set $X_F = \mathcal{H}(\pi, F)$. By Gentzen's mid-sequent theorem π induces a proof of $\bigvee_{F \in \Gamma} F^{X_F}$ and we are done. □

3 Proof grammars

It this section we define a class of grammars suitable for the analysis of Π_2-proofs. Our definition of a grammar somehow deviates from a standard one as it allows certain (controlled) bounded non-terminals to be re-written in the derivations. It is, nevertheless, possible to provide an equivalent definition of these grammars in terms of (standard) context-free tree grammars [Afshari et al., 2015]. The presentation given in this paper has the advantage that the types of non-terminals are fixed from the outset allowing the verification of (proof-specific) language properties such as reductions proved in Section 5 to be clearer.

3.1 Terms, positions and substitution

Fix a ranked alphabet Σ and let $\mathcal{V}$ be a fixed set of variable symbols distinct from Σ. Let $\mathrm{Term}(\Sigma)$ denote the set of terms in the simply-typed λ-calculus built from $\Sigma \cup \mathcal{V}$. The set of *positions* of a term $T \in \mathrm{Term}(\Sigma)$, denoted $\mathrm{Pos}(T)$, are the nodes of the underlying tree, i.e.

$$\mathrm{Pos}(T) = \{\langle\rangle\} \cup \begin{cases} \emptyset, & \text{if } T \in \Sigma \cup \mathcal{V}, \\ \{ip \mid i \in \{0,1\} \wedge p \in \mathrm{Pos}(T_i)\}, & \text{if } T = T_0 \cdot T_1, \\ \{0p \mid p \in \mathrm{Pos}(T_0)\}, & \text{if } T = (\lambda\alpha.T_0). \end{cases}$$

For a term T and $p \in \mathrm{Pos}(T)$, we write $T|p$ for the subterm of T at position p.

Given $T \in \mathrm{Term}(\Sigma)$ and $\alpha \in \mathcal{V}$, let $\mathrm{Free}(T, \alpha) \subseteq \mathrm{Pos}(T)$ be the collection of positions at which α appears free in T:

$$\begin{aligned} \mathrm{Free}(\alpha, \alpha) &= \{\langle\rangle\} \\ \mathrm{Free}(\gamma, \alpha) &= \emptyset \quad \text{if } \alpha \neq \gamma \in \mathcal{V} \\ \mathrm{Free}(a, \alpha) &= \emptyset \quad \text{if } a \in \Sigma \\ \mathrm{Free}(T_1 \cdot T_2, \alpha) &= \{ip \mid i \in \{0,1\} \text{ and } p \in \mathrm{Free}(T_i, \alpha)\} \\ \mathrm{Free}(\lambda\gamma.T, \alpha) &= \begin{cases} \emptyset, & \text{if } \gamma = \alpha \\ \{0p \mid p \in \mathrm{Free}(T, \alpha)\}, & \text{if } \gamma \neq \alpha \end{cases} \end{aligned}$$

We introduce two forms of term substitution. Let $S, T \in \mathrm{Term}(\Sigma)$.

- $T[a/S]$, where $a \in \Sigma \cup \mathcal{V}$, denotes the *safe* substitution of S for a in T. So in particular, $(\lambda\gamma.T)[\gamma/S] = \lambda\gamma.T$ and for a distinct from γ, $(\lambda\gamma.T)[a/S] = \lambda\delta.(T[\gamma/\delta][a/S])$ where δ is a variable symbol not occurring free in S.
- $T[p/S]$, where $p \in \mathrm{Pos}(T)$, is defined by recursion on p:

$$T[\langle\rangle/S] = S \qquad a[p/S] = S, \quad a \in \Sigma \cup \mathcal{V}$$
$$(T_0 \cdot T_1)[0p/S] = T_0[p/S] \cdot T_1 \qquad (\lambda\alpha.T)[0p/S] = \lambda\alpha.T[p/S]$$
$$(T_0 \cdot T_1)[1p/S] = T_0 \cdot (T_1[p/S])$$

  Note that this may involve an *unsafe* substitution.

3.2 Syntax and semantics of proof grammars

A *proof grammar* is a tuple $\mathcal{G} = \langle \mathcal{N}, \Sigma, \mathcal{S}, \mathrm{Pr}\rangle$ where $\mathcal{N}$ is a set of typed *non-terminals* of order at most 1, $\mathcal{S} \subseteq \mathcal{N}$ is a set of *starting symbols* (of base type), Σ is a ranked alphabet, called *terminals*, disjoint from $\mathcal{N}$, and Pr consists of pairs $(a, T) \subset \mathcal{N} \times \mathrm{Term}(\Sigma \cup \mathcal{N})$ (called *production rules* and written $a \to T$) such that a and T have the same type. Given a proof grammar $\mathcal{G}$ we assume $\mathcal{G} = \langle \mathcal{N}_\mathcal{G}, \Sigma_\mathcal{G}, \mathcal{S}_\mathcal{G}, \mathrm{Pr}_\mathcal{G}\rangle$.

Let d be a sequence $\langle \rho_i, p_i\rangle_{i<k}$ of pairs of production rules of a proof grammar $\mathcal{G}$ and positions, and S and T terms. We call d a *derivation from S to T*, written $d\colon S \to T$, if there exist terms $(N_i)_{i\leq k}$ such that $N_0 = S$, $N_k = T$, and for each $0 \leq i < k$,

1. ρ_i is a production rule of $\mathcal{G}$ and $p_i \in \mathrm{Pos}(N_i)$,
2. For $\rho_i = (a \to S)$, we have $N_i|p_i = a$ and $N_{i+1} = N_i[p_i/S]$.

The sequence of terms $(N_i)_{i\leq k}$ is uniquely determined by d and S, whence we may write $d(i)$ for N_i. The *length* of d, $\mathrm{lh}(d)$, is k. We write $\mathrm{Der}(\mathcal{G})$ for the set of derivations in $\mathcal{G}$, and say *T is derivable from S* if there exists a derivation $d\colon S \to T$. A derivation d *writes* a non-terminal a if there is a production rule of the form $a \to S$ for some S occurring in d.

The *language* of a proof grammar $\mathcal{G}$, $\mathcal{L}(\mathcal{G})$, is the set of terms not containing free occurrences of non-terminal symbols that are derivable from the starting symbols of the grammar:

$$Ł(\mathcal{G}) = \{T \mid \exists d\colon \sigma \to T \text{ s.t. } \sigma \in \mathcal{S}_\mathcal{G} \text{ and } T \text{ contains no free occurrence of a non-terminal}\}.$$

When comparing languages of proof grammars it is convenient to work modulo β-convertibility. For λ-terms S and T, we write $S \to_\beta T$ to abbreviate T is obtainable from S by one step β-reduction, and $\to_\beta^*$ for the reflexive transitive closure of this relation. Thus for grammars $\mathcal{G}_1$, $\mathcal{G}_2$, we write $\mathcal{L}(\mathcal{G}_1) \subseteq Ł(\mathcal{G}_2)$ to express that for every $S \in Ł(\mathcal{G}_1)$ there exists a $T \in Ł(\mathcal{G}_2)$ such that $T \to_\beta^* S$.

3.3 Rigidity

Rigid grammars were studied in [Hetzl, 2010, 2012a] for the construction of terms appearing in cut-elimination for first-order logic with Π_1-cuts. In this section we extend the notion of rigidity for the analysis of proofs with Π_2-cuts.

Let $\mathcal{G}$ be a proof grammar and suppose $\lhd$ is a transitive binary relation on $\mathcal{N}_\mathcal{G}$ and $\mathcal{R} \subseteq \mathcal{N}_\mathcal{G}$ is a designated set of non-terminals. A derivation $d = \langle (a_i \to S_i), q_i\rangle_{i<\mathrm{lh}(d)} : S \to T$ induces an equivalence relation on $\{i \mid i < \mathrm{lh}(d)\}$ corresponding to connectedness in parse trees: for $j_0, j_1 < \mathrm{lh}(d)$, set $j_0 \approx_d j_1$ iff there exist $i_0 \le j_0, j_1$ such that

1. $q_{i_0} \le q_{j_0}, q_{j_1}$,
2. for every $k \in \{0, 1\}$ and $i_0 < i < j_k < \mathrm{lh}(d)$, if $q_i \le q_{j_k}$ and $a_i \in \mathcal{R}$ then $a_{j_k} \ntriangleleft a_i$.

In other words, two non-terminals occurring in (the natural tree representation of) the derivation d are considered connected if there is no non-terminal of higher priority between them and their closest common ancestor. We write $j_0 \sim_d j_1$ if $j_0 \approx_d j_1$ and in addition $a_{j_0} = a_{j_1} \in \mathcal{R}$. Notice that $\sim_d$ may not be an equivalence relation on $\{i \mid i < \mathrm{lh}(d)\}$. For instance, if $a_j \notin \mathcal{R}$ then $j \nsim_d j$.

We consider derivations that *respect* the relation $\sim_d$ and we permit unsafe substitutions that are *controlled* by priority ordering $\lhd$:

Definition 3.1 (Rigid derivations). *Let $\mathcal{G}$ be a proof grammar and suppose $\lhd$ is a transitive binary relation on $\mathcal{N}_\mathcal{G}$ and $\mathcal{R} \subseteq \mathcal{N}_\mathcal{G}$. A derivation $d = \langle a_i \to T_i, q_i\rangle_{i<k} : S \to T$ in $\mathcal{G}$ is rigid with respect to $(\lhd, \mathcal{R})$ if*

1. *for every $i, j < k$, $i \sim_d j$ implies $T|q_i = T|q_j$, and*
2. *for every $i < k$ with $a_i \in \mathcal{R}$, if $d(i)|q = \lambda a_i.S_0$ for some position $q < q_i$ and term S_0, then there exist position $q < q' < q_i$, term S_1 and variable b such that $d(i)|q' = \lambda b.S_1$ and $a_i \lhd b$.*

A *rigid* proof grammar is a tuple $\mathcal{G} = \langle \mathcal{N}, \mathcal{R}, \lhd, \Sigma, \mathcal{S}, \mathrm{Pr}\rangle$ such that $\langle \mathcal{N}, \Sigma, \mathcal{S}, \mathrm{Pr}\rangle$ is a proof grammar, $\mathcal{R} \subseteq \mathcal{N}$ and $\lhd$ is a transitive relation on $\mathcal{N}$. $\mathcal{R} = \mathcal{R}_\mathcal{G}$ is the set of *rigid* non-terminals of $\mathcal{G}$ and $\lhd = \lhd_\mathcal{G}$ is the *priority ordering of* $\mathcal{G}$. $\mathcal{G}$ is *acyclic* if the ordering $\lhd_\mathcal{G}$ is acyclic, that is for all $a \in \mathcal{N}_\mathcal{G}$ it is not the case that $a \lhd_\mathcal{G} a$, and is *totally rigid* if $\mathcal{R}_\mathcal{G} = \mathcal{N}_\mathcal{G}$.

In a rigid proof grammar $\mathcal{G}$, a *derivation* is simply a derivation in the underlying proof grammar and a *rigid derivation* is a derivation which is rigid with respect to $(\lhd_{\mathcal{G}}, \mathcal{R}_{\mathcal{G}})$.

A rigid derivation $d\colon S \to T$ (in $\mathcal{G}$) is *full* if there is no extension of d that is a rigid derivation in $\mathcal{G}$. The *language* of a rigid proof grammar $\mathcal{G}$, $\mathcal{L}(\mathcal{G})$, is the collection of terms derivable from full rigid derivations starting from $\mathcal{S}_{\mathcal{G}}$:

$$Ł(\mathcal{G}) = \{T \mid \exists d\colon \sigma \to T \text{ s.t. } \sigma \in \mathcal{S}_{\mathcal{G}} \text{ and } d \text{ is full and } (\lhd_{\mathcal{G}}, \mathcal{R}_{\mathcal{G}})\text{-rigid}\}.$$

Note Definition 3.1 allows certain variables that are bounded (by an abstraction) to be re-written. Therefore in rigid derivations, arbitrary β-reductions are not permissible until the derivation is *fully* written in the sense given above.

Observation 3.2. *The language of a rigid proof grammar is a set of closed (well-typed) ground-type λ-terms.*

Observation 3.3 (Permutation). *Let d and d' be derivations in a rigid grammar $\mathcal{G}$ and suppose d' is a permutation of d. d' is rigid iff d is rigid.*

Example 3.4. *Let $\mathcal{G}$ be the rigid proof grammar with start symbol σ, the non-terminals σ, α and γ, rigid non-terminals $\mathcal{R}$, ordering $\lhd$, all other symbols terminals of appropriate arity, and production rules*

$$\sigma \to f(\alpha, \gamma, \gamma) \qquad\qquad \gamma \to g(\gamma) \mid a \qquad\qquad \alpha \to \gamma$$

If $\gamma \notin \mathcal{R}$ we have, unsurprisingly, $\mathcal{L}(\mathcal{G}) = \{f(g^m(a), g^n(a), g^o(a)) \mid m, n, o \geq 0\}$. For $\gamma \in \mathcal{R}$,
1. *If $\gamma \lhd \gamma \lhd \alpha$ and $\alpha \in \mathcal{R}$, $\mathcal{L}(\mathcal{G}) = \{f(g^m(a), g^n(a), g^n(a)) \mid m, n \geq 0\}$;*
2. *If $\gamma \lhd \gamma \ntriangleleft \alpha$, $\mathcal{L}(\mathcal{G}) = \{f(g^m(a), g^m(a), g^m(a)) \mid m \geq 0\}$;*
3. *If $\gamma \ntriangleleft \gamma \ntriangleleft \alpha$, $\mathcal{L}(\mathcal{G}) = \{f(a, a, a)\}$.*

Example 3.5. *Let $\mathcal{G}$ be a proof grammar and suppose $\mathcal{R} \subseteq \mathcal{N}_{\mathcal{G}}$. Further, let $d = \langle \rho_i, q_i \rangle_{i<k}\colon M \to N$ be a derivation in $\mathcal{G}$. Two simple choices of $\lhd$ are*
1. Global rigidity. *Set $\lhd_G = \emptyset$. Then d is $(\lhd_G, \mathcal{R})$-rigid iff for all $i, j < k$, $d(i)|q_i = d(j)|q_j \in \mathcal{R}$ implies $N|q_i = N|q_j$.*
2. Local rigidity. *$\lhd_L = \mathcal{N}_{\mathcal{G}} \times \mathcal{N}_{\mathcal{G}}$. In this form rigidity is treated only at the level of production rules: d is $(\lhd_L, \mathcal{R})$-rigid iff for every $i < k$, if $\rho_i = (a \to S)$ and $S|p = S|q \in \mathcal{R}$, then $N|q_i p = N|q_i q$.*

Notation 3.6. *Given a derivation $d = \langle \rho_i, q_i \rangle_{i<k}$ and a position p we define d^p to be the derivation $\langle \rho_i, pq_i \rangle_{i<k}$.*

Definition 3.7 (Subderivation). *Let $d = \langle \rho_i, q_i \rangle_{i<\mathrm{lh}(d)}$ be a derivation and suppose $k < \mathrm{lh}(d)$. The* subderivation of d from k *is the longest derivation $e = \langle \hat{\rho}_i, \hat{q}_i \rangle_{i<\mathrm{lh}(e)}$ such that $\hat{q}_0 = \langle\rangle$ and e^{q_k} is a subsequence of d.*

Definition 3.8 (Strong rigidity). *A derivation $d = \langle \rho_i, q_i \rangle_{i<\mathrm{lh}(d)}$ is strongly rigid iff for all $i_0, i_1 < \mathrm{lh}(d)$ if $i_0 \sim_d i_1$ then the subderivations of d from i_0 and i_1 are identical or a permutation of one another.*

Restricting to strongly-rigid derivations does not reduce the language of a rigid grammar:

Lemma 3.9. *Let $\mathcal{G}$ be a rigid grammar. For all M and N, if there is a rigid derivation from M to N in $\mathcal{G}$ then there exists a strongly rigid derivation from M to N in $\mathcal{G}$.*

Proof: By recursion through d: for each p and q such that $d(i)|p = d(j)|q \in \mathcal{N}_\mathcal{G}$ and $N|p = N|q$ replace the sub-derivation of d at q by a copy of the sub-derivation of d at p. The result will be strongly rigid iff d is rigid. □

3.4 Bounds on rigid proof grammars

Lemma 3.10 (Bounding). *Suppose $\mathcal{G}$ is a rigid proof grammar satisfying the following condition:*

$$\begin{gathered}\text{for every rigid derivation } d = \langle \rho_i, p_i \rangle_{i\le k} \text{ in } \mathcal{G} \text{ and every } i < j \le k, \\ \text{if } p_i \le p_j \text{ and } d(i)|p_i = d(j)|p_j \text{ then there exists } l \in (i,j) \text{ such that} \\ p_i \le p_l \le p_j \text{ and } d(j)|p_j \lhd_\mathcal{G} d(l)|p_l.\end{gathered} \tag{2}$$

If $\mathcal{G}$ is acyclic then $\mathcal{L}(\mathcal{G})$ is finite; if $\mathcal{G}$ is acyclic and totally rigid, $|\text{Ł}(\mathcal{G})| \le |\mathrm{Pr}_\mathcal{G}|^{2^{|\mathcal{N}_\mathcal{G}|-1}}$.

Proof: Assume $\mathcal{G}$ is a rigid proof grammar satisfying (2) and $\lhd_\mathcal{G}$ is acyclic. Fix a rigid derivation $d = \langle \rho_i, p_i \rangle_{i\le k}$ in $\mathcal{G}$ from an starting symbol. Call a derivation $\hat{d} = \langle \hat{\rho}_i, \hat{p}_i \rangle_{i\le\hat{k}}$ *a path through d* if $\hat{d}$ is a subsequence of d, $(\hat{\rho}_0, \hat{p}_0) = (\rho_0, p_0)$, and for every $i < j < \hat{k}$, $\hat{p}_i \le \hat{p}_j$.

Let $\hat{d}$ be a path through d and let $\mathcal{N}_d$ denote the set of non-terminals written by $\hat{d}$, i.e. $\mathcal{N}_d = \{a \in \mathcal{N}_\mathcal{G} \mid \exists i < \hat{k} \exists S\, \hat{\rho}_i = (a \to S)\}$. Suppose $\mathcal{N}_d = \{a_0, a_1, \dots, a_{|\mathcal{N}_d|-1}\}$ and for all $i < j < |\mathcal{N}_d|$, $a_i \not\lhd a_j$ which is possible as $\lhd_\mathcal{G}$ is acyclic. By the stated requirement there can be at most one $i \le \hat{k}$ such that $\hat{\rho}_i$ writes a_0, whence there are at most two production rules in $\hat{d}$ that write a_1, four production rules in $\hat{d}$ writing a_2, and in general no more than 2^i production rules writing a_i. So $\hat{k} < 2^{|\mathcal{N}_d|}$. The set of paths through d forms a tree which has branching degree bounded by some constant K independent of d and depth bounded by $2^{|\mathcal{N}_\mathcal{G}|}$, and it follows that $k \le K^{2^{|\mathcal{N}_\mathcal{G}|}}$. As d is arbitrary there are only finitely many rigid derivations and the first result follows.

Suppose $\mathcal{G}$ is acyclic, totally rigid and satisfies (2). Let $m = |\mathrm{Pr}_\mathcal{G}|$ and $n = |\mathcal{N}_\mathcal{G}|$. We argue, by induction on the number of non-terminals in $\mathcal{G}$, that the set of terms

$T \in \mathrm{Terms}(\Sigma_{\mathcal{G}})$ rigidly derivable from a starting symbol in $\mathcal{G}$ has size bounded by $m^{2^{n-1}}$.

The base case is $n = 1$. By the main assumption of the lemma every derivation $d\colon \sigma \to T$ with $\sigma \in \mathcal{S}_{\mathcal{G}}$ and $\in \mathrm{Terms}(\Sigma_{\mathcal{G}})$ has length 1, of which there are no more than m. For the induction step, suppose $n = n_0 + 1$. Let $\mathcal{N} = \mathcal{N}_{\mathcal{G}} \setminus \{a\}$ where $a \neq \sigma$ is chosen such that $a \not\prec b$ for all $b \in \mathcal{N}_{\mathcal{G}} \setminus \{\sigma\}$. Suppose $d\colon \sigma \to T \in \mathrm{Terms}(\Sigma_{\mathcal{G}})$ is a rigid derivation in $\mathcal{G}$. By the main assumption d can be re-ordered to have the form $d_0 d_1$ where $d_0\colon \sigma \to S$ and $d_1\colon S \to T = S[a/S']$ for appropriate terms S and S', such that the non-terminal a is not written in d_0 and not introduced by a production rule in d_1. The induction hypothesis implies there is no more than $m^{2^{n_0-1}}$ possibilities for each of S and S', whence there are $\leq m^{2^{n-1}}$ possibilities for T. □

3.5 Grammars for Π_2-proofs

In this section we associate to each simple Π_2-proof π a proof grammar $\mathcal{G}_\pi$ which will be used in the subsequent sections to prove the confluence result. We begin with a motivating example:

Example 3.11. *Let $\pi \vdash \exists x F$ be the proof given below in which there is a Π_2-cut on $\forall x \exists y A$, a Π_1-cut on $\forall x B$ and we assume these are the only cuts on quantified formulæ in π; α and γ the respectively eigenvariables of these cuts; t_i and u_i witness terms of the existential quantifier in, respectively, $\exists x \forall y \bar{A}$ and $\exists x \bar{B}$; and β_i the eigenvariable of the universal quantifier in $\forall y \bar{A}(t_i, y)$. For a formula C, we use C^* to distinguish between the two copies of C that may appear in the same sequent in the proof.*

$$
\dfrac{
  \dfrac{
    \dfrac{
      \dfrac{B(\gamma), A(\alpha, s(\alpha,\gamma)), F(g(\gamma,\alpha))}{B(\gamma), A(\alpha, s(\alpha,\gamma)), \exists x F}\,\exists
    }{\forall x B, \exists y A(\alpha, y), \exists x F}\,\exists
    \qquad
    \dfrac{
      \dfrac{
        \dfrac{\bar{B}(u_1), \bar{B}^*(u_2(\alpha))}{\bar{B}(u_1), \exists x \bar{B}^*}\,\exists
      }{\exists x \bar{B}, \exists x \bar{B}^*}\,\exists
    }{\exists x \bar{B}}\,c
  }{
    \dfrac{\exists y A(\alpha, y), \exists x F}{\forall x \exists y A(x,y), \exists x F}\,\forall
  }\,cut
  \qquad
  \pi' \vdash \exists x \forall y \bar{A}(x,y)
}{\pi \vdash \exists x F}\,cut
$$

where π' is given by

$$\dfrac{\dfrac{\dfrac{\dfrac{\dfrac{\bar{A}(t_1,\beta_1),\bar{A}^*(t_2(\beta_1),\beta_2)}{\bar{A}(t_1,\beta_1),\forall y\bar{A}^*(t_2(\beta_1),y)}\,\forall}{\bar{A}(t_1,\beta_1),\exists x\forall y\bar{A}^*(x,y)}\,\exists}{\forall y\bar{A}(t_1,y),\exists x\forall y\bar{A}^*(x,y)}\,\forall}{\exists x\forall y\bar{A}(x,y),\exists x\forall y\bar{A}^*(x,y)}\,\exists}{\pi'\vdash\exists x\forall y\bar{A}(x,y)}\,c$$

Having the cut reduction process in mind, it is natural to consider the following grammar. First we introduce a starting symbol $\sigma_{\exists xF}$ which will write to the (literal) witness of $\exists xF$ in π, i.e. the term $g(\gamma,\alpha)$. This production rule initiates the search for the witnesses to $\exists xF$. Next we add production rules that write the eigenvariables α and γ of the two cuts to the associated terms t_i and u_i respectively, mirroring the substitutions performed in Gentzen-style cut elimination. To capture the correct rules for eigenvaribles of the universal quantifier in $\exists x\forall y\bar{A}$, i.e. β_0 and β_1, we introduce a functional non-terminal σ which represents the existential quantifier in $\forall x\exists yA(x,y)$ and allows β_i to write (modulo β-reduction) to the term $s(t_i,\gamma)$.[2] More precisely, $\mathcal{G}=\langle\mathcal{N},\mathcal{R},\lhd,\Sigma,\mathcal{S},\mathrm{Pr}\rangle$ with $\mathcal{S}=\{\sigma_{\exists xF}\}$, $\mathcal{R}=\{\alpha,\gamma,\beta_1,\beta_2\}$, $\mathcal{N}=\mathcal{S}\cup\mathcal{R}\cup\{\sigma\}$ and Pr consisting of rules:

$$\begin{array}{lll}\sigma_{\exists xF}\to g(\gamma,\alpha) & \gamma\to u_1\mid u_2 & \\ \alpha\to t_1\mid t_2 & \beta_i\to\sigma\cdot t_i\quad \text{for } i=1,2 & \sigma\to\lambda\alpha.s\end{array}$$

The priority ordering is given by the order in which the eigenvariables are eliminated in the proof, increasing in priority from top down, and left to right through a Π_2-cut (which has the universal formula on the left subproof):

$$\gamma\lhd\alpha\lhd\beta_2\lhd\beta_1$$

We now calculate $\mathcal{L}(\mathcal{G})$. There are two possible starting derivations for $\sigma_{\exists xF}$:

$$\sigma_{\exists xF}\to g(\gamma,\alpha)\to g(u_1,\alpha) \tag{3}$$

$$\sigma_{\exists xF}\to g(\gamma,\alpha)\to g(u_2(\alpha),\alpha) \tag{4}$$

Extending (3) *we obtain the closed term $g(u_1,t_1)\in Ł(\mathcal{G})$ as well as*

$$\sigma_{\exists xF}\to^* g(u_1,\alpha)\to g(u_1,t_2(\beta_1))\to g(u_1,t_2(\sigma\cdot t_1))\to g(u_1,t_2(\lambda\alpha.s(\alpha,\gamma)\cdot t_1)).$$

2 Further motivation for the use of non-terminals of function type can be found in [Afshari et al., 2015].

Since $\gamma \lhd \beta_1$, the γ appearing in the last term above is not connected to its earlier occurrence in (3) and may freely write to either u_1 or u_2, yielding terms $g(u_1, t_2(s(t_1, u_1))$ and $g(u_1, t_2(s(t_1, u_2(t_1)))$ in $\mathcal{L}(\mathcal{G})$.

To extend (4), first note that the (possible) two occurrences of α in $g(u_2(\alpha), \alpha)$ must be written to the same result, of which the following are allowed.

$$g(u_2(t_1), t_1)$$
$$g(u_2(t_2(s(t_1, u_1))), t_2(s(t_1, u_1)))$$
$$g(u_2(t_2(s(t_1, u_2(t_1)))), t_2(s(t_1, u_2(t_1))))$$

Indeed, it is not hard to check that any cut-free proof π' obtainable from π via the reduction steps in Figures 1 and 2, $\mathcal{H}(\pi', \exists x F)$ consists of exactly the six closed terms derived above.

We now proceed with the definition of the grammar. Let $\pi \vdash \Gamma$ be a simple Π_2-proof of a set of prenex Π_2 and Σ_2 formulæ. The *proof grammar for π* is the grammar $\mathcal{G}_\pi = \langle \mathcal{N}_\pi, \mathcal{R}_\pi, \lhd_\pi, \Sigma_\pi, \mathcal{S}_\pi, \mathrm{Pr}_\pi \rangle$ where the components are defined as follows.

Symbols and their types

We will use symbols of the form τ_F^p or σ_F^p where p is a position and F is a formula occurring in π.[3] Each such symbol is assigned a *type*, either 0 (ground type) or $0 \to 0$ (function type):

- $\mathrm{type}(\tau_F^p) = \mathrm{type}(\sigma_F^p) = 0$ if $F \in \Pi_2$.
- $\mathrm{type}(\tau_F^p) = \mathrm{type}(\sigma_F^p) = (0 \to 0)$ otherwise (i.e. $F \in \Sigma_2 \setminus \Pi_2$).
- All eigenvariables in π and first-order terms are ground-type terms.

Alphabet and non-terminals

Σ_π consists of the function symbols, constants and variables occurring in π, and symbols $\tau_F^{\langle\rangle}$ for every $F \in \Gamma$. The set of non-terminals, $\mathcal{N}_\pi$, consists of

- rigid non-terminals $\mathcal{R}_\pi = \mathrm{EV}(\pi)$;
- starting symbols $\mathcal{S}_\pi = \{\sigma_F^{\langle\rangle} \mid F \in \Gamma\}$;
- symbols τ_A^p where $p \in \mathrm{Pos}(\pi) \setminus \{\langle\rangle\}$ and A is a formula occurring in the end-sequent of $\pi|p$;

3 For brevity we write τ_F for $\tau_F^{\langle\rangle}$.

- symbols σ^p_A where $p \in \mathrm{Pos}(\pi)$ and A is a formula occurring in the end-sequent of $\pi|p$.

Table 2. Production rules in $\mathcal{G}_\pi$

$\pi\|p \vdash \Gamma$	Production rules in $\mathcal{G}_\pi$ for position p, $\mathrm{Pr}_{\pi,p}$
$\pi\|p \vdash A, \bar{A}$	$\{\sigma^p_A \to \tau^p_{\bar{A}}, \sigma^p_{\bar{A}} \to \tau^p_A\}$
$\dfrac{\Delta, A_0, A_1}{\pi\|p \vdash \Delta, A_0 \vee A_1}\vee$	$\mathrm{prod}(p, \Delta, \emptyset)$
$\dfrac{\Pi, A_0 \quad \Delta, A_1}{\pi\|p \vdash \Pi, \Delta, A_0 \wedge A_1}\wedge$	$\mathrm{prod}(p, \Pi, \Delta)$
$\dfrac{\Delta, A[x/\alpha]}{\pi\|p \vdash \Delta, \forall xA}\forall$	$\mathrm{prod}(p, \Delta, \emptyset) \cup \{\alpha \to \tau^p_{\forall xA}, \sigma^p_{\forall xA} \to \lambda\alpha.\sigma^{p0}_{A[x/\alpha]}\}$
$\dfrac{\Delta, A[y/s]}{\pi\|p \vdash \Delta, \exists yA}\exists$	$\mathrm{prod}(p, \Delta, \emptyset) \cup \{\sigma^p_{\exists yA} \to s,\ \tau^{p0}_{A[y/s]} \to \tau^p_{\exists yA} \cdot s\}$
$\dfrac{\Delta, A \quad \Pi, \bar{A}}{\pi\|p \vdash \Delta, \Pi}\mathrm{cut}$	$\mathrm{prod}(p, \Delta, \Pi) \cup \{\tau^{p0}_A \to \sigma^{p1}_{\bar{A}}, \tau^{p1}_{\bar{A}} \to \sigma^{p0}_A\}$
$\dfrac{\Delta}{\pi\|p \vdash \Delta, \Pi}\mathrm{w}$	$\mathrm{prod}(p, \Delta, \emptyset)$
$\dfrac{\Delta, \Pi, \Pi^*}{\pi\|p \vdash \Delta, \Pi}\mathrm{c}$	$\mathrm{prod}(p, \Pi \cup \Delta, \emptyset) \cup \{\tau^{p0}_{F^*} \to \tau^p_F, \sigma^p_F \to \sigma^{p0}_{F^*} \mid F \in \Pi\}$
	$\mathrm{prod}(p, \Gamma, \Delta) = \mathrm{prod}_\tau(p, \Gamma, \Delta) \cup \mathrm{prod}_\sigma(p, \Gamma, \Delta)$
	$\mathrm{prod}_\tau(p, \Gamma_0, \Gamma_1) = \{\tau^{pj}_F \to \tau^p_F \mid j \in \{0,1\} \wedge F \in \Gamma_j\}$
	$\mathrm{prod}_\sigma(p, \Gamma_0, \Gamma_1) = \{\sigma^p_F \to \sigma^{pj}_F \mid j \in \{0,1\} \wedge F \in \Gamma_j\}$

Production rules and priority ordering

As the set of production rules in $\mathcal{G}_\pi$ we choose $\mathrm{Pr}_\pi := \bigcup_{p\in\mathrm{Pos}(\pi)} \mathrm{Pr}_{\pi,p}$, where the sets $\mathrm{Pr}_{\pi,p}$ (defined in Table 2) are determined by the rule of inference occurring at position p in π. Informally, non-terminals of the form σ^p_F represent the existential quantifier in F (if there is one) with their production rules 'searching' for witnesses within the sub-proof $\pi|p$. Dually, τ_F^p represents the universal quantifier in F (at p) and link eigenvariables on one side of a cut to the existential witnesses on the other side.

Like the production rules, the rigidity ordering, $\lhd_\pi$, is determined by the local structure of π and is the smallest transitive relation satisfying the following five conditions.

- Concerning production rules for axioms and side formulæ to all inference rules:
  1. For each production rule of the form $a \to \tau_B^q$ or $a \to \tau_B^q \cdot s$ where $a \in \mathcal{N}_\pi \setminus \mathrm{EV}(\pi)$ we have $a \lhd_\pi \tau_B^q$;
  2. For each rule of the form $\sigma_A^p \to \sigma_B^{pj}$ or $\sigma_A^p \to \lambda\alpha.\sigma_B^{pj}$ ($j \in \{0, 1\}$) we have $\sigma_A^p \lhd_\pi \sigma_B^{pj}$;
- For production rules introducing or eliminating eigenvariables:
  3. For each production rule $\alpha \to \tau_{\forall xA}^q$ where $\alpha \in \mathrm{EV}(\pi)$ we have $\alpha \lhd_\pi \tau_{\forall xA}^q$, and $\xi \lhd_\pi \alpha$ for every $\xi \in \mathrm{EV}(\pi|q0)$;
  4. For each rule of the form $\sigma_A^p \to s$ and each variable α appearing in s we set $\sigma_A^p \lhd_\pi \alpha$;
- And for production rules of cut formulæ:
  5. For a rule $\tau_A^{pi} \to \sigma_{\bar{A}}^{pj}$ with $i = 1 - j \in \{0, 1\}$, if $A = \forall x A_0$ for some Σ_1 formula A_0 then $\tau_A^{pi} \lhd_\pi \sigma_{\bar{A}}^{pj}$.

The first four conditions increase the priority of non-terminals as one follows either 'τ' production rules 'down' the proof towards the conclusion or principal cuts, or 'σ' productions 'upwards' towards witnesses of existential quantifiers. The final condition mediates the passage between the two paths over (one direction of) a Π_2-cut. The additional cases added by 3. capture, through the rigidity ordering, the duplication of eigenvariables that may occur when reducing a quantified cut.

Example 3.12. *Consider the proof π in Example 3.11. The grammar $\mathcal{G}_\pi$ is the tuple $\langle \mathcal{N}_\pi, \mathcal{R}_\pi, \lhd_\pi, \Sigma_\pi, \mathcal{S}_\pi, \mathrm{Pr}_\pi \rangle$ where $\mathcal{S}_\pi = \{\sigma_{\exists xF}^{\langle\rangle}\}$, $\mathcal{R}_\pi = \{\alpha, \gamma, \beta_1, \beta_2\}$, we have in particular $\gamma \lhd_\pi \alpha \lhd_\pi \tau_{\forall x \exists y A}^0 \lhd_\pi \sigma_{\exists x \forall y \bar{A}}^1 \lhd_\pi \beta_2 \lhd_\pi \beta_1$ and the production rules are:*

1. *For the starting symbol we have:*

$$\sigma_{\exists xF}^{\langle\rangle} \to \sigma_{\exists xF}^{0} \to \sigma_{\exists xF}^{00} \to \sigma_{\exists xF}^{000} \to \sigma_{\exists xF}^{0000} \to g(\gamma, \alpha)$$

2. *For the eigenvariable γ:*

$$\gamma \to \tau_{\forall xB}^{000} \to \sigma_{\exists x\bar{B}}^{001} \to \sigma_{\exists x\bar{B}}^{0010} \mid \sigma_{\exists x\bar{B}^*}^{0010}$$
$$\sigma_{\exists x\bar{B}}^{0010} \to u_1$$
$$\sigma_{\exists x\bar{B}^*}^{0010} \to \sigma_{\exists x\bar{B}^*}^{00100} \to u_2$$

3. *For eigenvariable α:*

$$\begin{aligned} \alpha \to \tau^0_{\forall x \exists y A} \to \sigma^1_{\exists x \forall y \bar{A}} &\to \sigma^{10}_{\exists x \forall y \bar{A}} \mid \sigma^{10}_{\exists x \forall y \bar{A}^*} \\ \sigma^{10}_{\exists x \forall y \bar{A}} &\to t_1 \\ \sigma^{10}_{\exists x \forall y \bar{A}^*} &\to \sigma^{100}_{\exists x \forall y \bar{A}^*} \to \sigma^{1000}_{\exists x \forall y \bar{A}^*} \to t_2 \end{aligned}$$

4. *For eigenvariables β_1 and β_2:*

$$\begin{aligned} \beta_1 \to \tau^{100}_{\forall y \bar{A}(t_1)} &\to \tau^{10}_{\exists x \forall y \bar{A}} \cdot t_1 \\ \tau^{10}_{\exists x \forall y \bar{A}} &\to \tau^{1}_{\exists x \forall y \bar{A}} \\ \beta_2 \to \tau^{10000}_{\forall y \bar{A}^*(t_2)} &\to \tau^{1000}_{\exists x \forall y \bar{A}^*} \cdot t_2 \\ \tau^{1000}_{\exists x \forall y \bar{A}^*} &\to \tau^{100}_{\exists x \forall y \bar{A}^*} \to \tau^{10}_{\exists x \forall y \bar{A}^*} \to \tau^{1}_{\exists x \forall y \bar{A}} \\ \tau^1_{\exists x \forall y \bar{A}} \to \sigma^0_{\forall x \exists y A} &\to \lambda\alpha.\sigma^{00}_{\exists y A} \\ \sigma^{00}_{\exists y A} &\to \sigma^{000}_{\exists y A} \to s \end{aligned}$$

Notice that the function type non-terminal $\sigma^0_{\forall x \exists y A}$ plays the same role as σ in grammar $\mathcal{G}$ of Example 3.11. Indeed, $\mathcal{L}(\mathcal{G}) = Ł(\mathcal{G}_\pi)$.

Shifting symbols, terms and rules

Following the definition above, it is clear that if π is a sub-proof of π', say $\pi = \pi'|p$, then $\mathcal{G}_\pi$ can be viewed as a sub-grammar of $\mathcal{G}_{\pi'}$ by 'shifting' the annotation of non-terminals in $\mathcal{G}_\pi$ by the position p. This action of 'shifting' a grammar relative to a position turns out to be a useful operation on grammars. We let $\mathcal{G}^p$ denote the result of shifting $\mathcal{G}$ relative to p, that is, $\mathcal{G}^p = \langle \mathcal{N}^p_{\mathcal{G}}, \mathcal{R}_{\mathcal{G}}, \lhd^p_\pi, \Sigma_{\mathcal{G}}, \mathcal{S}^p_{\mathcal{G}}, \mathrm{Pr}^p_{\mathcal{G}} \rangle$, where

$$\begin{aligned} \mathcal{N}^p_{\mathcal{G}} &= \{a^p \mid a \in \mathcal{N}_{\mathcal{G}}\} & \mathcal{S}^p_{\mathcal{G}} &= \{a^p \mid a \in \mathcal{S}_{\mathcal{G}}\} \\ \mathrm{Pr}^p_{\mathcal{G}} &= \{a^p \to T^p \mid (a \to T) \in \mathrm{Pr}_{\mathcal{G}}\} & \lhd^p_\pi &= \{(a^p, b^p) \mid a \lhd_\pi b\} \end{aligned}$$

given by

$$\begin{aligned} (\tau^q_F)^p &= \tau^{pq}_F & (\sigma^q_F)^p &= \sigma^{pq}_F & \gamma^p &= \gamma \quad \text{for } \gamma \in \mathcal{R}_{\mathcal{G}} \\ (T_1 \cdot T_2)^p &= T^p_1 \cdot T^p_2 & (\lambda\alpha.T)^p &= \lambda\alpha.(T^p) \end{aligned}$$

So in particular $\mathcal{G}^p_{\pi|p}$ is a sub-grammar of $\mathcal{G}_\pi$ whenever $p \in \mathrm{Pos}(\pi)$.

Lemma 3.13. *If π is a simple Π_2-proof then $\lhd_\pi$ is acyclic.*

Proof: By induction on the proof $\pi \vdash \Gamma$ noticing that no production rule in $\mathcal{G}_\pi$ writes $\tau^{\langle\rangle}_F$ or introduces $\sigma^{\langle\rangle}_F$ (for $F \in \Gamma$), and that if π ends in a cut on a simple

Π_2 formula $A = \forall x \exists y B$ (say $\pi|0 \vdash \Gamma, A$ and $\pi|1 \vdash \Gamma, \bar{A}$) then for all $a \in \mathcal{N}^0_{\pi|0}$ and $b \in \mathcal{N}^1_{\pi|1}$ we have $a \not\triangleleft_\pi b$. □

Theorem 3.14. *Let π be a simple Π_2-proof. Then $|\text{Ł}(\mathcal{G}_\pi)| < 2^{2^{2|\pi|}}$.*

Proof: First, note that requirement (2) of Lemma 3.10 is satisfied by all proof grammars arising from simple Π_2-proofs. We define a new grammar $\mathcal{G}'$ by making the following changes to $\mathcal{G}_\pi$: for every non-terminal $\nu^p_F \notin \mathcal{R}_\pi$ with $\nu \in \{\tau, \sigma\}$ and production rule $\nu^p_F \to S$, if there is a production rule $a \to \nu^p_F$ in $\mathcal{G}_\pi$ then remove $\nu^p_F \to S$ and instead add $a \to S$. Observe this process is well-defined as the term S cannot contain any occurrences of ν^p_F. Once all production rules writing ν^p_F are replaced, remove ν^p_F from the set of non-terminals. Let $\mathcal{G}_0$ be the resulting grammar, and $\mathcal{G}'$ the subsequent grammar obtained by indicating any remaining non-rigid non-terminals in $\mathcal{G}_0$ to be rigid.

$\mathcal{G}'$ is a totally rigid grammar and $|\mathcal{N}_{\mathcal{G}'}| + |\text{Pr}_{\mathcal{G}'}| \leq 2|\pi|$. Moreover, $\mathcal{L}(\mathcal{G}_\pi) = \text{Ł}(\mathcal{G}')$: it is easy to see that the 'compression' preserves rigid derivations and so $\mathcal{L}(\mathcal{G}_0) = \text{Ł}(\mathcal{G}_\pi)$. In $\mathcal{G}_0$ the only non-rigid non-terminals remaining are of the form $\tau^p_{\exists yA}$ or $\sigma^p_{\forall y\bar{A}}$ for some p and A. Observe that i) there is only one choice for writing such non-terminals, and ii) $\tau^p_{\exists yA}, \sigma^p_{\forall y\bar{A}} \triangleleft_\pi \alpha$ where α is the eigenvariable of $\forall y\bar{A}$. Thus any two non-terminals that become disconnected through $\tau^p_{\exists yA}$ or $\sigma^p_{\forall y\bar{A}}$ in $\mathcal{G}'$ were already disconnected in $\mathcal{G}_\pi$ through α. Indeed $\mathcal{G}'$ and $\mathcal{G}_\pi$ have the same language, and the comparability between their rigid derivations means $\mathcal{G}'$ also satisfies requirement (2).

Now Lemma 3.10 implies $|\text{Ł}(\mathcal{G}')| \leq |\text{Pr}_{\mathcal{G}'}|^{2^{|\mathcal{N}_{\mathcal{G}'}|-1}}$. We therefore have

$$|\text{Ł}(\mathcal{G}_\pi)| = |\text{Ł}(\mathcal{G}')| \leq 2^{2^{|\mathcal{N}_{\mathcal{G}'}|+|\text{Pr}_{\mathcal{G}'}|-1}} < 2^{2^{2|\pi|}}.$$

This concludes the proof. □

4 Technical lemmas

In Section 5 we will prove that for regular simple Π_2-proofs π and π' if $\pi \rightsquigarrow \pi'$ then $\mathcal{L}(\mathcal{G}_{\pi'}) \subseteq \text{Ł}(\mathcal{G}_\pi)$. Before considering the theorem, however, we require additional results concerning the fine structure of proof grammars. In Section 4.1 we introduce the notion of homomorphism between proof grammars as a means to test language containment. Section 4.2 highlights a number of properties relating to derivations in proof grammars and Section 4.3 prepares the ground-work for replacing sub-proofs by ones with comparable grammars.

4.1 Comparing grammars

An easy way to compare the languages induced by grammars is by providing a function mapping non-terminals of one grammar into another preserving rigid derivability. In the simplest form (which is all that is required here) this is given by a *homomorphism* as described below. First we need the following definitions.

Controlled β-reduction

We generalise the relation of β-reduction to obtain a finer relation between terms. Given two λ-terms S and T and a (possibly empty) set of variables $X \subseteq \mathcal{V}$, we write $S \rightarrowtail_X T$ if one of the following conditions hold.

- $S = T$;
- $S = S_0 \cdot S_1$, $T = T_0 \cdot T_1$ and $S_i \rightarrowtail_X T_i$;
- $S = \lambda\alpha.S_0$, $T = \lambda\alpha.T_0$ and $S_0 \rightarrowtail_X T_0$;
- $S = (\lambda\alpha.S_0) \cdot S_1$, $\alpha \in X$ and $T = S_0[\alpha/S_1]$.

Notice that $S \rightarrowtail_{\mathcal{V}} T$ is the usual 1-step parallel β-reduction. We write $S \rightarrowtail T$ iff $S \rightarrowtail_{\mathcal{V}} T$. This is not to be confused with $S \rightarrowtail_{\emptyset} T$ which holds iff $S = T$. Also let $S \rightarrowtail_X^* T$ denote the (reflexive) transitive closure of $S \rightarrowtail_X T$.

Following the definition of $S \rightarrowtail T$ we define a canonical function $r_{T,S}$ mapping positions in T to their corresponding position in S:

$$\begin{aligned}
r_{T,S}(\langle\rangle) &= \langle\rangle\\
r_{(T_0\cdot T_1),(S_0\cdot S_1)}(ip) &= i\, r_{T_i,S_i}(p), \quad \text{for } i \in \{0,1\}\\
r_{(\lambda\alpha.T_0),(\lambda\alpha.S_0)}(0p) &= 0\, r_{T_0,S_0}(p)\\
r_{S_0[\alpha/S_1],(\lambda\alpha.S_0)\cdot S_1}(p) &= \begin{cases} 1p', & \text{if } p = qp',\ q \in \mathrm{Free}(\alpha, S_0) \text{ and } p' \in \mathrm{Pos}(S_1)\\ 00p, & \text{otherwise}\end{cases}
\end{aligned}$$

where $\mathrm{Free}(\alpha, S_0)$ denotes the positions in S_0 marking free occurrences of α.

Homomorphism function

Let $\mathcal{F}$ and $\mathcal{G}$ be rigid proof grammars and $f\colon \mathcal{N}_{\mathcal{F}} \to \mathcal{N}_{\mathcal{G}}$. If $\Sigma_{\mathcal{F}} \subseteq \Sigma_{\mathcal{G}}$ then f naturally extends to a function $(.)^f\colon \mathrm{Terms}(\Sigma_{\mathcal{F}} \cup \mathcal{N}_{\mathcal{F}}) \to \mathrm{Terms}(\Sigma_{\mathcal{G}} \cup \mathcal{N}_{\mathcal{G}})$ given by

$$\begin{aligned}
\alpha^f &= f(\alpha) \quad \text{if } \alpha \in \mathcal{N}_{\mathcal{F}}, & (\lambda\alpha.S)^f &= \lambda\alpha^f.S^f\\
a^f &= a \quad \text{if } a \in \Sigma_{\mathcal{F}}, & (S\cdot T)^f &= S^f \cdot T^f
\end{aligned}$$

Definition 4.1 (Homomorphism). *Let $\mathcal{F}, \mathcal{G}$ be rigid grammars and suppose $\Sigma_{\mathcal{F}} \subseteq \Sigma_{\mathcal{G}}$ and $\mathcal{S}_{\mathcal{F}} \subseteq \mathcal{S}_{\mathcal{G}}$. A* homomorphism *from $\mathcal{F}$ into $\mathcal{G}$ is a function $f: \mathcal{N}_{\mathcal{F}} \to \mathcal{N}_{\mathcal{G}}$ such that*
1. *$f(\sigma) = \sigma$ for every $\sigma \in \mathcal{S}_{\mathcal{F}}$;*
2. *$f(a)$ has the same type as a for every $a \in \mathcal{N}_{\mathcal{F}}$;*
3. *$f(\alpha) \in \mathcal{R}_{\mathcal{G}}$ iff $\alpha \in \mathcal{R}_{\mathcal{F}}$;*
4. *$a \lhd_{\mathcal{F}} b$ implies $f(a) \lhd_{\mathcal{G}} f(b)$;*
5. *for each $p = (a \to S) \in \mathcal{F}$, there is a derivation $\rho^f = \langle \rho_i, p_i \rangle_{i<k}: f(a) \to S^f$ in $\mathcal{G}$ such that $\rho^f(i)|p_i \in \mathcal{R}_{\mathcal{G}}$ only if $i = 0$;*
6. *For every rigid derivation $d = \langle (a_i \to S_i), p_i \rangle_{i<l}: \sigma \to S$ in $\mathcal{F}$ with $\sigma \in \mathcal{S}_{\mathcal{F}}$ and every $j_0, j_1 < l$, if $f(a_{j_0}) = f(a_{j_1})$ and $j_0 \approx_d j_1$ then $a_{j_0} = a_{j_1}$.*

We write $f: \mathcal{F} \to \mathcal{G}$ to stipulate that $\Sigma_{\mathcal{F}} \subseteq \Sigma_{\mathcal{G}}$, $\mathcal{S}_{\mathcal{F}} \subseteq \mathcal{S}_{\mathcal{G}}$ and f is a homomorphism from $\mathcal{F}$ to $\mathcal{G}$.

Notice that condition 6 is trivially satisfied if f is injective, thus its only role is for in the case of Contraction Reduction.

A homomorphism $f: \mathcal{F} \to \mathcal{G}$ extends to a function $(.)^f: \mathrm{Der}(\mathcal{F}) \to \mathrm{Der}(\mathcal{G})$ in the obvious way: given a derivation $d = \langle \rho, p \rangle e: S \to T$ in $\mathcal{F}$, suppose $e^f: d(1)^f \to T^f$ is already defined and set $d^f := \langle \rho^f, p \rangle e^f: S^f \to T^f$ in $\mathcal{G}$.

Lemma 4.2. *Suppose $f: \mathcal{F} \to \mathcal{G}$ is a homomorphism. If $d: S \to T$ is a rigid derivation in $\mathcal{F}$ then $d^f: S^f \to T^f$ is a rigid derivation in $\mathcal{G}$.*

Proof: Let $\mathcal{F}$ and $\mathcal{G}$ be as above, $d = \langle (a_i \to S_i), p_i \rangle_{i<k}: S \to T$ a rigid derivation in $\mathcal{F}$ and $d^f = \langle (b_j \to M_j), q_j \rangle_{j<\mathrm{lh}(d^f)}: S^f \to T^f$ the derivation in $\mathcal{G}$ induced by f.

First suppose $j_0, j_1 < \mathrm{lh}(d^f)$ are such that $j_0 \approx_{d^f} j_1$ and $b_{j_0} = b_{j_1} \in \mathcal{R}_{\mathcal{G}}$. Condition 4 (combined with the definition of $d \mapsto d^f$) implies that there are $i_0, i_1 < \mathrm{lh}(d)$ such that $b_{j_l} = f(a_{i_l})$ and $q_{j_l} = p_{i_l}$ for $l = 0, 1$. Let $j^* < j_0, j_1$ be the index witnessing $j_0 \approx_{d^f} j_1$. Set $j \leq j^*$ be the greatest index such that $b_j = f(a_{i^*})$ for some $i^* < \mathrm{lh}(d)$. We now verify that i^* witnesses $i_0 \approx_d i_1$: To obtain a contradiction, suppose for $l = 0$ or $l = 1$ there is a non-terminal a_i with $i^* < i < i_l$ such that $p_i \leq p_l$ and $a_{i_l} \lhd_{\mathcal{F}} a_i$. By condition 4 we have $b_{j_l} = f(a_{i_l}) \lhd_{\mathcal{G}} f(a_i)$. Let $f(a_i) = b_{j'}$. By the definition of j we know for every $j^* \geq j' > j$, $b_{j'} \notin f(\{a_i \mid i \leq \mathrm{lh}(d)\})$. Therefore, $j' > j^*$, and since $b_{j_l} \lhd_{\mathcal{G}} b_{j'}$ we see that $b_{j'}$ violates $j_0 \approx_{d^f} j_1$, hence we are done. Since $i_0 \approx_d i_1$, from condition 6 we deduce $a_{i_0} = a_{i_1}$ so, since d is rigid, $T|p_{i_0} = T|p_{i_1}$. But then $T^f|q_{j_0} = T^f|q_{j_1}$ as required.

To check the second condition of rigidity for d^f, suppose $d^f(j)|x = \lambda\alpha.S_0^f$ for $x < q_j$ and $b_j = \alpha \in \mathcal{R}_{\mathcal{G}}$. Note that for any term T and position y, $T^f|y = (T|y)^f$. In particular, since α is also rigid there is $i < \mathrm{lh}(d)$ such that $d^f(j) = (d(i))^f, f(a_i) = b_j$ with $a_i \in \mathcal{R}_{\mathcal{F}}$, and $p_i = q_j$. Then we have $d(i)|x = \lambda a_i.S_0$ and so by rigidity of d there exists $x < y < p_i$ such that $d(i)|y = \lambda b.S_1$ and $a_i \lhd_{\mathcal{F}} b$. Let $\beta = f(b)$. Then $d^f(j)|y = \lambda\beta.S_1^f$ and by condition 4 $\alpha \lhd_{\mathcal{G}} \beta$. □

Since homomorphisms fix terms from $\Sigma_{\mathcal{F}}$ we conclude

Corollary 4.3. *Suppose $\mathcal{F}$ and $\mathcal{G}$ are rigid proof grammars and there is a homomorphism $f\colon \mathcal{F} \to \mathcal{G}$. Then $\mathcal{L}(\mathcal{F}) \subseteq Ł(\mathcal{G})$.*

4.2 Partitioning

Let π be a proof ending in a cut. The grammar $\mathcal{G}_\pi$ can be viewed as the union of the two disjoint subgrammars $\mathcal{G}^0_{\pi|0}$ and $\mathcal{G}^1_{\pi|1}$ and 'connecting' production rules (i.e. rules introduced by the cut). A full derivation d in $\mathcal{G}_\pi$ can therefore be permuted to the form $d_1 d_2 \cdots d_k$ where each d_i is a full derivation in either $\mathcal{G}^0_{\pi|0}$ or $\mathcal{G}^1_{\pi|1}$, plus one direction of the connecting rules. The next lemma demonstrates that in the case of cuts on (simple) Π_2-formulæ, there can be at most three such 'alternations' between the two subgrammars. This observation will be particularly relevant in the analysis of the Quantifier Reduction rule (Lemma 5.6).

Lemma 4.4 (Cut Partition). *Suppose π is the simple Π_2-proof given below where A is assumed to be a Π_2 formula.*

$$\frac{\pi_0 \vdash \Gamma, A \qquad \pi_1 \vdash \Delta, \bar{A}}{\pi \vdash \Gamma, \Delta}\,\mathit{cut}$$

For $F \in \Gamma$ and $N \in \mathrm{Term}(\Sigma_\pi)$, if $d\colon \sigma^0_F \to N$ is a rigid derivation in $\mathcal{G}_\pi$ then there is a permutation $d' = d_1 d_2 d_3 d_4\colon \sigma^0_F \to N$ of d such that

- *$d_1 : \sigma^0_F \to M_1$ is a full derivation in $\mathcal{G}^0_{\pi_0} \cup \{\tau^0_A \to \sigma^1_{\bar{A}}\}$;*
- *$d_2 : M_1 \to M_2$ is a rigid derivation $\mathcal{G}^1_{\pi_1} \cup \{\tau^1_{\bar{A}} \to \sigma^0_A\}$;*
- *$d_3 : M_2 \to M_3$ is a derivation in $\mathcal{G}^0_{\pi_0}$;*
- *$d_4 : M_3 \to N$ is a derivation using only the rules $\{\tau^0_F \to \tau^{\langle\rangle}_F \mid F \in \Gamma\} \cup \{\tau^1_F \to \tau^{\langle\rangle}_F \mid F \in \Delta\}$.*

Proof: The derivation d' is obtained through permuting the production rules so that we fully re-write the non-terminals of a subgrammar $\mathcal{G}^i_{\pi_i}$ before passing (through the cut on A) to those of $\mathcal{G}^{1-i}_{\pi_{1-i}}$. In this way it is clear how to obtain full derivations d_1 and d_2 satisfying the requirements such that $d_1 d_2$ is a subsequence of d. Note, as d_1 and d_2 are full, M_1 and M_2 are, respectively, terms in $\Sigma_\pi \cup \{\sigma^1_{\bar{A}}\} \cup \{\tau^0_F \mid F \in \Gamma\}$ and $\Sigma_\pi \cup \{\sigma^0_A\} \cup \{\tau^0_F \mid F \in \Gamma\} \cup \{\tau^1_F \mid F \in \Delta\}$. Finally, to obtain d_3 we rewrite (according to d) all occurrences of σ^0_A in M_2. Observe that in doing so the non-terminal τ^0_A will not be created: d_3 writes each occurrence of σ^0_A in M_2 to a term of the form $\lambda\alpha.S$ where α is the *unique* eigenvariable for the external universal quantifier in A; the definition of $\lhd_\pi$ and the restriction on rigid derivations implies d_3 may not utilise a production rule of the form $\alpha \to \tau^q_A$, so τ_A^0 cannot appear in M_3. Therefore, the only non-terminals from $\mathcal{G}_\pi$ in M_3 are $\{\tau_F^0 \mid F \in \Gamma\} \cup \{\tau_F^1 \mid F \in \Delta\}$ which will be written with the remaining of production rules from $\mathrm{prod}_\tau(\langle\rangle, \Gamma, \Delta)$. □

Theorem 4.5 (Partition). *Let π be a simple Π_2-proof and $p \in \mathrm{Pos}(\pi)$. Denote by $\mathcal{G}_p$ the sub-grammar of $\mathcal{G}$ comprising only those production rules writing non-terminals in $\{\sigma_F^q, \tau_F^q \mid p \leq q\} \cup \mathrm{EV}(\pi|p)$ with start symbols $\{\sigma_F^p \mid F$ in the end sequent of $\pi|p\}$. If $N \in Ł(\mathcal{G}_\pi)$, then there exists a rigid derivation $d\colon \sigma \to N$ in $\mathcal{G}_\pi$ of the form $d = d_0 d_1 \cdots d_L$ where for all $i \leq L$,*

- *$d_i\colon M_i \to M_{i+1}$ is a full derivation in $\mathcal{G}_\pi \setminus \mathcal{G}_p$ if i is even;*
- *$d_i\colon M_i \to M_{i+1}$ is a full derivation in $\mathcal{G}_p$ if i is odd.*

Proof: The proof is by induction on p. The base case $p = \langle\rangle$ holds trivially, choosing $d_0 = d$. Suppose $p = qj$ where $j \in \{0, 1\}$. Let L' and $\langle d_i' \rangle_{i \leq L'}$ be given by the induction hypothesis for $q \in \mathrm{Pos}(\pi)$. For each $i \leq L'$ therefore $d_i'\colon M_i \to M_{i+1}$ is a full derivation in $\mathcal{G}_\pi \setminus \mathcal{G}_q$, if i is even, and in $\mathcal{G}_q$ otherwise.

First suppose the final rule of $\pi|q$ is a cut. By Lemma 4.4 for each odd i, d_i' can be replaced by a derivation of the form $e_{i,0}e_{i,1}e_{i,2}\colon M_i \to M_{i+1}$ where $e_{i,0}$ and $e_{i,2}$ are full derivations in $\mathcal{G}_{q0} = \mathcal{G}_p$ and $e_{i,1}$ is a full derivation in $\mathcal{G}_{q1}$.

Suppose $j = 1$. Then we pick $L = L'$ and set

$$d_i = \begin{cases} e_{i-1,2}d_i'e_{i+1,0} & i \leq L \text{ even}, \\ e_{i,1}, & i \leq L \text{ odd}, \end{cases}$$

where, if necessary, $e_{L+1,0} = e_{-1,2} = \langle\rangle$. If instead $j = 0$ we set $L = 2L'$ and for each $2i \leq L'$ we define

$$d_{4i} = d_{2i}' \qquad d_{4i+2} = e_{2i+1,1}$$
$$d_{4i+1} = e_{2i+1,0} \qquad d_{4i+3} = e_{2i+1,2}.$$

Now suppose the rule at position q in π is not a cut. Notice that if $e\colon a \to S$ is a derivation in $\mathcal{G}_q$ with $a \in \mathcal{N}_{\mathcal{G}_q}$ then e is either a derivation wholly in $\mathcal{G}_{q0}$ or a derivation wholly in $\mathcal{G}_{q1}$. Thus for each odd $i \leq L'$, d_i' can be re-ordered as a derivation $e_{i,0}e_{i,1}\colon M_i \to M_{i+1}$ where $e_{i,0}$ is full in $\mathcal{G}_{q0}$ and $e_{i,1}$ is full in $\mathcal{G}_{q1}$. Moreover, observe that the two derivations are independent of one another, so $e_{i,1}e_{i,0}\colon M_i \to M_{i+1}$. Using this fact it is straightforward to alter the arrangement of derivations to obtain a satisfying sequence $\langle d_i \rangle_{i \leq L}$ with $L \leq L'$. □

4.3 Lifting

Another ingredient required in the analysis of the Quantifier Reduction rule is commuting derivations with β-reduction. More precisely, given a derivation

$d\colon T \to U$ and β-reduction $R \rightarrowtail_X T$, whether it is possible to 'lift' d to a derivation $e\colon R \to S$ such that $S \rightarrowtail_X U$ as in the figure below. This is not in general possible as demonstrated by the example in Figure 3. Lemma 4.8 pins down sufficient conditions for completing the square. Before this we define the reverse process which can be visualised as 'pushing' a derivation through β-reduction.

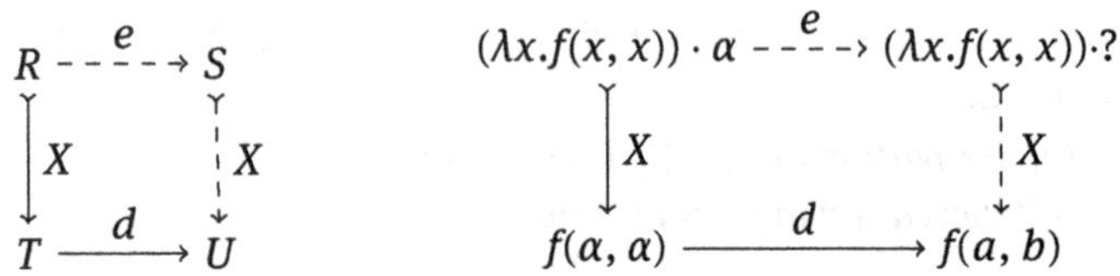

Fig. 3. Lifting

Definition 4.6. *Let $e = \langle \rho_i, p_i \rangle_{i<\mathrm{lh}(e)}\colon R \to S$ be a derivation in grammar $\mathcal{G}$ and suppose $R \rightarrowtail_X T$. The derivation e^T is defined by recursion on* $\mathrm{lh}(e)$.

- *For $\mathrm{lh}(e) = 0$, $e^T := \langle\rangle$;*
- *For $\mathrm{lh}(e) = n+1$, $e^T := e_0^T \langle \rho_n, q_i \rangle_{i\le k}$ where $e_0 = \langle \rho_i, p_i \rangle_{i<n}$ and $\{q_i \mid i \le k\}$ is an enumeration of $\{q \mid r_{T,R}(q) = p_n\}$.*

It is easily seen that

Lemma 4.7. *If $e\colon R \to S$, $R \rightarrowtail_X T$ and $e^T\colon T \to U$ then $S \rightarrowtail_X U$, and if d is strongly rigid, so is e^T.*

Lemma 4.8 (Lifting). *Let $d = \langle a_i \to S_i, p_i \rangle_{i<k}\colon T \to U$ be a strongly rigid derivation in grammar $\mathcal{G}$ and $R \rightarrowtail_X T$. Suppose*

1. *for all positions q_0, q_1, if $r_{T,R}(q_0) = r_{T,R}(q_1)$ then the subderivations of d from q_0 and q_1 are permutations of one another;*
2. *if $a_i \in X$ then there exists $p < p_i$ and non-terminal $\alpha \notin X$ such that $a_i \lhd \alpha$ and $U|p = \lambda\alpha.(U|p0)$.*

Then there exists a strongly rigid derivation $e\colon R \to S$ such that e^T is a permutation of d and $S \rightarrowtail_X U$.

Proof: Let $r_{T,R}$ be the canonical function mapping positions in T to their corresponding position in R. To each $p \in \mathrm{Pos}(S)$ we can associate a particular position in T corresponding to p. Thus let $r_{T,S}^{-1}$ be a function such that

$$r_{T,S}^{-1}(p) \in \{q \mid r_{T,S}(q) = p\}$$

for every position p. We now define $e = f(d)$ recursively:

- For $k = 0$, $f(d) = \langle\rangle$;
- For $k = l + 1$, if $p_l = r^{-1}_{T,R}(p)q$ for some maximal $p \in \mathrm{Pos}(R)$ and some q, set $f(d) = f(d_0)\langle a_l \to S_l, pq\rangle$ where $d_0 = \langle a_i \to S_i, p_i\rangle_{i<l}$, otherwise set $f(d) = f(d_0)$.

$f(d)$ is well-defined since condition 1. ensures that $f(d)$ is invariant under the choice of the inverse r^{-1}, and condition 2. guarantees that every applicable production rule in d will be also freely applicable in $f(d)$. □

5 Reductions

We begin with showing that the simplest reductions yield homomorphisms.

Lemma 5.1. *Suppose $\pi \rightsquigarrow \pi'$ is an instance of a reduction rule in Figure 1 or 2 except Contraction or Quantifier Reduction. Then $\mathcal{L}(\mathcal{G}_{\pi'}) \subseteq \mathcal{L}(\mathcal{G}_\pi)$. If the reduction is not a case of Weakening Reduction then in fact $\mathcal{L}(\mathcal{G}_{\pi'}) = \mathcal{L}(\mathcal{G}_\pi)$.*

Proof: Suppose $\pi \rightsquigarrow \pi'$ is and instance of Axiom Reduction and

$$\dfrac{\overset{\pi_0}{\Gamma, A} \qquad A, \bar{A}}{\pi|p \vdash \Gamma, A}\ \text{cut} \qquad\qquad \overset{\pi_0}{\pi'|p \vdash \Gamma, A}$$

A homomorphism $f\colon \mathcal{G}_{\pi'} \to \mathcal{G}_\pi$ can be easily given by a function which maps a non-terminal v^{pq}_F for $v \in \{\sigma, \tau\}$ and $pq \in \mathrm{Pos}(\pi')$ with $q \neq \langle\rangle$ to the corresponding ones in π at position $p0q$ and is identity on all other non-terminals. Consider a rule $\sigma^p_F \to S$ in π' then $\sigma^{0p} \to f(S)$ is a production rule in π. If $F \in \Gamma$, $\sigma^p_F \to \sigma^{p0}_F$ is also a production rules in $\mathcal{G}_\pi$; otherwise, $F = A$ and the sequence

$$\sigma^p_A \to \sigma^{p1}_A \to \tau^{p1}_{\bar{A}} \to \sigma^{p0}_A$$

is a derivation in π. In either case, we have a derivation $f(\sigma^p_F) \to f(S)$ in π as required. Conversely, one can define a homomorphism $f\colon \mathcal{G}_\pi \to \mathcal{G}_{\pi'}$ by the function below, hence $\mathcal{L}(\mathcal{G}_\pi) = \mathcal{L}(\mathcal{G}_{\pi'})$.

$$f(v^p_F) = v^p_F \qquad\qquad f(v^{p0q}_F) = v^{pq}_F$$
$$f(\sigma^{p1}_A) = f(\tau^{p1}_{\bar{A}}) = \sigma^p_A \qquad\qquad f(\tau^{p1}_A) = f(\sigma^{p1}_{\bar{A}}) = \tau^p_A$$

Now consider an instance of Weakening Reduction. Suppose

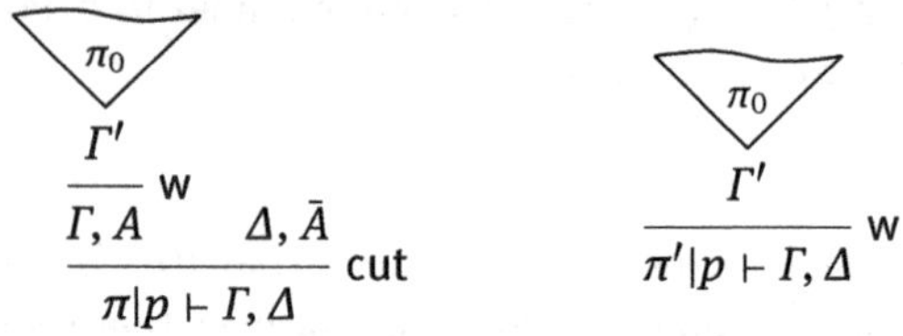

where $\Gamma' \subseteq \Gamma$. A homomorphism from $\mathcal{G}_{\pi'}$ to $\mathcal{G}_\pi$ is defined as follows. The function $f\colon \mathcal{N}_{\pi'} \to \mathcal{N}_\pi$ is determined by

$$f(\nu_F^{pq}) = \nu_F^{p0q} \qquad \text{for } \nu \in \{\sigma, \tau\} \text{ and every } q \neq \langle\rangle \text{ with } pq \in \mathrm{Pos}(\pi'|p);$$
$$f(\gamma) = \gamma \qquad \text{for all other non-terminals.}$$

Production rules from π' are readily mapped to derivation in π and the remaining conditions of d are trivially met.

Suppose instead that $\pi \rightsquigarrow \pi'$ is a case of rule permutation. Of particular interest is if the reduction is the permutation of two cuts, for example $\pi|p$ and $\pi'|p$ are of the form

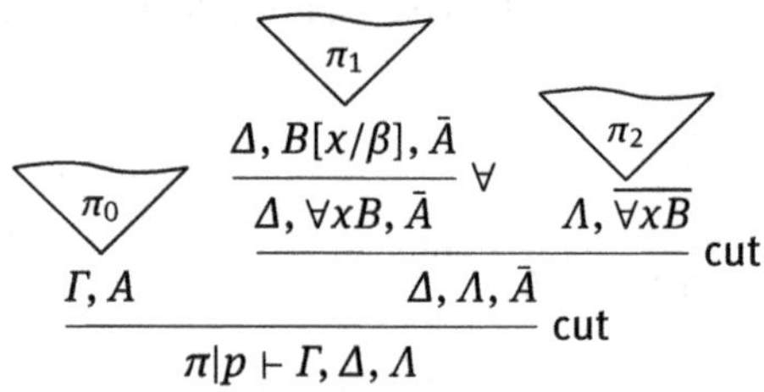

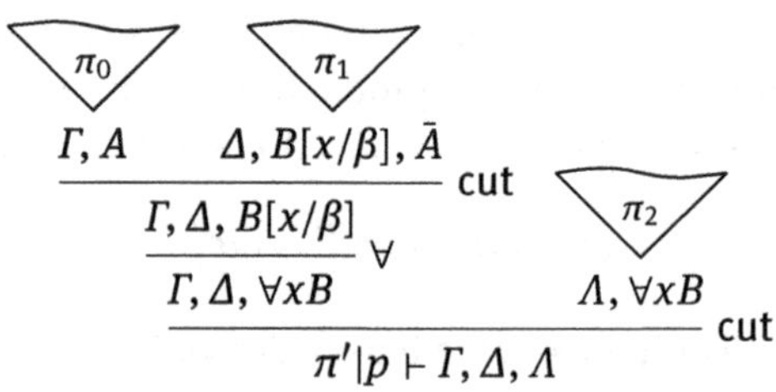

where A is a Π_2 formula and B is Σ_1.

We begin showing $\mathcal{L}(\mathcal{G}_{\pi'}) \subseteq \mathcal{L}(\mathcal{G}_\pi)$. This is achieved by considering the homomorphism that maps, for $i = 0, 1, 2$, the copy of $\mathcal{G}_{\pi_i}$ in $\mathcal{G}_{\pi'}$ to the copy in $\mathcal{G}_\pi$. Explicitly, we define $f\colon \mathcal{N}_{\pi'} \to \mathcal{N}_\pi$ as follows. For each position q, formula F and $\nu \in \{\sigma, \tau\}$ set

$$f(\nu_F^{p000q}) = \nu_F^{p0q} \qquad f(\nu_F^{p001q}) = \nu_F^{p100q} \qquad f(\nu_F^{p1q}) = \nu_F^{p11q}.$$

The non-terminals arising from positions $p0$ and $p00$ in π' are then appropriately mapped to those in π so connections between the sub-grammars are preserved:

$$f(v_F^{p0}) = \begin{cases} v_F^{p0} & \text{if } F \in \Gamma \\ v_F^{p10} & \text{if } F \in \Delta \cup \{\forall x B\} \end{cases} \qquad f(v_F^{p00}) = \begin{cases} v_F^{p0} & \text{if } F \in \Gamma \\ v_F^{p100} & \text{if } F \in \Delta \cup \{B[x/\beta]\} \end{cases}$$

In all other cases (i.e. a is a non-terminal of $\mathcal{G}_{\pi'}$ but not of $\mathcal{G}^p_{\pi'|p}$) we set $f(a) = a$.

We now argue that if A is not quantifier free and the last rule in π_0 is not weakening then f is a homomorphism. Since π is a simple proof, A has the form $\forall x A_0$ for some Σ_1 formula A_0 and there is an eigenvariable $\alpha \in \mathrm{EV}(\pi_0)$ such that the last inference rule of π_0 is

$$\frac{\Gamma, A_0[x/\alpha]}{\Gamma, A}\ \forall$$

By the definition of $\lhd_\pi$ notice that $\xi \lhd_\pi \alpha \lhd_\pi \tau_A^{p0}$ for every $\xi \in \mathrm{EV}(\pi_0) \setminus \{\alpha\}$. It is clear that f respects the ordering $\lhd'_\pi$ when restricted to non-terminals from either π_0, π_1 or π_2. The only other cases we need show are

$$f(\tau_A^{p000}) \lhd_\pi f(\sigma_{\bar{A}}^{p001}) \qquad\qquad \xi \lhd_\pi \beta \lhd_\pi f(\tau_{\forall x B}^{p0}) \lhd_\pi f(\sigma_{\overline{\forall x B}}^{p1})$$

for every $\xi \in \mathrm{EV}(\pi_0)$. The first inequation is given by the definition of $\lhd_\pi$. The second is determined by observing

$$\xi \lhd_\pi \tau_A^{p0} \lhd_\pi \sigma_{\bar{A}}^{p1} \lhd_\pi \sigma_{\bar{A}}^{p100} \lhd_\pi \beta \lhd_\pi \tau_{\forall x B}^{p10} \lhd_\pi \sigma_{\overline{\forall x B}}^{p11}.$$

As f is injective on eigenvariables in π' condition 6 is also satisfied and by Corollary 4.3 we are done.

In the other cases, namely that A is quantifier free or A has no eigenvariable associated to it in π_0, f need not be a homomorphism because for an eigenvariable $\xi \in \mathrm{EV}(\pi_0)$ we will have $\xi \not\lhd_\pi \beta$ but $\xi \lhd'_\pi \beta$. Nevertheless, by the Cut Partition Lemma 4.4 it is simple to show that f maps rigid derivations in $\mathcal{G}_{\pi'}$ to rigid derivations in $\mathcal{G}_\pi$.

The converse direction, $\mathcal{L}(\mathcal{G}_\pi) \subseteq Ł(\mathcal{G}_{\pi'})$, is witnessed by the function $f\colon \mathcal{N}_\pi \to \mathcal{N}_{\pi'}$ given by

$$f(v_F^{p0q}) = v_F^{p000q} \qquad f(v_F^{p100q}) = v_F^{p001q} \qquad f(v_F^{p11q}) = v_F^{p1q}$$

$$f(v_F^{p10}) = \begin{cases} v_F^{p001} & \text{if } F = \overline{A} \\ v_F^{p0} & \text{otherwise} \end{cases} \qquad f(v_F^{p1}) = \begin{cases} v_F^{p0} & \text{if } F \in \Delta \\ v_F^{p001} & \text{if } F = \overline{A} \\ v_F^{p1} & \text{otherwise} \end{cases}$$

for each q, F and $v \in \{\sigma, \tau\}$, and setting $f(a) = a$ in all other cases which in all cases is a homomorphism. □

We now proceed with Contraction Reduction. In this case the obvious renaming of non-terminals is not injective and additional properties are required in order to verify the preservation of rigidity i.e. condition 6 of homomorphism.

For the remainder of the section we use symbols x and y (possibly with indices) as meta-variables for positions in terms to contrast with positions in proofs which will be denoted by p, q, etc.

Lemma 5.2 (Contraction Reduction). *If π, π' are simple Π_2-proofs and $\pi \rightsquigarrow \pi'$ is an instance of Contraction Reduction, $\mathcal{L}(\mathcal{G}_{\pi'}) = Ł(\mathcal{G}_\pi)$.*

Proof: Suppose that the reduction $\pi \rightsquigarrow \pi'$ is an instance of Contraction Reduction to position p as follows:

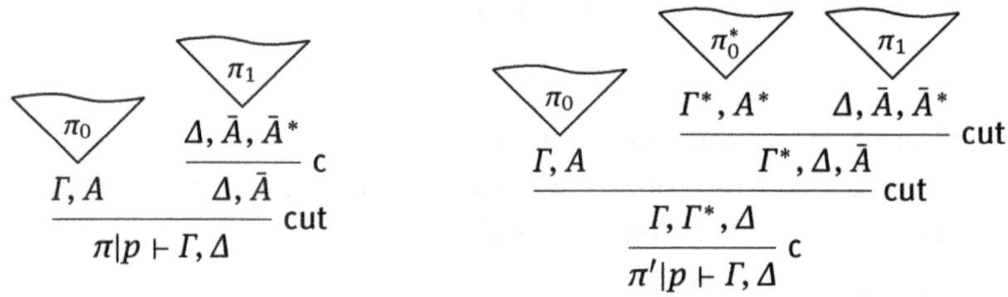

Note that we have assumed the contraction is on a single formula $\bar{A}$; the case for contraction of multiple formulæ is analogous. Consider the function $f\colon \mathcal{N}_{\pi'} \to \mathcal{N}_\pi$ defined by

$$
\begin{array}{rlrl}
f(\gamma) = f(\gamma^*) = \gamma & \text{for } \gamma \in \mathrm{EV}(\pi_0) & f(v_F^{p00q}) = f(v_{F^*}^{p010q}) = v_F^{p0q} & \\
f(\delta) = \delta & \text{for } \delta \in \mathrm{EV}(\pi_1) & f(v_F^{p011q}) = v_F^{p10q} & \\
f(v_F^{p01}) = v_F^{p10} & \text{for } F \in \Delta \cup \{\bar{A}\} & f(v_{F^*}^{p01}) = v_F^{p0} & \text{for } F \in \Gamma \\
f(v_F^{p0}) = v_F^{p10} & \text{for } F \in \Delta & f(v_{F/F^*}^{p0}) = v_F^{p0} & \text{for } F \in \Gamma \\
f(a) = a & & &
\end{array}
$$

It is easy to verify that f transforms production rules in $\mathcal{G}_{\pi'}$ to derivations in $\mathcal{G}_\pi$. For instance, the production rules $\tau_A^{p00} \to \sigma_{\bar{A}}^{p01}$ and $\tau_{A^*}^{p010} \to \sigma_{\bar{A}^*}^{p011}$, from the cuts in π' on A and A^*, are transformed under f to $\tau_A^{p0} \to \sigma_{\bar{A}}^{p1} \to \sigma_{\bar{A}}^{p10}$ and $\tau_A^{p0} \to \sigma_{\bar{A}}^{p1} \to \sigma_{\bar{A}^*}^{p10}$ respectively.

Condition 6 is the only non-trivial case to verify. Suppose, in search of a contradiction, $d = \langle (a_i \to S_i), y_i \rangle_{i<\mathrm{lh}(d)}\colon a_0 \to S$ is a strongly rigid derivation in $\mathcal{G}_{\pi'}$ and $i_0, i_1 < \mathrm{lh}(d)$ are such that i) $i_0 \approx_d i_1$, ii) $f(a_{i_0}) = f(a_{i_1}) \in R_\pi$ and iii) $a_{i_0} \neq a_{i_1}$. We may assume d is the shortest rigid derivation satisfying these conditions, so in particular for every $i < \mathrm{lh}(d)$, $y_i \leq y_{i_0}$ or $y_i \leq y_{i_1}$, both hold only if $i = 0$, and $\mathrm{lh}(d) = \max\{i_0 + 1, i_1 + 1\}$. By the definition of f, (ii) and (iii) we may assume $a_{i_0} = \gamma \in \mathrm{EV}(\pi_0)$ and $a_{i_1} = \gamma^* \in \mathrm{EV}(\pi_0^*)$. The argument breaks into three cases:

1. d writes one of the two non-terminals $\sigma_{A^*}^{p010}$ or σ_{A}^{p00};
2. $a_0 \in \mathrm{EV}(\pi_0) \cup \mathrm{EV}(\pi_0^*)$ or $a_0 \in \{\sigma_F^{p00q}, \sigma_F^{p010q}\}$ for some $q \in \mathrm{Pos}(\pi_0)$ and simple formula F;
3. there exist distinct $j_0, j_1 < \mathrm{lh}(d)$ and formulæ $F_0, F_1 \in \Gamma$ such that for each $k \in \{0,1\}$, $0 < j_k < i_k$, $y_{j_k} \leq y_{i_k}$, $a_{j_0} = \sigma_{F_0}^{p0}$ and $a_{j_1} = \sigma_{F_1^*}^{p0}$.

Notice that the failure of 1 implies either 3 or that a_0 is a non-terminal from one of π_0 or π_0^*, a condition that, by the minimality of d, is equivalent to 2.

First suppose 1 holds. Due to the minimality of d the derivation must pass through (from right to left) at least one of the two marked cuts, so there is $i \geq j$ such that either $a_i = \tau_{\bar{A}}^{p01}$ and $S_i = \sigma_A^{p00}$, or $a_i = \tau_{\bar{A}^*}^{p011}$ and $S_i = \sigma_{A^*}^{p010}$. This is only possible if $\bar{A}$ contains a universal quantifier and so, by simplicity of π, $\bar{A}$ cannot be in the class Π_2, that is, $A = \forall x \exists y A_0$ for some quantifier-free A_0. But then we may find an eigenvariable of $\mathrm{EV}(\pi_1)$ that is written by d and contradicts assumption (i).

Suppose 2 holds; in particular that $a_0 \in \mathrm{EV}(\pi_0)$ or $a_0 = \sigma_F^{p00q}$ for some $q \in \mathrm{Pos}(\pi_0)$ and simple formula F. Since $a_{i_1} \in \mathrm{EV}(\pi_0^*)$ and $y_0 < y_{i_1}$ there exists $j_0 < i_1$ such that $y_0 < y_{j_0} \leq y_{i_1}$ and either $a_{j_0} = \tau_A^{p00}$ or $a_{j_0} \in \{\tau_F^q \mid q \leq p \text{ and } F \text{ is simple}\}$ (if the production rule in d giving rise to a_{j_0} writes an eigenvariable from $\mathrm{EV}(\pi') \setminus \mathrm{EV}(\pi_0)$ then in fact $a_{j_0} = \tau_F^q$ for some $q < p$ and some F). Assuming 1 does not hold the latter case will hold for some choice of j_0 whence there also exists $j_0 \leq j_1 < i_1$ such that $a_{j_1} \in \{\sigma_F^p \mid F \in \Gamma\}$. But then there will be $j_0 < i < j_1$ such that $y_i \leq y_{j_1} \leq y_{i_1}$, $a_i \in \mathrm{EV}(\pi') \setminus \mathrm{EV}(\pi'|p)$ and $\xi \lhd_\pi' a_i$ for every $\xi \in \mathrm{EV}(\pi'|p)$, contradicting (i).

If 3 holds then similar to the above argument there must exist $i < \mathrm{lh}(d)$ such that $a_i \in \mathrm{EV}(\pi') \setminus \mathrm{EV}(\pi'|p)$ and $\xi \lhd_\pi' a_i$ for every $\xi \in \mathrm{EV}(\pi')$.

By Lemma 4.2 we conclude $\mathcal{L}(\mathcal{G}_{\pi'}) \subseteq \mathrm{Ł}(\mathcal{G}_\pi)$. The converse inclusion holds because f is, in a suitable sense, surjective. In particular, for every rigid derivation d in $\mathcal{G}_\pi$ there exists a derivation d' in $\mathcal{G}_{\pi'}$ such that d is a permutation of $f(d')$. d' is chosen by replacing each sub-derivation that resides wholly within π_0 by its counterpart in π_0^* if this sub-derivation is immediately preceded or succeeded by the production rules $\tau_{\bar{A}^*}^{p10} \to \tau_{\bar{A}}^{p1}$ or $\sigma_{\bar{A}}^{p1} \to \sigma_{\bar{A}^*}^{p10}$. This operation is guaranteed to yield a derivation in $\mathcal{G}_{\pi'}$ by the Partition Lemma for Cut which implies that there is no derivation in $\mathcal{G}_\pi$ that write σ_A^{p0} to a term containing τ_A^{p0} and involves only rules taken from π_0. □

Before we proceed with Quantifier Reduction we need the following lemmas.

Lemma 5.3. *If π is a simple Π_2-proof, $d\colon \tau_A^p \to S$ is a derivation in $\mathcal{G}_\pi$ and $S|x = \alpha \in \mathrm{EV}(\pi|q)$ for some $x \in \mathrm{Pos}(S)$ and $p < q \in \mathrm{Pos}(\pi)$ then there exist $y < x$ and $\beta \in \mathrm{EV}(\pi)$ such that $S|y = \lambda\beta.(S|y0)$ and $\alpha \lhd \beta$.*

Proof: By examination of the definition of $\mathcal{G}_\pi$. □

Lemma 5.4 (Substitution). *Let π be a regular proof, s a term whose free variables do not appear in* $\mathrm{EV}(\pi)$ *and* $\alpha \notin \mathrm{EV}(\pi)$. *Then for every $d\colon M \to N$ in $\mathcal{G}_{\pi[\alpha/s]}$ there exists a derivation $d'\colon M' \to N'$ in $\mathcal{G}_\pi$ such that $M = M'[\alpha/s]$ and $N = N'[\alpha/s]$ and d' is strongly rigid iff d is strongly rigid.*

Proof: Let f be the natural homomorphism from $\mathcal{G}_\pi$ to $\mathcal{G}_{\pi[\alpha/s]}$. Notice f is surjective on derivations up to permutation. □

Let $d = \langle (a_i \to S_i), y_i \rangle_{i<\mathrm{lh}(d)}\colon S \to T$ be a derivation. Recall the relation $\sim_d$ defined in Section 3.3. We could instead define $\sim_d$ on the set $\mathrm{Pos}(T)$. For $x_0, x_1 \in \mathrm{Pos}(T)$, $x_0 \sim_d x_1$ if and only if there exists $j_0, j_1 < \mathrm{lh}(d)$ such that

1. $x_0 = y_{j_0}$ and $x_1 = y_{j_1}$,
2. $a_{j_0} = a_{j_1} \in \mathcal{R}$,
3. for every $i_0 < i < \mathrm{lh}(d)$ and $k \in \{0, 1\}$, if $a_i \in R$ and $y_i \leq y_{j_k}$ then $a_{j_i} \ntriangleleft a_i$.

The relation $\sim_d$ on positions (as opposed to indices) is not in general transitive but it suffices to define rigidity and is a useful alternative. For example,

Lemma 5.5. *Let $d\colon S \to T$ be a derivation and suppose d' is a permutation of d. If d' is a derivation and $\sim$ is defined on positions then $\sim_d = \sim_{d'}$. In addition, $\sim_d$ is independent of the choice of S.*

Lemma 5.6 (Quantifier Reduction). *If π and π' are the simple Π_2-proofs and $\pi \rightsquigarrow \pi'$ is an instance of Quantifier Reduction* $\mathcal{L}(\mathcal{G}_{\pi'}) =$ Ł$(\mathcal{G}_\pi)$.

Proof: Suppose that the reduction $\pi \rightsquigarrow \pi'$ is an instance of Quantifier Reduction to position p as follows:

$$\dfrac{\dfrac{\overset{\pi_0}{\Gamma_0, A_\alpha}}{\Gamma_0, \forall x\, A_x}\,\forall \qquad \dfrac{\overset{\pi_1}{\Gamma_1, \bar{A}_s}}{\Gamma_1, \exists x\, \bar{A}_x}\,\exists}{\pi|p \vdash \Gamma_0, \Gamma_1}\,\mathrm{cut} \qquad\qquad \dfrac{\overset{\pi_0[\alpha/s]}{\Gamma_0, A_s} \qquad \overset{\pi_1}{\Gamma_1, \bar{A}_s}}{\pi'|p \vdash \Gamma_0, \Gamma_1}\,\mathrm{cut}$$

where A_t abbreviates $A[\alpha/t]$. Let $\mathcal{G} = \mathcal{G}_\pi$ and $\mathcal{G}' = \mathcal{G}_{\pi'}$. We begin showing $\mathcal{L}(\mathcal{G}_{\pi'}) \subseteq$ Ł$(\mathcal{G}_\pi)$.

Suppose $a \in N_{\mathcal{G}'}$ and let $d\colon a \to T$ be a full strongly rigid derivation in $\mathcal{G}'$. We may assume d is partitioned according to the Partition Theorem 4.5, whence $d = d_0 d_1 \cdots d_{2L}$ and for each $i \leq L$, $d_i\colon S_i \to S_{i+1}$ is a full derivation in $\mathcal{G}' \setminus \mathcal{G}'_p$, if i is even, and in $\mathcal{G}'_p$ otherwise. We determine a term $T' \rightarrowtail^* T$ and rigid derivation $d'\colon a \to T'$ in $\mathcal{G}$ by recursion on L.

If $L = 0$, $d = d_0$ is a derivation solely in $\mathcal{G}' \setminus \mathcal{G}'_p$ and hence also a derivation in $\mathcal{G}$. So suppose $L > 0$. The Cut Partition Lemma 4.4 determines a strongly rigid permutation of d_1 of the form $e_0e_1e_2e_3e_4$ where

- $e_0 : S_1 \to S_1^1$ is derivation using only rules in $\{\sigma_F^p \to \sigma_F^{pi} \mid i < 2 \wedge F \in \Gamma_i\}$;
- $e_1 : S_1^1 \to S_1^2$ is a full derivation in the grammar $\mathcal{G}'_{p0}$;
- $e_2 : S_1^2 \to S_1^3$ is a full derivation in $(\mathcal{G}'_{p1} \setminus \{\tau_{\bar{A}_s}^{p1}\})$;
- $e_3 = \langle(\tau_{\bar{A}_s}^{p1} \to \sigma_{A_s}^{p0}), x_i\rangle_{1\le i\le k} : S_1^3 \to S_1^4 = S_1^3[\tau_{\bar{A}_s}^{p1}/\sigma_{A_s}^{p0}]$;
- $e_4 = e_{4,1}^{x_1} \cdots e_{4,k}^{x_k} : S_1^4 \to S_2$ where for each $1 \le i \le k$, $e_{4,i}$ is a derivation of $S_2|x_i$ from $\sigma_{A_s}^{p0}$ using only rules in $\mathcal{G}'_{p0}$, and $e_{4,i}^{x_i}$ is the induced derivation of $S_1^4[x_i/S_2|x_i]$ from S_1^4 obtained by adjusting $e_{4,i}$ to S_1^4.

The Substitution Lemma 5.4 provides translations of e_1 and e_4 to derivations $\hat{e}_1 : S_1^1 \to \hat{S}_1^2$ and $\hat{e}_4 = \hat{e}_{4,1}^{x_1} \cdots \hat{e}_{4,k}^{x_k} : \hat{S}_1^4 \to \hat{S}_2$ in $\mathcal{G}_{p0}$ such that

$$S_1^2 = \hat{S}_1^2[\tau_{A_\alpha}^{p00}/\tau_{A_s}^{p0}][\alpha/s] \qquad S_1^4 = \hat{S}_1^4[\sigma_{A_s}^{p0}/\sigma_{A_\alpha}^{p00}] \qquad S_2 = \hat{S}_2[\alpha/s].$$

Within $\mathcal{G}_{\pi|p}$ there is a simple derivation of s from α (i.e., $\alpha \to \tau_{\forall x A_x}^{p0} \to \sigma_{\exists x \bar{A}_x}^{p1} \to s$) which, when combined with $\hat{e}_1$ and the natural translation of e_2 into $\mathcal{G}_{p10}$, yields a derivation $e = e_0\hat{e}_1\hat{e}_2 : S_1 \to \hat{S}_1^3 = S_1^3[\tau_{\bar{A}_s}^{p1}/\tau_{\bar{A}_s}^{p10}]$.

The derivation e_3, when translated to $\mathcal{G}_{\pi|p}$,[4] becomes

$$\hat{e}_3 : \hat{S}_1^3 \to S_1^3[\tau_{\bar{A}_s}^{p1}/(\lambda\alpha.\, \sigma_{A_\alpha}^{p00}) \cdot s]$$

and may be connected with e and $\hat{e}_{4,1}^{x_100} \cdots \hat{e}_{4,k}^{x_k00}$ to yield the derivation $\hat{e}$ given by

$$e_0\hat{e}_1\hat{e}_2\hat{e}_3\hat{e}_{4,1}^{x_100} \cdots \hat{e}_{4,k}^{x_k00} : S_1 \to S_2' = S_2[x_1/(\lambda\alpha.\hat{S}_2|x_1) \cdot s] \cdots [x_k/(\lambda\alpha.\hat{S}_2|x_k) \cdot s].$$

Clearly $S_2' \rightarrowtail_\alpha S_2$. All that remains is to show that the derivation $d_2 \cdots d_{2L} : S_2 \to T$ 'lifts' to a derivation $\hat{d}_2 : S_2' \to \hat{T}$ such that $d_0\hat{e}\hat{d}_2 : a \to \hat{T}$ is rigid and $\hat{T} \rightarrowtail T$ as then applying the induction hypothesis to $\hat{d}_2$ will determine a rigid derivation $d' : a \to T'$ in $\mathcal{G}$ such that $T' \rightarrowtail T$.

First we note the following. Suppose y_0 and y_1 are distinct positions in S_2 such that $S_2|y_0 = S_2|y_1 \in \mathcal{R}_{\mathcal{G}'}$ and $r_{S_2,S_2'}(y_0) \sim_{d_0\hat{e}} r_{S_2,S_2'}(y_1)$. If neither y_0 nor y_1 has the form x_i1y for any $i \le k$ and position y then by the choice of $\hat{e}$ and strong rigidity of d in fact $y_0 \sim_{d_0d_1} y_1$. If $y_0 \ge x_i1$ however, then since $S_2|y_0$ is an eigenvariable outside $\pi|p$, also $y_0 \sim_{d_0d_1} y_1$.

We can now verify that the lifting lemma can indeed be applied to $d_2 \cdots d_{2L}$ to obtain a term $\hat{T} \rightarrowtail T$ and derivation $\hat{d}_2 : S_2' \to \hat{T}$: The first condition for lifting

4 Each $(\tau_{\bar{A}_s}^{p1} \to \sigma_{A_s}^{p0})$ is readily translated to $\tau_{\bar{A}_s}^{p10} \to \tau_{\exists x \bar{A}_x}^{p1} \cdot s \to \sigma_{\forall x A_x}^{p0} \cdot s \to (\lambda\alpha.\sigma_{A_\alpha}^{p00}) \cdot s$.

is provided by the argument in the previous paragraph in the case $r_{S_2,S_2'}(y_0) = r_{S_2,S_2'}(y_1)$; the second condition follows from Lemma 5.3. The previous argument also suffices to deduce that $d_0\hat{e}\hat{d}_2$ is rigid. As applying the induction hypothesis to $\hat{d}_2$ does not affect rigidity, we are done.

The converse direction, namely $\mathcal{L}(\mathcal{G}_\pi) \subseteq \text{Ł}(\mathcal{G}_{\pi'})$, is more straightforward. Every derivation e starting from α with $\mathrm{lh}(e) \geq 3$ will have $e(1) = \tau^{p0}_{\forall x A_x}$, $e(2) = \sigma^{p1}_{\overline{\forall x A_x}}$ and $e(3) = s$. Similarly, for any derivation e' in $\mathcal{G}_\pi$, if $i < \mathrm{lh}(e')$ and $e'(i)|x$ has the form $(\lambda\alpha.S_0) \cdot S_1$ then $S_1 = s$. These observations allow a simple translation of derivations in $\mathcal{G}_\pi$ to derivations in $\mathcal{G}_{\pi'}$ preserving rigidity. Let $d\colon a \to S$ be a derivation in $\mathcal{G}_\pi$ where a and S contain neither $\tau^{p1}_{\overline{\forall x A}}$ nor $\sigma^{p0}_{\forall x A}$, and let $S \rightarrowtail^*_\alpha S'$ be the term in which all abstractions on α have been reduced. Then there exists a derivation $d'\colon f(a) \to f(S')$ where $f\colon \mathcal{N}_\pi \to \mathrm{Terms}(\Sigma_\pi \cup \mathcal{N}_{\pi'})$ is defined by

$$f(\alpha) = f(\tau^{p0}_{\forall x A_x}) = f(\sigma^{p1}_{\exists x \bar{A}_x}) = s \qquad f(v^{p00q_0}_F) = v^{p0q_0}_{F[\alpha/s]} \qquad f(v^{p10q_1}_F) = v^{p1q_1}_F$$

for $v \in \{\sigma, \tau\}$, $q_0 \in \mathrm{Pos}(\pi_0)$ and $q_1 \in \mathrm{Pos}(\pi_1)$, and $f(a) = a$ in all other cases. Finally we may deduce that d' is rigid whenever d is strongly rigid by an induction on $\mathrm{lh}(d)$ using Lemma 4.7. As f is the identity function on terms in $\mathrm{Terms}(\Sigma_\pi)$, we conclude $\mathcal{L}(\mathcal{G}_\pi) \subseteq \text{Ł}(\mathcal{G}_{\pi'})$. □

6 Conclusion

We have established the following theorems.

Theorem 6.1. *Let $\pi_0, \pi_1, \ldots, \pi_k$ be a sequence of simple Π_2-proofs such that π_{i+1} is obtained from π_i by the reduction steps outlined in Section 2.1. Then $\mathcal{L}(\mathcal{G}_{\pi_k}) \subseteq \text{Ł}(\mathcal{G}_{\pi_0})$. If, moreover, no step in the reduction is an instance of weakening reduction then $\mathcal{L}(\mathcal{G}_{\pi_0}) = \text{Ł}(\mathcal{G}_{\pi_k})$.*

Theorem 6.2. *Let $\pi \vdash \Gamma$ be a simple Π_2-proof where Γ is a set of Σ_1 formulæ. There exists a grammar $\mathcal{G}$ such that*
1. $|\mathrm{Pr}_\mathcal{G}| \leq |\pi| + |\Gamma|$,
2. $|\text{Ł}(\mathcal{G})| \leq 2^{2^{2|\pi|}}$,
3. *The formula $\bigvee_{F \in \Gamma} F^{\{t | (F,t) \in \text{Ł}(\mathcal{G})\}}$ is valid,*
4. *If π' is obtained from π by a sequence of simple Π_2-proofs using the reduction steps of Section 2.1 and all cuts in π' are on quantifier-free formulæ then the Herbrand set for π', $\mathcal{H}(\pi')$, is contained in $\mathcal{L}(\mathcal{G})$; if no application of weakening reduction is used then $\mathcal{H}(\pi') = \text{Ł}(\mathcal{G})$.*

Proof: Let $\mathcal{G}$ be the modification of $\mathcal{G}_\pi$ which contains a single (fresh) start symbol σ and production rules $\sigma \to (F, \sigma_F)$ for each $F \in \Gamma \cap \Sigma_1$. 1. is immediate from the definition of $\mathcal{G}_\pi$, 2. is a consequence of Lemma 3.14 and 4. is a corollary of the previous theorem and the observation that if π contains only cuts on quantifier-free formulæ then, up to β-equivalence of terms, $\mathcal{H}(\pi) = Ł(\mathcal{G})$. Finally, 3. is deducible from the first part of 4. by observing that standard cut elimination strategies, such as in [Troelstra and Schwichtenberg, 1996], preserve simplicity of proofs. □

These results may be extended to proofs involving prenex Π_2 and Σ_2 formulæ. This can be achieved by endowing the simply typed λ-calculus with product types and pairing functions in the usual way. In addition to the typing definition for simple Π_1 and Σ_1 formulæ defined in the paper, for each position q and prenex Π_2 formula $A = \forall x B$ the two non-terminals τ^q_A and $\sigma^q_{\bar{A}}$ are assigned type type$(\tau^q_A) = o \times \text{type}(\tau^q_B)$, and σ^q_A and $\tau^q_{\bar{A}}$ have the dual type $o \rightarrow \text{type}(\sigma^q_B)$ where o denotes the type of first-order terms. In other words, if $A = \forall x_1 \cdots \forall x_m \exists y_1 \cdots \exists y_n B$ and B is quantifier free then

$$\text{type}(\tau^q_A) = \text{type}(\sigma^q_{\bar{A}}) = o^{m+1} = \underbrace{o \times (o \times \cdots \times o)}_{m+1}$$
$$\text{type}(\sigma^q_A) = \text{type}(\tau^q_{\bar{A}}) = \underbrace{o \rightarrow o \rightarrow \cdots \rightarrow o}_{m} \rightarrow o^{n+1}.$$

The production rules for existential and universal introduction in Table 2 then become respectively

$$\text{prod}(p, \Pi, \emptyset) \cup \{\sigma^p_{\exists yA} \rightarrow (s, \sigma^{p0}_{A[y/s]}),\ \tau^{p0}_{A[y/s]} \rightarrow \tau^p_{\exists yA} \cdot s\},$$
$$\text{prod}(p, \Pi, \emptyset) \cup \{\alpha \rightarrow (\tau^p_{\forall xA})_0,\ \tau^{p0}_{A[x/\alpha]} \rightarrow (\tau^p_{\forall xA})_1,\ \sigma^p_{\forall xA} \rightarrow \lambda\alpha.\sigma^{p0}_{A[x/\alpha]}\}$$

where $(.,.)$ is a (polymorphic) binary pairing function with first and second projections $(.)_0$ and $(.)_1$. Aside from a few technical considerations all the lemmas and proofs presented here generalise to proofs with cuts and end-sequents of prenex Π_2 and Σ_2 formulæ.

6.1 Future work

We have shown the language of grammars defined for simple Π_2-proofs is invariant under most cut reduction rules. It would be interesting to investigate under which further transformations language invariance is maintained. For example, as part of Lemma 5.1 we have shown that permuting two cuts does not change the language of the proof grammar. In other words, composition by cut is an associative operation which, as another interesting line of future work, would allow the consideration of a category of proofs in the spirit of [Hyland, 2002]. Note, the results and techniques presented in this paper are independent of the exact syntactic variant of the sequent calculus used and apply to two-sided, additive, etc. calculi as well.

Further generalisation of this work would involve larger classes of cut-formulæ of which there are two natural directions to proceed. The first is to generalise to cuts on a wider class of Π_2 and Σ_2-formulæ. The simplest extension, which can already be realised, is to simple proofs with cuts on arbitrary prenex Π_2 formulæ (this generalisation is described in detail above); more complex will be considering boolean combinations of prenex Π_2 and Σ_2 formulæ and non-prenex formulæ. A second generalisation would see proof grammars extended to prenex Π_n-formulæ. The relationship between proofs and grammars established thus far suggests how to proceed: with Π_1-cuts inducing non-terminals of type level 0 and Π_2-cuts inducing, in addition, non-terminals if type level 1, the conjecture is that Π_{n+1}-cuts will be amenable to analysis using non-terminals of level $\leq n$ and order n recursion schemes (a natural extension of context-free tree grammars to higher-order). This is reminiscent to the relationship between the number of quantifier alternations in an induction and the type level of the functional obtained from Gödel's Dialectica interpretation [Gödel, 1958], see e.g. [Avigad and Feferman, 1999].

Acknowledgment: The authors' research was supported by the Wiener Wissenschafts-, Forschungs- und Technologiefonds (WWTF) project VRG12-04.

References

Bahareh Afshari, Stefan Hetzl, and Graham E. Leigh. Herbrand Disjunctions, Cut Elimination and Context-Free Tree Grammars. In Thorsten Altenkirch, editor, *13th International Conference on Typed Lambda Calculi and Applications (TLCA 2015)*, volume 38 of *Leibniz International Proceedings in Informatics (LIPIcs)*, pages 1–16, Dagstuhl, Germany, 2015. Schloss Dagstuhl–Leibniz-Zentrum fuer Informatik.

Jeremy Avigad and Solomon Feferman. Gödel's functional ("dialectica") interpretation. In Sam Buss, editor, *The Handbook of Proof Theory*, pages 337–405. North-Holland, 1999.

Matthias Baaz and Stefan Hetzl. On the non-confluence of cut-elimination. *Journal of Symbolic Logic*, 76(1):313–340, 2011.

Samuel R. Buss. The undecidability of k-provability. *Annals of Pure and Applied Logic*, 53: 75–102, 1991.

Samuel R. Buss. On Herbrand's Theorem. In *Logic and Computational Complexity*, volume 960 of *Lecture Notes in Computer Science*, pages 195–209. Springer, 1995.

Alessandra Carbone and Stephen Semmes. *A Graphic Apology for Symmetry and Implicitness.* Oxford Mathematical Monographs. Oxford University Press, 2000.

Jean-Yves Girard. Linear logic. *Theoretical Computer Science*, 50(1):1–101, 1987.

Kurt Gödel. Über eine noch nicht benützte Erweiterung des finiten Standpunktes. *Dialectica*, 12:280–287, 1958.

Willem Heijltjes. Classical proof forestry. *Annals of Pure and Applied Logic*, 161(11):1346–1366, 2010.

Jacques Herbrand. *Recherches sur la théorie de la démonstration*. PhD thesis, Université de Paris, 1930.

Stefan Hetzl. On the form of witness terms. *Archive for Mathematical Logic*, 49(5):529–554, 2010.

Stefan Hetzl. Applying Tree Languages in Proof Theory. In Adrian-Horia Dediu and Carlos Martín-Vide, editors, *Language and Automata Theory and Applications (LATA) 2012*, volume 7183 of *Lecture Notes in Computer Science*, pages 301–312. Springer, 2012a.

Stefan Hetzl. The Computational Content of Arithmetical Proofs. *Notre Dame Journal of Formal Logic*, 53(3):289–296, 2012b.

Stefan Hetzl and Lutz Straßburger. Herbrand-Confluence for Cut-Elimination in Classical First-Order Logic. In Patrick Cégielski and Arnaud Durand, editors, *Computer Science Logic (CSL) 2012*, volume 16 of *Leibniz International Proceedings in Informatics (LIPIcs)*, pages 320–334. Schloss Dagstuhl–Leibniz-Zentrum fuer Informatik, 2012.

Stefan Hetzl and Lutz Straßburger. Herbrand-confluence. *Logical Methods in Computer Science*, 9(4), 2013.

J. Martin E. Hyland. Proof theory in the abstract. *Annals of Pure and Applied Logic*, 114:43–78, 2002.

Richard McKinley. Proof nets for Herbrand's Theorem. *ACM Transactions on Computational Logic*, 14(1):5:1–5:31, 2013.

Anne S. Troelstra and Helmut Schwichtenberg. *Basic Proof Theory*. Cambridge: Cambridge University Press, 1996.

Marco Benini
Proof-Oriented Categorical Semantics

Abstract: The proposition-as-types correspondence applied to first-order logical theories and the associated type systems says that they are different although equivalent presentations of the same entity. But the nature of this entity is not clear, as these systems are interpreted in distinct categories and in different ways. In this contribution, an alternative framework is introduced so that the interpretations of the logical system and of the type theory identify the same entity, which acquires both, a logical and a computational meaning at the semantic level. It is shown that this semantics is sound and complete, both in the logical and in the type-theoretic sense. Also, it is shown that the standard models for interpreting first-order systems in category theory can be transformed in equivalent models in the presented framework.

Keywords: Categorical semantics, Propositions-as-types

Mathematics Subject Classification 2010: 03B20, 03C07, 03G30

1 Introduction

The proposition-as-types correspondence in its classic formulation [Howard, 1980] tells that the propositional intuitionistic logic and the simple theory of types are different but equivalent presentations of the same entity. Specifically, considering the logical fragment limited to conjunction and implication, and the type theory based on product and arrow types, the correspondence says that each logical formula can be seen as a type, and each proof as a typed term, and vice versa. The computational behaviour, that is, the reduction of typed terms, acquires a logical meaning, namely, normalisation of proofs; conversely, each proof acquires a computational meaning, being, thanks to the correspondence, a term apt to be reduced.

In [Johnstone, 2002b, Section D.4.2], the propositional fragment is naturally interpreted in a topos: each formula becomes a monomorphism, and the connectives are interpreted as the operations in the lattice of subobjects. Also, the simple theory of types is naturally interpreted in weak Cartesian closed categories: each type becomes an object, so that a product type becomes a product in the category, and an arrow type becomes an exponential object. Actually, the full power of a

Marco Benini: Dipartimento di Scienza e Alta Tecnologia, Università degli Studi dell'Insubria, via Valleggio 11, I-22100 Como, Italy, e-mail: marco.benini@uninsubria.it

topos is not needed to interpret the propositional fragment, or conversely, we can safely limit the interpretation of the simple theory of types to toposes, so that the world where the logical and type-theoretic models live is the same. Nevertheless, the difference between the models for logic and type theory is striking: in one case, formulae are arrows, in the other, objects[1]. So, in particular, these theories do not seem to speak about the same entities.

Moving to the richer first-order language, it is possible to construct a type theory that allows to establish the propositions-as-types correspondence as in the propositional case. But in the first-order case, the class of categories used to provide models for the logical side of the correspondence is very different from the one used to assign a meaning to type theory. In fact, quantifiers are modelled by suitable adjoint functors, according to Lavwere's seminal paper [Lavwere, 1969], while the nowadays prominent semantics of types interpret quantifiers as product or function spaces, see [The Univalent Foundations Program, 2013, Chapter 1]. Although both interpretations of quantifiers arise as adjoints to some elementary functor, they differ: the pullback functor in Lavwere's case, where the left and right adjoint provide the interpretations for the existential and universal quantifiers, respectively; functions spaces are right adjoint to the product functor $\_\times X$, as they are just the exponential objects in the category of types; product spaces arise as right adjoints to diagonal functors $X \mapsto (X, X)$. At least in principle, it is possible to use toposes as a unifying class of categories, but it is hard to imagine that a topos is the natural entity which is described by the first-order language. Also, as in the propositional case, the way to interpret formulae and types is radically distinct, giving rise to different models, even within the same class of categories.

In the present contribution, a novel class of categories is introduced, to give a semantics both to first-order intuitionistic theories and to their corresponding type theories. But, differently from the standard way to interpret logic and type theory, the models are the same to the point that formulae and types denote the same objects, and similarly proofs and typed terms denote the same arrows. In this way, the semantics poses the propositions-as-types correspondence as the

1 In a way, this is too simple: in fact, what is important is the *structure* of the entity interpreting a formula. For example, a conjunction $A \wedge B$ interpreted as a type becomes a product, which is defined together with its projections and universal arrows, coming from its construction as a limit. Conversely, $A \wedge B$ interpreted as a logical formula becomes a meet in the lattice of subobjects of the context of interpretation, thus an equivalence class of arrows with a common codomain, related to the other subobjects via a suitable notion of order. The important part is not the product or the equivalence class, but how the interpretation relates to other objects or other equivalence classes, respectively. It is this structure which should emerge in the semantics, and it is the objective of this contribution to make it explicit.

central notion, which justifies the "proof-oriented" attribute of the title. It will be shown that the semantics is sound and complete both in the logical and in the type-theoretical sense. Also, we will draw a brief sketch of the relations of the proposed semantics with other ways to assign a meaning to first-order logic.

The main feature of the proposed semantics, the one allowing to say that first-order logic and the corresponding type theory, are presentations of the same entity, is that the models are *point-free*. This means that logical terms do not denote elements in some universe, but rather they are used to construct the *glue* that keeps together the interpretation of formulae and proofs, so that both logic and computation will arise from the same structure. For example, considering arithmetic, logical terms denote natural numbers in standard set-theoretic semantics, like Kripke or Tarski models; in a standard categorical semantics, like Heyting categories, logical terms denote special arrows, which correspond to the natural numbers. On the contrary, in the framework that will be introduced in the following, logical terms are used as 'placeholders' to describe the action of the substitution operation on a suitable category, whose objects denote logical formulae and whose arrows denote logical proofs. It is really the algebraic action induced on the category by the substitution operation that forms the 'glue' which allows to define an interpretation for formulae and proofs. And, since any first-order theory in this framework has a classifying model, we will conclude that this model is exactly the entity presented by the logical theory and the corresponding type theory.

2 Preliminaries

Before introducing the semantics, it is useful to precisely fix the syntax of the logical systems we are going to use. This allows to introduce the type theories, so that the propositions-as-types correspondence will hold by construction.

2.1 Logic

The logical systems we are interested in are usually described as being theories in the multi-sorted, first-order, intuitionistic logic. Referring to [Troelstra and Schwichtenberg, 2000] for an extensive presentation, these theories are described by the following definitions.

Definition 1 (Logical signature). *A* logical signature $\langle S, F, R\rangle$ *is composed by a set* S *of* sort symbols, *a set* F *of* function symbols *each one of the form* $f : s_1 \times \cdots \times s_n \to s_0$ *where, for every* $0 \le i \le n$, $s_i \in S$, *and a set* R *of* relation symbols *each one of the* *form* $r{:}s_1{\times}\cdots{\times}s_n$ *where, for every* $1 \le i \le n$, $s_i \in S$. *When* $n = 0$, *a function symbol is usually called a* constant, *while a relation symbol is usually called a* propositional constant.

Definition 2 (Logical term). *Given a signature* $\langle S, F, R\rangle$, *assume that there is a denumerable set* V_s *of* logical variables *for each* $s \in S$. *Then, a* logical term *is inductively defined as follows, along with the set of its* free variables*:*

1. *if* $v \in V_s$ *then* $v : s$ *is a term and* $\mathsf{FV}(v : s) = \{v : s\}$;
2. *if* $f{:}s_1{\times}\cdots{\times}s_n \to s_0 \in F$ *and* $t_1{:}s_1, \dots, t_n{:}s_n$ *are terms, then so is* $f(t_1, \dots, t_n)$: s_0, *and* $\mathsf{FV}(f(t_1, \dots, t_n) : s_0) = \bigcup_{i=1}^n \mathsf{FV}(t_i : s_i)$.

Definition 3 (Logical formula). *Fixed a signature* $\langle S, F, R\rangle$, *a* logical formula *is inductively defined, along with its free variables, as:*

1. $\top$ *and* $\bot$ *are formulae, and* $\mathsf{FV}(\top) = \mathsf{FV}(\bot) = \emptyset$;
2. *if* $p : s_1 \times \cdots \times s_n \in R$ *and* $t_1 : s_1, \dots, t_n : s_n$ *are terms, then* $p(t_1, \dots, t_n)$ *is a formula and* $\mathsf{FV}(p(t_1, \dots, t_n)) = \bigcup_{i=1}^n \mathsf{FV}(t_i : s_i)$;
3. *if* A *and* B *are formulae, so are* $A \wedge B$, $A \vee B$, *and* $A \supset B$, *and* $\mathsf{FV}(A \wedge B) = \mathsf{FV}(A \vee B) = \mathsf{FV}(A \supset B) = \mathsf{FV}(A) \cup \mathsf{FV}(B)$;
4. *if* $x \in V_s$ *and* A *is a formula, so are* $\forall x : s\, A$ *and* $\exists x : s\, A$, *and* $\mathsf{FV}(\forall x : s\, A) = \mathsf{FV}(\exists x : s\, A) = \mathsf{FV}(A) \setminus \{x : s\}$.

A notion of *substitution* is defined in the standard way, see e.g. [Troelstra and Schwichtenberg, 2000, Section 1.1.2].

Definition 4 (Substitution). *Fixed a signature* $\langle S, F, R\rangle$, *the act of substituting* $x \in V_s$ *with* $r : s$ *in the term* $t : u$, *notation* $t[r/x] : u$, *is inductively defined as:*

1. *if* $v \in V_u$, *then* $v[r/x] : u = r : s$ *when* x *is identical to* v *and* u *is identical to* s, *while* $v[r/x] : u = v : u$ *otherwise;*
2. $f(t_1, \dots, t_n)[r/x] : u = f(t_1[r/x], \dots, t_n[r/x]) : u$.

The act of substituting $x \in V_s$ *with the term* $r : s$ *in the formula* A, *notation* $A[r/x]$, *is inductively defined as:*

1. $\top[r/x] = \top$, $\bot[r/x] = \bot$;
2. $p(t_1, \dots, t_n)[r/x] = p(t_1[r/x], \dots, t_n[r/x])$;
3. $(A{\wedge}B)[r/x] = A[r/x]{\wedge}B[r/x]$, $(A{\vee}B)[r/x] = A[r/x]{\vee}B[r/x]$, *and* $(A \supset B)[r/x] = A[r/x] \supset B[r/x]$;
4. $(\forall y{:}u\, A)[r/x] = \forall y{:}u\, A[r[z/y]/x]$, *with* $z{:}u \notin \mathsf{FV}(A)$, *when* $x{:}s$ *is different from* $y{:}u$, *and* $(\forall y{:}u\, A)[r/x] = \forall y{:}u\, A$ *otherwise;* $(\exists y{:}u\, A)[r/x] = \exists y{:}u\, A[r[z/y]/x]$, *with* $z{:}u \notin \mathsf{FV}(A)$, *when* $x{:}s$ *is different from* $y{:}u$, *and* $(\exists y{:}u\, A)[r/x] = \exists y{:}u\, A$ *otherwise, that is, when* x *is identical to* y *and* s *is identical to* u.

$$\frac{}{\Gamma \vdash A}\,\text{Assum where } A \in \Gamma \qquad \frac{}{\Gamma \vdash A}\,\text{Ax where } A \in T$$

$$\frac{\Gamma \vdash A \quad \Gamma \vdash B}{\Gamma \vdash A \wedge B}\,\wedge\text{I} \qquad \frac{\Gamma \vdash A \wedge B}{\Gamma \vdash A}\,\wedge\text{E}_1 \qquad \frac{\Gamma \vdash A \wedge B}{\Gamma \vdash B}\,\wedge\text{E}_2$$

$$\frac{\Gamma \vdash A}{\Gamma \vdash A \vee B}\,\vee\text{I}_1 \qquad \frac{\Gamma \vdash A}{\Gamma \vdash B \vee A}\,\vee\text{I}_2 \qquad \frac{\Gamma \vdash A \vee B \quad \Gamma, A \vdash C \quad \Gamma, B \vdash C}{\Gamma \vdash C}\,\vee\text{E}$$

$$\frac{\Gamma, A \vdash B}{\Gamma \vdash A \supset B}\,\supset\text{I} \qquad \frac{\Gamma \vdash A \supset B \quad \Gamma \vdash A}{\Gamma \vdash B}\,\supset\text{E}$$

$$\frac{}{\Gamma \vdash \top}\,\top\text{I} \qquad \frac{\Gamma \vdash \bot}{\Gamma \vdash B}\,\bot\text{E}$$

$$\frac{\Gamma \vdash A}{\Gamma \vdash \forall x : s\, A}\,\forall\text{I}\,^{(*)} \qquad \frac{\Gamma \vdash \forall x : s\, A}{\Gamma \vdash A[r/x]}\,\forall\text{E}$$

$$\frac{\Gamma \vdash A[r/x]}{\Gamma \vdash \exists x : s\, A}\,\exists\text{I} \qquad \frac{\Gamma \vdash \exists x : s\, A \quad \Gamma, A \vdash B}{\Gamma \vdash B}\,\exists\text{E}\,^{(**)}$$

Fig. 1. The logical calculus: in $^{(*)}$, the variable $x : s$ must be *fresh* in the corresponding proof, i.e., $x : s \notin \bigcup_{C \in \Gamma} \mathrm{FV}(C)$, with Γ the set of assumptions; also, in $^{(**)}$, the variable $x : s$ must be fresh, too, i.e., $x : s \notin \mathrm{FV}(B) \cup \bigcup_{C \in \Gamma} \mathrm{FV}(C)$.

It is worth remarking that substitution operates at the meta-level. Hence, if x is a variable and k a constant of the same sort, we do not distinguish the term $x[k/x]$ from k within the syntax.

Definition 5 (Logical theory). *A (logical)* theory *is a collection of formulae. A formula in a theory T is said to be an* axiom *of T.*

Definition 6 (Logical calculus). *Fixed a signature and a theory T, a* derivation *or* proof *$\Gamma \vdash_T A$ of a formula A from the finite set of formulae Γ, the* assumptions, *is inductively defined by the rules in Figure 1. These are the usual rules of natural deduction [Troelstra and Schwichtenberg, 2000, p. 36] for multi-sorted intuitionistic first-order logic, presented in the sequent notation.*

2.2 Type theory

The type theories are constructed to validate the propositions-as-types interpretation. To avoid clashes with the categorical notation, used to describe the semantics, and to emphasise the correspondence with the logical systems, we adopt the logical notation whenever appropriate.

Thus, a *type* is just a logical formula. According to the propositions-as-types correspondence, a *typed term* should be a logical derivation, but the definition of proofs should be made more precise than we did in the previous section. In the first place, assumptions and their use have to be made explicit and traceable: this is obtained using *contexts*.

Definition 7 (Context). *Fixed a signature $\langle S, F, R\rangle$, assume that, for each type t, there is a denumerable set W_t of* typed variables*; as usual, they are denoted as $x : t$. Then, a* context *is any finite list of typed variables such that each typed variable appears only once.*

Out of the formal definition, the meaning is simply that each context denotes a set of assumptions: so, $x : t$ in a context means that the type t, a formula in the propositions-as-types interpretation, is an assumption whose name is x. We required assumptions to form finite sets, and thus we require contexts to be finite lists. It is evident that, forgetting the sequence, each finite list generates a finite set. Conversely, having a sequence is handy to precisely define the semantics of proofs, but, as we are going to show, essentially irrelevant, since a change in the order of elements induces an isomorphic choice of objects and arrows for the interpretation of formulae and derivations, respectively.

Definition 8 (Typed theory). *A* typed theory *is a collection T of elements of the form $a : A$, with a a constant and A a type. We require that each constant appears only once, that is, a constant uniquely identify an axiom.*

Every logical theory can be seen as a typed theory by decorating each axiom with a distinct constant, and, vice versa, every typed theory generates a logical theory, by forgetting the constants. So the correspondence between the two notions allows to ignore their difference most of the times.

Definition 9 (Typed term). *Given a signature $\langle S, F, R\rangle$, a typed theory T, and a context Γ, a* typed term *is inductively defined as:*

1. *if $x : t \in \Gamma$ then $\Gamma.x : t$ is a typed term, and $\mathsf{FV}(\Gamma.x : t) = \{x : t\}$;*
2. *if $a : A \in T$ then $\Gamma.a : A$ is typed term, and $\mathsf{FV}(\Gamma.a : A) = \emptyset$;*
3. *if $\Gamma.s : A$ and $\Gamma.t : B$ are typed terms, then so is $\Gamma.\wedge\mathsf{I}(s, t) : A \wedge B$; moreover, $\mathsf{FV}(\Gamma.\wedge\mathsf{I}(s, t) : A \wedge B) = \mathsf{FV}(\Gamma.s : A) \cup \mathsf{FV}(\Gamma.t : B)$;*

4. *if* $\Gamma.t : A \wedge B$ *is a typed term, then so are* $\Gamma.\wedge\mathsf{E}_1(t) : A$ *and* $\Gamma.\wedge\mathsf{E}_2(t) : B$*; also,* $\mathsf{FV}(\Gamma.\wedge\mathsf{E}_1(t) : A) = \mathsf{FV}(\Gamma.\wedge\mathsf{E}_2(t) : B) = \mathsf{FV}(\Gamma.t : A \wedge B)$*;*
5. *if* $\Gamma.t : A$ *is a typed term, so are* $\Gamma.\vee\mathsf{I}_1(t) : A \vee B$ *and* $\Gamma.\vee\mathsf{I}_2(t) : B \vee A$*, and, moreover,* $\mathsf{FV}(\Gamma.\vee\mathsf{I}_1(t) : A \vee B) = \mathsf{FV}(\Gamma.\vee\mathsf{I}_2(t) : B \vee A) = \mathsf{FV}(\Gamma.t : A)$*;*
6. *if* $\Gamma.s : A \vee B$, $\Gamma, u : A.t : C$*, and* $\Gamma, v : B.r : C$ *are typed terms (and* $u : A$ *and* $v : B$ *cannot occur in* Γ*), then so is* $\Gamma.\vee\mathsf{E}(s, t, r, u : A, v : B) : C$*, and*

$$\begin{aligned}\mathsf{FV}(\Gamma.\vee\mathsf{E}(s, t, r, u : A, v : B) : C) = \quad & \mathsf{FV}(\Gamma.s : A \vee B) \cup \\ & \cup (\mathsf{FV}(\Gamma, u : A.t : C) \setminus \{u : A\}) \cup \\ & \cup (\mathsf{FV}(\Gamma, v : B.r : C) \setminus \{v : B\}) \quad ;\end{aligned}$$

7. *if* $x \in W_A$ *and* $\Gamma, x : A.t : B$ *is a typed term, then* $\Gamma.\supset\mathsf{I}(t, x : A) : A \supset B$ *is, too, and* $\mathsf{FV}(\Gamma.\supset\mathsf{I}(t, x : A) : A \supset B) = \mathsf{FV}(\Gamma.t : B) \setminus \{x : A\}$*;*
8. *if* $\Gamma.s : A \supset B$ *and* $\Gamma.t : A$ *are typed terms, then* $\Gamma.\supset\mathsf{E}(s, t) : B$ *is a typed term, too; moreover,* $\mathsf{FV}(\Gamma.\supset\mathsf{E}(s, t) : B) = \mathsf{FV}(\Gamma.s : A \supset B) \cup \mathsf{FV}(\Gamma.t : A)$*;*
9. $\Gamma.\top\mathsf{I} : \top$ *is a typed term, and* $\mathsf{FV}(\Gamma.\top\mathsf{I} : \top) = \emptyset$*;*
10. *if* $\Gamma.t : \bot$ *is a typed term, so is* $\Gamma.\bot\mathsf{E}(t) : B$*, and* $\mathsf{FV}(\Gamma.\bot\mathsf{E}(t) : B) = \mathsf{FV}(\Gamma.t : \bot)$*;*
11. *if* $x \in V_s$ *and* $\Gamma.t : A$ *is a typed term such that* $x : s \notin \mathsf{FV}^*(\Gamma.t : A)$*, then also* $\Gamma.\forall\mathsf{I}(t) : \forall x : s\, A$ *is a typed term, and* $\mathsf{FV}(\Gamma.\forall\mathsf{I}(t) : \forall x : s\, A) = \mathsf{FV}(\Gamma.t : A)$*;*
12. *if* $\Gamma.t : \forall x : s\, A$ *is a typed term and* $r : s$ *is a logical term, then* $\Gamma.\forall\mathsf{E}(t, r) : A[r/x]$ *is a typed term, and* $\mathsf{FV}(\Gamma.\forall\mathsf{E}(t, r) : A[r/x]) = \mathsf{FV}(\Gamma.t : \forall x : s\, A)$*;*
13. *if* $x \in V_s$, $r : s$ *is a logical term, and* $\Gamma.t : A[r/x]$ *is a typed term, then also* $\Gamma.\exists\mathsf{I}(t) : \exists x : s\, A$ *is a typed term, and* $\mathsf{FV}(\Gamma.\exists\mathsf{I}(t) : \exists x : s\, A) = \mathsf{FV}(\Gamma.t : A[r/x])$*;*
14. *if* $\Gamma.t : \exists x : s\, A$ *and* $\Gamma, u : A.r : B$ *are typed terms such that* $x : s \notin \mathsf{FV}^*(\Gamma.r : B) \cup \mathsf{FV}(B)$*, then* $\Gamma.\exists\mathsf{E}(t, r, u : A)) : B$ *is a typed term, and*

$$\begin{aligned}\mathsf{FV}(\Gamma.\exists\mathsf{E}(t, r, u : A)) : B) = \quad & \mathsf{FV}(\Gamma.t : \exists x : s\, A) \cup \\ & \cup (\mathsf{FV}(\Gamma, u : A.r : B) \setminus \{u : A\}) \quad .\end{aligned}$$

In the previous definition, $\mathsf{FV}^*(\Gamma.t : A)$ *is the set consisting of those logical variables* $x : s$ *such that there are a type* r *and a typed variable* $y \in W_r$ *such that* $x : s \in \mathsf{FV}(r)$ *and* $y : r \in \mathsf{FV}(\Gamma.t : A)$*. That is,* $\mathsf{FV}^*(\Gamma.t : A)$ *is the set of the free logical variables in the set of used*[2] *assumptions in* Γ*.*

---

2 Actually, this is an apparently more relaxed condition of freshness of a variable than the one in Figure 1. In essence, the present condition says that the variable used to introduce a universal quantification, or to eliminate an existential one, must not occur among the assumptions which are used via the assumption rule in the relevant subproofs. The other assumptions do not matter, as they are not relevant to establish the validity of the conclusion, which is exactly the statement that has to be proven in the meta-theory, and it should be evident.

A few remarks are due: the definition does not depend on the order of elements within Γ, but only on its underlying collection of elements; the term constructors have the same names as the logical inference rules and the same structure, thus establishing the proposition-as-types correspondence; discharged assumptions are rendered as typed variables occurring in a term, but not among its free typed variables; the conditions on variables to be fresh in the corresponding proof, see Figure 1, are reduced to require that a logical variable does not occur among the free variables in the undischarged assumptions. The mapping between the standard constructors in type theory, e.g., abstraction, application, pairing, etc., is immediate. Also, an evident notion of substitution is available for typed terms: each free typed variable can be replaced by any typed term in the same context provided the variable and the term share the same type.

As a computational system, the type theory operates by reducing typed terms. It is beyond the scope of this contribution to analyse the reduction process—the interested reader may look at normalisation in [Troelstra and Schwichtenberg, 2000, Chapter 6]. Instead, we present the reduction rules in the form of equalities.

Definition 10 (Typed equality). *Fixed a signature,* equality *of typed terms is inductively defined by the following rules: we will use the abbreviations* $\vec{x} : \vec{A}$ *for* $x_1 : A_1, \ldots, x_n : A_n$, $[\vec{s}/\vec{y}]$ *for* $[s_1/y_1, \ldots, s_m/y_m]$*, and similar where appropriate; also,* r, s, t*, are meta-variables for typed terms, while* u, v, x, y*, and* z *are proper variables, and* e *stands for a logical term.*

1. *if* $\vec{x}:\vec{A}.s : C = \vec{x}:\vec{A}.t : C$ *then* $\Gamma.s[\vec{r}/\vec{x}] : C = \Gamma.t[\vec{r}/\vec{x}] : C$ *where, for each* $1 \leq i \leq n$, $\Gamma.r_i : A_i$ *is a typed term;*
2. *if* $\Gamma.s_1 : B_1 = \Gamma.t_1 : B_1, \ldots, \Gamma.s_m : B_m = \Gamma.t_m : B_m$*, then* $\Gamma.r[\vec{s}/\vec{y}] : C = \Gamma.r[\vec{t}/\vec{y}] : C$*;*
3. $\Gamma.t : A = \Gamma.t : A$*;*
4. *if* $\Gamma.t : A = \Gamma.s : A$ *then* $\Gamma.s : A = \Gamma.t : A$*;*
5. *if* $\Gamma.t : A = \Gamma.s : A$ *and* $\Gamma.s : A = \Gamma.r : A$*, then* $\Gamma.t : A = \Gamma.r : A$*;*
6. *if* $\Gamma, u : A.s_1 : C = \Gamma, u : A.t_1[u/x] : C$ *and* $\Gamma, v : B.s_2 : C = \Gamma, v : B.t_2[v/y] : C$*, then* $\Gamma.\vee\mathsf{E}(r, s_1, s_2, u : A, v : B) : C = \Gamma.\vee\mathsf{E}(r, t_1, t_2, x : A, y : B) : C$*;*
7. *if* $\Gamma, x{:}B.s{:}C = \Gamma, x{:}B.t[x/y]{:}C$*, then also* $\Gamma.{\supset}\mathsf{I}(s, x{:}B){:}B \supset C = \Gamma.{\supset}\mathsf{I}(t, y{:}B){:}B \supset C$*;*
8. *similarly, if* $\Gamma, u{:}A.s{:}B = \Gamma, u{:}A.t[u/v]{:}B$*, then* $\Gamma.\exists\mathsf{E}(r, s, u{:}A){:}B = \Gamma.\exists\mathsf{E}(r, t, v{:}A) : B$*;*
9. $\Gamma.t : \top = \Gamma.\top\mathsf{I} : \top$*;*
10. $\Gamma.\wedge\mathsf{E}_1(\wedge\mathsf{I}(t, s)) : A = \Gamma.t : A$*;*
11. $\Gamma.\wedge\mathsf{E}_2(\wedge\mathsf{I}(t, s)) : A = \Gamma.s : A$*;*
12. $\Gamma.\wedge\mathsf{I}(\wedge\mathsf{E}_1(t), \wedge\mathsf{E}_2(t)) : A \wedge B = \Gamma.t : A \wedge B$*;*
13. $\Gamma.\vee\mathsf{E}(\vee\mathsf{I}_1(r), t, s, u : A, v : B) : C = \Gamma.t[r/u] : C$*;*

14. $\Gamma.\vee\mathsf{E}(\vee\mathsf{I}_2(r), t, s, u : A, v : B) : C = \Gamma.s[r/v] : C;$
15. $\Gamma.\vee\mathsf{E}(\vee\mathsf{E}(t_0, t_1, t_2, u_1 : A, u_2 : B), t_3, t_4, u_3 : C, u_4 : D) : E =$
    $= \Gamma.\vee\mathsf{E}(t_0, \vee\mathsf{E}(t_1, t_3, t_4, u_3 : C, u_4 : D),$
    $\vee\mathsf{E}(t_2, t_3, t_4, u_3 : C, u_4 : D), u_1 : A, u_2 : B) : E;$
16. $\Gamma, x : A.\bot\mathsf{E}(t) : A = \Gamma.x : A;$
17. $\Gamma.\supset\mathsf{E}(\supset\mathsf{I}(t, u : A), s) : B = \Gamma.t[s/u] : B;$
18. *if* $u : A \notin \mathsf{FV}(\Gamma.t : A \supset B), \Gamma.\supset\mathsf{I}(\supset\mathsf{E}(t, s), u : A) : A \supset B = \Gamma.t : A \supset B$;
19. $\Gamma.\forall\mathsf{E}(\forall\mathsf{I}(t), e) : B[e/z] = \Gamma.t[e/z] : B[e/z];$
20. *if* $\Gamma.\forall\mathsf{E}(t, e) : B[e/x] = \Gamma.\forall\mathsf{E}(s, e) : B[e/x]$ *for every logical term e of* , *then* $\Gamma.t :$ $\forall x : s\, B = \Gamma.s : \forall x : s\, B;$
21. $\Gamma.\exists\mathsf{E}(\exists\mathsf{I}(t : A[e/x]) : \exists x : s\, A, r, u : A) : B = \Gamma.r[e/x][t/u] : B;$
22. *if* $\Gamma.\exists\mathsf{E}(r, t, x : A)) : B = \Gamma.\exists\mathsf{E}(r, s, x : A)) : B, \Gamma, x : A.t : B = \Gamma, x : A.s : B;$
23. $\Gamma.r : B = \exists\mathsf{E}(t, r, x : A) : B$ *with* $x : A \notin \mathsf{FV}(r : B);$
24. $\Gamma.\exists\mathsf{E}(\exists\mathsf{E}(t, s, u : A), r, v : B) : C = \Gamma.\exists\mathsf{E}(t, \exists\mathsf{E}(s, r, v : B), u : A) : C.$

Rules (1)–(5) say that equality is an equivalence relation which is preserved by substituting equals for equals, and rules (6)–(8) say that bounded typed variables can be freely renamed. Rules (9)–(14), (16)–(19), and (21) are the standard reductions of product types, of sum types, β and η, and of universal and existential types, see [Girard et al., 1989]. Rules (15), (23) and (24) are standard permutation and simplification rules, see [Troelstra and Schwichtenberg, 2000, Chapter 6]. Rule (20) and (22) establish when the parametric subproofs of two equal proofs, are equal.

3 Semantics

The semantics of the logical systems and of the corresponding type theories is based on category theory: informally, a type is interpreted as an object and a logical proof as an arrow. The structure of the category allows to interpret all the connectives, while the logical terms generate, via substitution, the objects deputed to interpret quantifiers. The categorical structure is, essentially, the one of distributive categories, used by Taylor [1999] to interpret propositional intuitionistic logic plus a notion of substitution on the interpreted types.

Admitting the logical terms to form a groupoid [Mac Lane, 1998, p. 20], which is a natural extension of being a set—which is a discrete category, that is, a category whose only arrows are identities, while groupoids are categories whose arrows are all isomorphisms—it becomes possible to use the very same structure to generate a reasonable notion of equality at the semantic level, alongside quantifiers. Informally, an arrow in a groupoid indicates that two specific terms are equal by definition—which is evident when the arrow is the identity, but becomes meaningful when the arrow is an isomorphism. This equality is naturally mapped into an axiom of the logical theory or, equivalently, into a primitive arrow in the model of the theory inside the category of interpretation. Moreover, the structure of groupoid ensures that the generated equality is an equivalence relation, and the mapping into the category via the substitution functor forces equality to be a congruence. Actually, as it will be discussed in Section 5, using groupoids in the very beginning allows to abstract over the structure of terms, retaining only what is really needed to convey meaning to logic inside the semantics we are going to illustrate.

3.1 Categories

In short, the semantics is defined by the action the syntactical substitution process, see Definition 4, performs on a category $\mathbb{C}$ with enough structure. The action is represented as a family Σ of functors, each one modelling the substitution of a variable in an interpreted formula/type. Equivalently, the substitution functors are the way to interpret the syntactical substitution operation in the semantic level.

Definition 11 (Logical category). *Fixed a signature, a* logical category *is a triple* $\langle \mathbb{C}, \mathbb{G}, \Sigma \rangle$ *such that*

1. $\mathbb{C}$ *is a category with finite products, finite coproducts and exponentiation;*
2. $\mathbb{G}$ *is a groupoid of logical terms, that is, a category whose objects are all the logical terms on the signature, and such that every arrow is an isomorphism;*
3. $\Sigma = \{\Sigma_A(x:s)\colon \mathbb{G} \to \mathbb{C}\}_{A,x:s}$ *is a family of functors, called the* substitution functors, *indexed by all the types and all the logical variables;*
4. *each substitution functor* $\Sigma_A(x:s)$ *has a limit and a colimit;*
5. *substitution functors preserve the syntactical substitution, that is, for any* A, $x:s$, *and* $t:s$, $\Sigma_A(x:s)(t:s) = \Sigma_{A[t/x]}(x:s)(x:s)$;
6. *substitution functors are* stable *with respect to variables, that is, for any* $x:s$ *and* $y:t$, $\Sigma_A(x:s)(x:s) = \Sigma_A(y:t)(y:t)$;
7. *moreover, for any variable* $x:s$, *and any term* $t:s$,
   (a) $\Sigma_\bot(x:s)(t:s) = 0$, *the initial object of* $\mathbb{C}$;
   (b) $\Sigma_\top(x:s)(t:s) = 1$, *the terminal object of* $\mathbb{C}$;
   (c) *for any* A *and* B, $\Sigma_{A\wedge B}(x:s)(t:s) = \Sigma_A(x:s)(t:s) \times \Sigma_B(x:s)(t:s)$, *the binary product in* $\mathbb{C}$;
   (d) *for any* A *and* B, $\Sigma_{A\vee B}(x:s)(t:s) = \Sigma_A(x:s)(t:s) + \Sigma_B(x:s)(t:s)$, *the binary coproduct in* $\mathbb{C}$;

(e) *for any A and B, $\Sigma_{A \supset B}(x:s)(t:s) = \Sigma_B(x:s)(t:s)^{\Sigma_A(x:s)(t:s)}$, the exponential object in $\mathbb{C}$;*
(f) *for any A and $x : s$, $\Sigma_{\forall x:s\, A}(x:s)(x:s) = \lim \Sigma_A(x:s)$;*
(g) *for any A and $x : s$, $\Sigma_{\exists x:s\, A}(x:s)(x:s) = \mathsf{colim}\Sigma_A(x:s)$.*

When we refer to the *product of the category $\mathbb{C}$, we assume that a choice has been made among the possible isomorphic objects which may be considered products. And the same holds for the other categorical constructions we use, that is, coproduct, exponentiation and the limit and colimit in (4).*

It is worth remarking that, up to isomorphisms, the above mentioned choice does not matter: it is a basic fact of category theory that all the limits on the same functor are isomorphic, see, e.g., [Awodey, 2006].

There is an evident and natural way to assign an interpretation to any type:

Definition 12 (Interpretation of types). *Given a logical category $\langle \mathbb{C}, \mathbb{G}, \Sigma \rangle$ over a signature S, the interpretation of the type A is defined to be $MA = \Sigma_A(x:s)(x:s)$, for any variable $x : s$.*

Because of (6) in Definition 11, this definition is well-founded.

It is worth noticing that, fixed M, the requirement (5) in Definition 11 allows to uniquely define the family Σ. Nevertheless, we prefer to derive the interpretation of terms from the action of the substitution process over the category $\mathbb{C}$, since it is really this action which induces a meaning, as it is evident in (7) of the definition of logical category. Thus, it is the family Σ, which is the 'glue' referred to in the Introduction, the structure that keeps together the interpretation of formulae so that it can acquire a meaning allowing to use it as a model for logic.

Definition 13 (T-model). *Given a typed theory T, a T-*model *$\langle \mathbb{C}, \mathbb{G}, \Sigma, \alpha \rangle$ is such that $\langle \mathbb{C}, \mathbb{G}, \Sigma \rangle$ is a logical category on the language of T, and α is map from the theory T to* $\mathsf{Obj}\mathbb{C}$ *such that there exists an arrow $\alpha(a : A) \colon 1 \to MA$ in $\mathbb{C}$ for each axiom $a : A \in T$. This arrow will force all the axioms to be valid, see Definition 14, hence the use of the name 'model'.*

3.2 Interpretations

To define the semantics of typed terms, it has to be remarked that a logical category is *distributive*, i.e., $A \times (B + C) \simeq (A \times B) + (A \times C)$ for any $A, B, C \in \mathsf{Obj}\mathbb{C}$. In a logical category, it also holds that $MA \times M\exists x{:}s\, B \simeq M\exists x{:}s\, A \wedge B$, sometimes called the *Frobenius isomorphism*. These isomorphisms will be denoted by Δ and Δ_F, respectively, and they come for free from Cartesian closure. Then, the interpretation

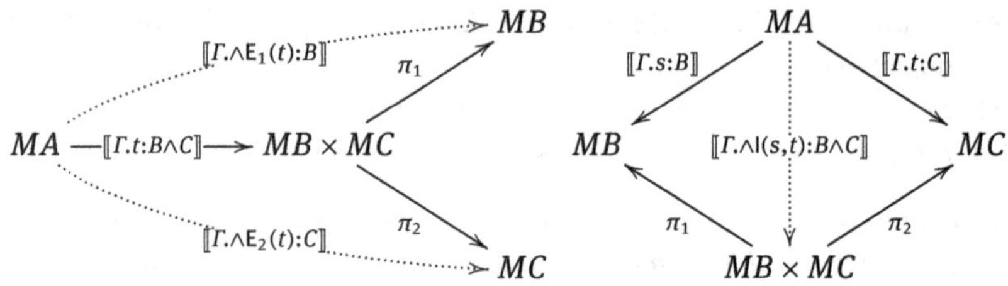

Fig. 2. The semantics of $\wedge\mathsf{E}_1$, $\wedge\mathsf{E}_2$ (left), and $\wedge\mathsf{I}$ (right).

of a typed term $x_1 : A_1, \dots, x_n : A_n.t : B$ will be an arrow from $MA_1 \times \cdots \times MA_n$ to MB. Since changing the order of elements in the context $x_1 : A_1, \dots, x_n : A_n$ induces an isomorphic product in the category, as it is immediate to check, interpretation will be automatically defined up-to-isomorphisms, in this very precise sense.

Definition 14 (Typed terms semantics). *Fixed a T-model $\langle \mathbb{C}, \mathbb{G}, \Sigma, \alpha \rangle$, the interpretation of a typed term $\Gamma.t : B$, notation $[\![\Gamma.t : B]\!]$, is inductively defined as follows: to simplify the notation, we pose Γ to be $x_1 : A_1, \dots, x_n : A_n$, and $A = A_1 \wedge \cdots \wedge A_n$,*

1. *if $x : A_i$ is a typed variable, $[\![\Gamma.x : A_i]\!] = \pi_i : MA \to MA_i$, the i-th projector of the product MA; this is the only point where we use the sequence of elements within the context, and the reader is invited to check that reordering the elements reduces to choose an isomorphic projector in the isomorphic product arising from the reordering;*
2. *if $a : B \in T$ is an axiom, $[\![\Gamma.a : B]\!] = \,! \circ \alpha(a : B)$, with $! : MA \to 1$ the universal arrow of the terminal object, and $\alpha(a : B)$ as in Definition 13;*
3. *$[\![\Gamma.\wedge\mathsf{I}(s,t) : B \wedge C]\!] = ([\![\Gamma.s : B]\!], [\![\Gamma.t : C]\!])$, with $(\_,\_)$ the universal arrow of the product $MB \times MC$, see Figure 2;*
4. *$[\![\Gamma.\wedge\mathsf{E}_1(t) : B]\!] = \pi_1 \circ [\![\Gamma.t : B \wedge C]\!]$ where π_1 is the first projector of the product $MB \times MC$; also, $[\![\Gamma.\wedge\mathsf{E}_2(t) : C]\!] = \pi_2 \circ [\![\Gamma.t : B \wedge C]\!]$ where π_2 is the second projector of the product $MB \times MC$, see Figure 2;*
5. *$[\![\Gamma.\vee\mathsf{I}_1(t) : B + C]\!] = \imath_1 \circ [\![\Gamma.t : B]\!]$, with $\imath_1$ the first injection of $MB + MC$; similarly, $[\![\Gamma.\vee\mathsf{I}_2(t) : B + C]\!] = \imath_2 \circ [\![\Gamma.t : C]\!]$, with $\imath_2$ the second injection of the coproduct $MB + MC$, , see Figure 3;*
6. *$[\![\Gamma.\vee\mathsf{E}(s,t,r,u : B, v : C) : D]\!] = [[\![\Gamma, u : B.t : D]\!], [\![\Gamma, v : C.r : D]\!]] \circ \Delta \circ (1_{MA}, [\![\Gamma.s : B \vee C]\!])$, where $[\_,\_]$ is the couniversal arrow of the coproduct $(MA \times MB) + (MA \times MC)$, $(\_,\_)$ is the universal arrow of the product $MA \times (MB + MC)$, and the arrow Δ is the distributive isomorphism, see Figure 3;*

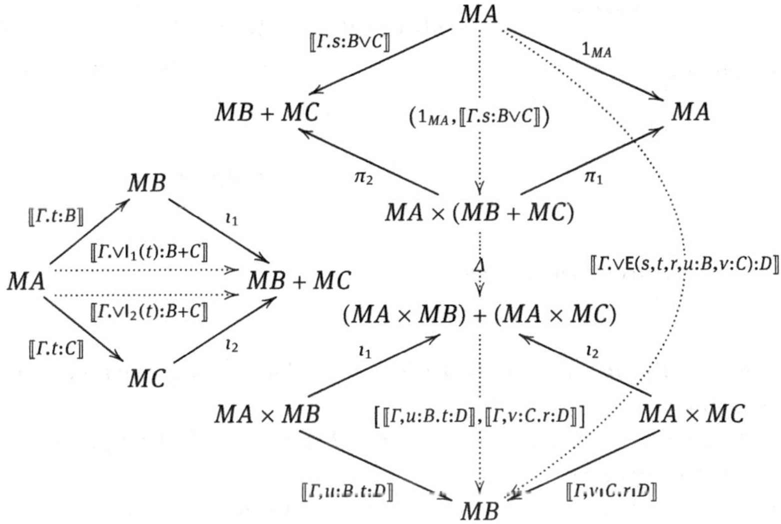

Fig. 3. The semantics of ∨I$_1$, ∨I$_2$ (left), and ∨E (right).

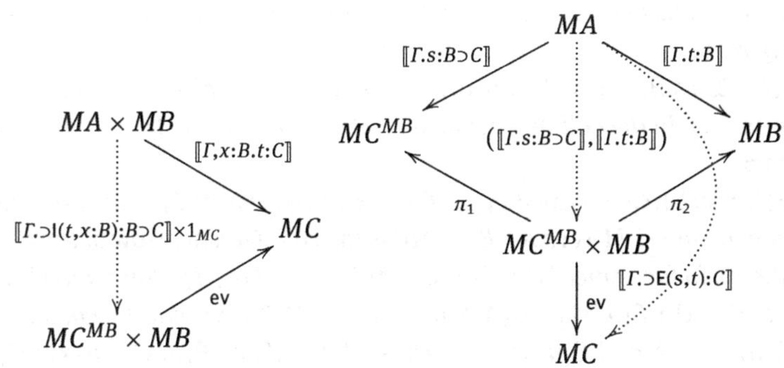

Fig. 4. The semantics of ⊃I (left) and ⊃E (right).

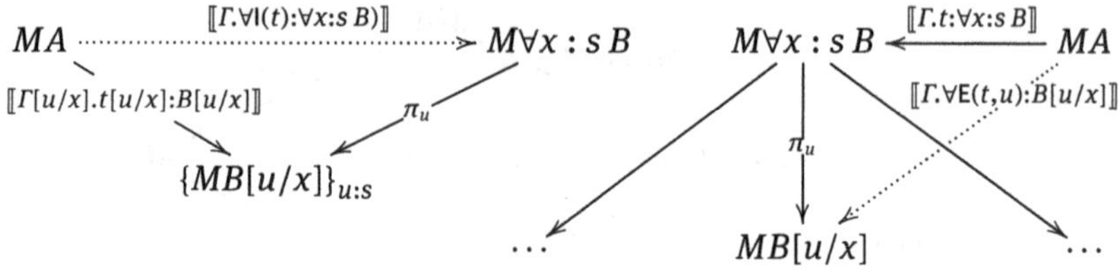

Fig. 5. Semantics of ∀I (left) and ∀E (right).

7. $[\![\Gamma.{\supset}\mathsf{I}(t, x : B) : B \supset C]\!] = \mathrm{tr}([\![\Gamma, x : B.t : C]\!])$, *with* tr *the exponential transpose, see Figure 4;*
8. $[\![\Gamma.{\supset}\mathsf{E}(s, t) : C]\!] = \mathrm{ev} \circ ([\![\Gamma.s : B \supset C]\!], [\![\Gamma.t : B]\!])$ *where* ev *is the exponential evaluation, see Figure 4;*
9. $[\![\Gamma.\top\mathsf{I} : \top]\!] = ! : MA \to 1$, *the universal arrow of the terminal object;*
10. $[\![\Gamma.\bot\mathsf{E}(t) : B]\!] = ! \circ [\![\Gamma.t : \bot]\!]$, *with* $! : 0 \to MB$ *the couniversal arrow of the initial object;*
11. $[\![\Gamma.\forall\mathsf{I}(t) : \forall x : s\, B)]\!]$ *is the universal arrow* $MA \to M\forall x : s\, B$ *of the limit on* $\Sigma_B(x : s)$. *In fact,* $[\![\Gamma.t : B]\!] : MA \to MB$ *and, by definition of* M, $[\![\Gamma[u/x].t[u/x] : B[u/x]]\!] : MA[u/x] \to MB[u/x]$ *for any logical term* $u : s$. *But, by construction of* ∀I, $x : s \notin \mathsf{FV}^*(\Gamma.t : B)$, *hence* $\Gamma[u/x] = \Gamma$ *and* $A[u/x] = A$. *Therefore,* $(MA, \{[\![\Gamma.t[u/x] : B[u/x]]\!]\}_{u:s})$ *is a cone on* $\Sigma_B(x : s)$ *and thus there is a unique universal arrow to* $\lim \Sigma_B(x : s) = M\forall x : s\, B$, *see Figure 5.*
12. $[\![\Gamma.\forall\mathsf{E}(t, u) : B[u/x]]\!] = \pi_u \circ [\![\Gamma.t : \forall x : s\, B]\!]$, *with* π_u *the projection corresponding to the logical term* $u : s$ *in the limit cone on* $\Sigma_B(x : s)$ *having* $M\forall x : s\, B$ *as vertex, see Figure 5;*
13. $[\![\Gamma.\exists\mathsf{I}(t) : \exists x : s\, B]\!] = \iota_u \circ [\![\Gamma.t : B[u/x]]\!]$, *where* ι_u *is the injection corresponding to the term* $u : s$ *in the colimit cocone on* $\Sigma_B(x : s)$ *having* $M\exists x : s\, B$ *as vertex, see Figure 6;*
14. $[\![\Gamma.\exists\mathsf{E}(t, r, v{:}B) : C]\!] = ! \circ \Delta_F \circ (1_{MA}, [\![\Gamma.t : \exists x : s\, B]\!])$, *where* Δ_F *is the Frobenius isomorphism, and* $! : M\exists x : s\, A \wedge B \to MC$ *is the couniversal arrow of the colimit on* $\Sigma_{A\wedge B}(x{:}s)$. *In fact, since* $[\![\Gamma, v{:}B.r{:}C]\!] : MA \wedge B \to MC$, *it follows that* $[\![\Gamma[u/x], v : B[u/x].r[u/x] : C[u/x]]\!] : M(A \wedge B)[u/x] \to MC[u/x]$, *but, by construction of* ∃E, $A[u/x] = A$ *and* $C[u/x] = C$. *Thus,* $(MC, \{[\![\Gamma, v : B[u/x].r[u/x] : C]\!]\}_{u:s})$ *is a cocone on* $\Sigma_{A\wedge B}(x : s)$ *and so there is unique couniversal arrow to it from* $\mathrm{colim}\Sigma_{A\wedge B}(x : s) = M\exists x : s\, A \wedge B$, *see Figure 6.*

Then, fixed a theory T and a T-model, two typed terms $\Gamma.t : A$ and $\Gamma.s : B$ are *semantically equal* whenever their interpretations are isomorphic in the evident co-slice category, $[\![\Gamma.t : A]\!] \cong [\![\Gamma.s : B]\!]$. Also, a formula A is *valid* in a T-model when

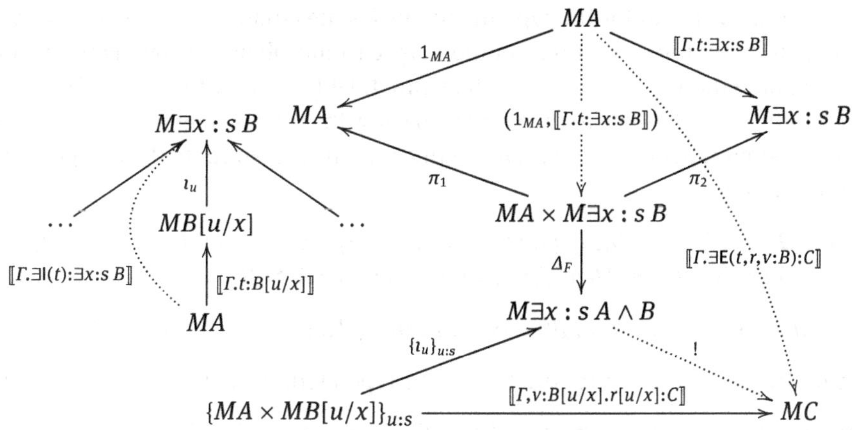

Fig. 6. The semantics of ∃I (left) and ∃E (right).

there is an arrow $p\colon 1 \to MA$. It is worth remarking that the axioms in T are valid in any T-model via the α map. Also, it is important to remark again that reordering the elements in a context, yields to isomorphic interpretations of types and typed terms.

4 Soundness and Completeness

Fixed a theory T, a soundness result for the logical system states that, whenever $\Gamma.p : A$ is a typed term, i.e., a logical proof in the proposition-as-types correspondence, any T-model in which all the assumptions are valid, makes A valid, too. A completeness result is the converse. Because of the proof-oriented nature of the semantics, this soundness result is almost immediate.

Theorem 1 (Logical soundness). *If the formula A is derivable from Γ in the theory T via the proof p, then A is valid in any T-model making all the assumptions valid.*

Proof: Let Γ be $x_1 : B_1, \ldots, x_n : B_n$, the sequence being arbitrary. Since each assumption is valid, there is $1 \to MB_i$ in the T-model for each $1 \leq i \leq n$, thus $\gamma\colon 1 \to MB_1 \times \cdots \times MB_n$ is the universal arrow of the product. But $[\![\Gamma.p : A]\!] : MB_1 \times \cdots \times MB_n \to MA$, so $[\![\Gamma.p : A]\!] \circ \gamma$ makes A valid. A different ordering of assumptions yields, evidently, to an isomorphic evidence of validity for A. □

The soundness result for the type theory on T states that, if two typed terms are equal, they have the same interpretation, up to isomorphisms. Showing this result is long and tedious, so we just sketch its proof, which can be found in full details in [Benini, 2013]. The key to the proof is to show that composing proofs by substituting assumptions with derivations, reduces in the semantics to the composition of interpretations.

Lemma 2. *Let $\vec{x} : \vec{A}$ be the context $x_1 : A_1, \ldots, x_n : A_n$, and let $\Gamma.r_i : A_i$ be a typed term for each $1 \le i \le n$. Then, for any typed term $\vec{x} : \vec{A}.t : B$,*

$$[\![\Gamma.t[r_1/x_1, \ldots, r_n/x_n] : B]\!] = [\![\vec{x} : \vec{A}.t : B]\!] \circ ([\![\Gamma.r_1 : A_1]\!], \ldots, [\![\Gamma.r_n : A_n]\!]) \ .$$

Proof: By induction on the structure of the typed term $\vec{x} : \vec{A}.t : B$. □

Theorem 3 (Soundness). *If a typed equality $\Gamma.s : A = \Gamma.t : A$ is derivable in the type theory based on T, then, in any T-model, $[\![\Gamma.s : A]\!] \cong [\![\Gamma.t : A]\!]$.*

Proof: By induction on the definition of typed equality. □

The logical interpretation of the soundness theorem for type theory is a weak form of normalisation: as far as we regard typed equality as a way to reduce proofs, equality preserves the semantics.

To show the completeness results for the type theories and for the logical systems, we proceed by constructing a syntactical model $\mathcal{M}$ of a theory T. Then, we will prove that $\mathcal{M}$ classifies all the T-models, which means the $\mathcal{M}$ model is, in a sense, the minimal one verifying the theory. The completeness results follow immediately.

Definition 15 (Syntactical equivalence). *Fixed a theory T, its equivalence classes of typed terms $[\Gamma.t : A]$ are defined as the minimal sets such that*

1. $\Gamma.t : A \in [\Gamma.t : A]$;
2. *if $\Gamma'.s : B = \Gamma'.r : B$ is derivable from T, and $\Gamma'.s : B \in [\Gamma.t : A]$, then also $\Gamma'.r : B \in [\Gamma.t : A]$;*
3. *if $\Gamma', z : D_1 \wedge D_2.s[\wedge\mathsf{E}_1(z)/y_1, \wedge\mathsf{E}_2(z)/y_2] : B \in [\Gamma.t : A]$ with $z : D_1 \wedge D_2 \notin \mathsf{FV}(\Gamma', y_1 : D_1, y_2 : D_2.s : B)$, then $\Gamma', y_1 : D_1, y_2 : D_2.s : B \in [\Gamma.t : A]$;*
4. *if $\Gamma', z : D.s : B \in [\Gamma.t : A]$ and $y : D \notin \mathsf{FV}(\Gamma', z : D.s : B)$, then it holds that $\Gamma', y : D.s[y/z] : B \in [\Gamma.t : A]$.*

Fact 1. *Each typed term $\Gamma.t : A$ appears in some equivalence class $[y : C.s : A]$ with the context consisting of a single typed variable.*

Proof: Let Γ be $x_1 : A_1, \ldots, x_n : A_n$. Pose C as $A_1 \wedge \cdots \wedge A_n$, with C be $\top$ when $n = 0$, and take $y : C$ to be new: then,

$$s = t[\wedge\mathsf{E}_1(y)/x_1, \wedge\mathsf{E}_1(\wedge\mathsf{E}_2(y))/x_2, \ldots, \wedge\mathsf{E}_2(\cdots(\wedge\mathsf{E}_2(y))\cdots)/x_n] \ .$$

The idea is that the first element of Γ can be obtained from y as $\wedge\mathsf{E}_1(y)$, and the rest of Γ is represented by $\wedge\mathsf{E}_2(y)$. Recursively dismounting the rest, we can access every element of Γ, which provides the formula s above. □

Fact 2. *If $\Gamma.t : B$ and $\Gamma.s : B$ are in the same equivalence class $[y : C.r : D]$ then B is identical to D and $\Gamma.t : B = \Gamma.s : B$.*

Proof: An immediate induction on the definition of syntactical equivalence proves B is the same as D, $\Gamma, y{:}C.t{:}B = \Gamma, y{:}C.r{:}B$ and $\Gamma, y{:}C.s{:}B = \Gamma, y{:}C.r{:}B$. □

So, syntactical equivalence induces a partition on the collection of typed terms, thus it is an equivalence relation.

Definition 16 (Syntactical category). *Given a theory T, its* syntactical category *$\langle \mathbb{C}_T, \mathbb{G}_T, \Sigma_T \rangle$ is defined as*

1. *the category $\mathbb{C}_T$ whose objects are all the types, the equivalence classes $[x{:}A.t{:}D]$ define the arrows $A \to D$, with $[x : A.x : A]$ as identities, and composition is given by substitution:*

$$[y : D.s : C] \circ [x : A.t : D] = [x : A.s[t/y] : C] \;\; ;$$

2. *the groupoid $\mathbb{G}_T$ is the discrete category of logical terms*[3]*.*
3. *the substitution functor Σ_T is generated, remember (5) in Definition 11, by the type interpretation M_T defined by $M_T A = A$.*

Proposition 4. *Fixed a theory T, its syntactical category $\langle \mathbb{C}_T, \mathbb{G}_T, \Sigma_T \rangle$ is logical.*

Proof: Products are formed by conjunction and $\top$; coproducts by disjunction and $\bot$; exponentiation is implication; the limit and the colimit of any substitution functor are the corresponding universally and existentially quantified formulae. Mere calculation allows to verify that these objects behave as expected. □

Proposition 5. *Given a theory T, and thus its syntactical category $\langle \mathbb{C}_T, \mathbb{G}_T, \Sigma_T \rangle$, the T-model $\langle \mathbb{C}_T, \mathbb{G}_T, \Sigma_T, \alpha_T \rangle$, where α_T maps the axiom $(a : A) \mapsto [x : \top.a : A]$, is called the* canonical model *and it satisfies exactly those typed equalities which are provable from T.*

Proof: Observing that the equivalence class $[x{:}\top.a{:}A]$ contains $.a{:}A$, the derivation of A via the axiom rule from the empty context, it follows that the map α_T

3 In the light of Section 5, when using abstract terms on the (abstract) signature $\langle \mathbb{G}, \Sigma, R \rangle$, $\mathbb{G}_T$ will become $\mathbb{G}$, thus constraining the canonical model to preserve the equalities which are part of the structure of abstract terms. It is worth remarking that this change has no impact on the results in this section, except for a technical assumption which will be discussed in Section 5.

induces the arrow $[.a : A]: 1 \to A$, thus the definition of T-model is correct. By induction on the structure of $x : C.t : B$, we prove that $[\![x : C.t : B]\!] = [x : C.t : B]$. Then, the conclusion follows immediately. □

Proposition 6. *Fixed a theory T, for every logical category $\langle \mathbb{C}, \mathbb{G}, \Sigma \rangle$, there is a bijection between equivalence classes, modulo natural equivalences, of structure-preserving functors $\mathbb{C}_T \to \mathbb{C}$ and equivalence classes, modulo isomorphisms, of T-models in $\mathbb{C}$, induced by the map $F \mapsto F(M_T)$.*

Proof: Essentially, this is Proposition 4.2.5 in Part D of [Johnstone, 2002b], see also [Benini, 2013]. □

Theorem 7 (Completeness). *If $\Gamma.t : A$ is semantically equal to $\Gamma.s : A$ in every T-model, then $\Gamma.t : A = \Gamma.s : A$ is derivable from T.*

Proof: It is safe to assume that Γ is $x : B$, thanks to syntactical equivalence. So, if $[\![x : B.t : A]\!] = [\![x : B.s : A]\!]$ is valid in every T-model, it holds in the canonical model as well. But, from the proof of Proposition 5, $[\![x : B.t : A]\!] = [x : B.t : A]$ and $[\![x : B.s : A]\!] = [x : B.s : A]$, so $[x : B.t : A] = [x : B.s : A]$ and the conclusion follows from Fact 2. □

Corollary 8. *If a formula A is valid in every T-model, then A is provable in the theory T.*

Proof: If A is valid in every T-model, $1 \to A$ is an arrow in the canonical model, thus $[.t : A]$ is inhabited. □

Hence, given a theory T, both the logical system and the type theory on T are sound and complete, and the canonical model captures exactly the valid formulae and the valid typed equalities: in the words of the Introduction, the canonical model is the entity the theory T really speaks of. And any other T-model may be obtained from the canonical one by transforming it via a suitable structure-preserving functor, which happens to be a *logical functor* [Mac Lane and Moerdijk, 1992, Section IV.2].

An inconsistent theory T allows to derive $\bot$: in the present semantics, T has a canonical model as well, but it is logically trivial, since $M\top \to M\bot$ is an arrow, that is, the initial and the terminal object are isomorphic. Of course, being the canonical model a classifier for other models via structure-preserving functors, all the model will have zero objects too. Oppositely, if a theory has at least a model with no zero object, it is consistent. It should be remarked that an inconsistent theory has a canonical model which still contains non-trivial computational information about how its proofs reduce: this can be useful for non-logical applications.

The reader may wonder why we have not explicitly considered equality in the completeness proof: in other semantics, the presence of equality modifies the way a canonical model is synthesised from the language. The answer is that there is no need to treat equality as special: since the semantics is point-free, i.e., logical terms are not part of it, and they do not get interpreted in some entity in the logical category, but rather they are only visible through the action of the substitution functor, equality is just another atomic relation.

4.1 Equalities

As we said, logical categories contain enough structure to model equality in the semantics. Although it is always possible to treat equality between logical terms by an axiomatic theory, which gets interpreted in the previously explained way, and thus give rise to a sound and complete theory, there are cases when a purely semantic approach is preferable.

The general idea is to consider the groupoid $\mathbb{G}$ of logical terms as the definition of the equality relation, and to use the machinery developed so far to project the groupoid structure onto the logical category, and to synthesise a theory for equality from this projection. To do so, assume the signature to contain a relation symbol $=_s : s \times s$ for some sort s. Then, we require the logical category $\langle \mathbb{C}, \mathbb{G}, \Sigma \rangle$ to satisfy

1. if $f \in \mathrm{Hom}_{\mathbb{G}}(t : s, r : s)$ then there is $\pi(f) \in \mathrm{Hom}_{\mathbb{C}}(1, Mt =_s r)$;
2. if $\mathrm{Hom}_{\mathbb{G}}(t : s, r : s) = \emptyset$ then there is $\varepsilon_{t,r} \in \mathrm{Hom}_{\mathbb{C}}(Mt =_s r, 0)$.

We want to show that these two requirements are enough to make $=_s$ a congruence, i.e., an equivalence relation such that, if $t =_s r$ then $MA[t/x] = MA[r/x]$ for each type A and variable $x : s$. In the following, we assume to work within a logical category $\langle \mathbb{C}, \mathbb{G}, \Sigma \rangle$ satisfying the requirements above.

Fact 3. *If $g \in \mathrm{Hom}_{\mathbb{C}}(MA, MB)$ then $A \supset B$ is valid.*

Proof: Consider the following exponentiation diagram:

$$\begin{array}{ccc} MB^{MA} \times MA & \xrightarrow{\text{ev}} & MB \\ \Big\uparrow {\scriptstyle \mathrm{tr}(g \circ \pi_2) \times 1_{MA}} & & \Big\uparrow {\scriptstyle g} \\ 1 \times MA & \xrightarrow[\pi_2]{} & MA \end{array}$$

Then, $\mathrm{tr}(g \circ \pi_2) \colon 1 \to MA \supset B$. □

Fact 4. *If, for all terms $t : s$, $f_t \in \mathrm{Hom}_{\mathbb{C}}\left(1, MB[t/x]^{MA[t/x]}\right)$, then $\forall x : s\, A \supset B$ is valid.*

Proof: Since $(1, \{f_t\}_{t:s})$ is a cone on $\Sigma_{A \supset B}(x : s)$, there exists the universal arrow $1 \to M\forall x : s\, A \supset B = \lim \Sigma_{A \supset B}(x : s)$. □

Proposition 9. *The formula $\forall x : s\, x =_s x$ is valid.*

Proof: Since, for each term $t{:}s$, $1_{t:s} \in \mathrm{Hom}_{\mathbb{G}}(t{:}s, t{:}s)$, $1 \to Mt =_s t$ is an arrow in $\mathbb{C}$. Thus, $1 = M\top$ is the vertex of a cone on $\Sigma_{x=_s x}(x : s)$. Then, the universal arrow of $\lim \Sigma_{x=_s x}(x : s)$ from 1 establishes the conclusion. □

Proposition 10. *The formula $\forall x : s\, \forall y : s\, x =_s y \supset y =_s x$ is valid.*

Proof: Let $t : s$ and $r : s$ be terms of sort s. Either there is $f \in \mathrm{Hom}_{\mathbb{G}}(t : s, r : s)$, or $\mathrm{Hom}_{\mathbb{G}}(t{:}s, r{:}s) = \emptyset$. In the former case, being $\mathbb{G}$ a groupoid, f has an inverse, and $\pi(f^{-1})\colon 1 \to Mr =_s t$ is in $\mathbb{C}$. Hence $\pi(f^{-1}) \circ (!\colon Mt =_s r \to 1) : Mt =_s r \to Mr =_s t$. In the latter case, $(!\colon 0 \to Mr =_s t) \circ \varepsilon_{t,r}\colon Mt =_s r \to Mr =_s t$. In both case, there is some $g\colon Mt =_s r \to Mr =_s t$. So, by Fact 3, $Mt =_s r \supset Mr =_s t$ is valid. Thus, by Fact 4 twice, the conclusion follows. □

Proposition 11. *The formula $\forall x : s\, \forall y : s\, \forall z : s\, x =_s y \wedge y =_s z \supset x =_s z$ is valid.*

Proof: Let $t : s$, $u : s$ and $v : s$ be terms. Either there are $f_1 \in \mathrm{Hom}_{\mathbb{G}}(t : s, u : s)$ and $f_2 \in \mathrm{Hom}_{\mathbb{G}}(u : s, v : s)$, or one of $\mathrm{Hom}_{\mathbb{G}}(t : s, u : s)$ and $\mathrm{Hom}_{\mathbb{G}}(u : s, v : s)$ is empty.

In the former case, $f_2 \circ f_1 \in \mathrm{Hom}_{\mathbb{G}}(t{:}, v : s)$, so $\pi(f_2 \circ f_1)\colon 1 \to Mt =_s v$ is in $\mathbb{C}$. Thus, $\pi(f_2 \circ f_1) \circ (!\colon Mt =_s u \times Mu =_s v \to 1) : Mt =_s v \times Mu =_s v \to Mt =_s v$. In the latter case, either $Mt =_s u \to 0$ or $Mu =_s v \to 0$ is in $\mathbb{C}$, that is $f\colon Mt =_s u \times Mu =_s v \to 0$ can be constructed by composing with the appropriate projection. Thus, $(!\colon 0 \to Mt =_s v) \circ f\colon Mt =_s u \times Mu =_s v \to Mt =_s v$. Summing up, in any case there is an arrow $g\colon Mt =_s u \times Mu =_s v \to Mt =_s v$. So, by Fact 3 and Fact 4, the conclusion follows. □

Proposition 12. *For any formula A, if $t =_s r$ is valid, so is $A[t/x] \supset A[r/x]$.*

Proof: Being $t =_s r$ valid, either there is $f \in \mathrm{Hom}_{\mathbb{G}}(t : s, r : s)$ or the category $\mathbb{C}$ is trivial, i.e., 0 is isomorphic to 1. In the latter case, the conclusion follows immediately, as 0 is initial. In the former case, since $\Sigma_A(x : s)(f)\colon MA[t/x] \to MA[r/x]$, Fact 3 yields the result. □

Thus, the natural typed theory for equality arising from the groupoid $\mathbb{G}$ is $T_=$, the typed theory consisting all the axioms $a_{\pi(f)} : t =_s r$ corresponding to the case (1) in the requirements on the logical category, and all the axioms $a_{\varepsilon_{t,r}} : \neg t =_s r$, corresponding to case (2). Then, soundness follows immediately from Propositions 9, 10, 11, and 12. And completeness comes for free.

5 Abstracting logical terms

A peculiar aspect of working within a point-free framework is that logical terms do not denote elements of some universe; rather, they link together formulae and proofs through the substitution functor. Precisely, terms and variables have very specific roles: terms are used as replacements for variables in the substitution pro cess. In turn, the semantic substitution process is modelled by the action of the substitution functors, and by the substitution rules in Definition 4 in the syntax. And, by clause (5) of Definition 11, these two instances of the process are coherent.

On the contrary, variables are used in a variety of ways: they act as terms; they can be replaced by terms in the substitution process; they are used to form quantified formulae; they constrain how inference rules may be applied—see clauses (11) and (14) in Definition 9; they index the substitution functors whose limits and colimits are used to interpret quantified formulae. All the uses of variables, as described so far, are deeply interrelated.

Since terms and variables do not get interpreted, although their roles are essential to build both the syntax and the semantics, it is natural to ask whether they are really needed. Or, oppositely, if they may be replaced by a simpler, more abstract structure. The answer is positive, although not yet completely satisfactory. In fact, as we will show in the following, we can use any groupoid as the structured collection of terms, obtaining essentially the same results, and, at the same time, allowing for an apparently more expressive syntax.

Substituting logical terms with the objects of a groupoid requires to modify the syntax. Specifically, the notion of *signature*, i.e., Definition 1 changes as follows: an abstract signature is a triple $\langle \mathbb{G}, \Sigma, R \rangle$ composed by a groupoid $\mathbb{G}$, a functor $\Sigma\colon \mathbb{G} \times \mathbb{G} \times \mathbb{G} \to \mathbb{G}$, and a set R of relation symbols, each one of the form $r : n$, where $n \in \mathbb{N}$ indicates its *arity*, that is, the number of arguments. The Σ functor models substitution of terms in $\mathbb{G}$: $\Sigma_t(x)(r)$ is the result of substituting x with r in t. We require Σ to be *stable*: $\Sigma_t(x)(x) = t$ for any $x, t \in \mathrm{Obj}\,\mathbb{G}$, compare with (6) in Definition 11.

Fact 5. *Fixed an abstract signature $\langle \mathbb{G}, \Sigma, R \rangle$, let $x_1, x_2, r_1, r_2, t_1, t_2 \in \mathrm{Obj}\,\mathbb{G}$ such that $x_1 \cong x_2$, $r_1 \cong r_2$, and $t_1 \cong t_2$. Then $\Sigma_{t_1}(x_1)(r_1) \cong \Sigma_{t_2}(x_2)(r_2)$.*

In an abstract signature, variables can be identified with those objects in $\mathbb{G}$ that significantly affect the substitution process: formally, $x \in \mathrm{Obj}\,\mathbb{G}$ is a *variable* when there are $t, r \in \mathrm{Obj}\,\mathbb{G}$ such that $\Sigma_t(x)(r) \not\cong t$. To conform to the standard notation, we call V the collection of all variables in an abstract signature, and we write $t[r/x]$ instead of $\Sigma_t(x)(r)$ when x is a variable.

Then, formulae, typed terms and the rest of the framework can be immediately redefined with no significant impact. The results, soundness and completeness in particular, can be derived in same way, except that an additional technical assumption is needed: in fact, it is easy to consider groupoids having a non-empty but still finite collection of variables. Since we require to freely form quantified formulae, and substitution requires to rename bounded variables at need, the storage of *new* variables may exhaust in those groupoids, preventing to reproduce the proofs of our results in them. Still, when a groupoid has an infinite number of non-isomorphic variables, or even when $V = \emptyset$, the proofs in this contribution are valid. The special case $V = \emptyset$ means to consider propositional systems, since quantified formulae cannot be formed.

Of course, introducing sorts is easy: it suffices to consider $\mathbb{G} = \bigsqcup_{s \in S} \mathbb{G}_s$, the coproduct of the groupoids $\mathbb{G}_s$, each one having infinite non-isomorphic variables. For a simple signature as in Definition 1, the groupoid $\mathbb{G}$ is the coproduct $\bigsqcup_{s \in S} \mathbb{G}_s$, that is, the disjoint union of the groupoids $\mathbb{G}_s$, with $\mathbb{G}_s$ the discrete category, i.e., the set of terms in the sort s.

An essential notion to model the constraints that variables introduce in the framework, as already illustrated, is that of *free variables in a term*: the free variables in $t \in \mathsf{Obj}\mathbb{G}$ is the collection $\mathsf{FV}(t)$ of variables x such that there is $r \in \mathsf{Obj}\mathbb{G}$ for which $t[r/x] \not\equiv t$. It is immediate to check that such a definition works as expected, because of Fact 5.

As an example, let us consider the category $\mathbb{P}$ of quasi orders and monotone maps. The category $\mathbb{G}_{\mathbb{P}}$ has as objects all the quasi orders but its arrows are just the isomorphisms of $\mathbb{P}$. Clearly, $\mathbb{G}_{\mathbb{P}}$ is a groupoid, and thus we can use it as the 'collection' of abstract terms in a logical language. Assuming to have equality as the only relation symbol, and interpreting it in the standard way, see Section 4.1, the theory will model quasi orders up to isomorphisms, forcing isomorphic orderings to be equal in the theory.

Let $\Sigma\colon \mathbb{G}_{\mathbb{P}} \times \mathbb{G}_{\mathbb{P}} \times \mathbb{G}_{\mathbb{P}} \to \mathbb{G}_{\mathbb{P}}$ be any functor such that

- $\Sigma_Q(S)(R) = Q$ when the finite quasi order S does not occur in Q or R does not occur in S,
- $\Sigma_Q(S)(R) = \langle (Q \setminus S) \sqcup R; \leq \rangle$ when the finite quasi order S occurs in Q and R occurs in S (notice how a choice has to be made, since S may occur in Q more than once, and R may occur in S in more than one way)
- the $\leq$ ordering relation above is defined as $x \leq y$ when
  - $x, y \in Q \setminus S$ and $x \leq_Q y$;
  - $x, y \in R$ and $x \leq_R y$;
  - $x \in Q \setminus S$, $y \in R$ and there is $z \in S$ such that $x \leq_Q z$ and $z \leq_S y$;
  - $x \in R$, $y \in Q \setminus S$ and there is $z \in S$ such that $x \leq_S z$ and $z \leq_Q y$.

Clearly, because it acts on a groupoid, Σ is a functor, and it is *stable*, since $\Sigma_Q(S)(S) = Q$, as it is immediate to verify.

Clearly, in this abstract signature, the 'terms' are the quasi orders, which are considered up-to-isomorphisms. The variables are the finite quasi orders, and the 'substitution' on terms replaces a finite quasi order S with a suborder R, in the obvious way. This is useful, for example, to show that chains can be compressed by identifying equivalent elements, thus reducing well quasi orders to well orders. This property is proved in a completely formal environment, by considering all the possible Σ functors above. Evidently, the advantage of this technique is to move the semantic way of reasoning, which is typical of algebraic arguments, like the ones used in the theory of quasi orders, to the syntactic level, naturally.

6 Comparison with other semantics

First-order intuitionistic systems have been intensively studied and many different ways to formally interpret these logical systems have been proposed and compared to each other. In this spirit, we will compare logical categories with similar frameworks, based on category theory, so to make the comparison precise. From our results, the reader may easily relate the semantics in this contribution with other systems based on different backgrounds.

The two major categorical frameworks to assign a meaning to first-order logic are Heyting categories and, as an especially important instance, elementary toposes. Both semantics are described in [Johnstone, 2002b], and we assume them to be well-known; in this respect, we adopt the notation of [Johnstone, 2002b] and its terminology. In fact, these semantics are variations over the same idea: to associate a subobject to each formula in a context. So, the context provides the object of reference and, specifically, the pseudo-complemented lattice of its subobjects. Formulae get a meaning by interpreting connectives as lattice operations, while, following Lavwere [1969]'s idea of interpreting quantifications as adjoints, quantifiers are rendered by relating the lattices of different objects through their connecting arrows. In fact, Heyting categories are, in a way, the minimal categories allowing for such an interpretation, while elementary toposes can be regarded as their natural algebraic and categorical completions.

So, we will show that each model $\mathcal{M}$ for a logical theory T in a Heyting category can be directly and uniformly transformed in a T-model $\mathcal{L}$ in a logical category, in such a way that the valid statements in $\mathcal{M}$ are exactly those in $\mathcal{L}$. Of course, since every elementary topos is also a Heyting category, the same transformation can be done on models in a topos. Thus, we will deduce that the expressive power of the semantics based on logical categories is at least the same as the one based on Heyting algebras, or elementary toposes, as far as we are only interested in what can be expressed in the first-order language and within the notion of validity.

Fixed[4] a signature Σ, we want to map each Σ-structure $\mathcal{M}$ in a Heyting category $\mathbb{C}$ to an essentially equivalent logical category $\mathsf{LH}(\mathcal{M})$.

Definition 17. *The category* $\mathsf{LH}(\mathcal{M}) = \langle \mathbb{D}, \mathbb{G}, \Xi \rangle$, *is defined as*

1. *The objects of* $\mathbb{D}$ *are* $[\![\mathsf{FV}(A).A]\!]_{\mathbb{C}}$, *the interpretation of the formula* A *in its minimal context*[5] *inside* $\mathbb{C}$, *for all the formulae* A. *It is worth noticing that the order of variables in the context denoted by* $\mathsf{FV}(A)$ *is not relevant since the finite products over the same objects are isomorphic;*
2. $\mathsf{Hom}_{\mathbb{D}}([\![\mathsf{FV}(A).A]\!]_{\mathbb{C}}, [\![\mathsf{FV}(B).B]\!]_{\mathbb{C}})$ *contains at most one arrow, and this happens exactly when* $[\![\mathsf{FV}(A) \cup \mathsf{FV}(B).A]\!]_{\mathbb{C}} \leq [\![\mathsf{FV}(A) \cup \mathsf{FV}(B).B]\!]_{\mathbb{C}}$ *in the lattice of subobjects of* $\mathcal{M}(\mathsf{FV}(A) \cup \mathsf{FV}(B))$;
3. *identities are the obvious arrows generated by reflexivity of* $\leq$;
4. *composition in* $\mathbb{D}$ *reduces to transitivity of* $\leq$;
5. *the groupoid* $\mathbb{G}$ *is the set, seen as a discrete category, of all terms in the language over* Σ;
6. Ξ *is the substitution functor generated by the interpretation map* $N \colon A \mapsto [\![\mathsf{FV}(A).A]\!]_{\mathbb{C}}$, *sending* A *to its interpretation in the minimal context.*

Proposition 13. *For any* Σ*-structure* $\mathcal{M}$ *in a Heyting category* $\mathbb{C}$, $\mathsf{LH}(\mathcal{M})$ *is a logical category.*

Proof: See [Benini, 2013]. It reduces to show that the lattice operations are mapped by N into the right categorical constructions, and that the adjoints seen through the N map construct the limit and the colimit of substitution functors[6]. □

4 Unfortunately, Johnstone [2002b] uses the Greek letter Σ for signatures, while we used it for the substitution functor: since there will be no double usage in this section, we conform to [Johnstone, 2002b] from now on.

5 Again, the notion of context here is different from the previously defined one, since the variables are logical, i.e., they are decorated with sorts instead of types. It is worth remarking that our contexts are the ones used in [Johnstone, 2002b] to define the interpretation of the simple theory of types.

6 In [Benini, 2013], an auxiliary device, the notion of *star* and *costar* subcategory, has been used. Subsequently, these auxiliary notions have turned out to be superfluous, and so they have been dropped from the current presentation. Nevertheless, proofs are, essentially, unchanged

Proposition 14. *Let $\mathcal{M}$ be a Σ-structure in a Heyting category $\mathbb{C}$. A sequent $\vec{x}:\vec{C}: A \vdash B$ is valid in $\mathcal{M}$ if and only if it is valid in* $\mathsf{LH}(\mathcal{M})$.

Proof: The 'only if' part is immediate by definition of LH. So, since the sequent is valid in the Heyting semantics, $[\![\vec{x}:\vec{C}.A]\!]_{\mathbb{C}} \leq [\![\vec{x}:\vec{C}.B]\!]_{\mathbb{C}}$ in the lattice of subobjects of $\mathcal{M}(\vec{x}:\vec{C})$. Then, it is easy to show [Benini, 2013] that $[\![\mathsf{FV}(A) \cup \mathsf{FV}(B).A]\!]_{\mathbb{C}} \leq [\![\mathsf{FV}(A) \cup \mathsf{FV}(B).B]\!]_{\mathbb{C}}$ in the lattice of subobjects of $\mathcal{M}(\mathsf{FV}(A) \cup \mathsf{FV}(B))$, thus, by definition of LH, there is an arrow $NA \to NB$ in $\mathbb{D}$, as required. □

Thus, given a model in a Heyting category, it is possible to transform it in a model in a logical category so that the valid formulae are exactly the same in both models. A natural question is whether it is possible to build a model in a Heyting category out of a T-model in a logical category. Up to present, the answer is negative and the intuitive reason is that the T-model is point-free: since logical terms and sorts are external to the core semantics, it seems unlikely that their interpretation can be reconstructed and/or generated.

But, supposing to have a T-model $\mathsf{LH}(\mathcal{M})$ in a logical category, which is known to come from a model $\mathcal{M}$ in a Heyting category, is it possible to recover $\mathcal{M}$? Again, as far as we know, the answer is negative, and for the same reason as above. Nevertheless, assuming that $\mathcal{M}$ is a model in a topos, we can, in some sense, invert the construction process leading to $\mathsf{LH}(\mathcal{M})$ by synthesising a very abstract model $\mathcal{N}$ in another topos which preserves validity of formulae as in Proposition 14. This result can be achieved by extending the map LH to a proper functor.

Definition 18 (Logical functor). *A functor $F\colon \mathfrak{E} \to \mathfrak{F}$ between the toposes $\mathfrak{E}$ and $\mathfrak{F}$ is said to be* logical, *see [Johnstone, 2002a, Section A.2.1] and [Mac Lane and Moerdijk, 1992, Section IV.2], when*
1. *F preserves finite limits;*
2. *for all $A \in \mathsf{Obj}\,\mathfrak{E}$, $F(\wp A)$ with the relation $F(\varepsilon_A)$ is a powerset object for FA, see [Goldblatt, 2006, p. 104].*

Definition 19. *Fixed a topos $\mathfrak{E}$ and a signature $\langle S, F, R\rangle$,* $\mathsf{Str}\,\mathfrak{E}$ *is the category whose objects are the models of the language in $\mathfrak{E}$, and whose arrows are the corresponding logical endo-functors preserving interpretations: $G\colon \mathcal{U} \to \mathcal{V} \in \mathsf{Str}\,\mathfrak{E}$ when G is a logical functor $\mathfrak{E} \to \mathfrak{E}$ such that $M_{\mathcal{V}}s = G(M_{\mathcal{U}}s)$ for any $s \in S$, $M_{\mathcal{V}}f = G(M_{\mathcal{U}}f)$ for any $f \in F$, and $M_{\mathcal{V}}r = G(M_{\mathcal{U}}r)$ for any $r \in R$.*

Fact 6. *If $F\colon \mathcal{U} \to \mathcal{V} \in \mathsf{Str}\,\mathfrak{E}$, then $M_{\mathcal{V}}t : s = F(M_{\mathcal{U}}t : s)$ for any logical term $t : s$, and $[\![\Gamma.A]\!]_{\mathcal{V}} = F([\![\Gamma.A]\!]_{\mathcal{U}})$ for any formula A in the context Γ.*

So, $\mathsf{Str}\,\mathfrak{E}$ is the *right* domain for extending LH to a functor: it contains the models in the topos $\mathfrak{E}$ and its arrows are exactly the functors between models preserving the logical structure, validity in particular.

Definition 20. *Fixed a signature S,* Str *is the category*

1. *whose objects are the logical categories $\langle \mathbb{C}, \mathbb{G}, \Sigma \rangle$ having as objects the formulae on S, $\mathbb{G}$ is a groupoid of terms over S, and Σ is generated by a surjective interpretation M;*
2. *whose arrows are the functors preserving interpretations, or, equivalently, the functors preserving limits and colimits.*

Because all the categories $\mathbb{C}$ in the above definition are small, the latter requirement can be rephrased by saying that the arrows are the functors having both a left and a right adjoint.

Fact 7. *For each model $\mathfrak{U}$ in some topos $\mathfrak{E}$,* $\mathsf{LH}(\mathfrak{U}) \in \mathsf{Str}$.

Fact 8. *If T is a theory over the signature S, and $\langle \mathbb{C}_T, \mathbb{G}_T, \Sigma_T \rangle$ is the syntactical category, see Definition 16, then* $\langle \mathbb{C}_T, \mathbb{G}_T, \Sigma_T \rangle \in \mathsf{ObjStr}$.

Thus, Str is the *right* codomain for extending LH to a functor: it contains all the images of models in Str $\mathfrak{E}$ and the characterising model for a theory.

Definition 21 (LH functor). *Given a topos $\mathfrak{E}$ and a signature S, then the functor* $\mathsf{LH}\colon \mathsf{Str}\,\mathfrak{E} \to \mathsf{Str}$ *extends the map* LH*: for any $F\colon \mathfrak{U} \to \mathfrak{V}$ in* Str $\mathfrak{E}$, $\mathsf{LH}(F)$ *is an arrow between* $\mathsf{LH}(\mathfrak{U}) = \langle \mathbb{D}_{\mathfrak{U}}, \mathbb{G}, \Sigma_{\mathfrak{U}} \rangle$ *and* $\mathsf{LH}(\mathfrak{V}) = \langle \mathbb{D}_{\mathfrak{V}}, \mathbb{G}, \Sigma_{\mathfrak{V}} \rangle$ *defined by*

1. $\mathsf{LH}(F)(\llbracket \mathsf{FV}(A).A \rrbracket_{\mathfrak{U}}) = F(\llbracket \mathsf{FV}(A).A \rrbracket_{\mathfrak{U}})$ *for any formula A on S;*
2. *for any $f\colon \llbracket \mathsf{FV}(A).A) \rrbracket_{\mathfrak{U}} \to \llbracket \mathsf{FV}(B).B) \rrbracket_{\mathfrak{U}}$, $\mathsf{LH}(F)(f)$ is the unique arrow such that $\Rightarrow_{\mathfrak{V}} \circ \langle F\llbracket \Gamma.A \rrbracket_{\mathfrak{U}}, F\llbracket \Gamma.B \rrbracket_{\mathfrak{U}} \rangle = F \top_{M_{\mathfrak{V}} \Gamma}$ with $\Gamma = \mathsf{FV}(A) \cup \mathsf{FV}(B)$.*

The definition is quite technical but its essence is that each logical functor preserving validity of models in a topos can be transformed in a canonical way into a homomorphism between logical categories. This happens exactly because, obviously, validity preserving arrows preserve validity, a fact that can be expressed in topos-theoretic terms, as in point (2) above.

We remind the following result, see [Awodey, 2006, Corollary 9.17]:

Proposition 15. *Let $F\colon \mathbb{C} \to \mathbb{D}$ be a functor between small categories. Then, the functor $F^*\colon [\mathbb{D}^{\mathrm{op}}, \mathbf{Set}] \to [\mathbb{C}^{\mathrm{op}}, \mathbf{Set}]$ given by $F^*(G) = G \circ F$, has both left and a right adjoints $F_! \dashv F^* \dashv F_*$. Moreover, there is a natural isomorphism $F_! \circ \mathbf{Y}_{\mathbb{C}} \cong \mathbf{Y}_{\mathbb{D}} \circ F$, where $\mathbf{Y}$ is the Yoneda embedding:*

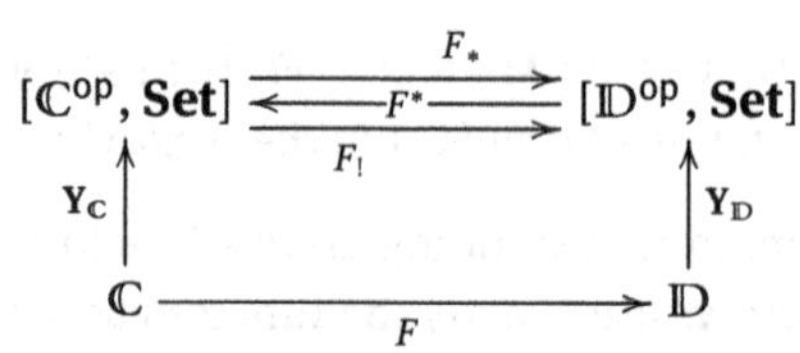

Hence, taking F to be LH, from a small topos $\mathfrak{E}$, it is possible to reconstruct topos-theoretic models from T-models, but they live in the bigger world of the presheaf topos $[(\mathsf{Str}\ \mathfrak{E})^{\mathsf{op}}, \mathbf{Set}]$. The fact that all the reconstructed models are equivalent in terms of validity is a consequence of the existence of adjoints.

7 Conclusions

This contribution illustrates a categorical semantics for the first-order, multi-sorted logical theories, together with the associated type theories. The semantics has been shown to be sound and complete, both in the logical and in the type-theoretical sense, and every theory has a classifying model.

Moreover, the models are point-free, since logical terms are not interpreted in some universe inside the models, but rather they are used to define the notion of semantic substitution, i.e., the action syntactical substitution performs over the category which is the 'universe' of the model, whose objects stand for formulae, and whose arrows stands for logical proofs. Hence, logical terms do not need a supporting universe to be studied via the logical language of the describing theory.

Although this semantics is equivalent to Heyting categories, as both makes valid the same collection of formulae,the correspondence between models is not direct: given a model in a Heyting category, it has been shown how a corresponding logical category is constructed. But, the point-free nature of logical categories prevents the construction of a suitable universe to interpret terms, which is needed to derive an associated Heyting category.

Acknowledgment: This work has been supported by the project *Correctness by Construction* (CORCON), EU 7th framework programme, grant n. PIRSES-GA-2013-612638, and by the project *Abstract Mathematics for Actual Computation: Hilbert's Program in the 21st Century*, John Templeton Foundation.

References

Steve Awodey. *Category Theory*, volume 49 of *Oxford Logic Guides*. Oxford University Press, 2006.

Marco Benini. Intuitionistic first-order logic: Categorical semantics via the Curry-Howard isomorphism. Technical report, ArXiv.org, 2013. http://arxiv.org/abs/1307.0108.

Jean-Yves Girard, Yves Lafont, and Paul Taylor. *Proofs and Types*, volume 7 of *Cambridge Tracts in Theoretical Computer Science*. Cambridge University Press, 1989.

Robert Goldblatt. *Topoi: The Categorical Analysis of Logic*. Dover Publications, 2006.

William Howard. The formulae-as-types notion of construction. In J. Roger Seldin and Jonathan P. Hindley, editors, *To H.B. Curry: Essays on Combinatory Logic, Lambda Calculus and Formalism*, pages 479–490. Academic Press, 1980. Original paper manuscript from 1969.

Peter T. Johnstone. *Sketches of an Elephant: A Topos Theory Compendium*, volume 1. Oxford University Press, 2002a.

Peter T. Johnstone. *Sketches of an Elephant: A Topos Theory Compendium*, volume 2. Oxford University Press, 2002b.

William F. Lavwere. Adjointness in foundations. *Dialectica*, 23:281–296, 1969. Reprinted in *Theory and Applications of Categories*, 16:1–16 (2006).

Saunders Mac Lane. *Categories for the Working Mathematician*. Springer, second edition, 1998.

Saunders Mac Lane and Ieke Moerdijk. *Sheaves in Geometry and Logic: A First Introduction to Topos Theory*. Springer, 1992.

Paul Taylor. *Practical Foundations of Mathematics*, volume 59 of *Cambridge studies in advanced mathematics*. Cambridge University Press, 1999.

The Univalent Foundations Program. *Homotopy Type Theory: Univalent Foundations of Mathematics*. Lulu press, 2013.

Anne S. Troelstra and Helmut Schwichtenberg. *Basic Proof Theory*, volume 43 of *Cambridge Tracts in Theoretical Computer Science*. Cambridge University Press, second edition, 2000.

Ulrich Berger, Kenji Miyamoto, Helmut Schwichtenberg, and Hideki Tsuiki

Logic for Gray-code Computation

Abstract: Gray-code is a well-known binary number system where neighboring values differ in one digit only. Tsuiki (2002) has introduced Gray code to the field of real number computation. He assigns to each number a unique $1\bot$-sequence, i.e., an infinite sequence of $\{-1, 1, \bot\}$ containing at most one copy of $\bot$ (meaning undefinedness). In this paper we take a logical and constructive approach to study real number computation based on Gray-code. Instead of Tsuiki's indeterministic multihead Type-2 machine, we use pre-Gray code, which is a representation of Gray-code as a sequence of constructors, to avoid the difficulty due to $\bot$ which prevents sequential access to a stream. We extract real number algorithms from proofs in an appropriate formal theory involving inductive and coinductive definitions. Examples are algorithms transforming pre-Gray code into signed digit code of real numbers, and conversely, the average for pre-Gray code and a translation of finite segments of pre-Gray code into its normal form. These examples are formalized in the proof assistant Minlog.

Keywords: Gray-code, Real number computation, Inductive and coinductive definitions, Program extraction

Mathematics Subject Classification 2010: 03D78, 03F60, 03B70, 03B35

1 Introduction

Gray-code (also called reflected binary code) is widely known in digital communication, due to its property that the Hamming distance between adjacent Gray-codes is always 1. Based on Gray-code, Di Gianantonio and Tsuiki studied independently an expansion of real numbers as infinite sequences of $\{0, 1, \bot\}$ each of which contains at most one $\bot$ standing for undefinedness [Gianantonio, 1999, Tsuiki, 2002]. Tsuiki called it *modified Gray expansion.* He also studied computability of real numbers, and presented several algorithms to do real number computation via Gray-code. The motivation of this paper is to shed light on the

Ulrich Berger: Swansea University, Swansea, SA2 8PP, Wales, UK, e-mail: u.berger@swansea.ac.uk
Kenji Miyamoto, Helmut Schwichtenberg: Ludwig-Maximilian Universität München, Munich, Germany, e-mail: kenji.miyamoto@math.lmu.de, schwicht@math.lmu.de
Hideki Tsuiki: Division of Mathematics, Faculty of Integrated Human Studies, Kyoto University, Kyoto 606, Japan, e-mail: tsuiki@i.h.kyoto-u.ac.jp

logical aspect of Gray-code computation from the constructive standpoint. We formalize Gray-code in the *Theory of Computable Functionals*, TCF in short, and also in the proof assistant Minlog[1], which is an implementation of TCF, by means of inductive and coinductive definitions [Schwichtenberg and Wainer, 2012]. In order to make use of Tsuiki's idea in TCF, we introduce pre-Gray code which is Gray-code represented as ordinary streams. Through the realizability interpretation we extract from proofs programs as terms in an extension T^+ of Gödel's T involving higher type recursion and corecursion operators. As case studies we extract real number algorithms in our setting of pre-Gray code. The correctness of the extracted programs is automatically ensured by the soundness theorem.

The rest of this paper is organized as follows. In Section 2 we investigate Gray-code and introduce pre-Gray code representation of real numbers. In Section 3 we describe realizability in our framework TCF, w.r.t. inductive and coinductive definitions. This provides a suitable setting to study logical aspects of signed digit streams and pre-Gray code. Section 4 presents proofs about coinductive representations that correspond to algorithms; the latter are described informally. 4.1 studies the average of two real numbers in signed digit code, and 4.4 directly for pre-Gray code. In 4.2 and 4.3 we give translators from pre-Gray into signed digit code, and vice versa. 4.5 studies a translation of finite segments of pre-Gray code into its normal form. Section 5 deals with the conversion of Gray-code to modified Gray expansion. In Section 6 we present and discuss the terms extracted from formalizations (in the proof assistant Minlog) of the proofs in Sections 4 and 5.

Related work

There are programming languages which can process modified Gray expansion directly. Tsuiki and Sugihara studied an extension of Haskell with the non-deterministic choice operator `gamb` which works as McCarthy's `amb` operator [Tsuiki and Sugihara, 2005]. Tsuiki studied a logic programming language with guarded clauses and committed choice [Tsuiki, 2005]. Terayama and Tsuiki studied an extension of PCF with parallelism [Terayama and Tsuiki, 2013]. In this paper we avoid using the above features by adopting pre-Gray code. Concerning stream based real arithmetic. Wiedmer [Wiedmer, 1977, 1980] used signed digit streams for real number computation. Its corecursive treatment was studied by Ciaffaglione and Di Gianantonio in Coq [Ciaffaglione and Gianantonio, 1999]. Berger and Seisenberger studied program extraction to obtain programs dealing

1 See http://www.minlog-system.de/

with signed digit streams [Berger and Seisenberger, 2010]. Some of their results are formalized by Miyamoto and Schwichtenberg in TCF and Minlog [Miyamoto, 2013, Miyamoto and Schwichtenberg, 2015]. Chuang studied the average and the multiplication of real numbers using coinduction in Agda via the Curry-Howard isomorphism [Chuang, 2011].

2 Gray-code and its variations

2.1 Expansions of real numbers

We define binary expansion of the unit interval as the expansion of the unit[2] interval $\mathbb{I} = [-1, 1]$ as infinite sequences of **PSD** $= \{-1, 1\}$ (proper signed digits) so that $v = a_1 a_2 \dots$ represents

$$\sum_{i=1}^{\infty} \frac{a_i}{2^i}. \tag{1}$$

With binary expansion, a finite sequence $a_1, a_2, \dots, a_n$ denotes the interval $f_{a_1}(f_{a_2} \dots (f_{a_n}(\mathbb{I}) \dots)$ for f_a the function

$$f_a(x) = \frac{x + a}{2},$$

and $a_1 a_2 \dots$ denotes the real number that belongs to the intersection of the intervals denoted by its finite truncations. Though binary expansion is simple and has little redundancy, it cannot be used for stream-based computation of the reals because, for example, the first digit of the number 0 cannot be determined by any arbitrary approximation information of the number. To remedy this, signed digit code is commonly used in real number computation. Signed digit code is a representation of the same interval with the same formula (1), but with three digits **SD** $= \{-1, 0, 1\}$. In this paper, we view finite sequences of **SD** as a free algebra **I** with a nullary constructor $\mathsf{nil}_\mathbf{I}$ and three unary constructors $\mathsf{C}_{-1}, \mathsf{C}_0, \mathsf{C}_1$ of type $\mathbf{I} \to \mathbf{I}$. That is,

$$\mathbf{I} = \mathsf{nil}_\mathbf{I} + \mathsf{C}_{-1}\,\mathbf{I} + \mathsf{C}_0\,\mathbf{I} + \mathsf{C}_1\,\mathbf{I}.$$

Signed digit code has a lot of redundancy, as $\bar{1}1$ and $0\bar{1}$ represent the same interval $[-1/2, 0]$ and $1\bar{1}$ and 01 represent the same interval $[0, 1/2]$. Here, $\bar{1}$ is the notation of -1 in a sequence. Modified Gray expansion is a unique representation

2 For simplicity we base our study on $[-1, 1]$ rather than $[0, 1]$.

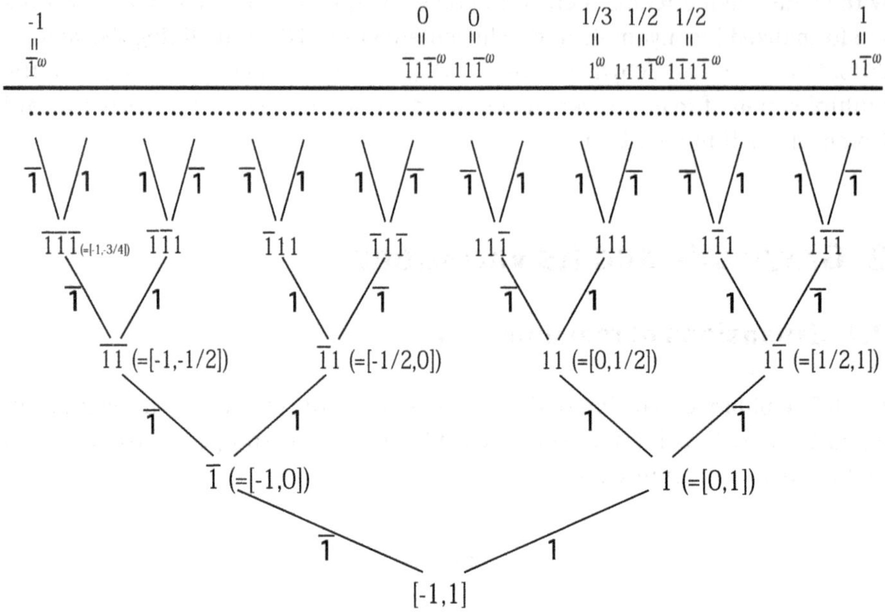

Fig. 1. Gray expansion.

of $\mathbb{I}$ that can be used for real number computation. It is based on Gray expansion which is another way of expanding $\mathbb{I}$ with **PSD**. In Gray expansion, the sequence is flipped after an appearance of 1. That is, let LR_a for $a \in \mathbf{PSD}$ be functions defined as

$$LR_a(x) = -a\frac{x-1}{2} \tag{2}$$

so that $LR_{-1} = f_{-1}$ but $LR_1(x) = f_1(-x)$ (Figure 1). Then, the number represented by a sequence $v = a_1a_2\dots$ is the limit of the shrinking intervals $LR_{a_1}(LR_{a_2}\dots(LR_{a_n}(\mathbb{I})\dots)$ which is equal to

$$\sum_{i=1}^{\infty}\frac{-\prod_{j\le i}(-a_j)}{2^i}. \tag{3}$$

With Gray expansion, each dyadic rational number (i.e., $k/2^i$ for integers $-2^i \le k \le 2^i$) other than -1 and 1 is represented in two ways as is the case for the binary expansion. For example, 0 is expanded as $\bar{1}1\bar{1}^\omega$ and $11\bar{1}^\omega$. However, the two expansions differ only at one digit and the sequence after the digit they differ is always $1\bar{1}^\omega$. *Modified Gray expansion* assigns the $1\bot$-sequence $s\bot 1\bar{1}^\omega$ to a dyadic rational number which has two Gray expansions $s11\bar{1}^\omega$ and $s\bar{1}1\bar{1}^\omega$ for $s \in \{1, \bar{1}\}^*$ [Tsuiki, 2002]. In this way, each real number in the unit interval is represented as a unique infinite $1\bot$-sequence, which is an infinite sequence of $\{-1, 1, \bot\}$ such that at most one copy of $\bot$ is contained in the sequence. In this paper, we consider its variant that assigns all three sequences $s11\bar{1}^{\omega}$, $s\bar{1}1\bar{1}^{\omega}$, and $s\bot 1\bar{1}^{\omega}$ to this dyadic rational number and simply call it the *(infinite) Gray-code*. Gray-codes of real numbers in $[-1, 1]$ range over the subset $\{1, \bar{1}\}^{\omega} \cup \{1, \bar{1}\}^{*}\bot 1\bar{1}^{\omega}$ of $\{-1, 1, \bot\}^{\omega}$. We will also call each $1\bot$-sequence in this set an (infinite) Gray-code. The following table shows the difference of the three representations according to the character a allowed when a dyadic rational number is represented as $sa1\bar{1}^{\omega}$.

Table 1. Variations of Gray-code

Gray expansion	$a \in \{-1, 1\}$,
Modified Gray expansion	$a = \bot$,
Gray-code	$a \in \{-1, 1, \bot\}$.

As we will study in Section 5, Gray-code and modified Gray expansion are equivalent in that a Gray-code can be coinductively converted to modified Gray expansion.

In order to define the meaning of Gray-codes more precisely, we introduce finite Gray-codes. A *finite* $1\bot$*-sequence* of length n is an infinite sequence $t = t_0 t_1 \ldots$ of $\{-1, 1, \bot\}$ such that $t_{n-1} \neq \bot$, $t_k = \bot$ for $k \geq n$, and $t_k = \bot$ for at most one $k < n$. We sometime omit the suffix $\bot^{\omega}$ of a finite $1\bot$-sequence and write it as a sequence of $\{-1, 1, \bot\}$ of length n. We call a finite $1\bot$-sequence in $\{1, \bar{1}\}^{*} \cup \{1, \bar{1}\}^{*}\bot 1\bar{1}^{*}$ a *finite Gray-code*. We define the order generated by $\bot \sqsubseteq -1$ and $\bot \sqsubseteq 1$ on $\{-1, 1, \bot\}$, and its product order on $\{-1, 1, \bot\}^{\omega}$. The set of finite/infinite $1\bot$-sequences form a Scott-Ershov domain BD with compact elements finite $1\bot$-sequences. Similarly, finite/infinite Gray-codes form a Scott-Ershov domain RD with compact elements finite Gray-codes. We say that a finite $1\bot$-sequence t approximates a $1\bot$-sequence s if $t \sqsubseteq s$.

We can define the meaning of Gray-code based on this domain structure. The meaning $[\![s]\!]$ of a finite Gray-code s is the same interval as the meaning of s with Gray expansion if $s \in \{\bar{1}, 1\}^{*}$, and is the union of $[\![s'\bar{1}1\bar{1}^{n}]\!]$ and $[\![s'11\bar{1}^{n}]\!]$ if s has the form $s'\bot 1\bar{1}^{n}$ for $s' \in \{\bar{1}, 1\}^{*}$. The meaning $[\![t]\!]$ of an infinite Gray-code t is the unique real number that belongs to the intersection of $[\![s]\!]$ for s finite Gray-codes that approximate t. The following proposition is immediate from the definition.

Proposition 2.1.
1. *For* $t \in \{-1, 1\}^{\omega}$, $[\![t]\!]$ *is the same as the value obtained by* (3).
2. *For* $s\bot 1\bar{1}^{\omega}$ *with* $s \in \{-1, 1\}^{*}$, $[\![s\bot 1\bar{1}^{\omega}]\!] = [\![s\bar{1}1\bar{1}^{\omega}]\!] = [\![s11\bar{1}^{\omega}]\!]$.

2.2 An algebra of $\bot$-sequences

Note that $\bot$ is not an ordinary character and a machine cannot read or write a $\bot$ on a tape. In [Tsuiki, 2002] an IM2-machine (indeterministic multihead Type-2 machine) was introduced to input and output $1\bot$-sequences. An IM2-machine has two heads on each input/output tape so that it can skip a $\bot$ and access the rest the sequence. In this paper, instead of such a direct manipulation of $1\bot$-sequences, we define pre-Gray code, which is a "representation" of Gray-code as sequences of constructors representing how an $1\bot$-sequence is obtained, and consider computation through usual stream programs instead of IM2-machines.

We introduce the algebra $\mathbf{OB} = (|\mathbf{OB}|, C \cup \{\mathsf{nil}\})$ of finite $1\bot$-sequences, before introducing pre-Gray code. The carrier set $|\mathbf{OB}|$ is the set of finite $1\bot$-sequences. It is generated by four unary constructors in $C := \{\mathsf{cons}_1, \mathsf{cons}_{-1}, \mathsf{ins}_1, \mathsf{ins}_{-1}\}$ as well as a nullary constructor nil. Recall that an ordinary binary sequence is a term of a free algebra with two unary constructors cons_a for $a \in \{-1, 1\}$ which prepend a to a sequence as well as nil. On the other hand, a $1\bot$-sequence is generated by two additional constructors ins_a for $a \in \{-1, 1\}$ which insert a as the second character to a sequence.

Example 2.2. *The term* $\mathsf{ins}_1(\mathsf{ins}_{-1}(\mathsf{cons}_{-1}(\mathsf{ins}_1\ \mathsf{nil})))$ *denotes* $\bar{1}1\bar{1}\bot 1$*:*

$$\begin{aligned}
\mathsf{nil} &\quad \textit{denotes} \quad \bot^{\omega}, \\
(\mathsf{ins}_1\ \mathsf{nil}) &\quad \textit{denotes} \quad \bot 1 \bot^{\omega}, \\
(\mathsf{cons}_{-1}(\mathsf{ins}_1\ \mathsf{nil})) &\quad \textit{denotes} \quad \bar{1}\bot 1 \bot^{\omega}, \\
(\mathsf{ins}_{-1}(\mathsf{cons}_{-1}(\mathsf{ins}_1\ \mathsf{nil}))) &\quad \textit{denotes} \quad \bar{1}\bar{1}\bot 1 \bot^{\omega}, \\
(\mathsf{ins}_1(\mathsf{ins}_{-1}(\mathsf{cons}_{-1}(\mathsf{ins}_1\ \mathsf{nil})))) &\quad \textit{denotes} \quad \bar{1}1\bar{1}\bot 1 \bot^{\omega}.
\end{aligned}$$

When writing a term of **OB**, we omit nil and write it as a sequence of C. Thus, we write $[\mathsf{ins}_1, \mathsf{ins}_{-1}, \mathsf{cons}_{-1}, \mathsf{ins}_1]$ for this term.

It is not hard to caluculate, that $[\mathsf{cons}_{-1}, \mathsf{cons}_1, \mathsf{cons}_{-1}, \mathsf{ins}_1]$ also denotes the same $1\bot$-sequence. We write $\varphi(p)$ for the $1\bot$-sequence denoted by $p \in C^*$. More precisely, $\varphi([c_1, \ldots, c_n]) = (c_1 \circ \cdots \circ c_n)(\bot^{\omega})$.

For coalgebraic computation, one needs to read sequences of constructors from left to right. If a sequence of C is read from left to right, then it can be considered as a procedure to construct a $1\bot$-sequence as follows. We start with an infinite tape with the state $\bot^{\omega}$. We view cons_a as the operation to fill the leftmost $\bot$ with a and ins_a as the operation to fill the second $\bot$ from the left with a. We write $\psi(p)$ for the $1\bot$-sequence obtained by this procedure. More precisely, if we define $c' : \{-1, 1, \bot\}^\omega \to \{-1, 1, \bot\}^\omega$ ($c \in C$) by

$$\begin{aligned}\mathsf{cons}'_a(s) &= \text{filling in } s \text{ the first bottom from the left by } a\\ \mathsf{ins}'_a(s) &= \text{filling in } s \text{ the second bottom from the left by } a\end{aligned}$$

then $\psi([c_1, \ldots, c_n]) = (c'_n \circ \cdots \circ c'_1)(\bot^\omega)$.

Example 2.3. *We construct* $\bar{1}1\bar{1}\bot 1$ *according to* $[\mathsf{cons}_{-1}, \mathsf{cons}_1, \mathsf{cons}_{-1}, \mathsf{ins}_1]$ *as* $\bot^\omega \to \bar{1}\bot^\omega \to \bar{1}1\bot^\omega \to \bar{1}1\bar{1}\bot^\omega \to \bar{1}1\bar{1}\bot 1\bot^\omega$ *and according to* $[\mathsf{ins}_1, \mathsf{ins}_{-1}, \mathsf{cons}_{-1}, \mathsf{ins}_1]$ *as* $\bot^\omega \to \bot 1\bot^\omega \to \bot 1\bar{1}\bot^\omega \to \bar{1}1\bar{1}\bot^\omega \to \bar{1}1\bar{1}\bot 1\bot^\omega$.

Proposition 2.4. $\varphi(p) = \psi(p)$ *for* $p \in C^*$.

Proof: We show that

$$(c_1 \circ \cdots \circ c_n)(\bot^\omega) = (c'_n \circ \cdots \circ c'_1)(\bot^\omega). \tag{4}$$

Note that c' satisfies the equations

$$\begin{aligned}c'(b : s) &= b : c'(s)\ (b \neq \bot),\\ \mathsf{cons}'_a(\bot : s) &= a : s,\\ \mathsf{ins}'_a(\bot : s) &= \bot : \mathsf{cons}'_a(s).\end{aligned}$$

Using the equations for c and c' one easily verifies that

$$c(\bot^\omega) = c'(\bot^\omega), \tag{5}$$

$$c \circ d' = d' \circ c. \tag{6}$$

From (6) one obtains

$$(c_1 \circ \cdots \circ c_n) \circ c' = c' \circ (c_1 \circ \cdots \circ c_n) \tag{7}$$

by induction on n. Now (7) and (5) yield (4), again by induction on n. □

Note that c' is increasing. That is, $s \sqsubseteq c'(s)$ for $c \in C$. Therefore, we can consider an infinite sequence $q \in C^\omega$ of the four constructors cons_1, $\mathsf{cons}_{\bar{1}}$, ins_1, $\mathsf{ins}_{\bar{1}}$ as representing an infinite $1\bot$-sequence which is obtained as the least upper bound of $\{\varphi(p)(= \psi(p)) \mid p \text{ is a finite prefix of } q\}$. For example, $[\mathsf{ins}_1, \mathsf{ins}_{-1}, \mathsf{ins}_{-1}, \mathsf{ins}_{-1}, \ldots]$ represents $\bot 1\bar{1}^\omega$, and we write $\varphi(q)$ for the $1\bot$-sequence represented by $q \in C^\omega$.

As we have noted, the algebra **OB** is not a free algebra and we have equations

$$\mathsf{ins}_a \circ \mathsf{cons}_b = \mathsf{cons}_b \circ \mathsf{cons}_a \tag{8}$$

for $a, b \in$ **PSD**. Actually, **OB** is the universal algebra in that the set of finite $1\bot$-sequences is equal to the quotient of C^* by these equations.

2.3 An algebra of Gray-code and an auxiliary algebra

Recall that finite Gray-codes form a subset $\{\bar{1}, 1\}^* \cup \{\bar{1}, 1\}^* \bot 1 \bar{1}^*$ of the set of finite $1\bot$-sequences. In order to represent only this set of finite Gray-codes, we define a subalgebra **G** of **OB** simultaneously with another subalgebra **H**. The carrier set of **G** is the set of finite Gray-codes. A naive attempt is to define them as follows.

$$\begin{aligned}\mathbf{G} &= (\{\bar{1}, 1\}^* \cup \{\bar{1}, 1\}^* \bot 1 \bar{1}^*,\\ &\qquad \{\mathsf{cons}_a : \mathbf{G} \to \mathbf{G} \mid a \in \mathbf{PSD}\} \cup \{\mathsf{ins}_1 : \mathbf{H} \to \mathbf{G}, \mathsf{nil}_{\mathbf{G}} : \mathbf{G}\})\\ \mathbf{H} &= (\bot \bar{1}^*,\ \{\mathsf{ins}_{-1} : \mathbf{H} \to \mathbf{H}, \mathsf{nil}_{\mathbf{H}} : \mathbf{H}\})\end{aligned}$$

However, this definition does not allow filling a bottom with a digit by the cons_a constructor in the coinductive treatment of an $1\bot$-sequence. For this purpose, we need to add the constructors $\mathsf{cons}_a : \mathbf{G} \to \mathbf{H}$ for $a \in \mathbf{PSD}$ to the above definition. In order to distinguish the two constructors cons_a of types $\mathbf{G} \to \mathbf{G}$ and $\mathbf{G} \to \mathbf{H}$, we give them different names LR_a and Fin_a. We also rename ins_1 and ins_{-1} to U and D, respectively, and define the two algebras **G** and **H** with carrier sets $|\mathbf{G}| = \{\bar{1}, 1\}^* \cup \{\bar{1}, 1\}^* \bot 1 \bar{1}^*$ and $|\mathbf{H}| = \{\bar{1}, 1\}^+ \cup \{\bar{1}, 1\}^+ \bot 1 \bar{1}^* \cup \bot \bar{1}^*$ mutually recursively as follows.

$$\begin{aligned}\mathbf{G} &= (|\mathbf{G}|,\ \{\mathrm{LR}_a : \mathbf{G} \to \mathbf{G} \mid a \in \mathbf{PSD}\} \cup \{\mathrm{U} : \mathbf{H} \to \mathbf{G}, \mathsf{nil}_{\mathbf{G}} : \mathbf{G}\})\\ \mathbf{H} &= (|\mathbf{H}|,\ \{\mathrm{Fin}_a : \mathbf{G} \to \mathbf{H} \mid a \in \mathbf{PSD}\} \cup \{\mathrm{D} : \mathbf{H} \to \mathbf{H}, \mathsf{nil}_{\mathbf{H}} : \mathbf{H}\})\end{aligned}$$

Note that the carrier sets of both algebras are generated (but not freely) by their constructors. We call a term of type **G** a *finite pre-Gray code.*

In the coinductive treatment of an $1\bot$-sequence, $\mathrm{U} : \mathbf{H} \to \mathbf{G}$ means to leave the current cell *U*ndefined and fill the next cell with 1, $\mathrm{D} : \mathbf{H} \to \mathbf{H}$ means to *D*elay the determination of the value of the unfilled cell and add $\bar{1}$ to the end of the sequence, and Fin_a means to *Fin*ally fill the unfilled cell with a. Thus, both $\mathrm{U}(\mathrm{D}(\mathrm{Fin}_{-1}(\mathrm{U}(\mathsf{nil}_{\mathbf{H}}))))$ and $\mathrm{LR}_{-1}(\mathrm{LR}_1(\mathrm{LR}_{-1}(\mathrm{U}(\mathsf{nil}_{\mathbf{H}}))))$ are terms of type **G** representing the sequence $\bar{1}1\bar{1}\bot 1$.

We call an infinite sequence of these constructors all of whose finite truncations are term of type **G** an infinite term of type **G**, and similary, define an infinite term of **H**. An infinite term p of type **G** is representing an infinite Gray-code $\varphi(p)$ and thus representing a real number $[\![p]\!] \in \mathbb{I}$ defined as $[\![\varphi(p)]\!]$. For example, for $p = [\mathrm{U}, \mathrm{D}, \mathrm{D}, \ldots]$, $\varphi(p) = \bot 1 \bar{1}^\omega$ and $[\![p]\!] = 0$. We call an infinite term of type **G** an *(infinite) pre-Gray code.*

Since cons_a, and ins_a satisfy (8), the constructors of **G** and **H** satisfy the following equations for $a \in \mathbf{PSD}$.

$$\mathrm{U} \circ \mathrm{Fin}_a = \mathrm{LR}_a \circ \mathrm{LR}_1, \tag{9}$$

$$\mathrm{D} \circ \mathrm{Fin}_a = \mathrm{Fin}_a \circ \mathrm{LR}_{-1}. \tag{10}$$

We show that the set of finite Gray-codes is the quotient of the term algebra of **G** with these equations.

Proposition 2.5. *Let p be a term of type* **G** *and* $a_i \in \mathbf{PSD}$ $(1 \le i \le m)$.

1. *If* $\varphi(p) = a_1 \dots a_m \bot 1 \bar{1}^l$, *the equation* $p = [\mathrm{LR}_{a_1}, \dots, \mathrm{LR}_{a_m}, \mathrm{U}, \mathrm{D}^l]$ *can be derived from* (9) *and* (10).
2. *If* $\varphi(p) = a_1 \dots a_m$, *the equation* $p = [\mathrm{LR}_{a_1}, \dots, \mathrm{LR}_{a_m}]$ *can be derived from* (9) *and* (10).

Proof: Let $p = [c_1, \dots, c_n]$. We have $n = l + m$ because each constructor adds one digit to a sequence. Suppose that the argument type of c_i is **H** for $i \ge k$ and the return type of c_k is **G**. Then, from the definition of **G** and **H**, we have $k = m + 1$ and $c_k, \dots, c_n$ are uniquely determined as $c_k = \mathrm{U}$ and $c_i = \mathrm{D}$ for $i > k$. Therefore, (1) is immediately derived from (2). We prove (2) by induction on m. If $m = 0$, then $p = \mathsf{nil}_\mathbf{G}$ and this statement holds. Suppose that $c_m = \mathrm{LR}_b$. Since $\varphi(p) = \psi(p)$ by Proposition 2.4, $a_m = b$ and $\varphi([c_1, \dots, c_{m-1}]) = a_1 \dots a_{m-1}$. Therefore, it holds by the induction hypothesis. Suppose that $c_m = \mathrm{Fin}_b$. Since the argument type of c_{m-1} is **H**, p has the form $[c_1, \dots, c_{m-k-2}, \mathrm{U}, \mathrm{D}^k, \mathrm{Fin}_b]$. By induction hypothesis, $[c_1, \dots, c_{m-k-2}] = [\mathrm{LR}_{a_1}, \dots, \mathrm{LR}_{a_{m-k-2}}]$ is derived. On the other hand, $[\mathrm{U}, \mathrm{D}^k, \mathrm{Fin}_b] = [\mathrm{LR}_b, \mathrm{LR}_1, \mathrm{LR}_{-1}^k]$ is derived by applying (10) k times and then applying (9). Thus, (2) is proved. □

2.4 Pre-Gray code

As we defined, Gray-codes are representations of $\mathbb{I}$ as $\{-1, 1, \bot\}$-sequences and pre-Gray codes are terms of the algebra **G** of Gray-codes. For our study of real number computation based on pre-Gray code, we redefine **G** and **H** as free algebras and assign affine functions f_c to unary constructors c of **G** and **H** so that one can directly define meanings of pre-Gray codes.

First, since $\mathsf{nil}_\mathbf{G}$ and $\mathsf{nil}_\mathbf{H}$ express the empty $1\bot$-sequence, they denote the unit interval $\mathbb{I}$. It is natural to define f_{LR_a} and f_U as

$$f_{\mathrm{LR}_a} = -a\frac{x-1}{2} \ (= LR_a \text{ in (2)}), \tag{11}$$

$$f_\mathrm{U}(x) = \frac{x}{2}. \tag{12}$$

Since (9) and (10) hold, f_{Fin_a} and f_D should satisfy

$$f_\mathrm{U} \circ f_{\mathrm{Fin}_a} = f_{\mathrm{LR}_a} \circ f_{\mathrm{LR}_1}, \tag{13}$$

$$f_\mathrm{D} \circ f_{\mathrm{Fin}_a} = f_{\mathrm{Fin}_a} \circ f_{\mathrm{LR}_{-1}}. \tag{14}$$

From (13), we have

$$f_{\mathrm{Fin}_a}(x) = a\frac{x+1}{2} = f_{\mathrm{LR}_a}(-x), \tag{15}$$

and therefore from (14), we have

$$f_{\mathrm{D}}(x) = \frac{x}{2}. \tag{16}$$

Equations (11), (12), (15), and (16) define the meanings of constructors as affine functions, and they define the meaning $F_{\mathbf{G}}(p)$ of a term $p = [c_1, \ldots, c_n]$ of type **G** as

$$F_{\mathbf{G}}(p) = f_{c_1}(f_{c_2}(\ldots f_{c_n}(\mathbb{I})\ldots)). \tag{17}$$

They also define the meaning $F_{\mathbf{H}}(q)$ of a term q of type **H** similarly. Therefore, they define meanings $F_{\mathbf{G}}(p)$ and $F_{\mathbf{H}}(q)$ of infinite terms p of type **G** and q of type **H**, respectively, as the unique elements which belong to intersections of the meanings of their finite prefixes (Figure 2).

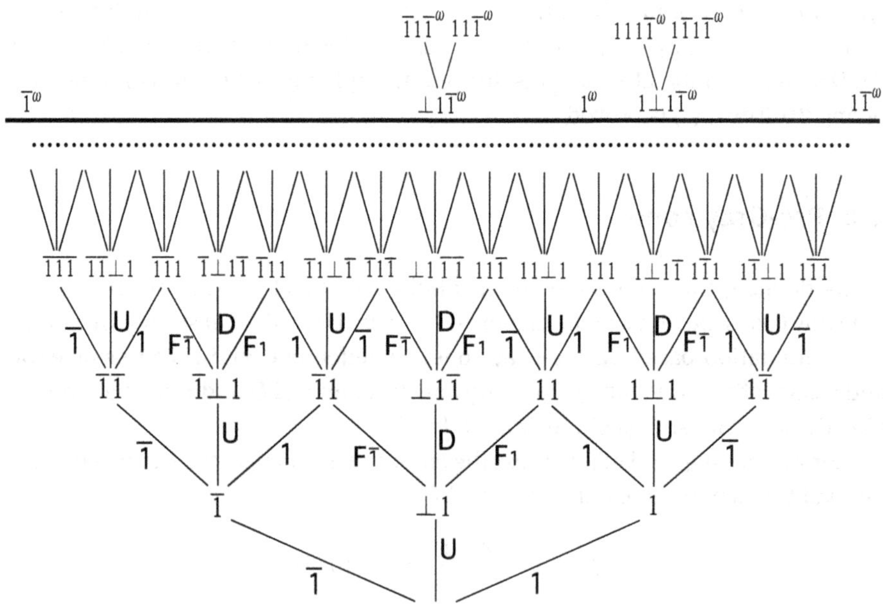

Fig. 2. The domain *RD* and constructors of pre-Gray code. Here, $\mathrm{LR}_{-1}, \mathrm{LR}_1, \mathrm{U}, \mathrm{Fin}_{-1}, \mathrm{Fin}_1$, and D are written as $\bar{1}, 1, \mathrm{U}, \mathrm{F}\bar{1}, \mathrm{F}1$, and D, respectively.

Proposition 2.6. *For a (possibly infinite) term p of type* **G**, *we have* $[\![p]\!] = F_{\mathbf{G}}(p)$.

Proof: We first prove the statement for the case that p is finite. Since equations (13) and (14) hold, we only need to consider the cases $p = [\mathrm{LR}_{a_1}, \ldots, \mathrm{LR}_{a_m}, \mathrm{U}, \mathrm{D}^n]$ and $p = [\mathrm{LR}_{a_1}, \ldots, \mathrm{LR}_{a_m}]$ by Proposition 2.5. The latter case is immediate from the definition. In the former case, we have $\varphi(p) = a_1 \ldots a_m \bot 1 \bar{1}^n$. Let $g = f_{\mathrm{LR}_{a_1}} \circ \cdots \circ f_{\mathrm{LR}_{a_m}}$. We have

$$\begin{aligned}
[\![p]\!] &= [\![a_1 \ldots a_m \bar{1} 1 \bar{1}^n]\!] \cup [\![a_1 \ldots a_m 1 1 \bar{1}^n]\!] \\
&= g(f_{\mathrm{LR}_{-1}} \circ f_{\mathrm{LR}_1} \circ f^n_{\mathrm{LR}_{-1}}(\mathbb{I})) \cup g(f_{\mathrm{LR}_1} \circ f_{\mathrm{LR}_1} \circ f^n_{\mathrm{LR}_{-1}}(\mathbb{I})) \\
&= g([-1/2^{n+1}, 0]) \cup g([0, 1/2^{n+1}]) \\
&= g([-1/2^{n+1}, 1/2^{n+1}]) \\
&= (g \circ f_{\mathrm{U}} \circ f^n_{\mathrm{D}})(\mathbb{I}) \\
&= F_{\mathbf{G}}([\mathrm{LR}_{a_1}, \ldots, \mathrm{LR}_{a_m}, \mathrm{U}, \mathrm{D}^n]).
\end{aligned}$$

The case p is infinite is immediately derived from the finite case. □

Note that the meaning $F_{\mathbf{H}}(p)$ of a term p of type **H** is also defined. If p is a term of type **H**, then $\varphi(p)$ may not be a finite Gray-code and even if it is, $F_{\mathbf{H}}(p)$ is different from $[\![\varphi(p)]\!]$ in general. For example, $\varphi([\mathrm{D}]) = \bot\bar{1}$ and $[\![\bar{1}\bar{1}]\!] \cup [\![1\bar{1}]\!] = [-1, -1/2] \cup [1/2, 1]$ is not an interval, and $[\![\varphi([\mathrm{Fin}_1, \mathrm{LR}_1])]\!] = [\![11]\!] = [0, 1/2]$ whereas $F_{\mathbf{H}}([\mathrm{Fin}_1, \mathrm{LR}_1]) = [1/2, 1]$.

The meaning $F_{\mathbf{H}}$ of **H** defines another representation of $\mathbb{I}$ which is obtained by flipping the second digit of Gray code (Figure 3). We extract conversion programs between these two representations from the proof of Lemma 4.10 in Section 4.4.

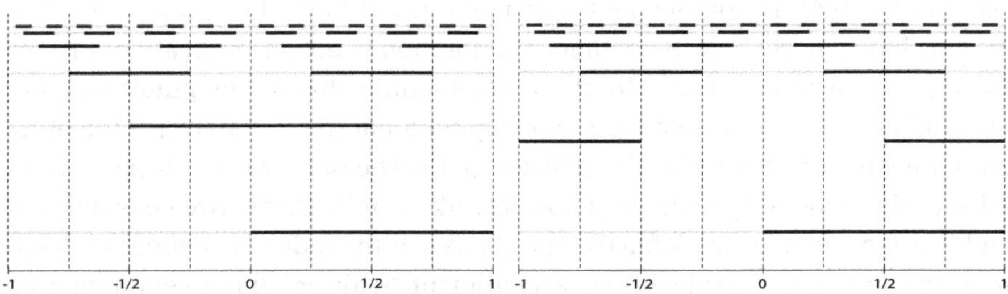

Fig. 3. Expansion by G (i.e., Gray expansion) and expansion by H.

We defined an infinite term of type **G** as an infinite sequence of constructors of **G** and **H** such that any truncation of the sequence forms a finite term of type **G**. In TCF, infinite structures like this can be treated as cototal ideals. Each algebra definition of TCF generates a basic domain of the Scott-Ershov model of partial continuous functionals. Among the ideals of such a domain we single out the total and cototal ones, which are our well-founded and non-well-founded objects, respectively. For the details, please consult [Schwichtenberg and Wainer, 2012]. For our algebras **I**, **G**, and **H**, every total ideal is a finite term and every cototal ideal is an infinite term of the algebra.

The notion of a cototal ideal also makes sense when the underlying algebra does *not* have nullary constructors. Since we will only be concerned with cototal ideals we take advantage of this fact and from now on omit the nullary constructor from our algebras. This will simplify the arguments below considerably (for instance in comparison with [Miyamoto and Schwichtenberg, 2015]). We also redefine our free algebras **I**, **G** and **H** so that **I** has a binary constructor C of type $\mathbf{SD} \to \mathbf{I} \to \mathbf{I}$ and so on; the intention is $\mathrm{LR}_a(p) = \mathrm{LR}(a, p)$. To sum up, our algebras have the following definitions.

$$\mathbf{I} = \mathrm{C}\,\mathbf{SD}\,\mathbf{I},$$
$$\mathbf{G} = \mathrm{LR}\,\mathbf{PSD}\,\mathbf{G} + \mathrm{U}\,\mathbf{H},$$
$$\mathbf{H} = \mathrm{Fin}\,\mathbf{PSD}\,\mathbf{G} + \mathrm{D}\,\mathbf{H}.$$

3 Coinductive representation of Gray-code via realizability

A constructive proof of a formula A can be viewed as a solution to the problem posed by A [Kolmogorov, 1932]. Such a solution is a (computable) function of a certain type $\tau(A)$ determined by the formula A. For example, $\forall_n \exists_m(\mathrm{Prime}(m) \wedge m > n)$ has type $\mathbf{N} \to \mathbf{N}$. Sometimes the solution is only a verification, like for $\forall_{n_1,n_2,n_3>0,m>2}(n_1^m + n_2^m \neq n_3^m)$. In such cases a solution has no computational content and the formula A is called non-computational (n.c., or Harrop); the other ones are called computationally relevant (c.r.). The only way c.r. formulas can arise is via inductively defined predicates, like I or ${}^{\mathrm{co}}I$ below (we consider $\exists_x A$ and $A \vee B$ as inductively defined). The clauses of the inductive definition determine the data type (free algebra) of a solution or “realizer”. It is essential that we allow non-computational universal quantifiers $\forall^{\mathrm{nc}}$ [Berger, 1993] to obtain the desired data type. For instance, in the clause $\forall_x^{\mathrm{nc}} \forall_d (I(x) \to I(\frac{x+d}{2}))$ $(d \in \{-1, 0, 1\})$ for I one is not interested in the real number x as input, but only in how the digit d gives rise to a new element of I. Here we work in such a constructive arithmetical theory with realizability (called TCF in [Schwichtenberg and Wainer, 2012]).

We want to extract algorithms for real number computation from proofs in an appropriate formal theory involving coinductive definitions. The idea is to leave infinite streams implicit, as realizers of atomic propositions on reals. For example, consider the problem to compute the average of two real numbers coded by infinite streams. We will coinductively define a unary predicate ${}^{\mathrm{co}}I$ and prove

$$\forall^{\mathrm{nc}}_{x,x'}({}^{\mathrm{co}}I(x) \to {}^{\mathrm{co}}I(x') \to {}^{\mathrm{co}}I(\frac{x+x'}{2})) \tag{18}$$

(recall that $\forall^{\mathrm{nc}}$ indicates that the reals x, x' have no computational significance, only the assumptions ${}^{\mathrm{co}}I(x)$, ${}^{\mathrm{co}}I(x')$ have). Associated with ${}^{\mathrm{co}}I$ is its "realizability extension" $({}^{\mathrm{co}}I)^{\mathbf{r}}$, a relation between streams v of signed digits and real numbers x. We can understand $({}^{\mathrm{co}}I)^{\mathbf{r}}(v, x)$ as saying that v is a stream representation of x witnessing ${}^{\mathrm{co}}I(x)$. The soundness theorem gives

$$({}^{\mathrm{co}}I)^{\mathbf{r}}(v,x) \to ({}^{\mathrm{co}}I)^{\mathbf{r}}(v',x') \to ({}^{\mathrm{co}}I)^{\mathbf{r}}(f(v,v'), \frac{x+x'}{2})$$

for some function f extracted from the proof. The function is the stream transformer for the average, and it is obtained (together with a proof of its correctness) from the proof of (18), which never mentions streams.

Now what is the predicate ${}^{\mathrm{co}}I$? Consider the operator

$$\Phi(X) := \{\, x \mid \exists^{\mathrm{r}}_{x'\in X}\exists_d (x = \frac{x'+d}{2})\,\},$$

where d ranges over **SD** $:= \{-1, 0, 1\}$. The ^{r} in $\exists^{\mathrm{r}}$ (not to be confused with the $^{\mathbf{r}}$ in $({}^{\mathrm{co}}I)^{\mathbf{r}}$) indicates that the quantified variable x' has no computational significance, only the kernel of the existential formula has. Since $\Phi(X)$ is stricly positive in X, our underlying theory provides us with unary predicates (or sets; they are not distinguished) I and ${}^{\mathrm{co}}I$ for the least and greatest fixed point of Φ:

$$\begin{aligned} I &:= \mu_X \Phi(X) && \text{least fixed point} \\ {}^{\mathrm{co}}I &:= \nu_X \Phi(X) && \text{greatest fixed point} \end{aligned}$$

satisfying the (strengthened) axioms

$$\begin{aligned} \Phi(I \cap X) \subseteq X \to I \subseteq X & \qquad \text{induction} \\ X \subseteq \Phi({}^{\mathrm{co}}I \cup X) \to X \subseteq {}^{\mathrm{co}}I & \qquad \text{coinduction} \end{aligned}$$

(they are called "strengthened" because their hypotheses are weaker than the fixed point property $\Phi(X) = X$).

The realizability extensions $I^{\mathbf{r}}$ and $({}^{\mathrm{co}}I)^{\mathbf{r}}$ are binary predicates on streams v of signed digits (coming from $\exists_d$ in the definition of $\Phi(X)$) and real numbers x. Consider the operator

$$\Phi^{\mathbf{r}}(Y) := \{\, (v,x) \mid \exists^{\mathrm{nc}}_{(v',x')\in Y}\exists_d (x = \frac{x'+d}{2} \wedge v = \mathrm{C}_d(v'))\,\}$$

(the ^{u} in $\exists^{\mathrm{nc}}$ indicates that neither the quantified variable nor the kernel has computational significance). Since $\Phi^{\mathbf{r}}(Y)$ is strictly positive in Y, again our underlying theory provides us with binary predicates (or relations) $I^{\mathbf{r}}$ and $({}^{\mathrm{co}}I)^{\mathbf{r}}$ for the least and greatest fixed point of $\Phi^{\mathbf{r}}$:

$$I^{\mathbf{r}} := \mu_Y \Phi^{\mathbf{r}}(Y) \qquad \text{least fixed point}$$
$$({}^{\mathrm{co}}I)^{\mathbf{r}} := \nu_Y \Phi^{\mathbf{r}}(Y) \qquad \text{greatest fixed point}$$

satisfying the (strengthened) axioms

$$\Phi^{\mathbf{r}}(I^{\mathbf{r}} \cap Y) \subseteq Y \to I^{\mathbf{r}} \subseteq Y \qquad \text{induction}$$
$$Y \subseteq \Phi^{\mathbf{r}}(({}^{\mathrm{co}}I)^{\mathbf{r}} \cup Y) \to Y \subseteq ({}^{\mathrm{co}}I)^{\mathbf{r}} \qquad \text{coinduction.}$$

The following proposition states that the definition of ${}^{\mathrm{co}}I$ is correct in the sense that the realizers of ${}^{\mathrm{co}}I(x)$ are exactly the signed digit representations of x.

Proposition 3.1. *For $v = a_1 a_2 \ldots \in \mathbf{SD}^{\omega}$ and $x \in \mathbb{I}$*

$$({}^{\mathrm{co}}I)^{\mathbf{r}}(v, x) \leftrightarrow x \in \bigcap_{n=1}^{\infty} f_{a_1}(f_{a_2}(\ldots f_{a_n}(\mathbb{I}) \ldots))$$

Proof: For the direction from left to right we show

$$\forall_{v,x}(({}^{\mathrm{co}}I)^{\mathbf{r}}(v, x) \to v = a_1 a_2 \ldots \wedge x \in f_{a_1}(f_{a_2}(\ldots f_{a_n}(\mathbb{I}) \ldots)))$$

by induction on n. For $n = 0$ this holds since $x \in \mathbb{I}$. For $n+1$ suppose that $({}^{\mathrm{co}}I)^{\mathbf{r}}(v, x)$ holds. Then, since $({}^{\mathrm{co}}I)^{\mathbf{r}}$ is a fixed point of $\Phi^{\mathbf{r}}$,

$$\exists_{v',x',a_1}(({}^{\mathrm{co}}I)^{\mathbf{r}}(v', x') \wedge x = f_{a_1}(x') \wedge v = \mathrm{C}_{a_1}(v')).$$

Let $v' = a_2 a_3 \ldots$. By induction hypothesis, $x' \in f_{a_2}(f_{a_3}(\ldots f_{a_{n+1}}(\mathbb{I}) \ldots))$. We have $v = \mathrm{C}_{a_1}(v') = a_1 a_2 \ldots$ and $x \in f_{a_1}(f_{a_2} \ldots (f_{a_{n+1}}(\mathbb{I})) \ldots)$.

The direction from right to left is shown by coinduction. Setting

$$\{ (v, x) \mid v = a_1 a_2 \ldots,\ x \in f_{a_1}(f_{a_2}(\ldots f_{a_n}(\mathbb{I}) \ldots)) \text{ for every } n \}$$

it suffices to show $P \subseteq \Phi^{\mathbf{r}}(P)$. Assume $(v, x) \in P$. Set $x' := 2x - a_1$ and $v' := a_2 a_3 \ldots$. Then clearly $P(v', x')$, $x = \frac{x'+a_1}{2}$ and $v = a_1 v' = \mathrm{C}_{a_1}(v')$. Hence $(v, x) \in \Phi^{\mathbf{r}}(P)$. □

For Gray-code we proceed similarly; for brevity only for the infinite case. We now need two predicates ${}^{\mathrm{co}}G$ and ${}^{\mathrm{co}}H$ instead of ${}^{\mathrm{co}}I$. The corresponding operators Γ, Δ are defined by

$$\Gamma(X, Y) := \{ y \mid \exists^{\mathrm{r}}_{x \in X} \exists_a (y = -a\frac{x-1}{2}) \vee \exists^{\mathrm{r}}_{x \in Y}(y = \frac{x}{2}) \},$$
$$\Delta(X, Y) := \{ y \mid \exists^{\mathrm{r}}_{x \in X} \exists_a (y = a\frac{x+1}{2}) \vee \exists^{\mathrm{r}}_{x \in Y}(y = \frac{x}{2}) \}$$

and we define $({}^{\mathrm{co}}G, {}^{\mathrm{co}}H) := \nu_{(X,Y)}(\Gamma(X,Y), \Delta(X,Y))$. This is understood as the greatest fixed point of (Γ, Δ), expressed by the (strengthened) simultaneous coinduction axiom

$$(X,Y) \subseteq (\Gamma({}^{\mathrm{co}}G \cup X, {}^{\mathrm{co}}H \cup Y), \Delta({}^{\mathrm{co}}G \cup X, {}^{\mathrm{co}}H \cup Y)) \to (X,Y) \subseteq ({}^{\mathrm{co}}G, {}^{\mathrm{co}}H),$$

where inclusion $\subseteq$ is meant component-wise.

For later use we note immediate consequences of the fact that $({}^{\mathrm{co}}G, {}^{\mathrm{co}}H)$ is a (simultaneous) fixed point of (Γ, Δ), CoGClause and CoGClauseInv:

$$\forall^{\mathrm{nc}}_x({}^{\mathrm{co}}G(x) \to \exists^{\mathrm{r}}_{x' \in {}^{\mathrm{co}}G}\exists_a(x = -a\frac{x'-1}{2}) \vee \exists^{\mathrm{r}}_{x' \in {}^{\mathrm{co}}H}(x = \frac{x'}{2})), \tag{19}$$

$$\forall^{\mathrm{nc}}_x(\exists^{\mathrm{r}}_{x' \in {}^{\mathrm{co}}G}\exists_a(x = -a\frac{x'-1}{2}) \vee \exists^{\mathrm{r}}_{x' \in {}^{\mathrm{co}}H}(x = \frac{x'}{2}) \to {}^{\mathrm{co}}G(x)). \tag{20}$$

The realizability extensions $({}^{\mathrm{co}}G)^{\mathbf{r}}$ and $({}^{\mathrm{co}}H)^{\mathbf{r}}$ are binary predicates on cototal ideals p in $\mathbf{G}$ or q in $\mathbf{H}$ (respectively) and real numbers x. Consider the operators

$$\begin{aligned}\Gamma^{\mathbf{r}}(Z,W) := \{\,(p,x) \mid{}& \exists_{(p',x') \in Z}\exists_a(x = -a\frac{x'-1}{2} \wedge p = \mathrm{LR}_a(p')) \vee^{\mathrm{nc}} \\ & \exists_{(q',x') \in W}(x = \frac{x'}{2} \wedge p = \mathrm{U}(q'))\,\}, \\ \Delta^{\mathbf{r}}(Z,W) := \{\,(q,x) \mid{}& \exists_{(p',x') \in Z}\exists_a(x = a\frac{x'+1}{2} \wedge q = \mathrm{Fin}_a(p')) \vee^{\mathrm{nc}} \\ & \exists_{(q',x') \in W}(x = \frac{x'}{2} \wedge q = \mathrm{D}(q'))\,\}\end{aligned}$$

(the ^{nc} in $\vee^{\mathrm{nc}}$ indicates that the disjunction has no computational significance). Since both $\Gamma^{\mathbf{r}}(Z,W)$ and $\Delta^{\mathbf{r}}(Z,W)$ are strictly positive in Z, W, our underlying theory provides us with a pair of binary predicates $({}^{\mathrm{co}}G)^{\mathbf{r}}$, $({}^{\mathrm{co}}H)^{\mathbf{r}}$ for the greatest fixed point of $(\Gamma^{\mathbf{r}}, \Delta^{\mathbf{r}})$:

$$(({}^{\mathrm{co}}G)^{\mathbf{r}}, ({}^{\mathrm{co}}H)^{\mathbf{r}}) := \nu_{(Z,W)}(\Gamma^{\mathbf{r}}(Z,W), \Delta^{\mathbf{r}}(Z,W))$$

satisfying the (strengthened) simultaneous coinduction axiom

$$\begin{aligned}&(Z,W) \subseteq (\Gamma^{\mathbf{r}}(({}^{\mathrm{co}}G)^{\mathbf{r}} \cup Z, ({}^{\mathrm{co}}H)^{\mathbf{r}} \cup W), \Delta^{\mathbf{r}}(({}^{\mathrm{co}}G)^{\mathbf{r}} \cup Z, ({}^{\mathrm{co}}H)^{\mathbf{r}} \cup W)) \to \\ &(Z,W) \subseteq (({}^{\mathrm{co}}G)^{\mathbf{r}}, ({}^{\mathrm{co}}H)^{\mathbf{r}})\end{aligned}$$

where again inclusion $\subseteq$ is meant component-wise.

Similar to Proposition 3.1, we show that ${}^{\mathrm{co}}G$ is correct in the sense that the realizers of ${}^{\mathrm{co}}G(x)$ are exactly the pre-Gray codes of x.

Proposition 3.2. *For $x \in \mathbb{I}$ and cototal ideals p in $\mathbf{G}$ and q in $\mathbf{H}$*

$$\begin{aligned}({}^{\mathrm{co}}G)^{\mathbf{r}}(p,x) &\leftrightarrow x = F_{\mathbf{G}}(p), \\ ({}^{\mathrm{co}}H)^{\mathbf{r}}(q,x) &\leftrightarrow x = F_{\mathbf{H}}(q).\end{aligned}$$

Proof: Recall that if $p = [c_1, c_2, \ldots]$ is a cototal ideal in **G**, then

$$F_{\mathbf{G}}(p) = \bigcap_{n=1}^{\infty} f_{c_1}(f_{c_2}(\ldots f_{c_n}(\mathbb{I})\ldots))$$

and similary for $F_{\mathbf{H}}(q)$. The proof is similar to the case of signed digits, but slightly more involved because of the simultaneous definition of $({}^{\mathrm{co}}G)^{\mathbf{r}}$ and $({}^{\mathrm{co}}H)^{\mathbf{r}}$. □

Remark 3.3 (Nested definition). *As an alternative to the above simultaneous definition of* ${}^{\mathrm{co}}G$ *and* ${}^{\mathrm{co}}H$*, we can take the nested definition* ${}^{\mathrm{co}}G' = \nu_X \Gamma(X, {}^{\mathrm{co}}H'(X))$ *where* ${}^{\mathrm{co}}H'_X = \nu_Y \Delta(X, Y)$*. The witnessing algebras would also be changed. In this paper we adopt the simultaneous one, since the extracted programs are simpler.*

4 Proofs about coinductive representations that correspond to algorithms

In each of the examples below, after the proof we state the rules (equations) expressing the algorithm implicit in this proof. Such an "informal program extraction" can be difficult and error-prone. In Section 6 this will be done precisely, using Minlog to extract a program (i.e., a term in an extension of Gödel's T) from a formalization of this proof.

4.1 Average for signed digit code

As a warm-up we prove the average property (18), following [Berger and Seisenberger, 2010]. Consider two sets of averages, the second one with a "carry" $i \in \mathbf{SD}_2 := \{-2, -1, 0, 1, 2\}$:

$$P := \{\frac{x+y}{2} \mid x, y \in {}^{\mathrm{co}}I\}, \qquad Q := \{\frac{x+y+i}{4} \mid x, y \in {}^{\mathrm{co}}I, i \in \mathbf{SD}_2\},$$

where $\mathbf{SD}_2$ are the "extended signed digits" $\{-2, -1, 0, 1, 2\}$; let i, j range over $\mathbf{SD}_2$. Recall that ${}^{\mathrm{co}}I$ is a fixed point of Φ. Hence ${}^{\mathrm{co}}I \subseteq \Phi({}^{\mathrm{co}}I)$, i.e.

$$\forall^{\mathrm{nc}}_{x\in {}^{\mathrm{co}}I} \exists^{\mathrm{r}}_{x'\in {}^{\mathrm{co}}I} \exists_d (x = \frac{x'+d}{2}) \qquad {}^{\mathrm{co}}I\text{-clause.} \tag{21}$$

It suffices to show that Q satisfies (21), for then by the greatest-fixed-point axiom for ${}^{\mathrm{co}}I$ we have $Q \subseteq {}^{\mathrm{co}}I$. Since we also have $P \subseteq Q$ we then obtain $P \subseteq {}^{\mathrm{co}}I$, which is our claim.

Lemma 4.1 (CoIAvToAvc).

$$\forall^{\mathrm{nc}}_{x,y\in{}^{\mathrm{co}}I}\exists^{\mathrm{r}}_{x',y'\in{}^{\mathrm{co}}I}\exists_i\Big(\frac{x+y}{2}=\frac{x'+y'+i}{4}\Big).$$

Proof: Immediate from (21). □

Lemma 4.2 (CoIAvcSatCoICl).

$$\forall_i\forall^{\mathrm{nc}}_{x,y\in{}^{\mathrm{co}}I}\exists^{\mathrm{r}}_{x',y'\in{}^{\mathrm{co}}I}\exists_{j,d}\Big(\frac{x+y+i}{4}=\frac{\frac{x'+y'+j}{4}+d}{2}\Big).$$

Proof: We need functions $J\colon \mathbf{SD}\to\mathbf{SD}\to\mathbf{SD}_2\to\mathbf{SD}_2$ and $K\colon \mathbf{SD}\to\mathbf{SD}\to\mathbf{SD}_2\to\mathbf{SD}$ such that $d+e+2i=J(d,e,i)+4K(d,e,i)$. They can be defined easily by cases on d, e and i. Using these we can relate the functions $\frac{x+d}{2}$ and $\frac{x+y+i}{4}$ by

$$\frac{\frac{x+d}{2}+\frac{y+e}{2}+i}{4}=\frac{\frac{x+y+J(d,e,i)}{4}+K(d,e,i)}{2}. \tag{22}$$

Now (21) gives the claim. □

By coinduction from Lemma 4.2 we obtain

Lemma 4.3 (CoIAvcToCoI).

$$\forall^{\mathrm{nc}}_z\Big(\exists^{\mathrm{r}}_{x,y\in{}^{\mathrm{co}}I}\exists_i\Big(z=\frac{x+y+i}{4}\Big)\to{}^{\mathrm{co}}I(z)\Big).$$

Proposition 4.4 (CoIAverage).

$$\forall^{\mathrm{nc}}_{x,y}\Big({}^{\mathrm{co}}I(x)\to{}^{\mathrm{co}}I(y)\to{}^{\mathrm{co}}I\Big(\frac{x+y}{2}\Big)\Big).$$

Proof: Immediate from Lemmata 4.1 and 4.3. □

Implicit algorithm. *Lemma 4.1 computes the first "carry" $i\in\mathbf{SD}_2$ and the tails of the inputs. Then $f\colon \mathbf{SD}_2\times\mathbf{I}\times\mathbf{I}\to\mathbf{I}$ defined corecursively by*

$$f(i,\mathrm{C}_d(v),\mathrm{C}_e(w))=\mathrm{C}_{K(d,e,i)}(f(J(d,e,i),v,w))$$

is called repeatedly in order to compute the average step by step.

4.2 From pre-Gray to signed digit code

We prove ${}^{\mathrm{co}}G\subseteq{}^{\mathrm{co}}I$. However, to be able to do this by coinduction we need to generalize our goal to

Lemma 4.5 (CoGToCoIAux). $\forall^{\mathrm{nc}}_x(\exists_a({}^{\mathrm{co}}G(ax)\vee{}^{\mathrm{co}}H(ax))\to{}^{\mathrm{co}}Ix)$.

Proof: For $P := \{ x \mid \exists_a ({}^{co}G(ax) \vee {}^{co}H(ax)) \}$ we must show $P \subseteq {}^{co}I$. By coinduction it suffices to prove $P \subseteq \Phi({}^{co}I \cup P)$. Let $x_1 \in P$. We show $x_1 \in \Phi({}^{co}I \cup P)$:

$$\exists^{r}_{x \in {}^{co}I \cup P} \exists_d \Big(x_1 = \frac{x+d}{2}\Big). \tag{23}$$

Since $x_1 \in P$ we have a such that ${}^{co}G(ax_1) \vee {}^{co}H(ax_1)$.

Case ${}^{co}G(ax_1)$. The ${}^{co}G$-clause ${}^{co}G \subseteq \Gamma({}^{co}G, {}^{co}H)$ applied to $ax_1 \in {}^{co}G$ gives us

$$\exists^{r}_{x \in {}^{co}G} \exists_b \Big(ax_1 = -b\frac{x-1}{2}\Big) \vee \exists^{r}_{x \in {}^{co}H} \Big(ax_1 = \frac{x}{2}\Big).$$

If the left hand side holds, we have $x_2 \in {}^{co}G$ and b such that $ax_1 = -b\frac{x_2-1}{2}$. Then (23) holds for $x := -abx_2$ and $d := ab$, since $-abx_2 \in P$ (by $x_2 \in {}^{co}G$ and the definition of P), and

$$x_1 = a^2 x_1 = -ab\frac{x_2-1}{2} = \frac{-abx_2+ab}{2} = \frac{x+d}{2}.$$

If the right hand side holds, we have $x_2 \in {}^{co}H$ such that $ax_1 = \frac{x_2}{2}$. Then (23) holds for $x := ax_2$ and $d := 0$, since $ax_2 \in P$ (by $x_2 \in {}^{co}H$ and the definition of P), and

$$x_1 = a^2 x_1 = a\frac{x_2}{2} = \frac{ax_2+0}{2} = \frac{x+d}{2}.$$

Case ${}^{co}H(ax_1)$. The ${}^{co}H$-clause ${}^{co}H \subseteq \Delta({}^{co}G, {}^{co}H)$ applied to $ax_1 \in {}^{co}H$ gives

$$\exists^{r}_{x \in {}^{co}G} \exists_b \Big(ax_1 = b\frac{x+1}{2}\Big) \vee \exists^{r}_{x \in {}^{co}H} \Big(ax_1 = \frac{x}{2}\Big).$$

If the left hand side holds, we have $x_2 \in {}^{co}G$ and b such that $ax_1 = b\frac{x_2+1}{2}$. Then (23) holds for $x := abx_2$ and $d := ab$, since $abx_2 \in P$ (by $x_2 \in {}^{co}G$ and the definition of P), and

$$x_1 = a^2 x_1 = ab\frac{x_2+1}{2} = \frac{abx_2+ab}{2} = \frac{x+d}{2}.$$

If the right hand side holds, we have $x_2 \in {}^{co}H$ such that $ax_1 = \frac{x_2}{2}$. Then (23) holds for $x := ax_2$ and $d := 0$, since $ax_2 \in P$ (by $x_2 \in {}^{co}H$ and the definition of P), and

$$x_1 = a^2 x_1 = a\frac{x_2}{2} = \frac{ax_2+0}{2} = \frac{x+d}{2}.$$ □

Implicit algorithm. $[f, g]$: **PSD** × **G** + **PSD** × **H** → **I** *defined by*

$$\begin{aligned} f(a, \mathrm{LR}_b(p)) &= \mathrm{C}_{ab}(f(-ab, p)), & g(a, \mathrm{Fin}_b(p)) &= \mathrm{C}_{ab}(f(ab, p)), \\ f(a, \mathrm{U}(q)) &= \mathrm{C}_0(g(a, q)), & g(a, \mathrm{D}(q)) &= \mathrm{C}_0(g(a, q)). \end{aligned}$$

An immediate consequence is

Proposition 4.6 (CoGToCoI). $\forall^{nc}_x ({}^{co}G(x) \to {}^{co}I(x))$.

4.3 From signed digit to pre-Gray code

Conversely we also have ${}^{\mathrm{co}}I \subseteq {}^{\mathrm{co}}G$. Again, to be able to prove this by coinduction we need to generalize our goal to

Lemma 4.7 (CoIToCoGAux).

$$\forall_x^{\mathrm{nc}}(\exists_a {}^{\mathrm{co}}I(ax) \to {}^{\mathrm{co}}Gx),$$
$$\forall_x^{\mathrm{nc}}(\exists_a {}^{\mathrm{co}}I(ax) \to {}^{\mathrm{co}}Hx).$$

Proof: For $P := \{x \mid \exists_a(ax \in {}^{\mathrm{co}}I)\}$ we show $P \subseteq {}^{\mathrm{co}}G$ simultaneously with $P \subseteq {}^{\mathrm{co}}H$. By coinduction it suffices to prove (i) $P \subseteq \Gamma({}^{\mathrm{co}}G \cup P, {}^{\mathrm{co}}H \cup P)$ and (ii) $P \subseteq \Delta({}^{\mathrm{co}}G \cup P, {}^{\mathrm{co}}H \cup P)$. For (i), let $x_1 \in P$. We show $x_1 \in \Gamma({}^{\mathrm{co}}G \cup P, {}^{\mathrm{co}}H \cup P)$:

$$\exists^{\mathrm{r}}_{x \in {}^{\mathrm{co}}G \cup P} \exists_a (x_1 = -a\frac{x-1}{2}) \vee \exists^{\mathrm{r}}_{x \in {}^{\mathrm{co}}H \cup P}(x_1 = \frac{x}{2}). \tag{24}$$

Since $x_1 \in P$ we have a_1 such that ${}^{\mathrm{co}}I(a_1x_1)$. The ${}^{\mathrm{co}}I$-clause ${}^{\mathrm{co}}I \subseteq \Phi({}^{\mathrm{co}}I)$ applied to $a_1x_1 \in {}^{\mathrm{co}}I$ gives us

$$\exists^{\mathrm{r}}_{x \in {}^{\mathrm{co}}I} \exists_d (a_1x_1 = \frac{x+d}{2}).$$

Hence we have $x_2 \in {}^{\mathrm{co}}I$ and d such that $a_1x_1 = \frac{x_2+d}{2}$.

Case $d = -1$. Then the left hand of (24) holds for $x := x_2$ and $a := -a_1$, since $x_2 \in P$ (by $x_2 \in {}^{\mathrm{co}}I$ and the definition of P), and

$$x_1 = a_1a_1x_1 = a_1\frac{x_2+d}{2} = a_1\frac{x_2-1}{2}.$$

Case $d = 1$. Then the left hand of (24) holds for $x := -x_2$ and $a := a_1$, since $-x_2 \in P$ (by $x_2 \in {}^{\mathrm{co}}I$ and the definition of P), and

$$x_1 = a_1a_1x_1 = a_1\frac{x_2+d}{2} = a_1\frac{x_2+1}{2} = -a_1\frac{-x_2-1}{2}.$$

Case $d = 0$. Then the right hand of (24) holds for $x := a_1x_2$, since $a_1x_2 \in P$ (by $x_2 \in {}^{\mathrm{co}}I$ and the definition of P), and

$$x_1 = a_1a_1x_1 = a_1\frac{x_2+d}{2} = \frac{a_1x_2}{2}.$$

This finishes the proof of (i). The proof of (ii) is similar, and we omit it. □

Implicit algorithm. $g\colon \tau \to \mathbf{G}$ *and* $h\colon \tau \to \mathbf{H}$ *with* $\tau := \mathbf{PSD} \times \mathbf{I}$, *defined by*

$$\begin{aligned} g(b, \mathrm{C}_{-1}(v)) &= \mathrm{LR}_{-b}(g(1, v)), & h(b, \mathrm{C}_{-1}(v)) &= \mathrm{Fin}_{-b}(g(-1, v)),\\ g(b, \mathrm{C}_{1}(v)) &= \mathrm{LR}_{b}(g(-1, v)), & h(b, \mathrm{C}_{1}(v)) &= \mathrm{Fin}_{b}(g(1, v)),\\ g(b, \mathrm{C}_{0}(v)) &= \mathrm{U}(h(b, v)), & h(b, \mathrm{C}_{0}(v)) &= \mathrm{D}(h(b, v)). \end{aligned}$$

An immediate consequence is

Proposition 4.8 (CoIToCoG). $\forall_x^{\mathrm{nc}}({}^{\mathrm{co}}I(x) \to {}^{\mathrm{co}}G(x))$.

4.4 Average for pre-Gray code

We consider the problem to compute the average of two real numbers given in pre-Gray code directly, without going via signed digit code.

As a preparation we treat the unary minus function. Here we make use of the fact that our coinduction axioms are in strengthened form (that is $X \subseteq \Phi({}^{co}I \cup X) \to X \subseteq {}^{co}I$ instead of $X \subseteq \Phi(X) \to X \subseteq {}^{co}I$, for example).

Lemma 4.9 (CoGMinus).

$$\forall^{nc}_x({}^{co}G(-x) \to {}^{co}Gx),$$
$$\forall^{nc}_x({}^{co}H(-x) \to {}^{co}Hx).$$

Proof: For $P := \{x \mid -x \in {}^{co}G)\}$ and $Q := \{x \mid -x \in {}^{co}H)\}$ we show $P \subseteq {}^{co}G$ simultaneously with $Q \subseteq {}^{co}H$. By coinduction it suffices to prove (i) $P \subseteq \Gamma({}^{co}G \cup P, {}^{co}H \cup Q)$ and (ii) $Q \subseteq \Delta({}^{co}G \cup P, {}^{co}H \cup Q)$. For (i), let $x_1 \in P$. We show $x_1 \in \Gamma({}^{co}G \cup P, {}^{co}H \cup Q)$:

$$\exists^r_{x \in {}^{co}G \cup P} \exists_a (x_1 = -a\frac{x-1}{2}) \vee \exists^r_{x \in {}^{co}H \cup Q}(x_1 = \frac{x}{2}). \tag{25}$$

The ${}^{co}G$-clause applied to $-x_1 \in {}^{co}G$ gives us

$$\exists^r_{x \in {}^{co}G} \exists_a (-x_1 = -a\frac{x-1}{2}) \vee \exists^r_{x \in {}^{co}H}(-x_1 = \frac{x}{2}).$$

In the first case we have $x_2 \in {}^{co}G$ and a with $-x_1 = -a\frac{x_2-1}{2}$. Then the left hand side of (25) holds for x_2 and $-a$ (here we use that our coinduction axiom is in strengthened form). In the second case we have $x_2 \in {}^{co}H$ with $-x_1 = \frac{x_2}{2}$. Then the right hand side of (25) holds for $-x_2$. This finishes the proof of (i). The proof of (ii) is similar, and we omit it. □

Implicit algorithm. *f : **G** → **G** and f' : **H** → **H** defined by*

$$f(\mathrm{LR}_a(p)) = \mathrm{LR}_{-a}(p), \qquad f'(\mathrm{Fin}_a(p)) = \mathrm{Fin}_{-a}(p),$$
$$f(\mathrm{U}(q)) = \mathrm{U}(f'(q)), \qquad f'(\mathrm{D}(q)) = \mathrm{D}(f'(q)).$$

Using Lemma 4.9 we prove that ${}^{co}G$ and ${}^{co}H$ are in fact equivalent.

Lemma 4.10 (CoHToCoG).

$$\forall^{nc}_x({}^{co}Hx \to {}^{co}Gx),$$
$$\forall^{nc}_x({}^{co}Gx \to {}^{co}Hx).$$

Proof: We show ${}^{\mathrm{co}}H \subseteq {}^{\mathrm{co}}G$ simultaneously with ${}^{\mathrm{co}}G \subseteq {}^{\mathrm{co}}H$. By coinduction it suffices to prove (i) ${}^{\mathrm{co}}H \subseteq \Gamma({}^{\mathrm{co}}G \cup {}^{\mathrm{co}}H, {}^{\mathrm{co}}H \cup {}^{\mathrm{co}}G)$ and (ii) ${}^{\mathrm{co}}G \subseteq \Delta({}^{\mathrm{co}}G \cup {}^{\mathrm{co}}H, {}^{\mathrm{co}}H \cup {}^{\mathrm{co}}G)$. For (i), let $x_1 \in {}^{\mathrm{co}}H$. We show $x_1 \in \Gamma({}^{\mathrm{co}}G \cup {}^{\mathrm{co}}H, {}^{\mathrm{co}}H \cup {}^{\mathrm{co}}G)$:

$$\exists^{\mathrm{r}}_{x \in {}^{\mathrm{co}}G \cup {}^{\mathrm{co}}H} \exists_a (x_1 = -a\frac{x-1}{2}) \vee \exists^{\mathrm{r}}_{x \in {}^{\mathrm{co}}H \cup {}^{\mathrm{co}}G} (x_1 = \frac{x}{2}). \tag{26}$$

The ${}^{\mathrm{co}}H$-clause applied to $x_1 \in {}^{\mathrm{co}}H$ gives us

$$\exists^{\mathrm{r}}_{x \in {}^{\mathrm{co}}G} \exists_a (x_1 = a\frac{x+1}{2}) \vee \exists^{\mathrm{r}}_{x \in {}^{\mathrm{co}}H} (x_1 = \frac{x}{2}).$$

In the first case we have $x_2 \in {}^{\mathrm{co}}G$ and a with $x_1 = a\frac{x_2+1}{2}$. Then the left hand side of (26) holds for $-x_2$ and a, using Lemma 4.9 and (again) that our coinduction axiom is in strengthened form. In the second case we have $x_2 \in {}^{\mathrm{co}}H$ with $x_1 = \frac{x_2}{2}$. Then the right hand side of (25) holds for x_2. This finishes the proof of (i). The proof of (ii) is similar, and we omit it. □

Implicit algorithm. *$g\colon \mathbf{H} \to \mathbf{G}$ and $h\colon \mathbf{G} \to \mathbf{H}$:*

$$g(\mathrm{Fin}_a(p)) = \mathrm{LR}_a(f^-(p)), \qquad h(\mathrm{LR}_a(p)) = \mathrm{Fin}_a(f^-(p)),$$
$$g(\mathrm{D}(q)) = \mathrm{U}(q), \qquad h(\mathrm{U}(q)) = \mathrm{D}(q)$$

where $f^- :=$ cCoGMinus (cL denotes the function extracted from the proof of a lemma L). Notice that no corecursive call is involved.

The direct proof of the existence of the average w.r.t. Gray-coded reals is similar to the proof in Section 4.1 of the existence of the average w.r.t. signed digit stream coded reals. It proceeds as follows. To prove

$$\forall^{\mathrm{nc}}_{x,y}({}^{\mathrm{co}}G(x) \to {}^{\mathrm{co}}G(y) \to {}^{\mathrm{co}}G(\frac{x+y}{2}))$$

consider again two sets of averages, the second one with a "carry":

$$P := \{\frac{x+y}{2} \mid x, y \in {}^{\mathrm{co}}G\}, \qquad Q := \{\frac{x+y+i}{4} \mid x, y \in {}^{\mathrm{co}}G,\ i \in \mathbf{SD}_2\}.$$

It suffices to show that Q satisfies the clause coinductively defining ${}^{\mathrm{co}}G$, for then by the greatest-fixed-point axiom for ${}^{\mathrm{co}}G$ we have $Q \subseteq {}^{\mathrm{co}}G$. Since we also have $P \subseteq Q$ we then obtain $P \subseteq {}^{\mathrm{co}}G$, which is our claim.

Lemma 4.11 (CoGAvToAvc).

$$\forall^{\mathrm{nc}}_{x,y \in {}^{\mathrm{co}}G} \exists^{\mathrm{r}}_{x',y' \in {}^{\mathrm{co}}G} \exists_i (\frac{x+y}{2} = \frac{x'+y'+i}{4}).$$

Proof: Immediate from CoGClause (19). □

Implicit algorithm. *We use f^* for cCoGPsdTimes and s for cCoHToCoG.*

$$\begin{aligned} f(\mathrm{LR}_a(p), \mathrm{LR}_{a'}(p')) &= (a + a', f^*(-a, p), f^*(-a', p')), \\ f(\mathrm{LR}_a(p), \mathrm{U}(q)) &= (a, f^*(-a, p), s(q)), \\ f(\mathrm{U}(q), \mathrm{LR}_a(p)) &= (a, s(q), f^*(-a, p)), \\ f(\mathrm{U}(q), \mathrm{U}(q')) &= (0, s(q), s(q')). \end{aligned}$$

Lemma 4.12 (CoGAvcSatCoICl).

$$\forall_i \forall^{\mathrm{nc}}_{x,y \in {}^{\mathrm{co}}G} \exists^{\mathrm{r}}_{x',y' \in {}^{\mathrm{co}}G} \exists_{j,d} \Big(\frac{x+y+i}{4} = \frac{\frac{x'+y'+j}{4} + d}{2}\Big).$$

Proof: As in Lemma 4.2 we need the functions J, K with their property (22). Then (19) gives the claim. □

Implicit algorithm.

$$\begin{aligned} f(i, \mathrm{LR}_a(p), \mathrm{LR}_{a'}(p')) &= (J(a, a', i), K(a, a', i), f^*(-a, p), f^*(-a', p')), \\ f(i, \mathrm{LR}_a(p), \mathrm{U}(q)) &= (J(a, 0, i), K(a, 0, i), f^*(-a, p), s(q)), \\ f(i, \mathrm{U}(q), \mathrm{LR}_a(p)) &= (J(0, a, i), K(0, a, i), s(q), f^*(-a, p)), \\ f(i, \mathrm{U}(q), \mathrm{U}(q')) &= (J(0, 0, i), K(0, 0, i), s(q), s(q')). \end{aligned}$$

Lemma 4.13 (CoGAvcToCoG).

$$\begin{aligned} &\forall^{\mathrm{nc}}_z \Big(\exists^{\mathrm{r}}_{x,y \in {}^{\mathrm{co}}G} \exists_i \Big(z = \frac{x+y+i}{4}\Big) \to {}^{\mathrm{co}}G(z)\Big), \\ &\forall^{\mathrm{nc}}_z \Big(\exists^{\mathrm{r}}_{x,y \in {}^{\mathrm{co}}G} \exists_i \Big(z = \frac{x+y+i}{4}\Big) \to {}^{\mathrm{co}}H(z)\Big). \end{aligned}$$

Proof: We show $Q \subseteq {}^{\mathrm{co}}G$ simultaneously with $Q \subseteq {}^{\mathrm{co}}H$. By coinduction it suffices to prove (i) $Q \subseteq \Gamma({}^{\mathrm{co}}G \cup Q, {}^{\mathrm{co}}H \cup Q)$ and (ii) $Q \subseteq \Delta({}^{\mathrm{co}}G \cup Q, {}^{\mathrm{co}}H \cup Q)$. For (i), let $z_1 \in Q$. We show $z_1 \in \Gamma({}^{\mathrm{co}}G \cup Q, {}^{\mathrm{co}}H \cup Q)$:

$$\exists^{\mathrm{r}}_{z \in {}^{\mathrm{co}}G \cup Q} \exists_a \Big(z_1 = -a\frac{z-1}{2}\Big) \vee \exists^{\mathrm{r}}_{z \in {}^{\mathrm{co}}H \cup Q} \Big(z_1 = \frac{z}{2}\Big). \tag{27}$$

Lemma 4.12 applied to $z_1 \in Q$ gives us $x_1, y_1 \in {}^{\mathrm{co}}G$ and i_1, d_1 such that

$$z_1 = \frac{\frac{x_1+y_1+i_1}{4} + d_1}{2}.$$

Case $d_1 = 0$. Go for the right hand side of (27) with $z := (x_1 + y_1 + i_1)/4 \in Q$. *Case* $d_1 = \pm 1$. Go for the left hand side of (27) with $a := d_1$ and $z := (-ax_1 - ay_1 - ai_1)/4 \in Q$. Then

$$-a\frac{z-1}{2} = -a\frac{4z-4}{8} = \frac{x_1 + y_1 + i_1 + 4a}{8} = z_1.$$

This finishes the proof of (i). The proof of (ii) is similar, and we omit it. □

Implicit algorithm. *In the proof we used SdDisj:* $\forall_d(d = 0 \vee \exists_a(d = a))$.

$$\begin{aligned} g(i, p, p') = {} & \text{let } (i_1, d, p_1, p'_1) = \text{cCoGAvcSatCoICl}(i, p, p') \text{ in} \\ & \text{case cSdDisj}(d) \text{ of} \\ & \quad 0 \to \text{U}(h(i, p_1, p'_1)) \\ & \quad a \to \text{LR}_a(g(-ai, f^*(-a, p_1), f^*(-a, p'_1))), \end{aligned}$$

$$\begin{aligned} h(i, p, p') = {} & \text{let } (i_1, d, p_1, p'_1) = \text{cCoGAvcSatCoICl}(i, p, p') \text{ in} \\ & \text{case cSdDisj}(d) \text{ of} \\ & \quad 0 \to \text{D}(h(i, p_1, p'_1)) \\ & \quad a \to \text{Fin}_a(g(-ai, f^*(-a, p_1), f^*(-a, p'_1))). \end{aligned}$$

Proposition 4.14 (CoGAverage).

$$\forall^{\text{nc}}_{x,y}({}^{\text{co}}G(x) \to {}^{\text{co}}G(y) \to {}^{\text{co}}G(\frac{x+y}{2})).$$

Proof: Compose Lemmata 4.11 and 4.13. □

4.5 A bounded translation from pre-Gray code to its normal form

For pre-Gray code there are many ways of expressing the same real number as we noted in Section 2. In particular, the terms $\text{U}(\text{D}^k(\text{Fin}_a\, p))$ and $\text{LR}_a(\text{LR}_1(\text{LR}^k_{-1}\, p))$ denote the same number as Proposition 2.5 says (D^k and LR^k_{-1} denote k-times repetition of the same constructor). Here we extract a program which transfers the former pattern in the first n elements of a pre-Gray code into the latter pattern.

Similar to G we inductively define a binary relation ${}^{\text{z}}G$ between real and natural numbers (used as bounds), this time with an initial clause. The definition is no longer simultaneous with H, but the latter can be defined independently in advance:

$${}^{\text{z}}\Delta(Z) := \{\, (y, m) \mid m = 0 \vee \exists^{\text{r}}_{(x,n)\in Z}(y = \frac{x}{2} \wedge m = n + 1)\,\}.$$

With ${}^{\text{z}}H = \mu_Z {}^{\text{z}}\Delta(Z)$ we can now define

$$\begin{aligned} {}^{\text{z}}\Gamma(X) := \{\, (y, m) \mid m = 0 \vee \exists^{\text{r}}_{(x,n)\in X}\exists_a(y = -a\frac{x-1}{2} \wedge m = n + 1) \vee \\ \exists^{\text{r}}_{(x,n)\in {}^{\text{z}}H}(y = \frac{x}{2} \wedge m = n + 1)\,\} \end{aligned}$$

and ${}^{\text{z}}G = \mu_X {}^{\text{z}}\Gamma(X)$. From a proof of $\{\, (x, n) \mid {}^{\text{co}}G(x)\,\} \subseteq {}^{\text{z}}G$, we extract the desired program to compute a prefix of a pre-Gray code of x of length n which does not contain a subsequence of the form $\mathrm{UD}^k\mathrm{Fin}_a$ from a pre-Gray code of x. The associated algebra for ${}^{\mathrm{z}}H$ it is just the natural numbers $\mathbf{N}$, and for ${}^{\mathrm{z}}G$ it is ${}^{\mathrm{z}}\mathbf{G}$ with constructors

$$\mathrm{Nz}\colon {}^{\mathrm{z}}\mathbf{G}, \qquad \mathrm{LRz}\colon \mathbf{PSD} \to {}^{\mathrm{z}}\mathbf{G} \to {}^{\mathrm{z}}\mathbf{G}, \qquad \mathrm{Uz}\colon \mathbf{N} \to {}^{\mathrm{z}}\mathbf{G}.$$

Lemma 4.15 (GenCoGLR). $\forall_x^{\mathrm{nc}}\forall_a({}^{\mathrm{co}}G(x) \to {}^{\mathrm{co}}G(-a\frac{x-1}{2}))$.

Proof: Easy by coinduction. □

Lemma 4.16 (CoGToBGAux).

$$\forall_n\forall_x^{\mathrm{nc}}({}^{\mathrm{co}}G(x) \to {}^{\mathrm{z}}G(x,n)),$$

$$\forall_n\forall_x^{\mathrm{nc}}({}^{\mathrm{co}}H(x) \to {}^{\mathrm{z}}H(x,n) \vee \exists^{\mathrm{r}}_{y\in{}^{\mathrm{co}}G}\exists_a({}^{\mathrm{z}}G(y,n-1) \wedge x = a\frac{y+1}{2})).$$

Proof: We prove both statements simultaneously by induction on n. The case $n = 0$ is trivial. For the step case, we first assume ${}^{\mathrm{co}}G(x_1)$ and prove $(x_1, n+1) \in {}^{\mathrm{z}}G$. We have

$$\exists^{\mathrm{r}}_{x_2\in{}^{\mathrm{co}}G}\exists_a(x_1 = -a\frac{x_2-1}{2}) \vee \exists^{\mathrm{r}}_{x_2\in{}^{\mathrm{co}}H}(x_1 = \frac{x_2}{2}).$$

(Case A) Suppose that the left hand side holds. Then, by induction hypothesis applied to x_2, we have ${}^{\mathrm{z}}G(x_2, n)$. Therefore $(x_1, n+1) \in {}^{\mathrm{z}}\Gamma({}^{\mathrm{z}}G) = {}^{\mathrm{z}}G$ because

$$\exists^{\mathrm{r}}_{(x_2,n)\in{}^{\mathrm{z}}G}\exists_a(x_1 = -a\frac{x_2-1}{2}).$$

(Case B) Suppose that the right hand side holds. Then, $x_1 = \frac{x_2}{2}$ for $x_2 \in {}^{\mathrm{co}}H$. Therefore, by induction hypothesis,

$${}^{\mathrm{z}}H(x_2, n) \vee \exists^{\mathrm{r}}_{x_3\in{}^{\mathrm{co}}G}\exists_a({}^{\mathrm{z}}G(x_3, n-1) \wedge x_2 = a\frac{x_3+1}{2}).$$

(Case B1) Suppose that the left hand side holds. Then, since ${}^{\mathrm{z}}H(x_2, n)$ and $x_1 = \frac{x_2}{2}$, ${}^{\mathrm{z}}G(x_1, n+1)$ holds.

(Case B2) Suppose that the right hand side holds. Then,

$$x_1 = \frac{x_2}{2} = a_3\frac{x_3+1}{4} = -a_3\frac{-\frac{x_3-1}{2}-1}{2} = -a_3\frac{x_4-1}{2}$$

for some $a_3, x_3 \in {}^{\mathrm{co}}G$ and $x_4 := -\frac{x_3-1}{2}$. Since $x_1 = -a_3\frac{x_4-1}{2}$, for our goal ${}^{\mathrm{z}}G(x_1, n+1)$ it suffices to prove ${}^{\mathrm{z}}G(x_4, n)$. In case $n = 0$ this follows from the initial clause for ${}^{\mathrm{z}}G$, and in case $n = m+1$ it follows from ${}^{\mathrm{z}}G(x_3, n-1)$ by the first generating clause for ${}^{\mathrm{z}}G$, since $x_4 = -\frac{x_3-1}{2}$.

Next, we suppose that ${}^{\mathrm{co}}H(x_1)$ and prove

$${}^{\mathrm{z}}H(x_1, n+1) \vee \exists^{\mathrm{r}}_{y\in{}^{\mathrm{co}}G}\exists_a({}^{\mathrm{z}}G(y, n) \wedge x_1 = a\frac{y+1}{2}).$$

The argument is almost the same as above. Since ${}^{\mathrm{co}}H(x_1)$, we have

$$\exists^{\mathrm{r}}_{x_2 \in {}^{\mathrm{co}}G} \exists_a (x_1 = a\frac{x_2+1}{2}) \vee \exists^{\mathrm{r}}_{x_2 \in {}^{\mathrm{co}}H}(x_1 = \frac{x_2}{2}).$$

(Case A) Suppose that the left hand side holds. We have $x_1 = a_2\frac{x_2+1}{2}$ for a_2 and $x_2 \in {}^{\mathrm{co}}G$. By induction hypothesis, ${}^{\mathrm{z}}G(x_2, n)$. Therefore

$$\exists^{\mathrm{r}}_{y \in {}^{\mathrm{co}}G} \exists_a ({}^{\mathrm{z}}G(y, n) \wedge x_1 = a\frac{y+1}{2}).$$

(Case B) Suppose that the right hand side holds. We have $x_1 = \frac{x_2}{2}$ for $x_2 \in {}^{\mathrm{co}}H$. By induction hypothesis,

$${}^{\mathrm{z}}H(x_2, n) \vee \exists^{\mathrm{r}}_{x_3 \in {}^{\mathrm{co}}G} \exists_a ({}^{\mathrm{z}}G(x_3, n-1) \wedge x_2 = a\frac{x_3+1}{2}).$$

(Case B1) Suppose that the left hand side holds. Then, since ${}^{\mathrm{z}}H(x_2, n)$ and $x_1 = \frac{x_2}{2}$, we have ${}^{\mathrm{z}}H(x_1, n+1)$.

(Case B2) Suppose that the right hand side holds. Then,

$$x_1 = \frac{x_2}{2} = a\frac{x_3+1}{4} = a\frac{\frac{x_3-1}{2}+1}{2} = a\frac{x_4+1}{2}$$

for $x_4 := \frac{x_3-1}{2}$. We prove the right hand side of our goal for x_4 and a. Since $x_1 = a\frac{x_4+1}{2}$ it suffices to prove $x_4 \in {}^{\mathrm{co}}G$ and ${}^{\mathrm{z}}G(x_4, n)$. From $x_3 \in {}^{\mathrm{co}}G$ we obtain $x_4 \in {}^{\mathrm{co}}G$ by Lemma 4.15. To prove ${}^{\mathrm{z}}G(x_4, n)$ we argue by cases on n. In case $n = 0$ this follows from the initial clause for ${}^{\mathrm{z}}G$, and in case $n = m + 1$ it follows from ${}^{\mathrm{z}}G(x_3, n-1)$ by the first generating clause for ${}^{\mathrm{z}}G$, since $x_4 = \frac{x_3-1}{2}$. □

Implicit algorithm. *$f\colon \mathbf{N} \to \mathbf{G} \to {}^{\mathrm{z}}\mathbf{G}$ and $g\colon \mathbf{N} \to \mathbf{H} \to \mathbf{N} + \mathbf{PSD} \times \mathbf{G} \times {}^{\mathrm{z}}\mathbf{G}$ are defined by simultaneous recursion*

$$\begin{aligned}
&f(0,p) = 0 \qquad g(0,q) = 0\\
&f(n+1, \mathrm{LR}_a(p)) = \mathrm{LRz}_a(f(n,p))\\
&f(n+1, \mathrm{U}(q)) = \text{case } g(n,q) \text{ of}\\
&\qquad\qquad m \to m\\
&\qquad\qquad (a,p,r) \to \mathrm{LRz}_a(\text{case } n \text{ of}\\
&\qquad\qquad\qquad 0 \to 0\\
&\qquad\qquad\qquad m+1 \to \mathrm{LRz}_1(r))\\
&g(n+1, \mathrm{Fin}_a(p)) = (a, p, f(n,p))\\
&g(n+1, \mathrm{D}(q)) = \text{case } g(n,q) \text{ of}\\
&\qquad\qquad m \to m+1\\
&\qquad\qquad (a,p,r) \to (a, \mathrm{LR}_{-1}(p), f(n, \mathrm{LR}_{-1}(p)))
\end{aligned}$$

An immediate consequence is

Proposition 4.17 (CoGToBG). $\forall_n \forall_x^{\mathrm{nc}}({}^{\mathrm{co}}G(x) \to {}^{\mathrm{Z}}G(x,n))$.

5 Conversion from Gray-code to modified Gray expansion

As we studied in Section 2.1, each dyadic rational number has three representations of the forms $s11\bar{1}^\omega$, $s\bar{1}1\bar{1}^\omega$ and $s\bot 1\bar{1}^\omega$ in Gray-code, and only the last one in modified Gray expansion. We show that Gray-code can be converted to modified Gray expansion. We denote by $K(RD)$ and $L(RD)$ the sets of compact and non-compact elements of *RD*, which coincide with the sets of finite Gray-codes and infinite Gray-codes, respectively. We also denote by $M(L(RD))$ the set of minimal elements of $L(RD)$, which coincides with the set of modified Gray expansions.

We say that s is a predecessor of t if $s \sqsubseteq t$ and no $u \in K(RD)$ satisfies $s \sqsubseteq u \sqsubseteq t$. $\bot^\omega$ have no predecessor, $s'\bot 1\bar{1}^k$, $\bar{1}^k$ and $1\bar{1}^k$ have one predecessor, and the other elements of $K(RD)$ have two predecessors (see Figure 2). We define a function ρ on $K(RD)$ so that $\rho(s)$ is the meet of the predecessors of s for $s \neq \bot$. In the following definition, $a \in \{\bar{1}, 1\}$, $k \geq 0$, $s' \in \{\bar{1}, 1\}^*$, and $a\bar{1}^{k-1}$ means $\bot^\omega$ if $k = 0$.

$$\rho(s) = \begin{cases} \bot^\omega & (s = \bot^\omega) \\ s'\bot 1\bar{1}^{k-1} & (s = s'\bot 1\bar{1}^k) \\ a\bar{1}^{k-1} & (s = a\bar{1}^k) \\ s'\bot 1\bar{1}^{k-1} & (s = s'a1\bar{1}^k) \end{cases}$$

Since $\rho(s) \sqsubseteq s$, we have $[\![\rho(s)]\!] \supseteq [\![s]\!]$. Moreover, $[\![\rho(s)]\!]$ is the smallest standard interval whose interior contains $[\![s]\!]$.

One can verify that ρ is monotonic. Therefore, ρ can be extended to a continuous function from *RD* to *RD* because *RD* is a Scott-Ershov domain. It is obvious that $\rho(L(RD)) \subseteq L(RD)$. The following proposition says that ρ is a conversion function from Gray-code to modified Gray expansion.

Proposition 5.1. *ρ is a retract function from $L(RD)$ to $M(L(RD))$. That is, $\rho(t) \in M(L(RD))$ and $\rho(t) \sqsubseteq t$ for $t \in L(RD)$. In particular, $\rho(t) = t$ for $t \in M(L(RD))$.*

Proof: Since $\rho(s) \sqsubseteq s$ for $s \in K(RD)$, $\rho(t) \sqsubseteq t$ for $t \in L(RD)$, and therefore $\rho(t) = t$ for $t \in M(L(RD))$. We show $\rho(L(RD)) \subseteq M(L(RD))$. Suppose that $t \in L(RD) \setminus M(L(RD))$ and let $t = sa1\bar{1}^\omega$ for some $s \in \{\bar{1}, 1\}^*$ and $a \in \{\bar{1}, 1\}$. Since $\rho(sa1\bar{1}^k) = s\bot 1\bar{1}^{k-1}$ for every $k \geq 0$, $\rho(t) = s\bot 1\bar{1}^\omega \in M(L(RD))$. □

We develop an algorithm to compute the function ρ at the level of pre-Gray code, i.e. we transform a pre-Gray code p to a pre-Gray code p' such that $\varphi(p') = \rho(\varphi(p))$, in particular p' will be the modified Gray-expansion of the real number denoted by p.

In the following, we sometimes write a for LR_a for simplicity. Since $\rho(a1\bar{1}) = \bot 1$ for $a \in \mathbf{PSD}$, if the sequence begins with $[\mathrm{LR}_a, \mathrm{LR}_1, \mathrm{LR}_{-1}]$, then we apply Equation (9) from right to left and replace it with $[\mathrm{U}, \mathrm{Fin}_a, \mathrm{LR}_{-1}]$ and fix U. We write this rule simply as

$$a\,1\,\bar{1} \;\mapsto\; \mathrm{U} \mid \mathrm{Fin}_a\,\bar{1}.$$

On the other hand, since $\rho(a11)=a$, if the sequence begins with $[\mathrm{U}, \mathrm{Fin}_a, \mathrm{LR}_1]$, we apply Equation (9) from left to right and replace it with $[\mathrm{LR}_a, \mathrm{LR}_1, \mathrm{LR}_1]$ and fix LR_a. Therefore, we have

$$\mathrm{U}\,\mathrm{Fin}_a\,1 \;\mapsto\; a\,1\,1.$$

Similarly, we have the following rules

$$\mathrm{Fin}_a\,\bar{1}\,\bar{1} \;\mapsto\; \mathrm{D} \mid \mathrm{Fin}_a\,\bar{1},$$
$$\mathrm{D}\,\mathrm{Fin}_a\,1 \;\mapsto\; \mathrm{Fin}_a \mid \bar{1}\,1.$$

If the sequence does not match to these four patterns, then we fix the first character. We repeat this procedure to the rest of the sequence. One can verify that the implicit algorithm extracted from the proof of Proposition 5.5 behaves in this way.

We extract a program that converts Gray-code to modified Gray expansion. To this end we define variants ${}^{\mathrm{co}}M$ of ${}^{\mathrm{co}}G$ and ${}^{\mathrm{co}}N$ of ${}^{\mathrm{co}}H$. Recall that the predicate $({}^{\mathrm{co}}G)^{\mathbf{r}}(p, y)$ expresses that p is a pre-Gray code of y by Proposition 3.2, and it is defined (as greatest fixed point) to mean that $p = \mathrm{LR}_a(p')$, $y = -a\frac{x-1}{2}$ and $({}^{\mathrm{co}}G)^{\mathbf{r}}(p', x)$ or else $p = \mathrm{U}(q)$, $y = \frac{x}{2}$ and $({}^{\mathrm{co}}H)^{\mathbf{r}}(q, x)$. In this definition, $p = \mathrm{LR}_{-1}(p')$ happens only if $y \le 0$, $p = \mathrm{LR}_1(p')$ happens only if $y \ge 0$ and $p = \mathrm{U}(q)$ happens only if $-\frac{1}{2} \le y \le \frac{1}{2}$. Modified Gray expansion is obtained by restricting these three cases to $y < 0$, $y > 0$, and $-\frac{1}{2} < y < \frac{1}{2}$. Therefore, ${}^{\mathrm{co}}M$ is defined so that the left clause of ${}^{\mathrm{co}}G$ is restricted to $y \ne 0$ and the right clause of ${}^{\mathrm{co}}G$ is restricted to $y \ne \pm\frac{1}{2}$. A similar restriction must be imposed on ${}^{\mathrm{co}}H$. Accordingly we define variants Γ', Δ' of the operators Γ, Δ by

$$\Gamma'(X, Y) := \{y \mid \exists^{\mathrm{r}}_{x\in X}\exists_a(y = -a\frac{x-1}{2} \wedge y \ne 0) \vee \exists^{\mathrm{r}}_{x\in Y}(y = \frac{x}{2} \wedge y \ne \pm\frac{1}{2})\},$$
$$\Delta'(X, Y) := \{y \mid \exists^{\mathrm{r}}_{x\in X}\exists_a(y = a\frac{x+1}{2} \wedge y \ne 0) \vee \exists^{\mathrm{r}}_{x\in Y}(y = \frac{x}{2} \wedge y \ne \pm\frac{1}{2})\}$$

and we define $({}^{\mathrm{co}}M, {}^{\mathrm{co}}N) := \nu_{(X,Y)}(\Gamma'(X, Y), \Delta'(X, Y))$.

The corresponding realizability predicates are defined by the operators

$$\begin{aligned}(\Gamma')^{\mathbf{r}}(Z, W) := \{ (p, x) \mid \exists_{(p',x')\in Z}\exists_a(x = -a\frac{x'-1}{2} \wedge p = \mathrm{LR}_a(p') \wedge x \neq 0) \vee^{\mathrm{nc}} \\ \exists_{(q',x')\in W}(x = \frac{x'}{2} \wedge p = \mathrm{U}(q')) \wedge x \neq \pm\frac{1}{2} \}, \\ (\Delta')^{\mathbf{r}}(Z, W) := \{ (q, x) \mid \exists_{(p',x')\in Z}\exists_a(x = a\frac{x'+1}{2} \wedge q = \mathrm{Fin}_a(p') \wedge x \neq 0) \vee^{\mathrm{nc}} \\ \exists_{(q',x')\in W}(x = \frac{x'}{2} \wedge q = \mathrm{D}(q')) \wedge x \neq \pm\frac{1}{2} \}\end{aligned}$$

as $(({}^{\mathrm{co}}M)^{\mathbf{r}}, ({}^{\mathrm{co}}N)^{\mathbf{r}}) := \nu_{(Z,W)}((\Gamma')^{\mathbf{r}}(Z, W), (\Delta')^{\mathbf{r}}(Z, W))$.

The following proposition shows that ${}^{\mathrm{co}}M$ is correct in the sense that the realizers of ${}^{\mathrm{co}}M(x)$ are exactly the pre-Gray codes of x that are mapped by φ to a modified Gray-expansion of x.

Proposition 5.2. *For cototal ideals p in $\mathbf{G}$ and $x \in \mathbb{I}$*

$$({}^{\mathrm{co}}M)^{\mathbf{r}}(p, x) \leftrightarrow \varphi(p) \text{ is a modified Gray-expansion of } x.$$

Proof: This is a direct consequence of the following lemma, because the modified Gray expansion of -1 and 1 are $\bar{1}^{\omega}$ and $1\bar{1}^{\omega}$, respectively, and they are the only cases modified Gray expansion has the form $s\bar{1}^{\omega}$ for $s \in \{\bar{1}, 1\}^*$. □

Lemma 5.3. *For $x \in \mathbb{I}$ and cototal ideals p in $\mathbf{G}$ and q in $\mathbf{H}$*

$$\begin{aligned}({}^{\mathrm{co}}M)^{\mathbf{r}}(p, x) &\leftrightarrow x = F_{\mathbf{G}}(p) \wedge (x \in \{-1, 1\} \vee \varphi(p) \neq s\bar{1}^{\omega} \text{ for } s \in \{\bar{1}, 1\}^*), \\ ({}^{\mathrm{co}}N)^{\mathbf{r}}(q, x) &\leftrightarrow x = F_{\mathbf{H}}(q) \wedge (x \in \{-1, 1\} \vee \varphi(q) \neq s\bar{1}^{\omega} \text{ for } s \in \{\bar{1}, 1\}^*).\end{aligned}$$

Proof: (From left to right). First, obviously, $({}^{\mathrm{co}}M)^{\mathbf{r}}(p, x) \to ({}^{\mathrm{co}}G)^{\mathbf{r}}(p, x)$ and $({}^{\mathrm{co}}N)^{\mathbf{r}}(q, x) \to ({}^{\mathrm{co}}H)^{\mathbf{r}}(q, x)$. Therefore, $({}^{\mathrm{co}}M)^{\mathbf{r}}(p, x) \to x = F_{\mathbf{G}}(p)$ and $({}^{\mathrm{co}}N)^{\mathbf{r}}(q, x) \to x = F_{\mathbf{H}}(q)$ holds by Proposition 3.2. We show

$$\forall_{s,p,x}(({}^{\mathrm{co}}M)^{\mathbf{r}}(p, x) \to x \in \{-1, 1\} \vee \varphi(p) \neq s\bar{1}^{\omega}) \tag{28}$$

$$\forall_{s,q,x}(({}^{\mathrm{co}}N)^{\mathbf{r}}(q, x) \to x \in \{-1, 1\} \vee \varphi(q) \neq s\bar{1}^{\omega}) \tag{29}$$

by induction on the length $|s|$ of $s \in \{\bar{1}, 1\}^*$. As the base case, (28) holds for $|s| = 0$ because $\varphi(p) = \bar{1}^{\omega} \wedge x = F_{\mathbf{G}}(p)$ implies $x = -1$. We study (29) for $|s| = 0$. We show that there is no pair (q, x) such that $\varphi(q) = \bar{1}^{\omega}$ and $({}^{\mathrm{co}}N)^{\mathbf{r}}(q, x)$. Suppose that such a pair exists. We have $x = F_{\mathbf{H}}(q)$ and $F_{\mathbf{H}}(q) = 0$ (cf. Figure 3 on page 79). Since $({}^{\mathrm{co}}N)^{\mathbf{r}}$ is a fixed point of $(\Delta')^{\mathbf{r}}$,

$$\exists_{p',x',a}(({}^{\mathrm{co}}M)^{\mathbf{r}}(p', x') \wedge x = a\frac{x'+1}{2} \wedge p = \mathrm{Fin}_a(p') \wedge x \neq 0)$$

or

$$\exists_{q',x'}((^{\mathrm{co}}N)^{\mathbf{r}}(q',x') \wedge x = \frac{x'}{2} \wedge p = \mathrm{D}(q') \wedge x \neq \pm\frac{1}{2}).$$

Since $x = 0$, we have the latter case and $p = \mathrm{D}(q')$ and $(^{\mathrm{co}}N)^{\mathbf{r}}(q', 0)$. Again for q', we have $q' = \mathrm{D}(q'')$ and $(^{\mathrm{co}}N)^{\mathbf{r}}(q'', 0)$. In this way, we have $q = \mathrm{D}^{\omega}$ and $\varphi(q) = \bot\bar{1}^{\omega}$, and we have contradiction. Thus, (29) holds for $|s| = 0$.

Suppose that (28) and (29) hold for $|s| = n$ and prove (28) for $|s| = n + 1$. Suppose that $(^{\mathrm{co}}M)^{\mathbf{r}}(p, x)$. Since $(^{\mathrm{co}}M)^{\mathbf{r}}$ is a fixed point of $(\Gamma')^{\mathbf{r}}$,

$$\exists_{p',x',a}((^{\mathrm{co}}M)^{\mathbf{r}}(p',x') \wedge x = -a\frac{x'-1}{2} \wedge p = \mathrm{LR}_a(p') \wedge x \neq 0)$$

or

$$\exists_{q',x'}((^{\mathrm{co}}N)^{\mathbf{r}}(q',x') \wedge x = \frac{x'}{2} \wedge p = \mathrm{U}(q') \wedge x \neq \pm\frac{1}{2}.$$

In the former case, by induction hypothesis, $x' \in \{-1, 1\}$ or else $\varphi(p') \neq s\bar{1}^{\omega}$ for any $s \in \{\bar{1}, 1\}^n$. The case $x' = 1$ does not happen because $x \neq 0$. If $x' = -1$, then $x \in \{-1, 1\}$. If $\varphi(p') \neq s\bar{1}^{\omega}$ for any $s \in \{\bar{1}, 1\}^n$, then $\varphi(p) = \varphi(\mathrm{LR}_a(p')) = a : \varphi(p') \neq s'\bar{1}^{\omega}$ for any $s' \in \{\bar{1}, 1\}^{n+1}$.

In the latter case, by induction hypothesis, $x' \in \{-1, 1\}$ or else $\varphi(q') \neq s\bar{1}^{\omega}$ for any $s \in \{\bar{1}, 1\}^n$. The case $x' \in \{-1, 1\}$ does not happen because $x \neq \pm\frac{1}{2}$. Suppose that $\varphi(q') \neq s\bar{1}^{\omega}$ for any $s \in \{\bar{1}, 1\}^n$. We have, for $a : t = \varphi(q')$, $\varphi(p) = \varphi(\mathrm{U}(q')) = a : 1 : t$ and $a : 1 : t \neq s'\bar{1}^{\omega}$ for any $s' \in \{\bar{1}, 1\}^{n+1}$.

The step case of (29) is similar and we omit it.

(From right to left). Easily proved by coinduction. □

Remark 5.4. *From* $^{\mathrm{co}}G = \Gamma(^{\mathrm{co}}G, {}^{\mathrm{co}}H)$ *and* $^{\mathrm{co}}H = \Delta(^{\mathrm{co}}G, {}^{\mathrm{co}}H)$ *we know that* $\gamma_a(x) := -a\frac{x-1}{2} \in {}^{\mathrm{co}}G$ $(x \in {}^{\mathrm{co}}H)$ *and* $\delta_a(x) := a\frac{x+1}{2} \in {}^{\mathrm{co}}H$ $(x \in {}^{\mathrm{co}}G)$.

Proposition 5.5 (CoGToCoM).

$$\forall_x^{\mathrm{nc}}(^{\mathrm{co}}G(x) \to {}^{\mathrm{co}}M(x)),$$
$$\forall_x^{\mathrm{nc}}(^{\mathrm{co}}H(x) \to {}^{\mathrm{co}}N(x)).$$

Proof: For $P := {}^{\mathrm{co}}G$ and $Q := {}^{\mathrm{co}}H$ we show $P \subseteq {}^{\mathrm{co}}M$ simultaneously with $Q \subseteq {}^{\mathrm{co}}N$. By coinduction it suffices to prove (i) $P \subseteq \Gamma'(^{\mathrm{co}}M \cup P, {}^{\mathrm{co}}N \cup Q)$ and (ii) $Q \subseteq \Delta'(^{\mathrm{co}}M \cup P, {}^{\mathrm{co}}N \cup Q)$. For (i), let $x_0 \in P$. We show $x_0 \in \Gamma'(^{\mathrm{co}}M \cup P, {}^{\mathrm{co}}N \cup Q)$:

$$\exists^{\mathrm{r}}_{x \in {}^{\mathrm{co}}M \cup P} \exists_a (x_0 = -a\frac{x-1}{2} \wedge x_0 \neq 0) \vee \exists^{\mathrm{r}}_{x \in {}^{\mathrm{co}}N \cup Q} (x_0 = \frac{x}{2} \wedge x_0 \neq \pm\frac{1}{2}). \tag{30}$$

The $^{\mathrm{co}}G$-clause applied to $x_0 \in {}^{\mathrm{co}}G$ gives us

$$\exists^{\mathrm{r}}_{x \in {}^{\mathrm{co}}G} \exists_a (x_0 = -a\frac{x-1}{2}) \vee \exists^{\mathrm{r}}_{x \in {}^{\mathrm{co}}H} (x_0 = \frac{x}{2}). \tag{31}$$

Case ga. The lhs of (31) holds. We have $x_1 \in {}^{\mathrm{co}}G$ and a_1 with $x_0 = -a_1\frac{x_1-1}{2}$. The ${}^{\mathrm{co}}G$-clause applied to $x_1 \in {}^{\mathrm{co}}G$ gives us

$$\exists^{\mathrm{r}}_{x\in{}^{\mathrm{co}}G}\exists_a(x_1 = -a\frac{x-1}{2}) \vee \exists^{\mathrm{r}}_{x\in{}^{\mathrm{co}}H}(x_1 = \frac{x}{2}). \tag{32}$$

Case gaa. The lhs of (32) holds. We have $x_2 \in {}^{\mathrm{co}}G$ and a_2 with $x_1 = -a_2\frac{x_2-1}{2}$.

Case ga$\bar{1}$. Assume $a_2 = -1$. Go for the lhs of (30) with $x_1 \in P$ and a_1. The goal $x_0 = -a_1\frac{x_1-1}{2}$ holds by the choice of x_1, a_1. Since $x_2 \in [-1,1]$, $x_1 = \frac{x_2-1}{2} \neq 1$. Thus, $x_0 = -a_1\frac{x_1-1}{2} \neq 0$.

Case ga1. Assume $a_2 = 1$. The ${}^{\mathrm{co}}G$-clause applied to $x_2 \in {}^{\mathrm{co}}G$ gives us

$$\exists^{\mathrm{r}}_{x\in{}^{\mathrm{co}}G}\exists_a(x_2 = -a\frac{x-1}{2}) \vee \exists^{\mathrm{r}}_{x\in{}^{\mathrm{co}}H}(x_2 = \frac{x}{2}). \tag{33}$$

Case ga1a. The lhs of (33) holds. We have $x_3 \in {}^{\mathrm{co}}G$ and a_3 with $x_2 = -a_3\frac{x_3-1}{2}$.

Case ga1$\bar{1}$. Assume $a_3 = -1$. Go for the rhs of (30) with $x = \delta_{a_1}(x_2) := a_1\frac{x_2+1}{2} \in Q$ (since $x_2 \in {}^{\mathrm{co}}G$ implies $\delta_a(x_2) \in {}^{\mathrm{co}}H$). The goal $x_0 = \frac{x}{2}$ holds since

$$x_0 = -a_1\frac{x_1-1}{2} = -a_1\frac{-a_2\frac{x_2-1}{2}-1}{2} = a_1\frac{x_2-1+2}{4} = \frac{x}{2}.$$

On the other hand, since $x_3 \in [-1,1]$, $x_2 = \frac{x_3-1}{2} \in [-1,0]$ and therefore, $x_0 = a_1\frac{x_2+1}{4} \in [-\frac{1}{4}, \frac{1}{4}]$. Thus, $x_0 \neq \pm\frac{1}{2}$.

Case ga11. Assume $a_3 = 1$. Go for the lhs of (30) with $x_1 \in P$ and a_1. The goal $x_0 = -a_1\frac{x_1-1}{2}$ holds by the choice of x_1, a_1. Since $x_3 \in [-1,1]$, $x_2 = -\frac{x_3-1}{2} \in [0,1]$ and hence $x_1 = -\frac{x_2-1}{2} \neq 1$. Thus, $x_0 = -a_1\frac{x_1-1}{2} \neq 0$.

Case ga1U. The rhs of (33) holds. We have $x_3 \in {}^{\mathrm{co}}H$ with $x_2 = \frac{x_3}{2}$. Go for the lhs of (30) with $x = x_1 \in Q$ and a_1. The goal $x_0 = -a_1\frac{x_1-1}{2}$ holds by the choice of x_1, a_1. Since $x_3 \in [-1,1]$, $x_2 \in [-\frac{1}{2}, \frac{1}{2}]$ and therefore $x_1 = -\frac{x_2-1}{2} \neq 1$. Thus, $x_0 = -a_1\frac{x_1-1}{2} \neq 0$.

Case gaU. The rhs of (32) holds. We have $x_2 \in {}^{\mathrm{co}}H$ with $x_1 = \frac{x_2}{2}$. Go for the lhs of (30) with $x = x_1 \in Q$ and a_1. The goal $x_0 = -a_1\frac{x_1-1}{2}$ holds by the choice of x_1, a_1. Since $x_2 \in [-1,1]$, $x_1 = \frac{x_2}{2} \neq 1$. Thus, $x_0 = -a_1\frac{x_1-1}{2} \neq 0$.

Case gU. The rhs of (31) holds. We have $x_1 \in {}^{\mathrm{co}}H$ with $x_0 = \frac{x_1}{2}$. We now proceed as above, applying the ${}^{\mathrm{co}}H$-clause to $x_1 \in {}^{\mathrm{co}}H$, and complete the proof of (i). The proof for (ii) is similar, and we omit it. □

Modified Gray expansion is a more desirable representation of real numbers than Gray-code in that it gives the unique code to each real number. However, a program which inputs and outputs modified Gray expansion is usually not easy to write, as the following conversion program indicates.

Implicit algorithm. $g\colon \mathbf{G} \to \mathbf{G}$ *and* $h\colon \mathbf{H} \to \mathbf{H}$*, defined by (with* a *for* LR_a*)*

$$
\begin{aligned}
&g(a(\bar{1}(p))) = a(g(\bar{1}(p))) && h(\mathrm{Fin}_a(\bar{1}(\bar{1}(p)))) = \mathrm{D}(h(\mathrm{Fin}_a(\bar{1}(p))))\\
&g(a(1(\bar{1}(p)))) = \mathrm{U}(h(\mathrm{Fin}_a(\bar{1}(p)))) && h(\mathrm{Fin}_a(\bar{1}(1(p)))) = \mathrm{Fin}_a(g(\bar{1}(1(p))))\\
&g(a(1(1(p)))) = a(g(1(1(p)))) && h(\mathrm{Fin}_a(\bar{1}(\mathrm{U}(q)))) = \mathrm{Fin}_a(g(\bar{1}(\mathrm{U}(q))))\\
&g(a(1(\mathrm{U}(q)))) = a(g(1(\mathrm{U}(q)))) && h(\mathrm{Fin}_a(1(p))) = \mathrm{Fin}_a(g(1(p)))\\
&g(a(\mathrm{U}(q))) = a(g(\mathrm{U}(q))) && h(\mathrm{Fin}_a(\mathrm{U}(q))) = \mathrm{Fin}_a(g(\mathrm{U}(q)))\\
&g(\mathrm{U}(\mathrm{Fin}_a(\bar{1}(p)))) = \mathrm{U}(h(\mathrm{Fin}_a(\bar{1}(p)))) && h(\mathrm{D}(\mathrm{Fin}_a(\bar{1}(p)))) = \mathrm{D}(h(\mathrm{Fin}_a(\bar{1}(p))))\\
&g(\mathrm{U}(\mathrm{Fin}_a(1(p)))) = a(g(1(1(p)))) && h(\mathrm{D}(\mathrm{Fin}_a(1(p)))) = \mathrm{Fin}_a(g(\bar{1}(1(p))))\\
&g(\mathrm{U}(\mathrm{Fin}_a(\mathrm{U}(q)))) = \mathrm{U}(h(\mathrm{Fin}_a(\mathrm{U}(q)))) && h(\mathrm{D}(\mathrm{Fin}_a(\mathrm{U}(q)))) = \mathrm{D}(h(\mathrm{Fin}_a(\mathrm{U}(q))))\\
&g(\mathrm{U}(\mathrm{D}(q))) = \mathrm{U}(h(\mathrm{D}(q))) && h(\mathrm{D}(\mathrm{D}(q))) = \mathrm{D}(h(\mathrm{D}(q)))
\end{aligned}
$$

When the above program cCoGToModCoG is composed with a program which inputs and outputs Gray-code, one obtains a program that inputs and outputs modified Gray expansion since a modified Gray expansion is itself a Gray-code. For example, cCoGAverage ∘ cCoGToModCoG is an average program on modified Gray expansion. Therefore, by constructing a program which inputs and outputs Gray-code, one automatically obtains a program which inputs and outputs modified Gray expansion.

6 Minlog and program extraction

Minlog is a proof assistant designed to study constructive proofs and their realizers, or more precisely the theory TCF [Schwichtenberg and Wainer, 2012]. All proofs in Sections 4 and 5 have been formalized in Minlog[3] and their realizers extracted, as terms in an extension of Gödel's T. In this section we present the extracted terms and discuss how they operate. They involve recursion and corecursion operators where the original proofs used induction or coinduction axioms, and the conversion rules for these operators determine how the extracted terms can be used as programs. The results of such an analysis have been shown in Sections 4 and 5 under the label "implicit algorithm".

3 See http://www.minlog-system.de/, which gives instructions on how to download (or clone) the system and the necessary software (Scheme in this case). The formalizations can be found in the directory minlog/examples/analysis/gray.scm.

6.1 Corecursion

Recall the type of the corecursion operator for **I**:

$$ {}^{\mathrm{co}}\mathcal{R}_{\mathbf{I}}^{\tau}\colon \tau \to (\tau \to \mathbf{SD} \times (\mathbf{I} + \tau)) \to \mathbf{I}. \tag{34} $$

The type $\mathbf{SD}\times(\mathbf{I}+\tau)$ appears since **I** has the single constructor C of type $\mathbf{SD} \to \mathbf{I} \to \mathbf{I}$. The meaning of ${}^{\mathrm{co}}\mathcal{R}_{\mathbf{I}}^{\tau} NM$ is defined by the conversion rule

$$ {}^{\mathrm{co}}\mathcal{R}_{\mathbf{I}}^{\tau} NM \mapsto \mathrm{C}_{\pi_1(MN)}([\mathrm{id}^{\mathbf{I}\to\mathbf{I}}, \lambda_y({}^{\mathrm{co}}\mathcal{R}_{\mathbf{I}}^{\tau} yM)]\pi_2(MN)). $$

We have used π_1, π_2 for the two projections of type $\rho \times \sigma$, and the notation $[f, g]\colon \rho + \sigma \to \tau$ (for $f\colon \rho \to \tau$ and $g\colon \sigma \to \tau$) defined by

$$ [f, g](z) := \begin{cases} f(x) & \text{if } z = \mathrm{inl}(x), \\ g(y) & \text{if } z = \mathrm{inr}(y). \end{cases} $$

We will also need the simultaneous corecursion operators ${}^{\mathrm{co}}\mathcal{R}_{\mathbf{G}}^{(\mathbf{G},\mathbf{H}),(\sigma,\tau)}$ and ${}^{\mathrm{co}}\mathcal{R}_{\mathbf{H}}^{(\mathbf{G},\mathbf{H}),(\sigma,\tau)}$ for **G**, **H**, of type

$$ \begin{aligned} {}^{\mathrm{co}}\mathcal{R}_{\mathbf{G}}^{(\mathbf{G},\mathbf{H}),(\sigma,\tau)} &\colon \sigma \to \delta_{\mathbf{G}} \to \delta_{\mathbf{H}} \to \mathbf{G} \\ {}^{\mathrm{co}}\mathcal{R}_{\mathbf{H}}^{(\mathbf{G},\mathbf{H}),(\sigma,\tau)} &\colon \tau \to \delta_{\mathbf{G}} \to \delta_{\mathbf{H}} \to \mathbf{H} \end{aligned} \tag{35} $$

with step types

$$ \begin{aligned} \delta_{\mathbf{G}} &:= \sigma \to \mathbf{PSD} \times (\mathbf{G} + \sigma) + (\mathbf{H} + \tau), \\ \delta_{\mathbf{H}} &:= \tau \to \mathbf{PSD} \times (\mathbf{G} + \sigma) + (\mathbf{H} + \tau). \end{aligned} $$

The type $\mathbf{PSD}\times(\mathbf{G}+\sigma)+(\mathbf{H}+\tau)$ appears since **G** has the two constructors $\mathrm{LR}\colon \mathbf{PSD} \to \mathbf{G} \to \mathbf{G}$ and $\mathrm{U}\colon \mathbf{H} \to \mathbf{G}$, and **H** has the two constructors $\mathrm{Fin}\colon \mathbf{PSD} \to \mathbf{G} \to \mathbf{H}$ and $\mathrm{D}\colon \mathbf{H} \to \mathbf{H}$. Omitting the upper indices of ${}^{\mathrm{co}}\mathcal{R}$, the terms ${}^{\mathrm{co}}\mathcal{R}_{\mathbf{G}} NMM'$ and ${}^{\mathrm{co}}\mathcal{R}_{\mathbf{H}} N'MM'$ are defined by the conversion rules

$$ {}^{\mathrm{co}}\mathcal{R}_{\mathbf{G}} NMM' \mapsto \begin{cases} \mathrm{LR}_{\pi_1(u)}([\mathrm{id}, \lambda_y({}^{\mathrm{co}}\mathcal{R}_{\mathbf{G}} yMM')]\pi_2(u)) & \text{if } MN = \mathrm{inl}(u) \\ \mathrm{U}([\mathrm{id}, \lambda_z({}^{\mathrm{co}}\mathcal{R}_{\mathbf{H}} zMM')]v) & \text{if } MN = \mathrm{inr}(v) \end{cases} $$

$$ {}^{\mathrm{co}}\mathcal{R}_{\mathbf{H}} N'MM' \mapsto \begin{cases} \mathrm{Fin}_{\pi_1(u)}([\mathrm{id}, \lambda_y({}^{\mathrm{co}}\mathcal{R}_{\mathbf{G}} yMM')]\pi_2(u)) & \text{if } M'N' = \mathrm{inl}(u) \\ \mathrm{D}([\mathrm{id}, \lambda_z({}^{\mathrm{co}}\mathcal{R}_{\mathbf{H}} zMM')]v) & \text{if } M'N' = \mathrm{inr}(v) \end{cases} $$

6.2 Notational conventions of Minlog

Table 2. Notational conventions of Minlog

	Heading 1	Heading 2
Types	`iv`, `ag`, `ah`, `bg`	base types for the algebra **I**, **G**, **H**, ${}^{2}\mathbf{G}$
	`rho=>sigma`	function typ
	`rho@@sigma`	product type
	`rho ysum sigma`	sum type
Variables (with fixed types)	`v`, `p`, `q`	of type **I**, **G** and **H**
	`d`, `a`, `i`	of type **SD**, **PSD**, $\mathbf{SD}_2$
	`ivw`	of type $\mathbf{SD}_2 \times \mathbf{I} \times \mathbf{I}$
	`jdvw`	of type $\mathbf{SD}_2 \times \mathbf{SD} \times \mathbf{I} \times \mathbf{I}$
	`ap`	of type $\mathbf{PSD} \times \mathbf{G}$
	`bv`	of type $\mathbf{PSD} \times \mathbf{I}$
	`ipp`	of type $\mathbf{SD}_2 \times \mathbf{G} \times \mathbf{G}$
	`idpp`	of type $\mathbf{SD}_2 \times \mathbf{SD} \times \mathbf{G} \times \mathbf{G}$
	`psf`	of type $(\mathbf{G} \to {}^{2}\mathbf{G}) \times (\mathbf{H} \to \mathbf{N} + \mathbf{PSD} \times \mathbf{G} \times {}^{2}\mathbf{G})$
	`apbg`	of type $\mathbf{PSD} \times \mathbf{G} \times {}^{2}\mathbf{G}$
Constants	`Rec`, `CoRec`	recursion, corecursion
	`Des`	destructor
	`PsdToSd`	embedding of **PSD** into **SD**
	`plus`, `times`, `inv`	arithmetic in **SD**
	`cL`	realizer for lemma L
Terms	`[x]r`	lambda abstraction $\lambda_x r$
	`r@s`	product term
	`left r`,`right r`	components (prefix, binding strongest)
	`InL`, `InR`	injections into a sum type

6.3 ColAverage

We analyze the term in Figure 4 extracted from ColAverage. The first argument N of the corecursion operator destructs `v`, `v0` into their components (d, v), (e, w) and forms $(d+e, v, w)$. The step function M, when applied to an argument `ivw` of type $\tau = \mathbf{SD}_2 \times \mathbf{I} \times \mathbf{I}$, M gives a result of type $\mathbf{SD} \times (\mathbf{I} + \tau)$, as follows. Destruct `ivw` into the form $(i, (d, v), (e, w))$, and let `jdvw` be the quadruple $(J(d, e, i), K(d, e, i), v, w)$. Return $(K(d, e, i), \mathrm{inr}(J(d, e, i), v, w))$.

```
[v,v0](CoRec sdtwo@@iv@@iv=>iv)
 (left Des v plus left Des v0@right Des v@right Des v0)
 ([ivw][let jdvw
     (J left Des left right ivw
        left Des right right ivw
        left ivw@
     K left Des left right ivw
       left Des right right ivw
       left ivw@
     right Des left right ivw@
     right Des right right ivw)
     (left right jdvw@InR(left jdvw@right right jdvw))])
```

Fig. 4. Extracted term for CoIAverage.

Hence we can write $\lambda_y{}^{\mathrm{co}}\mathcal{R}^{\tau}_{\mathbf{I}} yM$ as a function $f: \tau \to \mathbf{I}$ defined by

$$f(i, \mathrm{C}_d(v), \mathrm{C}_e(w)) = \mathrm{C}_{K(d,e,i)}(f(J(d, e, i), v, w)).$$

6.4 CoGToCoI

Consider the term in Figure 5 extracted from Lemma 4.5 (CoGToCoIAux).

```
[apq](CoRec psd@@(ag ysum ah)=>iv)apq
 ([apq0][case (right apq0)
     (InL p -> [case (Des p)
       (InL ap ->
       PsdToSd(left apq0 times left ap)@
       InR(inv(left apq0 times left ap)@InL right ap))
       (InR q -> Mid@InR(left apq0@InR q))])
     (InR q -> [case (Des q)
       (InL ap ->
       PsdToSd(left apq0 times left ap)@
       InR(left apq0 times left ap@InL right ap))
       (InR q0 -> Mid@InR(left apq0@InR q0))])])
```

Fig. 5. Extracted term for CoGToCoIAux.

We analyze the second argument M of the corecursion operator (the “step term”). When applied to an argument N of type $\tau = \mathbf{PSD} \times (\mathbf{G} + \mathbf{H})$, M returns a result of type $\mathbf{SD} \times (\mathbf{I} + \tau)$; it will be in the right part of $\mathbf{I} + \tau$ (i.e., here we do not use the fact that our coinductive definitions are in “strengthened” form). Consider the right hand side N_2 of N, of type $\mathbf{G} + \mathbf{H}$.

Case 1. If N_2 is of the form $\mathrm{inl}(p)$ with p of type $\mathbf{G}$, destruct p. Recall that $\mathbf{G}$ has two constructors, LR and U. If p is of the form $\mathrm{LR}_b(p')$, the result is $(ab, \mathrm{inr}(-ab, \mathrm{inl}(p')))$. If p is of the form $\mathrm{U}(q)$, the result is $(0, \mathrm{inr}(a, \mathrm{inr}(q)))$.

Case 2. If N_2 is of the form $\mathrm{inr}(q)$ with q of type $\mathbf{H}$, destruct q. Recall that $\mathbf{H}$ has two constructors, Fin and D. If q is of the form $\mathrm{Fin}_b(p)$, the result is $(ab, \mathrm{inr}(ab, \mathrm{inl}(p)))$. If q is of the form $\mathrm{D}(q')$, the result is $(0, \mathrm{inr}(a, \mathrm{inr}(q')))$. Hence $\lambda_y {}^{\mathrm{co}}\mathcal{R}_{\mathbf{I}}^{\tau} yM$ is a function $[f, g]\colon \mathbf{PSD} \times \mathbf{G} + \mathbf{PSD} \times \mathbf{H} \to \mathbf{I}$ defined by

$$
\begin{aligned}
f(a, \mathrm{LR}_b(p)) &= \mathrm{C}_{ab}(f(-ab, p)), & g(a, \mathrm{Fin}_b(p)) &= \mathrm{C}_{ab}(f(ab, p)),\\
f(a, \mathrm{U}(q)) &= \mathrm{C}_0(g(a, q)), & g(a, \mathrm{D}(q)) &= \mathrm{C}_0(g(a, q)).
\end{aligned}
$$

6.5 CoIToCoG

For Lemma 4.7 (CoIToCoGAux) we obtain the extracted term in Figure 6.

```
[bv](CoRec psd@@iv=>ag psd@@iv=>ah)bv
 ([bv0][case (left Des right bv0)
     (Lft -> InL(inv left bv0@InR(PRht@right Des right bv0)))
     (Rht -> InL(left bv0@InR(PLft@right Des right bv0)))
     (Mid -> InR(InR(left bv0@right Des right bv0)))])
 ([bv0][case (left Des right bv0)
     (Lft -> InL(inv left bv0@InR(PLft@right Des right bv0)))
     (Rht -> InL(left bv0@InR(PRht@right Des right bv0)))
     (Mid -> InR(InR(left bv0@right Des right bv0)))])
```

Fig. 6. Extracted term for CoIToCoGAux.

To understand this term recall the type (35) of the simultaneous corecursion operators ${}^{\mathrm{co}}\mathcal{R}_{\mathbf{G}}^{(\mathbf{G},\mathbf{H}),(\tau,\tau)}$ and ${}^{\mathrm{co}}\mathcal{R}_{\mathbf{H}}^{(\mathbf{G},\mathbf{H}),(\tau,\tau)}$, or shortly ${}^{\mathrm{co}}\mathcal{R}_{\mathbf{G}}$ and ${}^{\mathrm{co}}\mathcal{R}_{\mathbf{H}}$, with $\tau := \mathbf{PSD} \times \mathbf{I}$ and step types $\delta := \tau \to \mathbf{PSD} \times (\mathbf{G} + \tau) + (\mathbf{H} + \tau)$. We again analyze the particular step functions M, M' extracted from our proof. When applied to an argument N of type $\tau = \mathbf{PSD} \times \mathbf{I}$, M returns a result of type $\mathbf{PSD} \times (\mathbf{G} + \tau) + (\mathbf{H} + \tau)$, in the right part of $\mathbf{G} + \tau$ or $\mathbf{H} + \tau$. Let $N = (b, v)$ with v of type $\mathbf{I}$, of the form $\mathrm{C}_d(v')$. The result is

$$\begin{cases} \mathrm{inl}(-b, \mathrm{inr}(1, v')) & \text{if } d = -1, \\ \mathrm{inl}(b, \mathrm{inr}(-1, v')) & \text{if } d = 1, \\ \mathrm{inr}(\mathrm{inr}(b, v')) & \text{if } d = 0. \end{cases}$$

Similarly, when applied to an argument N of type $\tau = \mathbf{PSD} \times \mathbf{I}$, M' returns a result of type $\mathbf{PSD} \times (\mathbf{G} + \tau) + (\mathbf{H} + \tau)$. Let $N = (b, v)$ with v of type $\mathbf{I}$, of the form $\mathrm{C}_d(v')$. The result is

$$\begin{cases} \mathrm{inl}(-b, \mathrm{inr}(-1, v')) & \text{if } d = -1, \\ \mathrm{inl}(b, \mathrm{inr}(1, v')) & \text{if } d = 1, \\ \mathrm{inr}(\mathrm{inr}(b, v')) & \text{if } d = 0. \end{cases}$$

Hence we can write the two functions $\lambda_y {}^{\mathrm{co}}\mathcal{R}_{\mathbf{G}} y M M'$ and $\lambda_y {}^{\mathrm{co}}\mathcal{R}_{\mathbf{H}} y M M'$ as $g\colon \tau \to \mathbf{G}$ and $h\colon \tau \to \mathbf{H}$ defined by

$$\begin{aligned} g(b, \mathrm{C}_{-1}(v)) &= \mathrm{LR}_{-b}(g(1, v)), & h(b, \mathrm{C}_{-1}(v)) &= \mathrm{Fin}_{-b}(g(-1, v)), \\ g(b, \mathrm{C}_{1}(v)) &= \mathrm{LR}_{b}(g(-1, v)), & h(b, \mathrm{C}_{1}(v)) &= \mathrm{Fin}_{b}(g(1, v)), \\ g(b, \mathrm{C}_{0}(v)) &= \mathrm{U}(h(b, v)), & h(b, \mathrm{C}_{0}(v)) &= \mathrm{D}(h(b, v)). \end{aligned}$$

6.6 CoGAverage

For Lemma 4.9 (CoGMinus) the extracted term is shown in Figure 7.

```
[p](CoRec ag=>ag ah=>ah)p
 ([p0][case (Des p0)
     (InL ap -> InL(inv left ap@InL right ap))
     (InR q -> InR(InR q))])
 ([q][case (Des q)
     (InL ap -> InL(inv left ap@InL right ap))
     (InR q0 -> InR(InR q0))])
```

Fig. 7. Extracted term for CoGMinus.

We need simultaneous corecursion operators

$${}^{\mathrm{co}}\mathcal{R}_{\mathbf{G}}^{(\mathbf{G},\mathbf{H}),(\sigma,\tau)}, \quad \text{and} \quad {}^{\mathrm{co}}\mathcal{R}_{\mathbf{H}}^{(\mathbf{G},\mathbf{H}),(\sigma,\tau)}$$

of type (35). By analyzing the particular step functions M, M' extracted from our proof we see that we can write

$$\lambda_y {}^{\mathrm{co}}\mathcal{R}_{\mathbf{G}} y M M' \quad \text{and} \quad \lambda_z {}^{\mathrm{co}}\mathcal{R}_{\mathbf{H}} z M M'$$

as functions $f\colon \sigma \to \mathbf{G}$ and $f'\colon \tau \to \mathbf{H}$ defined by

$$\begin{aligned} f(\mathrm{LR}_a(p)) &= \mathrm{LR}_{-a}(p), & f'(\mathrm{Fin}_a(p)) &= \mathrm{Fin}_{-a}(p), \\ f(\mathrm{U}(q)) &= \mathrm{U}(f'(q)), & f'(\mathrm{D}(q)) &= \mathrm{D}(f'(q)). \end{aligned}$$

Lemma 4.9 (CoGMinus) gave us Lemma 4.10 (CoHToCoG).

```
[q](CoRec ah=>ag ag=>ah)q
 ([q0][case (Des q0)
     (InL ap -> InL(left ap@InL(cCoGMinus right ap)))
     (InR q1 -> InR(InL q1))])
 ([p][case (Des p)
     (InL ap -> InL(left ap@InL(cCoGMinus right ap)))
     (InR q0 -> InR(InL q0))])
```

Fig. 8. Extracted term for CoHToCoG.

The extracted term in Figure 8 clearly represents the functions shown as implicit algorithm in Section 4.4.

We now come to the average for Gray-code. As a preparation we need an easy consequence of CoGMinus, a lemma CoGPsdTimes with extracted term

```
[a,p][case a (PLft -> cCoGMinus p) (PRht -> p)].
```

For Lemma 4.11 (CoGAvToAvc) the extracted term in Figure 9 again clearly represents the function shown as implicit algorithm in Section 4.4.
For Lemma 4.12 (CoGAvcSatCoICl) the extracted term is shown in Figure 10. It is rather easy to parse into how it is written in Section 4.4.
For Lemma 4.13 (CoGAvcToCoG) we need as a preparation an easy lemma SdDisj: $\forall_d(d = 0 \vee \exists_a(d = a))$, with extracted term

```
[d][case d (Lft -> Inr PLft) (Rht -> Inr PRht) (Mid -> DummyL)]
```

It is easy to see that the extracted term for Lemma 4.13 (in Figure 11) gives the algorithm in Section 4.4.

```
[p,p0][case (Des p)
   (InL ap ->
   [case (Des p0)
     (InL ap0 -> left ap plus left ap0@
                 cCoGPsdTimes inv left ap right ap@
                 cCoGPsdTimes inv left ap0 right ap0)
     (InR q -> left ap plus Mid@
               cCoGPsdTimes inv left ap right ap@
               cCoHToCoG q)])
   (InR q ->
   [case (Des p0)
     (InL ap -> Mid plus left ap@
                cCoHToCoG q@
                cCoGPsdTimes inv left ap right ap)
     (InR q0 -> MT@cCoHToCoG q@cCoHToCoG q0)])]
```

Fig. 9. Extracted term for CoGAvToAvc.

```
[i,p,p0][case (Des p)
   (InL ap ->
   [case (Des p0)
     (InL ap0 -> J(PsdToSd left ap)(PsdToSd left ap0)i@
                 K(PsdToSd left ap)(PsdToSd left ap0)i@
                 cCoGPsdTimes inv left ap right ap@
                 cCoGPsdTimes inv left ap0 right ap0)
     (InR q ->   J(PsdToSd left ap)Mid i@
                 K(PsdToSd left ap)Mid i@
                 cCoGPsdTimes inv left ap right ap@
                 cCoHToCoG q)])
   (InR q ->
   [case (Des p0)
     (InL ap -> J Mid(PsdToSd left ap)i@
                K Mid(PsdToSd left ap)i@
                cCoHToCoG q@
                 cCoGPsdTimes inv left ap right ap)
     (InR q0 -> J Mid Mid i@K Mid Mid i@
                cCoHToCoG q@cCoHToCoG q0)])]
```

Fig. 10. Extracted term for CoGAvcSatCoICl.

```
[ipp](CoRec sdtwo@@ag@@ag=>ag sdtwo@@ag@@ag=>ah)ipp
 ([ipp0]
   [let idpp (cCoGAvcSatCoICl
               left ipp0 left right ipp0 right right ipp0)
     [case (cSdDisj left right idpp)
      (DummyL -> InR(InR(left idpp@right right idpp)))
      (Inr a ->
      InL(a@InR
       (a times inv left idpp@
        cCoGPsdTimes inv a left right right idpp@
        cCoGPsdTimes inv a right right right idpp)))]])
 ([ipp0][let idpp (cCoGAvcSatCoICl
                    left ipp0 left right ipp0 right right ipp0)
     [case (cSdDisj left right idpp)
      (DummyL -> InR(InR(left idpp@right right idpp)))
      (Inr a ->
      InL(a@InR
       (a times left idpp@
        cCoGPsdTimes a left right right idpp@
        cCoGPsdTimes a right right right idpp)))]])
```

Fig. 11. Extracted term for CoGAvcToCoG.

```
[n](Rec nat=>(ag=>bg)@@(ah=>nat ysum psd@@ag@@bg))n
 (([p]Nz)@([q]InL Zero))
 ([n0,psf]
   ([p][case (Des p)
       (InL ap -> LRz left ap(left psf right ap))
       (InR q -> [case (right psf q)
         (InL n -> Uz n)
         (InR apbg ->
         LRz left apbg
         [case n0
           (Zero -> Nz)
           (Succ n1 -> LRz PRht right right apbg)])])])@
```

Fig. 12. Extracted term for CoGToBGAux, part I

For Proposition 4.14 (CoGAverage) the extracted term is obtained by composition of those for Lemmata 4.11 and 4.13: `[p,p0]cCoGAvcToCoG(cCoGAvToAvc p p0)`.

6.7 CoGToBG

For Lemma 4.16 again the extracted term (see Figure 12 above and Figure 13 below) represents the algorithms given in Section 4.5

```
([q][case (Des q)
    (InL ap -> InR(left ap@right ap@left psf right ap))
    (InR q0 -> [case (right psf q0)
      (InL n1 -> InL(Succ n1))
      (InR apbg ->
      InR
      (left apbg@
       LR PLft left right apbg@
       left psf(LR PLft left right apbg)))])]))
```

Fig. 13. Extracted term for CoGToBGAux, part II

6.8 CoGToCoM

Finally Figure 15 gives the term extracted from our proof of Proposition 5.5.

```
[p](CoRec ag=>ag ah=>ah)p
 ([p0][case (Des p0)
     (InL ap -> [case (Des right ap)
       (InL ap0 -> [case (left ap0)
         (PLft -> InL(left ap@InR right ap))
         (PRht -> [case (Des right ap0)
           (InL ap1 -> [case (left ap1)
             (PLft -> InR(InR(cCoHClauseInv
                                  (InL(left ap@right ap0)))))
```

Fig. 14. Extracted term for CoGToCoM, to be continued on the next page

```
            (PRht -> InL(left ap@InR right ap))])
          (InR q -> InL(left ap@InR right ap))])])
      (InR q -> InL(left ap@InR right ap))])
    (InR q -> [case (Des q)
      (InL ap -> [case (Des right ap)
        (InL ap0 -> [case (left ap0)
          (PLft -> InR(InR(cCoHClauseInv(InL ap))))
          (PRht -> InL(left ap@InR(cCoGClauseInv
                   (InR(cCoHClauseInv(InL(PRht@right ap0)))))))])
        (InR q0 -> InR(InR(cCoHClauseInv(InL ap))))])
      (InR q0 -> InR(InR q))])])
([q][case (Des q)
    (InL ap -> [case (Des right ap)
      (InL ap0 -> [case (left ap0)
        (PLft -> [case (Des right ap0)
          (InL ap1 -> [case (left ap1)
            (PLft -> InR(InR(cCoHClauseInv
                                 (InL(left ap@right ap0)))))
            (PRht -> InL(left ap@InR right ap))])
          (InR q0 -> InL(left ap@InR right ap))])
        (PRht -> InL(left ap@InR right ap))])
      (InR q0 -> InL(left ap@InR right ap))])
    (InR q0 -> [case (Des q0)
      (InL ap -> [case (Des right ap)
        (InL ap0 -> [case (left ap0)
          (PLft -> InR(InR(cCoHClauseInv(InL ap))))
          (PRht -> InL(left ap@InR(cCoGClauseInv
                   (InR(cCoHClauseInv(InL(PLft@right ap0)))))))])
        (InR q1 -> InR(InR(cCoHClauseInv(InL ap))))])
      (InR q1 -> InR(InR q0))])])
```

Fig. 15. Extracted term for CoGToCoM, part II

Acknowledgment: This work was supported by the International Research Staff-Exchange Scheme (IRSES) Nr. 612638 CORCON and Nr. 294962 COMPUTAL of the European Commission, the JSPS Core-to-Core Program, A. Advanced research Networks and JSPS KAKENHI Grant Number 15K00015.

References

Ulrich Berger. Program extraction from normalization proofs. In M. Bezem and J.F. Groote, editors, *Typed Lambda Calculi and Applications*, volume 664 of *LNCS*, pages 91–106. Springer, Berlin, Heidelberg, New York, 1993.

Ulrich Berger and Monika Seisenberger. Proofs, programs, processes. In F. Ferreira et al., editors, *Proceedings CiE 2010*, volume 6158 of *LNCS*, pages 39–48. Springer, Berlin, Heidelberg, New York, 2010.

Chi Ming Chuang. *Extraction of Programs for Exact Real Number Computation Using Agda*. PhD thesis, Swansea University, Wales, UK, 2011.

Alberto Ciaffaglione and Pietro Di Gianantonio. A co-inductive approach to real numbers. In *Proc. of the workshop "Types 1999"*, volume 1956 of *LNCS*, pages 114–130. Springer, Berlin, Heidelberg, New York, 1999.

Pietro Di Gianantonio. An abstract data type for real numbers. *Theoretical Computer Science*, 221(1-2):295–326, 1999.

Andrey N. Kolmogorov. Zur Deutung der intuitionistischen Logik. *Math. Zeitschr.*, 35:58–65, 1932.

Kenji Miyamoto. *Program extraction from coinductive proofs and its application to exact real arithmetic*. PhD thesis, Mathematisches Institut der Universität München, 2013.

Kenji Miyamoto and Helmut Schwichtenberg. Program extraction in exact real arithmetic. *Mathematical Structures in Computer Science*, 25:1692–1704, 2015.

Shmuel Sagiv, editor. *Programming Languages and Systems, 14th European Symposium on Programming,ESOP 2005, Held as Part of the Joint European Conferences on Theory and Practice of Software, ETAPS 2005, Edinburgh, UK, April 4-8, 2005, Proceedings*, volume 3444 of *Lecture Notes in Computer Science*, 2005. Springer.

Helmut Schwichtenberg and Stanley S. Wainer. *Proofs and Computations*. Perspectives in Logic. Association for Symbolic Logic and Cambridge University Press, 2012.

Kei Terayama and Hideki Tsuiki. A stream calculus of bottomed sequences for real number computation. *Electr. Notes Theor. Comput. Sci.*, 298:383–402, 2013. 10.1016/j.entcs.2013.09.023. URL http://dx.doi.org/10.1016/j.entcs.2013.09.023.

Hideki Tsuiki. Real number computation through Gray code embedding. *Theoretical Computer Science*, 284:467–485, 2002.

Hideki Tsuiki. Real number computation with committed choice logic programming languages. *J. Log. Algebr. Program.*, 64(1):61–84, 2005. 10.1016/j.jlap.2004.07.005. URL http://dx.doi.org/10.1016/j.jlap.2004.07.005.

Hideki Tsuiki and Keiji Sugihara. Streams with a bottom in functional languages. In Sagiv [2005], pages 201–216. 10.1007/978-3-540-31987-0_15. URL http://dx.doi.org/10.1007/978-3-540-31987-0_15.

Edwin Wiedmer. *Exaktes Rechnen mit reellen Zahlen und anderen unendlichen Objekten*. PhD thesis, ETH Zürich, 1977.

Edwin Wiedmer. Computing with infinite objects. *Theoretical Computer Science*, 10:133–155, 1980.

Douglas S. Bridges

The Continuum Hypothesis Implies Excluded Middle

Abstract: Within the framework of Bishop-style constructive mathematics, it is first proved that the Continuum Hypothesis (CH) implies the law of excluded middle. A more explicit Brouwerian counterexample to CH is then discussed.

Keywords: Continuum hypothesis, Constructive, Excluded middle

Mathematics Subject Classification 2010: 03F60, 03E50

It is a well-known result of Goodman & Myhill [Goodman and Myhill, 1978][1] that in Bishop-style constructive mathematics (BISH)[2]—essentially, in their and our case, mathematics with intuitionistic logic and a constructive set theory such as CZF [Aczel and Rathjen] [3]—the full axiom of choice AC implies the law of excluded middle.[4] Given the classical independence of AC relative to ZF, and the similar status of Cantor's Continuum Hypothesis (which, save for the intuitionistic paper [Gielen et al., 1981], has largely been ignored by the constructive mathematics community), it is reasonable to ask whether the following holds in BISH:

Theorem 1. *The Continuum Hypothesis implies the law of excluded middle.*

We prove this theorem shortly. But first we need to clarify what we mean by 'the Continuum Hypothesis'. This can be handled without dealing directly with cardinals: for, classically, it is equivalent to the statement that if the cardinality of a set S lies between that of ω (the set of natural numbers) and that of $\mathcal{P}\omega$ (the power set of ω) then either S is cardinally similar to ω or else it is cardinally similar to $\mathcal{P}\omega$; in turn, this is equivalent classically to the what we take as the constructive version of the Continuum Hypothesis: namely,

1 based on work of Diaconescu [Diaconescu, 1975].

2 For more on BISH see [Bishop, 1967, Bishop and Bridges, 1985, Troelstra and van Dalen, 1988].

3 CZF is not the only set-theoretic foundation for BISH. Examples of others are constructive Morse set theory [Bridges and Alps], intuitionistic ZF [Friedman, 1977], and Myhill's CST [Myhill, 1975]. A popular foundation of a different sort is Martin-Löf's type theory [Martin-Löf, 1975].

4 Note that although the axiom of dependent choice, and the consequent one of countable choice, are acceptable to most constructive mathematicians, some—notably Richman [Richman, 2001]—have been developing mathematics constructively without even countable choice.

Douglas S. Bridges: School of Mathematics & Statistics, University of Canterbury, Private Bag 4800, Christchurch, New Zealand, e-mail: d.bridges@math.canterbury.ac.nz

CH: *If* $\omega \preccurlyeq S \preccurlyeq \mathcal{P}\omega$ *and* $\neg(S \preccurlyeq \omega)$, *then* $S \approx \mathcal{P}\omega$.

Here, '$\preccurlyeq$' is the usual 'is cardinally smaller than' predicate and '$\approx$' the 'cardinally similar' or 'equinumerous with' one; thus: '$S \preccurlyeq T$' means that there exists an injection (that is, one-one mapping) of S into T, and '$S \approx T$' that we have both $S \preccurlyeq T$ and $T \preccurlyeq S$. Both of these predicates are reflexive and transitive predicates, and '$\approx$' is also symmetric. If

$$S \preccurlyeq T \wedge \neg(S \approx T),$$

then we say that S is *cardinally strictly smaller* than T, and that T is *cardinally strictly greater* than S.

Our proof of Theorem 1 requires two preliminaries.

Lemma 2. *Let A be a set, and f a mapping of A into $\mathcal{P}A$. Then there exists $C \subset A$ that is not in the range of f.*

Proof: The argument is a positive re-phrasing of a familiar one in classical ZF. Define

$$C \equiv \{x \in A : x \notin f(x)\}.$$

Then $C \in \mathcal{P}A$. Suppose there exists $x \in A$ with $f(x) = C$. Then

$$x \in C \leftrightarrow x \notin f(x) = C,$$

which is absurd. Hence C is not in the range of f. □

Corollary 3. *An inhabited set is cardinally strictly smaller than its power set.*

We now give the proof of Theorem 1.

Proof: Let p be an arbitrary statement,[5] and define

$$S \equiv \{\{n\} : n \in \omega\} \cup \{E \subset \omega : p \vee \neg p\}.$$

Using the canonical injections f of ω into S, and g of S into $\mathcal{P}\omega$, we see that $\omega \preccurlyeq S \preccurlyeq \mathcal{P}\omega$. Moreover, if $S \approx \omega$, then, by Corollary 3, $\neg(S \approx \mathcal{P}\omega)$ and therefore $\neg(p \vee \neg p)$, which is absurd; whence $\neg(S \approx \omega)$. Assuming CH, we obtain $S \approx \mathcal{P}\omega$. It follows from this and Lemma 2 that there exists $C \subset \omega$ that is not in the range of f; whence

$$C \in \{E \subset \omega : p \vee \neg p\}$$

and therefore $p \vee \neg p$. □

5 What we present here is an informal proof that can be formalised in Constructive Morse Set Theory [Bridges and Alps]. To formalise the proof in Constructive ZF [Aczel and Rathjen], we would need to add the restriction that p be a bounded statement, in which case the conclusion is that CH implies the restricted law of excluded middle.

Theorem 3 embodies a Brouwerian counterexample to the Continuum Hypothesis. If we allow ourselves to invoke the classical results of Gödel [Gödel, 1940] and Cohen [Cohen, 1963], then we can use the ideas in the proof of Theorem 1 to beef up that example as follows.

First, consider the set

$$T \equiv \{E \subset \omega : \text{CH} \vee \neg\text{CH}\}.$$

If $T = \varnothing$, then we have the contradiction $\neg(\text{CH} \vee \neg\text{CH})$; hence T is nonempty. On the other hand, if T is inhabited—that is, we can construct an element of T—then CH $\vee\neg$CH holds. But the *constructive* interpretation of 'CH $\vee\neg$CH' is that either we have a proof of CH or we have a proof of $\neg$CH, which runs counter to the classical Gödel-Cohen proof that CH is independent of ZF set theory. Since BISH is consistent with classical mathematics, we see that T is an explicit example of a nonempty set which cannot possibly be proved inhabited.

This, in itself, is not uninteresting; but it also gives a new insight into the constructive status of CH. Consider the set

$$S \equiv \{\{n\} : n \in \omega\} \cup T,$$

which (see the proof of Theorem 1) satisfies

$$\omega \preccurlyeq S \preccurlyeq \mathcal{P}\omega \wedge \neg(S \approx \omega).$$

If $\neg(S \approx \mathcal{P}\omega)$, then $T = \varnothing$, which we have seen is absurd; hence $\neg\neg(S \approx \mathcal{P}\omega)$. On the other hand, if $S \approx \mathcal{P}\omega$, then—again see the proof of Theorem 1—T is inhabited, which, as we have just shown, cannot be proved. Thus S is an explicit example of a set that has the properties

$$\omega \preccurlyeq S \preccurlyeq \mathcal{P}\omega \wedge \neg(S \approx \omega) \wedge \neg\neg(S \approx \mathcal{P}\omega)$$

and that, within BISH, cannot possibly be proved cardinally similar to $\mathcal{P}\omega$.

Incidentally, the proof in [Goodman and Myhill, 1978] can be adapted similarly, to provide an explicit example of a set A and a subset S of $A \times \{0, 1\}$ such that

$$\forall_{x\in A}\exists_{y\in B}\,((x, y) \in S)$$

but it is impossible to prove, constructively, that there exists a function $f : A \to B$ such that $(x, f(x)) \in S$ for each $x \in A$.

Acknowledgment: I am grateful to Fred Richman and Maarten McKubre-Jordens for their comments on the first draft of this paper.

References

P. Aczel and M. Rathjen. Constructive set theory. forthcoming.

E. Bishop. *Foundations of Constructive Analysis*. McGraw-Hill, New York, 1967.

E. Bishop and D. S. Bridges. *Constructive Analysis*. Grundlehren der Math. Wiss. **279**, Springer, Heidelberg, 1985.

D. S. Bridges and R. A. Alps. Morse set theory as a foundation for constructive mathematics. preprint, University of Canterbury, New Zealand.

P. J. Cohen. The independence of the continuum hypothesis. *Proc. Nat. Acad. Sciences of the U.S.A.*, 51:1143–1148, 1963.

R. Diaconescu. Axiom of choice and complementation. *Proc. Amer. Math. Soc.*, 51:176–178, 1975.

H. Friedman. Set-theoretic foundations for constructive analysis. *Ann. Math.*, 105:1–28, 1977.

W. Gielen, H. de Swart, and W. Veldman. The continuum hypothesis in intuitionism. *J. Symb. Logic*, 46:121–136, 1981.

K. Gödel. *The Consistency of the Continuum Hypothesis*. Princeton. Univ. Press, Princeton, NJ, 1940.

N. Goodman and J. Myhill. Choice implies excluded middle. *Zeit. Math. Logik Grundlagen Math.*, 23:461, 1978.

P. Martin-Löf. An intuitionistic theory of types, predicative part. In H. E. Rose and J. C. Shepherdson, editors, *Logic colloquium '73, Proceedings of the logic colloquium, Bristol, July 1973*, Studies in logic and the foundations of mathematics, pages 73–118. North-Holland, 1975.

J. Myhill. Constructive set theory. *J. Symbolic Logic*, 40:347–382, 1975.

Fred Richman. Constructive mathematics without choice. In Peter Schuster, Ulrich Berger, and Horst Osswald, editors, *Reuniting the Antipodes: Constructive and Nonstandard Views of the Continuum*, volume 306 of *Synthese Library*, pages 207–210. Kluwer A. P., 2001.

A. S. Troelstra and D. van Dalen. *Constructivism in mathematics I*, volume 121 of *Studies in Logic and the Foundations of Mathematics*. North-Holland, 1988.

Ulrik Buchholtz, Gerhard Jäger, and Thomas Strahm

Theories of Proof-Theoretic Strength $\psi(\Gamma_{\Omega+1})$

Abstract: The purpose of this article is to present a range of theories with proof-theoretic ordinal $\psi(\Gamma_{\Omega+1})$. This ordinal parallels the ordinal of predicative analysis, Γ_0, and our theories are parallel to classical theories of strength Γ_0 such as $\widehat{\mathsf{ID}}_{<\omega}$, FP_0, ATR_0, Σ^1_1-DC_0 + (SUB), and Σ^1_1-AC_0 + (SUB). We also relate these theories to the unfolding of ID_1 which was already presented in the PhD thesis of the first author as a system of strength $\psi(\Gamma_{\Omega+1})$

Keywords: Subsystems of second order arithmetic and inductive definitions, Proof-theoretic ordinals, Unfolding

Mathematics Subject Classification 2010: 03F03, 03F05, 03F015, 03F35

1 Introduction

The ordinal $\psi(\Gamma_{\Omega+1})$ appeared first in [Bachmann, 1950], there it is denoted by $\varphi_{F_{\omega_2+1}(1)+1}(1)$.[1] This was the paper where Bachmann introduced the idea of using assigned fundamental sequences to ordinals of the third number class in order to define large countable ordinals, and this is what Howard [Howard, 1972] uses in his original ordinal analysis of ID_1. ID_1 is the theory of one generalized positive inductive definition, and its proof-theoretic ordinal is now known as the Bachmann-Howard ordinal.

Miller [Miller, 1976] proposed that $\psi(\Gamma_{\Omega+1})$ should be the proof-theoretic ordinal of a theory that relates to ID_1 as predicative analysis relates to first order arithmetic. Feferman's unfolding program [Feferman, 1996] provides a way to identity such a system because the unfolding of first order arithmetic is proof-theoretically equivalent to predicative analysis with proof-theoretic ordinal Γ_0 (cf. [Feferman and Strahm, 2000]).

For a history of the Bachmann method of describing constructive ordinals, and how it gave way to the more modern approach, we refer to [Crossley and Bridge Kister, 1986/87] and the preface of [Buchholz et al., 1981].

1 See section 2 for details on the notation system.

Ulrik Buchholtz: Department of Philosophy, Carnegie Mellon University, Pittsburgh, PA 15213-3890, USA, e-mail: ulrikb@andrew.cmu.edu
Gerhard Jäger and Thomas Strahm: Institut für Informatik, Universität Bern, Neubrückstrasse 10, 3012 Bern, Switzerland, e-mail: {jaeger,strahm}@iam.unibe.ch

Buchholtz [Buchholtz, 2014] recently proved that the unfolding of ID_1 has proof-theoretic ordinal $\psi(\Gamma_{\Omega+1})$, which indeed relates to $\psi(\varepsilon_{\Omega+1})$ (the ordinal of ID_1) as Γ_0 relates to ε_0 (the ordinal of first order Peano arithmetic).

In this paper we survey a range of further systems which also have proof-theoretic strength $\psi(\Gamma_{\Omega+1})$, for example Σ^1_1-$\mathsf{DC}^\bullet_0$ + $(\mathsf{SUB}^\bullet)$, $\mathsf{ATR}^\bullet_0$ + $(\mathsf{SUB}^\bullet)$, $\mathsf{FP}^\bullet_0$ and $\widehat{\mathsf{ID}}^\bullet_{<\omega}$.

Hancock [Hancock, 2000] separately conjectured that $\psi(\Gamma_{\Omega+1})$ is the ordinal of a certain kind of Martin-Löf type theory. This is made precise and verified in a companion article, which also identifies a system of explicit mathematics of strength $\psi(\Gamma_{\Omega+1})$.

The remainder of this paper is organized as follows: In the next section we set up some ordinal-theoretic preliminaries, including the definition of $\psi(\Gamma_{\Omega+1})$ and a review of derivation operators in the sense of Buchholz. Section 3 is centered around subsystems of second order arithmetic. Namely, we introduce systems Σ^1_1-$\mathsf{DC}^\bullet_0$ + $(\mathsf{SUB}^\bullet)$, $\mathsf{ATR}^\bullet_0$, and $\mathsf{FP}^\bullet_0$ resulting from their well-known relatives Σ^1_1-DC_0 + (SUB), ATR_0, and FP_0 by admitting least fixed points of arithmetical operators in the base language of second order arithmetic. In Section 4 we review the unfolding of ID_1; it is employed in Section 5 to establish the lower bound $\psi(\Gamma_{\Omega+1})$ of the above-mentioned systems via a formalized inductive model construction. Section 6 is devoted to the definition of finitely iterated fixed point theories $\widehat{\mathsf{ID}}^\bullet_n$ for $n < \omega$ and a reduction of $\mathsf{FP}^\bullet_0$ to the union of these theories, $\widehat{\mathsf{ID}}^\bullet_{<\omega}$. In Section 7 we sketch the main lines of the ordinal analysis of $\widehat{\mathsf{ID}}^\bullet_{<\omega}$, determining its proof-theoretic upper bound $\psi(\Gamma_{\Omega+1})$. The paper concludes with a final discussion on related systems of strength $\psi(\Gamma_{\Omega+1})$ in the setting of Martin-Löf type theory as well as Feferman's explicit mathematics.

2 Ordinal notations

In this section we try to give an account of the ordinal-theoretic environment and the ordinal-theoretic tools needed for putting the results of this article into perspective. We assume that the reader is familiar with the basic ordinal theory, the Veblen hierarchy of normal functions and collapsing functions à la Buchholz. A full exposition can be found in [Buchholz and Schütte, 1988], [Pohlers, 1989, 2009], and [Schütte, 1977].

Let *On* be the collection of all ordinals, Ω the least uncountable ordinal, and *AP* the collection of all *additive principal numbers*, meaning that $\alpha \in AP$ iff $(\forall \eta, \xi < \alpha)(\eta + \xi < \alpha)$. By $\alpha =_{NF} \alpha_1 + \cdots + \alpha_n$ we express that

$$\alpha = \alpha_1 + \cdots + \alpha_n \quad \text{and} \quad \alpha_1, \ldots, \alpha_n \in AP \quad \text{and} \quad \alpha_n \leq \cdots \leq \alpha_1 < \alpha.$$

Then the following result about the existence of the *Cantor normal form* is standard.

Lemma 2.1. *For every ordinal α with $0 < \alpha$ and $\alpha \notin AP$ there exist uniquely determined ordinals $\alpha_1, \ldots, \alpha_n$ such that $\alpha =_{NF} \alpha_1 + \cdots + \alpha_n$.*

The binary *Veblen function* φ is inductively defined by $\varphi 0\xi := \omega^\xi$ for all ordinals ξ and by choosing $\varphi\alpha$ to be the enumeration function of the closed and unbounded collection $\{\xi \in On : \forall\beta < \alpha(\varphi\beta\xi = \xi)\}$ if $\alpha > 0$. An ordinal α is called *strongly critical* iff $\alpha = \varphi\alpha 0$, and we let SC be the collection of all strongly critical ordinals. Now we set

$$\alpha =_{NF} \varphi\beta\gamma \quad :\Longleftrightarrow \quad \alpha = \varphi\beta\gamma \quad \text{and} \quad \beta, \gamma < \alpha$$

and obtain the following normal form property. For a proof see, for example, [Pohlers, 1989] or [Schütte, 1977].

Lemma 2.2. *For every ordinal $\alpha \in AP \setminus SC$ there exist uniquely determined ordinals β and γ such that $\alpha =_{NF} \varphi\beta\gamma$.*

It is common to write Γ_α for the α-th strongly critical ordinal, hence Γ is the normal function enumerating SC. Since $\Gamma_\Omega = \Omega$, it follows that $\Gamma_{\Omega+1}$ is the least strongly critical ordinal greater than Ω.

We now introduce for all ordinals α and β sets of ordinals $\mathbf{C}(\alpha, \beta)$ and ordinals $\psi\alpha$ following [Buchholz, 1992].

Definition 2.3. *The sets of ordinals $\mathbf{C}(\alpha, \beta)$ and the ordinals $\psi\alpha$ are defined for all ordinals α and β by induction on α.*
1. $\{0, \Omega\} \cup \beta \subseteq \mathbf{C}(\alpha, \beta)$.
2. *If $\eta, \xi \in \mathbf{C}(\alpha, \beta)$, then $\eta + \xi \in \mathbf{C}(\alpha, \beta)$ and $\varphi\eta\xi \in \mathbf{C}(\alpha, \beta)$.*
3. *If $\xi < \alpha$ and $\xi \in \mathbf{C}(\alpha, \beta)$, then $\psi\xi \in \mathbf{C}(\alpha, \beta)$.*
4. $\psi\alpha := \min\{\beta \in On : \mathbf{C}(\alpha, \beta) \cap \Omega \subseteq \beta\}$.

The sets $\mathbf{C}(\alpha, \beta)$ and the ordinals $\psi\alpha$ have a series of important properties whose proofs are not difficult and can be found in the references mentioned above.

Lemma 2.4. *For all ordinals $\alpha, \alpha_1, \alpha_2, \beta, \gamma_1, \ldots, \gamma_n$, we have:*
1. *If β is a limit ordinal, then $\mathbf{C}(\alpha, \beta) = \bigcup\{\mathbf{C}(\alpha, \xi) : \xi < \beta\}$.*
2. $\mathbf{C}(\alpha, \psi\alpha) \cap \Omega = \psi\alpha$.
3. $\psi\alpha \in SC$.
4. *If $\gamma =_{NF} \gamma_1 + \cdots + \gamma_n$ and $\gamma \in \mathbf{C}(\alpha, \beta)$, then $\gamma_1, \ldots, \gamma_n \in \mathbf{C}(\alpha, \beta)$.*
5. *If $\gamma =_{NF} \varphi\gamma_1\gamma_2$ and $\gamma \in \mathbf{C}(\alpha, \beta)$, then $\gamma_1, \gamma_2 \in \mathbf{C}(\alpha, \beta)$.*
6. *If $\alpha_1 < \alpha_2$ and $\alpha_1 \in \mathbf{C}(\alpha_2, \psi\alpha_2)$, then $\psi\alpha_1 < \psi\alpha_2$.*
7. *If $\alpha_1 \leq \alpha_2$, then $\psi\alpha_1 \leq \psi\alpha_2$ and $\mathbf{C}(\alpha_1, \psi\alpha_1) \subseteq \mathbf{C}(\alpha_2, \psi\alpha_2)$.*
8. $\mathbf{C}(\alpha, 0) = \mathbf{C}(\alpha, \psi\alpha)$.

So ψ is a weakly monotone function from On to the strongly critical ordinals less than or equal to $\psi(\Gamma_{\Omega+1})$. It also follows that $\psi(\Gamma_{\Omega+1})$ is the largest segment of ordinals in $\mathbf{C}(\Gamma_{\Omega+1}, \psi\Gamma_{\Omega+1})$, i.e., the least ordinal that cannot be generated by closing $\{0, \Omega\}$ under addition, ω-exponentiation, the binary Veblen function φ, and the function ψ.

The function ψ is weakly monotone but not strictly monotone: for example, if $\alpha = \min\{\xi \in On : \Gamma_\xi = \xi\}$, then $\psi\beta = \Gamma_\beta$ for all $\beta \leq \alpha$ and $\psi\gamma = \alpha$ for all γ such that $\alpha \leq \gamma \leq \Omega$. In order to obtain unique representations of the ordinals in $\mathbf{C}(\Gamma_{\Omega+1}, \psi\Gamma_{\Omega+1})$ we introduce a further normal form. Given ordinals α and β we define

$$\alpha =_{NF} \psi\beta \quad :\Longleftrightarrow \quad \alpha = \psi\beta \quad \text{and} \quad \beta \in \mathbf{C}(\beta, \psi\beta).$$

A detailed proof of the following normal form theorem for the function ψ can be found in [Pohlers, 1989].

Lemma 2.5. *For every strongly critical ordinal $\alpha \in \mathbf{C}(\Gamma_{\Omega+1}, \psi\Gamma_{\Omega+1})$ there exists a uniquely determined ordinal β such that $\alpha =_{NF} \psi\beta$.*

We end this section with some remarks about ordinal operators that will used to define *operator controlled derivations* in the sense of Buchholz Buchholz [1992].

Definition 2.6. *Let Pow(On) denote the collection of all sets of ordinals.*

1. *A class function*

$$\mathcal{H} : Pow(On) \to Pow(On)$$

   is called a derivation operator *iff it is monotone, inclusive plus idempotent and satisfies the following properties for all $X \in Pow(On)$ and all ordinals $\alpha, \alpha_1, \ldots, \alpha_n$:*
   (a) $\{0, \Omega\} \subseteq \mathcal{H}(X)$.
   (b) *If $\alpha =_{NF} \alpha_1 + \ldots \alpha_n$, then*

$$\alpha \in \mathcal{H}(X) \leftrightarrow \{\alpha_1, \ldots, \alpha_n\} \subseteq \mathcal{H}(X).$$

   (c) *If $\alpha =_{NF} \varphi\alpha_1\alpha_2$, then*

$$\alpha \in \mathcal{H}(X) \leftrightarrow \{\alpha_1, \alpha_2\} \subseteq \mathcal{H}(X).$$

2. *If $\mathcal{H}$ is a derivation operator, we define for all finite sets of ordinals $\mathfrak{m}$ and all ordinals σ operators*

$$\mathcal{H}[\mathfrak{m}],\ \mathcal{H}[\sigma],\ \mathcal{H}_\sigma : Pow(On) \to Pow(On)$$

by setting for all $X \in Pow(On)$:

$$\mathcal{H}[\mathfrak{m}](X) := \mathcal{H}(X \cup \mathfrak{m}),$$

$$\mathcal{H}[\sigma](X) := \mathcal{H}(X \cup \{\sigma\}),$$

$$\mathcal{H}_\sigma(X) := \bigcap\{\mathbf{C}(\alpha, \beta) : X \subseteq \mathbf{C}(\alpha, \beta) \text{ and } \sigma < \alpha\}.$$

Buchholz [Buchholz, 1992] provides a detailed analysis of derivation operators from which, in particular, we get all the properties summarized in the next lemma.

Lemma 2.7. *If $\mathcal{H}$ is a derivation operator, then we have for all finite sets of ordinals $\mathfrak{m}$, all ordinals σ, and all $X \in Pow(On)$:*

1. $\mathcal{H}[\mathfrak{m}]$, $\mathcal{H}[\sigma]$, *and* $\mathcal{H}_\sigma$ *are derivation operators.*
2. $\mathfrak{m} \subseteq \mathcal{H}(\emptyset) \implies \mathcal{H}[\mathfrak{m}] = \mathcal{H}$.
3. $\sigma \in \mathcal{H}(\emptyset) \implies \mathcal{H}[\sigma] = \mathcal{H}$.

3 Subsystems of second order arithmetic

Our language $\mathcal{L}_2$ of second order arithmetic contains number variables $a, b, c, u, v, w, x, y, z, \dots$ and set variables $U, V, W, X, Y, Z, \dots$ (both possibly with subscripts), function symbols for all primitive recursive functions, relation symbols for all primitive recursive relations, the relation symbol $\in$ for the element relation between natural numbers and sets of natural numbers as well as the standard logical connectives and auxiliary symbols. In addition, we have a distinguished anonymous unary relation symbol R that we use to define proof-theoretic ordinals and also plays a special role in the unfolding systems (see below). The *number terms* and *formulas* of $\mathcal{L}_2$ are defined as usual, and the *arithmetic* formulas of $\mathcal{L}_2$ are those without bounded set quantifiers; number and set parameters are permitted in arithmetic formulas.

Moreover, we frequently make use of the vector notation $\vec{\mathfrak{e}}$ as shorthand for a finite string $\mathfrak{e}_1, \dots, \mathfrak{e}_n$ of expressions whose length is not important or is evident from the context. Suppose now that $\vec{a}$ is the string of variables $a_1, \dots, a_n$ and $\vec{r}$ the string of number terms $r_1, \dots, r_n$. Then $A[\vec{r}/\vec{a}]$ is the formula that is obtained from A by simultaneously replacing all free occurrences of the variables $\vec{a}$ by the number terms $\vec{r}$; in order to avoid collision of variables, a renaming of bounded variables may be necessary. If the formula A is written as $B[\vec{a}]$, then we often simply write $B[\vec{r}]$ instead of $A[\vec{r}/\vec{a}]$; further variants of this notation below will be obvious.

The binary (infix) relation symbol = stands for the primitive recursive equality relation, the binary (infix) relation symbol < for the primitive recursive less-than relation, and t' for the successor of t. Very often we also write the same expression for a primitive recursive function (relation) as for the associated function (relation) symbol. Equality is only taken as basic symbol between numbers; equality between sets of numbers and functions is defined as

$$(U = V) \quad := \quad \forall a(a \in U \leftrightarrow a \in V).$$

In the following we make use of the standard primitive recursive coding machinery in $\mathcal{L}_2$: $\langle r_1, \ldots, r_n \rangle$ stands for the primitive recursively formed n-tuple of the number terms $r_1, \ldots, r_n$; *Seq* is the primitive recursive set of sequence numbers; $lh(r)$ denotes the length of (the sequence number coded by) r; if $i < lh(r)$, then $(r)_i$ is the i-th component of (the sequence coded by) r, i.e., $r = \langle (r)_0, \ldots, (r)_{lh(r) \dot{-} 1} \rangle$ provided that r is a sequence number.

The first order language $\mathcal{L}_1$ is the sub-language of $\mathcal{L}_2$ in which only formulas of $\mathcal{L}_2$ without set variables are permitted. Now we pick a fresh unary relation symbol P and write $\mathcal{L}_1(P)$ for the extension of $\mathcal{L}_1$ by P, i.e., expressions of the form $P(t)$ are permitted as atomic formulas of $\mathcal{L}_1(P)$. An $\mathcal{L}_1(P)$ formula is called *P-positive* if each occurrence of P in this formula is positive. We call P-positive formulas that contain at most u free *inductive operator forms* and let $\mathfrak{A}[P, u]$ range over such forms. If $\mathfrak{A}[P, u]$ does not contain the anonymous relation symbol R, it is called a *pure* inductive operator form.

Next we extend the language $\mathcal{L}_2$ to a new second order language $\mathcal{L}_2^\bullet$ by adding a fresh unary relation symbol $P_\mathfrak{A}$ for every pure inductive operator form $\mathfrak{A}[P, u]$; the number terms of $\mathcal{L}_2^\bullet$ are, of course, the number terms of $\mathcal{L}_2$. An $\mathcal{L}_2^\bullet$ formula is called *elementary* in case it does not contain bounded set variables. As syntactic variables we use $r, s, t, r_0, s_0, t_0, \ldots$ for number terms and $A, B, C, A_0, B_0, C_0, \ldots$ for formulas of $\mathcal{L}_2^\bullet$.

In the following we introduce a series of theories of second order arithmetic. The weakest of those, the theory ACA_0 is formulated in $\mathcal{L}_2$ and has the usual axioms and rules of inference of two-sorted logic with equality for the first sort, the axioms of primitive recursive arithmetic PRA for the primitive recursive functions and relations plus the axiom schema of *arithmetic comprehension*, i.e.,

$$\exists X \forall a(a \in X \leftrightarrow A[a])$$

for all arithmetic formulas $A[u]$ of $\mathcal{L}_2$, and the *induction axiom*

$$\forall X(0 \in X \land \forall a(a \in X \rightarrow a' \in X) \rightarrow \forall a(a \in Y)).$$

Well-known extensions of ACA_0 are obtained by adding axioms about comprehension and choice, for example

$$\forall a(\exists X A[X,a] \leftrightarrow \forall X B[X,a]) \rightarrow \exists Y \forall a(a \in Y \leftrightarrow \exists X A[X,a]), \qquad (\Delta^1_1\text{-CA})$$

$$\forall a \exists X C[a,X] \rightarrow \exists Y \forall a C[a,(Y)_a], \qquad (\Sigma^1_1\text{-AC})$$

$$\forall a \forall X \exists Y D[a,X,Y] \rightarrow \exists Z \forall a D[a,(Z)^a,(Z)_a], \qquad (\Sigma^1_1\text{-DC})$$

where $A[U,v]$, $B[U,v]$, $C[u,V]$, and $D[u,V,W]$ are arithmetic formulas of $\mathcal{L}_2$. In these formulations we are using the abbreviations

$$r \in (U)_s \quad := \quad \langle r,s \rangle \in U,$$

$$r \in (U)^s \quad := \quad r \in U \wedge r = \langle (r)_0,(r)_1 \rangle \wedge (r)_1 < s.$$

We write Δ^1_1-CA_0, Σ^1_1-AC_0, and Σ^1_1-DC_0 for the theories ACA_0 + (Δ^1_1-CA), ACA_0 + (Σ^1_1-AC), and ACA_0 + (Σ^1_1-DC), respectively, and recall that Δ^1_1-CA_0 and Σ^1_1-AC_0 are conservative extensions of Peano arithmetic PA, whereas Σ^1_1-DC_0 has proof-theoretic ordinal $\varphi\omega 0$. For details see [Barwise and Schlipf, 1975], [Buchholz et al., 1981], and [Cantini, 1986].

Before turning to the next principle, we introduce some notation: If A and $B[v]$ are $\mathcal{L}_2$ formulas, then $A_U[\{a : B[a]\}]$ indicates the result of substituting $B[r]$ for each occurrence of $(r \in U)$ in A. The *substitution rule* is the rule of inference

$$\frac{\forall X A}{A_X[\{a : B[a]\}]} \qquad (\mathsf{SUB})$$

for arithmetic $\mathcal{L}_2$ formulas $A[U]$ and arbitrary $\mathcal{L}_2$ formulas $B[v]$. Obviously, the *bar rule*

$$\frac{\forall X TI[\lhd, X]}{TI[\lhd, \{a : B[a]\}]}$$

for binary primitive recursive relations $\lhd$ is a special case of (SUB). Here $TI[\lhd, U]$ stands for the formula

$$\forall x(\forall y(y \lhd x \rightarrow y \in U) \rightarrow x \in U) \rightarrow \forall x(x \in U).$$

From [Feferman and Jäger, 1983] we know that Δ^1_1-CA_0+(SUB), Σ^1_1-AC_0+(SUB), and Σ^1_1-DC_0 + (SUB) are proof-theoretically equivalent and of proof-theoretic strength Γ_0.

In the later considerations two further theories in $\mathcal{L}_2$ will play an important role: The first is the theory ATR_0 – the fourth system of Friedman's program of reverse mathematics – that extends ACA_0 by the schema of *arithmetic transfinite recursion*; a standard reference is [Simpson, 2009]. The second is the fixed point theory FP_0, resulting from ACA_0 by adding the fixed point axioms

$$\exists X \forall a(a \in X \leftrightarrow A[X,a]) \qquad (\mathsf{FP})$$

for all U-positive arithmetic formulas $A[U, v]$. As shown in [Avigad, 1996], ATR_0 and FP_0 are equivalent.

Theorem 3.1. *An $\mathcal{L}_2$ formula is provable in ATR_0 if and only if it is provable in FP_0.*

After these preliminary remarks we now turn to the theories that interest us most in this article. They are all formulated in the language $\mathcal{L}_2^\bullet$ and comprise the following *least fixed point axioms*

$$\forall a(\mathfrak{A}[P_\mathfrak{A}, a] \to P_\mathfrak{A}(a)), \tag{ID.1}$$

$$\forall X(\forall a(\mathfrak{A}[X, a] \to a \in X) \to \forall a(P_\mathfrak{A}(a) \to a \in X)). \tag{ID.2}$$

for all inductive operator forms $\mathfrak{A}[P, u]$. Please observe that (ID.2) only claims minimality with respect to sets, not with respect to $\mathcal{L}_2^\bullet$ definable classes.

The theory $\mathsf{ACA}_0^\bullet$ is the $\mathcal{L}_2^\bullet$ system that contains the axioms of ACA_0 (formulated for $\mathcal{L}_2^\bullet$), all least fixed point axioms (ID.1) and (ID.2) plus the comprehension schema

$$\exists X \forall a(a \in X \leftrightarrow A[a]), \tag{E-CA}$$

for elementary $\mathcal{L}_2^\bullet$ formulas $A[u]$. As a consequence, any $P_\mathfrak{A}$ defines a set in $\mathsf{ACA}_0^\bullet$.

It is an easy exercise to show that $\mathsf{ACA}_0^\bullet$ is a conservative extension of the famous theory ID_1 of non-iterated positive inductive definitions; for details about ID_1 cf., for example, [Buchholz et al., 1981] or [Pohlers, 2009].

The schemas (Δ_1^1-CA$^\bullet$), (Σ_1^1-AC$^\bullet$), and (Σ_1^1-DC$^\bullet$) are the analogues of (Δ_1^1-CA), (Σ_1^1-AC), and (Σ_1^1-DC) with the arithmetic $\mathcal{L}_2$ formulas replaced by elementary $\mathcal{L}_2^\bullet$ formulas. Accordingly, the theories Δ_1^1-$\mathsf{CA}_0^\bullet$, Σ_1^1-$\mathsf{AC}_0^\bullet$, and Σ_1^1-$\mathsf{DC}_0^\bullet$ are defined to be the theories $\mathsf{ACA}_0^\bullet$ + (Δ_1^1-CA$^\bullet$), $\mathsf{ACA}_0^\bullet$ + (Σ_1^1-AC$^\bullet$), and $\mathsf{ACA}_0^\bullet$ + (Σ_1^1-DC$^\bullet$).

Of course, there is also an analogue of the substitution rule for the language $\mathcal{L}_2^\bullet$. Simply consider

$$\frac{\forall X A}{A_X[\{a : B[a]\}]} \tag{SUB$^\bullet$}$$

for all elementary $\mathcal{L}_2^\bullet$ formulas $A[U]$ and arbitrary $\mathcal{L}_2^\bullet$ formulas $B[v]$. Thus the bar rule for $\mathcal{L}_2^\bullet$ reads as

$$\frac{\forall X TI[\lhd, X]}{TI[\lhd, \{a : B[a]\}]}$$

for binary relations $\lhd$ that are primitive recursive in the least fixed points $P_\mathfrak{A}$ and arbitrary $\mathcal{L}_2^\bullet$ formulas $B[v]$, and is a special case of (SUB$^\bullet$).[2]

2 It is because of (SUB$^\bullet$) that we restrict ourselves to *pure* operator forms in $\mathcal{L}_2^\bullet$.

In the following we shall prove that Δ^1_1-$\mathsf{CA}^\bullet_0$ + (SUB$^\bullet$), Σ^1_1-$\mathsf{AC}^\bullet_0$ + (SUB$^\bullet$), and Σ^1_1-$\mathsf{DC}^\bullet_0$ + (SUB$^\bullet$) are theories with proof-theoretic ordinal $\psi(\Gamma_{\Omega+1})$. Two other interesting systems of the same strength are $\mathsf{ATR}^\bullet_0$ and $\mathsf{FP}^\bullet_0$, obtained from ATR_0 and FP_0, respectively, by relativizing them to the language $\mathcal{L}^\bullet_2$.

More precisely, let the schema of *elementary transfinite recursion* be as the schema of arithmetic transfinite recursion but with elementary $\mathcal{L}^\bullet_2$ formulas instead of arithmetic $\mathcal{L}_2$ formulas. Then $\mathsf{ATR}^\bullet_0$ is the extension of $\mathsf{ACA}^\bullet_0$ by elementary transfinite recursion. Similarly, the fixed point axioms of $\mathcal{L}_2$ are lifted to

$$\exists X \forall a(a \in X \leftrightarrow A[X, a]) \qquad (\mathsf{FP}^\bullet)$$

for arbitrary U-positive elementary formulas $\mathfrak{A}[U, v]$ of $\mathcal{L}^\bullet_2$, and $\mathsf{FP}^\bullet_0$ is the $\mathcal{L}^\bullet_2$ theory $\mathsf{ACA}^\bullet_0$ + (FP$^\bullet$).

There exists a close relationship between our theories formulated in $\mathcal{L}_2$ and their counterparts in $\mathcal{L}^\bullet_2$. Consider an inductive operator form $\mathfrak{A}[U, v]$ together with the axiom schema

$$\exists X\Big(\forall a(\mathfrak{A}[X, a] \to a \in X) \land \forall Y(\forall a(\mathfrak{A}[Y, a] \to a \in Y) \to X \subseteq Y)\Big) \qquad (\mathsf{LFP})$$

for all inductive operator forms $\mathfrak{A}[U, v]$. Added to ACA_0 it implies that every inductive operator form has a least fixed point, where "least" means least with respect to all sets that are fixed points.

Theorem 3.2. *Let T be one of the theories* ACA_0, Δ^1_1-CA_0, Σ^1_1-AC_0, Σ^1_1-DC_0, ATR_0, *or* FP_0, *which are all formulated in $\mathcal{L}_2$. Then we have:*

1. *$T^\bullet$, which is formulated in $\mathcal{L}^\bullet_2$, is a conservative extension of T +* (LFP) *with respect to all $\mathcal{L}_2$ sentences.*
2. $\mathsf{ATR}^\bullet_0$ *and* $\mathsf{FP}^\bullet_0$ *prove the same $\mathcal{L}^\bullet_2$ formulas.*

Proof: Since every relation constant $P_{\mathfrak{A}}$ for an inductive operator form $\mathfrak{A}[U, v]$ defines a set, all instances of (LFP) are provable in the theories $T^\bullet$. Therefore, T+(LFP) $\subseteq T^\bullet$. To establish conservativity, we simply fix for each inductive operator form $\mathfrak{A}[U, v]$ the uniquely determined least fixed point U, which exists according to (LFP), and interpret $P_{\mathfrak{A}}(t)$ as $(t \in U)$. These considerations together with Theorem 3.1 also yield the second assertion. □

We write $\mathfrak{M}[U]$ to express that U is (the range of the sets of) a *countable coded ω-model of* Σ^1_1-DC_0 in the sense of [Simpson, 2009]. Elementhood in such an U is then abbreviated by

$$V \mathrel{\dot\in} U \quad := \quad \exists a(V = (U)_a),$$

and $\vec{V} \mathrel{\dot\in} U$ means that all components of $\vec{V}$ belong to U. Moreover, given an arbitrary $\mathcal{L}_2$ formula A, its relativization A^U to U is obtained from A by replacing all quantifiers $\exists X(\ldots X \ldots)$ and $\forall X(\ldots X \ldots)$ by $\exists x(\ldots (U)_x \ldots)$ and $\forall x(\ldots (U)_x \ldots)$, respectively. Note that A^U is always arithmetic.

The theory ATR_0 has the following important property; cf. [Simpson, 2009] for a detailed proof and a discussion of the general context.

Theorem 3.3. *The theory* ATR_0 *proves that*

$$\forall X \exists Y (X \dot{\in} Y \wedge \mathfrak{M}[Y]).$$

For the subsequent considerations we let $\mathfrak{A}_0[U,v], \mathfrak{A}_1[U,v], \mathfrak{A}_2[U,v], \ldots$ be an arbitrary (but fixed) enumeration of all inductive operator forms and write

$$\mathfrak{F}_i[X] := \forall a(\mathfrak{A}_i[U,a] \rightarrow a \in U) \wedge \forall Y(\forall a(\mathfrak{A}_i[Y,a] \rightarrow a \in Y) \rightarrow X \subseteq Y),$$

expressing that set X is the least fixed point of $\mathfrak{A}_i[U,v]$. For any natural number n we write $For(n)$ for the collection of all $\mathcal{L}_2^\bullet$ formulas that do not contain relation symbols $P_{\mathfrak{A}_i}$ with $n < i$.

Lemma 3.4. *Let n be an arbitrary natural number. Under the assumptions*

1. $A[\vec{U}]$ *is an* $\mathcal{L}_2$ *formula whose free set variables are from the list* $\vec{U}$,
2. Σ^1_1-$\mathsf{DC}^\bullet_0 + (\mathsf{SUB}^\bullet)$ *proves* $A[\vec{U}]$ *by a proof* $\mathcal{P}$ *such that all formulas occurring in* $\mathcal{P}$ *belong to* $For(n)$,
3. *the set variables* $W_0, \ldots, W_n$ *do not occur in* $\mathcal{P}$, *and* $B[\vec{U}]$ *is the formula obtained from* $A[\vec{U}]$ *by substituting* $(t \in W_i)$ *for each subformula* $P_{\mathfrak{A}_i}(t)$, $i = 0, \ldots, n$,

the theory ATR_0 *proves*

$$\forall Z\Big(\Big(\bigwedge_{i=0}^{n}(\mathfrak{F}_i[W_i] \wedge W_i \dot{\in} Z) \wedge \vec{U} \dot{\in} Z \wedge \mathfrak{M}[Z]\Big) \rightarrow B^Z[\vec{U}]\Big).$$

Proof: We proceed by induction on $\mathcal{P}$. If $A[\vec{U}]$ is an axiom of Σ^1_1-$\mathsf{DC}^\bullet_0$, then our assertion is obvious since we relativize with respect to countable coded ω-models of Σ^1_1-DC_0; if $A[\vec{U}]$ is the conclusion of a rule of inference different from $(\mathsf{SUB}^\bullet)$, then our assertion follows directly from the induction hypothesis. So it only remains to discuss the case that $A[\vec{U}]$ is the conclusion of $(\mathsf{SUB}^\bullet)$. Then this inference has the form

$$\frac{\forall X C[\vec{U}, X]}{C[\vec{U}, \{x : D[\vec{U}, x]\}]},$$

where $A[\vec{U}]$ is the formula $C[\vec{U}, \{x : D[\vec{U}, x]\}]$, $C[\vec{U}, V]$ is an elementary $\mathcal{L}_2^\bullet$ formula, and $D[\vec{U}, v]$ an arbitrary $\mathcal{L}_2^\bullet$ formula. Let $E[\vec{U}, V]$and $F[\vec{U}, v]$ be the $\mathcal{L}_2$ formulas obtained from $C[\vec{U}, V]$ and $D[\vec{U}, v]$, respectively, by substituting $(t \in W_i)$ for each subformula $P_{\mathfrak{A}_i}(t)$, $i = 0, \ldots, n$. In view of the induction hypothesis ATR_0

proves

$$\forall Z\Big(\big(\bigwedge_{i=0}^{n}(\mathfrak{F}_i[W_i] \land W_i \mathrel{\dot\in} Z) \land \vec{U} \mathrel{\dot\in} Z \land \mathfrak{M}[Z]\big) \to (\forall X \mathrel{\dot\in} Z)E[\vec{U}, X]\Big), \qquad (*)$$

and we have to show in ATR_0 that

$$\forall Z\Big(\big(\bigwedge_{i=0}^{n}(\mathfrak{F}_i[W_i] \land W_i \mathrel{\dot\in} Z) \land \vec{U} \mathrel{\dot\in} Z \land \mathfrak{M}[Z]\big) \to E[\vec{U}, \{x : D^Z[\vec{U}, x]\}]\Big).$$

Working within ATR_0, pick a Z such that

$$\bigwedge_{i=0}^{n}(\mathfrak{F}_i[W_i] \ \land \ W_i \mathrel{\dot\in} Z) \ \land \ \vec{U} \mathrel{\dot\in} Z \ \land \ \mathfrak{M}[Z].$$

By Theorem 3.3 there exists an Y_0 for which

$$\bigwedge_{i=0}^{n}(\mathfrak{F}_i[W_i] \ \land \ W_i \mathrel{\dot\in} Y_0) \ \land \ \vec{U} \mathrel{\dot\in} Y_0 \ \land \ Z \mathrel{\dot\in} Y_0 \ \land \ \mathfrak{M}[Y_0].$$

Hence $\{x : D^Z[\vec{U}, x]\}$ is a set in Y_0, i.e., there exists an X_0 with

$$X_0 \mathrel{\dot\in} Y_0 \ \land \ \forall a(a \in X_0 \ \leftrightarrow \ D^Z[\vec{U}, a]).$$

Now we go back to (*), and obtain (by inserting Y_0 for Z and X_0 for X) that $E[\vec{U}, X_0]$, hence $E[\vec{U}, \{x : D^Z[\vec{U}, x]\}]$. This is what we had to show. □

Theorem 3.5. *Let $A[U]$ be an arithmetic formula of $\mathcal{L}_2$ with no set variables besides U. Then we have that*

$$\Sigma^1_1\text{-}\mathsf{DC}^\bullet_0 + (\mathsf{SUB}^\bullet) \vdash \forall X A[X] \quad \Longrightarrow \quad \mathsf{ATR}_0 + (\mathsf{LFP}) \vdash \forall X A[X].$$

Proof: From our assumption we obtain that there exists a natural number n and a proof of $A[\vec{U}]$ in Σ^1_1-$\mathsf{DC}^\bullet_0$ + $(\mathsf{SUB}^\bullet)$ such that all formulas in this proof belong to $For(n)$. We choose fresh set variables $W_0, \ldots, W_n$ not occurring in this proof and see by the previous lemma that ATR_0 proves

$$\forall Z\Big(\big(\bigwedge_{i=0}^{n}(\mathfrak{F}_i[W_i] \land W_i \mathrel{\dot\in} Z) \land U \mathrel{\dot\in} Z \land \mathfrak{M}[Z]\big) \to A[U]\Big),$$

hence also

$$\exists X_0 \ldots \exists X_n \exists Z\big(\bigwedge_{i=0}^{n}(\mathfrak{F}_i[X_i] \land X_i \mathrel{\dot\in} Z) \land U \mathrel{\dot\in} Z \land \mathfrak{M}[Z]\big) \to A[U].$$

In view of the schema (LFP) and Theorem 3.3, this means that $A[U]$ is provable in ATR_0 + (LFP). Consequently, ATR_0 + (LFP) proves $\forall X A[X]$. □

Corollary 3.6. *Let $A[U]$ be an arithmetic formula of $\mathcal{L}_2$ with no set variables besides U. Then we have that*

$$\Sigma^1_1\text{-DC}^\bullet_0 + (\text{SUB}^\bullet) \vdash \forall X A[X] \quad \Longrightarrow \quad \text{ATR}^\bullet_0 \vdash \forall X A[X].$$

Let us conclude this section with some remarks on the anonymous relation symbol R. Clearly, in the context of $\mathcal{L}_2$ and $\mathcal{L}^\bullet_2$ it plays the same role as any unspecified free set variable and would have been superfluous. More specifically: if T is one of the $\mathcal{L}_2$ or $\mathcal{L}^\bullet_2$ theories considered so far, then T proves A if and only if it proves $\forall X A_\mathsf{R}[X]$, where $A_\mathsf{R}[U]$ is obtained from A by replacing all occurrences of $\mathsf{R}(t)$ by $(t \in U)$.

In the next sections we shall turn to several first order theories, and then it is convenient to work within a syntax that provides for a place holder for arbitrary non-specified properties. A typical example is the first order definition of the proof-theoretic ordinal of a theory. For any primitive recursive relation $\lhd$ we set

$$TI[\lhd, \mathsf{R}] \quad := \quad \forall x(\forall y(y \lhd x \to \mathsf{R}(y)) \to \mathsf{R}(x)) \to \forall x \mathsf{R}(x).$$

Then if T is a theory formulated in a language containing the first order part of $\mathcal{L}_2$, the ordinal α is called *provable in T* if and only if there exists a primitive recursive well ordering $\lhd$ of order type α such that T proves $TI[\lhd, \mathsf{R}]$. The *proof-theoretic ordinal* $|T|$ of T is defined to be the least ordinal not provable in T.

Furthermore, it is more or less obvious that in the presence of $(\text{SUB}^\bullet)$ the following variant of the substitution rule is available. It will be needed for interpreting the substitution rule of the unfolding of ID_1, described in the next section.

Lemma 3.7. *Let A be an elementary $\mathcal{L}^\bullet_2$ formula and $B[u]$ an arbitrary $\mathcal{L}^\bullet_2$ formula. If $\Sigma^1_1\text{-DC}^\bullet_0 + (\text{SUB}^\bullet)$ proves A, then $\Sigma^1_1\text{-DC}^\bullet_0 + (\text{SUB}^\bullet)$ also proves $A_\mathsf{R}[\{a : B[a]\}]$, where $A_\mathsf{R}[\{a : B[a]\}]$ here indicates the result of substituting $B[t]$ for each occurrence of $\mathsf{R}(t)$ in A.*

4 Unfolding of ID_1

In this section we define the unfolding of ID_1 as in [Buchholtz, 2014]. This is an instance of Feferman's unfolding program [Feferman, 1996], for a general exposition, see also [Buchholtz, 2014]. This is defined in two steps; first the operational unfolding $\mathcal{U}_0(\mathsf{ID}_1)$ is introduced, and then the full unfolding $\mathcal{U}(\mathsf{ID}_1)$ is defined as an extension.

Let $\mathcal{L}^\bullet_1$ denote the fragment of $\mathcal{L}^\bullet_2$ without free or bound set variables (and thus also without the $\in$ relation). We use here a version of the unfolding in which the operational structure is given by a partial combinatory algebra. The language of $\mathfrak{U}_0(\mathsf{ID}_1)$ is the language $\mathcal{L}_1^\bullet$ extended with new constants k and s (combinators), p, p_0 and p_1 (pairing and projection), d (definition by cases), tt (true) and ff (false), e (equality), and the binary function symbol $\cdot$ (application). Terms are built in the usual way using variables and constants and closing under application and the function symbols of $\mathcal{L}_2^\bullet$. Further, we add a unary relation symbol N (natural numbers), and to account for partiality of application we also add a unary relation symbol $\downarrow$ (defined; expressing that a term has a value). When writing terms we drop the symbol for application and use the convention that application is left-associative to leave out parentheses. We often write $f(a_1, \ldots, a_n)$ for $f\, a_1 \ldots a_n$ and $\langle\!\langle a, b \rangle\!\rangle$ for $\mathsf{p}\, a\, b$.

For $\mathfrak{U}_0(\mathsf{ID}_1)$ we use Beeson's Logic of Partial Terms with strictness and equality, see [Beeson, 1985]. The non-logical axioms of $\mathfrak{U}_0(\mathsf{ID}_1)$ are:

1. The usual axioms of arithmetic, relativized to N, with the schematic form of complete induction on the natural numbers,

$$\mathsf{R}(0) \wedge \forall x(\mathsf{N}(x) \wedge \mathsf{R}(x) \rightarrow \mathsf{R}(x')) \rightarrow \forall x(\mathsf{N}(x) \rightarrow \mathsf{R}(x)).$$

2. The least fixed point axioms for each $P_{\mathfrak{A}}$ in schematic form relativized to N,[3]

$$\forall a(\mathsf{N}(a) \wedge \mathfrak{A}^{\mathsf{N}}[P_{\mathfrak{A}}, a] \rightarrow P_{\mathfrak{A}}(a)),$$
$$(\forall a(\mathsf{N}(a) \wedge \mathfrak{A}^{\mathsf{N}}[\mathsf{R}, a] \rightarrow \mathsf{R}(a)) \rightarrow \forall a(\mathsf{N}(a) \wedge P_{\mathfrak{A}}(a) \rightarrow \mathsf{R}(a))).$$

3. Partial combinatory algebra (PCA) axioms with pairing and definition by cases:
   (a) $\mathsf{k}\, a\, b = a$.
   (b) $\mathsf{s}\, a\, b\downarrow \wedge\, \mathsf{s}\, a\, b\, c \simeq a\, c\, (b\, c)$.
   (c) $\mathsf{p}_0 \langle\!\langle a, b \rangle\!\rangle = a \wedge \mathsf{p}_1 \langle\!\langle a, b \rangle\!\rangle = b$.
   (d) $\mathsf{d}\, a\, b\, \mathsf{tt} = a \wedge \mathsf{d}\, a\, b\, \mathsf{ff} = b$.
4. Decidable equality on natural numbers:
   (a) $\forall x, y(\mathsf{N}(x) \wedge \mathsf{N}(y) \rightarrow \mathsf{e}\, x\, y = \mathsf{tt} \vee \mathsf{e}\, x\, y = \mathsf{ff})$.
   (b) $\forall x, y(\mathsf{N}(x) \wedge \mathsf{N}(y) \rightarrow (\mathsf{e}\, x\, y = \mathsf{tt} \leftrightarrow x = y))$.

In addition, $\mathfrak{U}_0(\mathsf{ID}_1)$ includes the unrestricted substitution rule,

$$\frac{A}{A_{\mathsf{R}}[\{a : B[a]\}]}, \qquad \text{(SUB)}$$

where, because of partiality, $A_{\mathsf{R}}[\{a : B[a]\}]$ indicates the result of substituting $(r\downarrow \wedge B[r])$ for each occurrence of $\mathsf{R}(r)$ in A.

3 As usual, $\mathfrak{A}^{\mathsf{N}}$ denotes the formula $\mathfrak{A}$ with all quantifiers relativized to N.

Abstraction terms $\lambda x.t$ can be defined as usual, and from the PCA axioms we can show in $\mathfrak{U}_0(\mathsf{ID}_1)$:

1. $(\lambda x.t)\downarrow \wedge (\lambda x.t)\, x \simeq t$
2. $s\downarrow \rightarrow (\lambda x.t)\, s \simeq t[s/x]$

Here, $t \simeq s$ is an abbreviation for $t\downarrow \vee s\downarrow \rightarrow t = s$. Note that we use the notation of the λ-calculus even though the conversion relation is not exactly the same (in particular, it does not validate the (ξ)-rule of the λ-calculus).

The PCA axioms allow us to introduce a fixed point operator, but we cannot prove that it produces *least* fixed points.

Theorem 4.1 (Fixed point). *There is a closed term* fix *of* $\mathfrak{U}_0(\mathsf{ID}_1)$ *such that*

$$\mathfrak{U}_0(\mathsf{ID}_1) \vdash \mathsf{fix} f\downarrow \wedge \mathsf{fix} f\, x \simeq f\, (\mathsf{fix} f)\, x.$$

The language of the full unfolding $\mathfrak{U}(\mathsf{ID}_1)$ extends the language by additional constants to reflect the predicates of $\mathfrak{U}_0(\mathsf{ID}_1)$: nat (natural number), $\mathsf{i}_{\mathfrak{A}}$ (inductive set), eq (equality), pr_R (anonymous relation symbol R), inv (inverse image), conj (conjunction), neg (negation), un (universal quantification over the natural numbers), join (join, that is, disjoint union). In addition, we add the unary relation symbol Π (predicates) and the binary relation symbol $\in$ (predication). The axioms of $\mathfrak{U}(\mathsf{ID}_1)$ extend the ones of $\mathfrak{U}_0(\mathsf{ID}_1)$ by

5. Basic axioms about predicates:
   (a) $\Pi(\mathsf{nat}) \wedge \forall x(x \in \mathsf{nat} \leftrightarrow \mathsf{N}(x))$.
   (b) $\Pi(\mathsf{i}_{\mathfrak{A}}) \wedge \forall x(x \in \mathsf{i}_{\mathfrak{A}} \leftrightarrow P_{\mathfrak{A}}(x))$.
   (c) $\Pi(\mathsf{eq}) \wedge \forall x(x \in \mathsf{eq} \leftrightarrow \exists y(x = \langle\!\langle y, y\rangle\!\rangle))$.
   (d) $\Pi(\mathsf{pr}_\mathsf{R}) \wedge \forall x(x \in \mathsf{pr}_\mathsf{R} \leftrightarrow \mathsf{R}(x))$.
   (e) $\Pi(a) \rightarrow \Pi(\mathsf{inv}(a, f)) \wedge \forall x(x \in \mathsf{inv}(a, f) \leftrightarrow f\, x \in a)$.
   (f) $\Pi(a) \wedge \Pi(b) \rightarrow \Pi(\mathsf{conj}(a, b)) \wedge \forall x(x \in \mathsf{conj}(a, b) \leftrightarrow x \in a \wedge x \in b)$.
   (g) $\Pi(a) \rightarrow \Pi(\mathsf{neg}\, a) \wedge \forall x(x \in \mathsf{neg}(a) \leftrightarrow \neg(x \in a))$.
   (h) $\Pi(a) \rightarrow \Pi(\mathsf{un}\, a) \wedge \forall x(x \in \mathsf{un}(a) \leftrightarrow \forall y(\mathsf{N}(y) \rightarrow \langle\!\langle x, y\rangle\!\rangle \in a))$.
6. The dependent join axiom:

$$\Pi(a) \wedge (\forall y \in a)\Pi(f\, y) \rightarrow \Pi(\mathsf{join}(f, a)) \\ \wedge \forall x(x \in \mathsf{join}(f, a) \leftrightarrow \exists y, z(x = \langle\!\langle y, z\rangle\!\rangle \wedge y \in a \wedge z \in f(y))).$$

Finally, $\mathfrak{U}(\mathsf{ID}_1)$ contains the restricted substitution rule

$$\frac{A}{A_\mathsf{R}[\{a : B[a]\}]}, \qquad \text{(SUB)}$$

where A is any formula in the language of $\mathfrak{U}_0(\mathsf{ID}_1)$ and B is any formula in the language of $\mathfrak{U}(\mathsf{ID}_1)$ (with the same convention as for the substitution rule for $\mathfrak{U}_0(\mathsf{ID}_1)$).

5 Lower proof-theoretic bound

We define an interpretation of $\mathcal{U}(\mathsf{ID}_1)$ into Σ^1_1-$\mathsf{AC}^\bullet_0$ + $(\mathsf{SUB}^\bullet)$ in which we interpret the operational constants using indices of partial recursive functions. The predicates are then interpreted via a fixed point of an elementary positive operator form $\mathcal{A}[Q^+, x, y, z]$ where Q is a new ternary relation symbol. For the interpretation we need only consider a particular proof in $\mathcal{U}(\mathsf{ID}_1)$, and since such a proof refers only to finitely many of the least fixed points $P_{\mathfrak{A}}$ we can for simplicity (and because there is a universal such case), assume that $\mathcal{A}$ refers to a single inductive predicate $P_{\mathfrak{A}}$ (which we fix throughout this section).

The fixed point $Q_{\mathcal{A}}[x, y, z]$ of $\mathcal{A}[Q^+, x, y, z]$ is obtained in Σ^1_1-$\mathsf{AC}^\bullet_0$ using, as usual, Aczel's trick (cf. Feferman Feferman [1982]). In particular, we can consider the class $\mathfrak{C}$ of formulas $\exists X A[X, \vec{x}]$, where A is an elementary formula with the same restrictions as for the operator $\mathcal{A}$. There is a quinary $\mathfrak{C}$ formula $E[z, x_1, x_2, x_3, x_4]$ that enumerates the quaternary $\mathfrak{C}$ formulas. Using diagonalization we can then obtain a ternary $\mathfrak{C}$ formula $P_{\mathcal{A}}$ that is our desired fixed point of the operator $\mathcal{A}$. We record this as a lemma.

Lemma 5.1. *There is a $\mathfrak{C}$ formula $Q_{\mathcal{A}}[x, y, z]$ such that Σ^1_1-$\mathsf{AC}^\bullet_0$ proves*

$$\forall x, y, z(\mathcal{A}_Q[Q_{\mathcal{A}}, x, y, z] \leftrightarrow Q_{\mathcal{A}}[x, y, z])$$

where $\mathcal{A}_Q[Q_{\mathcal{A}}, x, y, z]$ denotes the formula obtained from $\mathcal{A}[Q^+, x, y, z]$ by replacing each occurrence of $Q(s, t, u)$ with $Q_{\mathcal{A}}[s, t, u]$.

The operational unfolding $\mathcal{U}_0(\mathsf{ID}_1)$ is interpreted in the usual way using its model in the partial recursive functions. See for example [Feferman and Strahm, 2000]. In particular, $a\ b$ is interpreted as $\{a\}(b)$ in the sense of ordinary recursion theory.

In order to interpret predicates we need codes of the following forms:

Table 1. Predicates and their codes

Predicate	Code
nat	$\langle 0, 0\rangle$
eq	$\langle 1, 0\rangle$
$\mathsf{i}_{\mathfrak{A}}$	$\langle 2, 0\rangle$
pr_R	$\langle 3, 0\rangle$
neg(a)	$\langle 4, a\rangle$
un(a)	$\langle 5, a\rangle$
conj(a, b)	$\langle 6, a, b\rangle$
inv(a, f)	$\langle 7, a, f\rangle$
join(f, a)	$\langle 8, f, a\rangle$

Thus, for example, the constant neg is interpreted as the index $\widehat{\mathsf{neg}}$ for a partial recursive function such that $\{\widehat{\mathsf{neg}}\}(a) = \langle 4, a\rangle$. We use the ternary fixed point $Q_{\mathcal{A}}[x, y, z]$ with following informal interpretation:

$Q_{\mathcal{A}}[x, 0, 0]$	x is a predicate
$Q_{\mathcal{A}}[x, y, 1]$	$y \in x$
$Q_{\mathcal{A}}[x, y, 2]$	$y \notin x$

The operator form $\mathcal{A}[Q, a, b, c]$ is now defined to be the disjunction of the following 26 clauses:

1. $a = \langle 0, 0\rangle \wedge b = 0 \wedge c = 0$,
2. $a = \langle 0, 0\rangle \wedge c = 1$,
3. $a = \langle 1, 0\rangle \wedge b = 0 \wedge c = 0$,
4. $a = \langle 1, 0\rangle \wedge \exists x(b = \langle x, x\rangle) \wedge c = 1$,
5. $a = \langle 1, 0\rangle \wedge \forall x(b \neq \langle x, x\rangle) \wedge c = 2$,
6. $a = \langle 2, 0\rangle \wedge b = 0 \wedge c = 0$,
7. $a = \langle 2, 0\rangle \wedge P_{\mathfrak{A}}(b) \wedge c = 1$,
8. $a = \langle 2, 0\rangle \wedge \neg P_{\mathfrak{A}}(b) \wedge c = 2$,
9. $a = \langle 3, 0\rangle \wedge b = 0 \wedge c = 0$,
10. $a = \langle 3, 0\rangle \wedge \mathsf{R}(b) \wedge c = 1$,
11. $a = \langle 3, 0\rangle \wedge \neg\mathsf{R}(b) \wedge c = 2$,
12. $\exists x(a = \langle 4, x\rangle \wedge Q(x, 0, 0)) \wedge b = 0 \wedge c = 0$,
13. $\exists x(a = \langle 4, x\rangle \wedge Q(x, 0, 0) \wedge Q(x, b, 2)) \wedge c = 1$,
14. $\exists x(a = \langle 4, x\rangle \wedge Q(x, 0, 0) \wedge Q(x, b, 1)) \wedge c = 2$,
15. $\exists x(a = \langle 5, x\rangle \wedge Q(x, 0, 0)) \wedge b = 0 \wedge c = 0$,
16. $\exists x(a = \langle 5, x\rangle \wedge Q(x, 0, 0) \wedge \forall y Q(x, \langle b, y\rangle, 1) \wedge c = 1$,
17. $\exists x(a = \langle 5, x\rangle \wedge Q(x, 0, 0) \wedge \exists y Q(x, \langle b, y\rangle, 2) \wedge c = 2$,
18. $\exists x, y(a = \langle 6, x, y\rangle \wedge Q(x, 0, 0) \wedge Q(y, 0, 0)) \wedge b = 0 \wedge c = 0$,
19. $\exists x, y(a = \langle 6, x, y\rangle \wedge Q(x, 0, 0) \wedge Q(y, 0, 0) \wedge Q(x, b, 1) \wedge Q(y, b, 1)) \wedge c = 1$.
20. $\exists x, y(a = \langle 6, x, y\rangle \wedge Q(x, 0, 0) \wedge Q(y, 0, 0) \wedge (Q(x, b, 2) \vee Q(y, b, 2))) \wedge c = 2$.
21. $\exists x, f(a = \langle 7, x, f\rangle \wedge Q(x, 0, 0)) \wedge b = 0 \wedge c = 0$,
22. $\exists x, f(a = \langle 7, x, f\rangle \wedge Q(x, 0, 0) \wedge Q(x, \{f\}(b), 1)) \wedge c = 1$,
23. $\exists x, f(a = \langle 7, x, f\rangle \wedge Q(x, 0, 0) \wedge (\{f\}(b)\uparrow \vee Q(x, \{f\}(b), 2)) \wedge c = 2$,
24. $\exists f, x(a = \langle 8, f, x\rangle \wedge Q(x, 0, 0) \wedge \forall y(Q(x, y, 2) \vee Q(\{f\}(y), 0, 0)))$
$\wedge\, b = 0 \wedge c = 0$,
25. $\exists f, x(a = \langle 8, f, x\rangle \wedge Q(x, 0, 0) \wedge \forall y(Q(x, y, 2) \vee Q(\{f\}(y), 0, 0))$
$\wedge\, \exists u, v(b = \langle u, v\rangle \wedge Q(x, u, 1) \wedge Q(\{f\}(u), v, 1)) \wedge c = 1$,
26. $\exists f, x(a = \langle 8, f, x\rangle \wedge Q(x, 0, 0) \wedge \forall y(Q(x, y, 2) \vee Q(\{f\}(y), 0, 0))$
$\wedge\, \forall u, v(b \neq \langle u, v\rangle \vee Q(x, u, 2) \vee Q(\{f\}(u), v, 2)) \wedge c = 2$.

We can now define

$$\Pi(x) := Q_{\mathcal{A}}[x, 0, 0] \wedge \forall y(Q_{\mathcal{A}}[x, y, 2] \leftrightarrow \neg Q_{\mathcal{A}}[x, y, 1]),$$
$$y \in x := \Pi(x) \wedge Q_{\mathcal{A}}[x, y, 1].$$

A similar trick was used by Feferman in [Feferman, 1982] in order to model universes in type theory and explicit mathematics. It is now a matter of routine to verify that this defines an interpretation * of $\mathfrak{U}(\mathsf{ID}_1)$ into $\Sigma^1_1\text{-}\mathsf{AC}^\bullet_0 + (\mathsf{SUB}^\bullet)$. Note that according to this interpretation, the premise of the substitution rule translates into an elementary formula of $\mathcal{L}^\bullet_2$.

Theorem 5.2. *The system $\mathfrak{U}(\mathsf{ID}_1)$ is contained in the system $\Sigma^1_1\text{-}\mathsf{AC}^\bullet_0 + (\mathsf{SUB}^\bullet)$ via the translation* *.

Hence, using [Buchholtz, 2014], we get following:

Corollary 5.3. $\psi(\Gamma_{\Omega+1}) \leq |\mathfrak{U}(\mathsf{ID}_1)| \leq |\Sigma^1_1\text{-}\mathsf{AC}^\bullet_0 + (\mathsf{SUB}^\bullet)|$.

In fact, the lower bound proof in [Buchholtz, 2014] can also be carried through in $\Delta^1_1\text{-}\mathsf{CA}^\bullet_0 + (\mathsf{SUB}^\bullet)$. As usual, jump hierarchies of elementary operators can be built using Δ^1_1 comprehension, see for example [Schütte, 1977].

Theorem 5.4. $\psi(\Gamma_{\Omega+1}) \leq |\Delta^1_1\text{-}\mathsf{CA}^\bullet_0 + (\mathsf{SUB}^\bullet)|$.

6 Finitely iterated fixed point theories

The aim of this section is to introduce first order theories $\widehat{\mathsf{ID}}^\bullet_n$, for all natural numbers $n \geq 1$, and to reduce $\mathsf{FP}^\bullet_0$ to the union of those. In the next section we shall then carry through the ordinal analysis of the theories $\widehat{\mathsf{ID}}^\bullet_n$ and thus determine the upper proof-theoretic bound of $\mathsf{FP}^\bullet_0$ and the systems equivalent to $\mathsf{FP}^\bullet_0$.

The theories $\widehat{\mathsf{ID}}^\bullet_n$ are the analogues of the well-known fixed point theories $\widehat{\mathsf{ID}}_n$, see [Feferman, 1982] or [Jäger et al., 1999], but with ID_1 rather than PA as the base theory. The languages $\mathcal{L}^\bullet(n)$ are defined by induction on n as follows:

1. $\mathcal{L}^\bullet(0)$ is the first order part $\mathcal{L}^\bullet_1$ of the language $\mathcal{L}^\bullet_2$.
2. Given $\mathcal{L}^\bullet(n-1)$, we first determine the collection $\mathfrak{C}(n-1)$ of all (not necessarily pure) inductive operator forms $\mathfrak{A}[P, u]$ formulated in $\mathcal{L}^\bullet(n-1)$, then select a fresh unary relation symbol $P^{(n)}_{\mathfrak{A}}$ for each $\mathfrak{A}[P, u]$ from $\mathfrak{C}(n-1)$, and let $\mathcal{L}^\bullet(n)$ be the extension of $\mathcal{L}^\bullet(n-1)$ by these new relation symbols, i.e.,

$$\mathcal{L}^\bullet(n) := \mathcal{L}^\bullet(n-1) \cup \{P^{(n)}_{\mathfrak{A}} : \mathfrak{A}[P, u] \in \mathfrak{C}(n-1)\}.$$

For any natural number $n \geq 1$, the the theory $\widehat{\mathsf{ID}}^\bullet_n$ is formulated in the language $\mathcal{L}^\bullet(n)$, its logic is the usual first order predicate logic with equality. The non-logical axioms of $\widehat{\mathsf{ID}}^\bullet_n$ are:

(A1) All axioms of primitive recursive arithmetic PRA plus the schema of complete induction on the natural numbers for all formulas of $\mathcal{L}^\bullet(n)$.

(A2) The *least fixed point axioms*

$$\forall a(\mathfrak{A}[P_{\mathfrak{A}}, a] \to P_{\mathfrak{A}}(a)), \tag{1}$$

$$\forall a(\mathfrak{A}[\{x : B[x]\}, a] \to B[a]) \to \forall a(P_{\mathfrak{A}}(a) \to B[a]) \tag{2}$$

for all inductive operator forms $\mathfrak{A}[P, u]$ of $\mathcal{L}_1$ and all formulas $B[u]$ of $\mathcal{L}^\bullet(n)$.

(A3) The *fixed point axioms*

$$\forall a(\mathfrak{A}[P^{(m)}_{\mathfrak{A}}, a] \leftrightarrow P^{(m)}_{\mathfrak{A}}(a))$$

for all natural numbers m with $1 \leq m \leq n$ and all inductive operator forms $\mathfrak{A}[P, u]$ from $\mathfrak{C}(m-1)$.

$\mathcal{L}^\bullet(< \omega)$ is the union of the languages $\mathcal{L}^\bullet(n)$ and $\widehat{\mathsf{ID}}^\bullet_{<\omega}$ is the union of the theories $\widehat{\mathsf{ID}}^\bullet_n$,

$$\widehat{\mathsf{ID}}^\bullet_{<\omega} := \bigcup \{\widehat{\mathsf{ID}}^\bullet_n : 1 \leq n < \omega\}.$$

The following theorem is the analogue of Avigad's reduction of FP_0 to the theory $\widehat{\mathsf{ID}}_{<\omega}$ of finitely iterated fixed points above PA; for details see Avigad [1996].

Theorem 6.1. *$\mathsf{FP}^\bullet_0$ is a conservative extension of $\widehat{\mathsf{ID}}^\bullet_{<\omega}$ with respect to all formulas of $\mathcal{L}^\bullet(0)$. In other words, if $\mathsf{FP}^\bullet_0$ proves the $\mathcal{L}^\bullet(0)$ formula A, then $\widehat{\mathsf{ID}}^\bullet_{<\omega}$ proves A as well.*

Proof: Taking up the strategy of Avigad [1996], we can establish this theorem by a model-theoretic argument. All we have to show is that any (first order) model of $\widehat{\mathsf{ID}}^\bullet_{<\omega}$ can be extended to a (second order) model of $\mathsf{FP}^\bullet_0$ that validates the same formulas of $\mathcal{L}^\bullet(0)$.

So let $\mathbb{M}$ be a model of $\widehat{\mathsf{ID}}^\bullet_{<\omega}$, write $|\mathbb{M}|$ for the universe of $\mathbb{M}$, and denote the $\mathbb{M}$-interpretations of the relation symbols $P_{\mathfrak{A}}$ for least fixed points and $P^{(n)}_{\mathfrak{A}}$ for arbitrary fixed points by $\mathbb{M}(P_{\mathfrak{A}})$ and $\mathbb{M}(P^{(n)}_{\mathfrak{A}})$, respectively. Then we define $\mathbb{S}_{\mathbb{M}}$ to be the collection of all these sets $\mathbb{M}(P_{\mathfrak{A}})$ and $\mathbb{M}(P^{(n)}_{\mathfrak{A}})$ plus their projections. Finally, $(\mathbb{M}, \mathbb{S}_{\mathbb{M}})$ is the second order extension of $\mathbb{M}$ where $\mathbb{S}_{\mathbb{M}}$ takes care of the second order part. Clearly, we have

$$\mathbb{M} \models A \leftrightarrow (\mathbb{M}, \mathbb{S}_{\mathbb{M}}) \models A$$

for all formulas A of $\mathcal{L}^\bullet(0)$. It remains to show that $(\mathbb{M}, \mathbb{S}_{\mathbb{M}})$ is a model of $\mathsf{FP}^\bullet_0$. For dealing with the fixed point axioms, consider an U-positive elementary formula

$A[U, V_1, \ldots, V_m, x, y_1, \ldots, y_n]$ with at most the indicated free set and number variables. To simplify notation we assume $m = n = 1$. We have to show that

$$(\mathbb{M}, \mathbb{S}_{\mathbb{M}}) \models \forall X \forall a \exists Y \forall b (b \in Y \leftrightarrow A[Y, X, b, a]). \tag{*}$$

To do so, choose an element $p \in |\mathbb{M}|$ and a set $M \in \mathbb{S}_{\mathbb{M}}$, given, for example, as

$$M = \{i \in |\mathbb{M}| : \langle i, q \rangle \in \mathbb{M}(P_{\mathfrak{B}}^{(k)})\}$$

for some fixed point relation symbol $P_{\mathfrak{B}}^{(k)}$ and some $q \in |\mathbb{M}|$. Now we define the formula $\mathfrak{C}[P, u]$ to be

$$u = \langle (u)_0, (u)_1, (u)_2 \rangle \ \wedge$$
$$A[\{x : P(\langle x, (u)_1, (u)_2 \rangle)\}, \{x : P_{\mathfrak{B}}^{(k)}(\langle x, (u)_2 \rangle)\}, (u)_0, (u)_1].$$

and observe that $\mathfrak{C}[P, u]$ is an inductive operator form with respect to the language $\mathcal{L}^{\bullet}(k)$. For the set

$$N := \{i \in |\mathbb{M}| : \langle i, p, q \rangle \in \mathbb{M}(P_{\mathfrak{C}}^{(k+1)})\}$$

and all $i \in |\mathbb{M}|$ we thus have

$$\begin{aligned} i \in N &\leftrightarrow \langle i, p, q \rangle \in \mathbb{M}(P_{\mathfrak{C}}^{(k+1)}), \\ &\leftrightarrow (\mathbb{M}, \mathbb{S}_{\mathbb{M}}) \models \mathfrak{C}[P_{\mathfrak{C}}^{(k+1)}, \langle i, p, q \rangle], \\ &\leftrightarrow (\mathbb{M}, \mathbb{S}_{\mathbb{M}}) \models A[N, M, i, p]. \end{aligned}$$

Hence N is the required fixed point, and (*) has been validated. All other cases are straightforward or treated similarly. □

By methods similar to those in [Avigad, 1996], the previous theorem can also be proved in a purely syntactic and proof-theoretic manner. We also conjecture that the speed-up result of [Avigad, 1996] carries over to $\mathsf{FP}_0^{\bullet}$ and $\widehat{\mathsf{ID}}_{<\omega}^{\bullet}$.

7 Upper proof-theoretic bound

To establish the upper proof-theoretic bounds on the theories $\widehat{\mathsf{ID}}_n^{\bullet}$ we shall combine methods of predicative and impredicative cut-elimination. To this end we first extend the languages $\mathcal{L}^{\bullet}(n)$ to languages $\mathcal{L}^{\bullet\infty}(n)$ by adding for inductive operator form $\mathfrak{A}[P, u]$ and each ordinal $\alpha < \Omega$ a new unary relation symbol $P_{\mathfrak{A}}^{<\alpha}$. These relation symbols are used to represent the stages of the least fixed points $P_{\mathfrak{A}}$. Then we restrict ourselves to the fragment of closed $\mathcal{L}^{\bullet\infty}(n)$ formulas A in negation-normal form and define $\neg A$ by de Morgan's rules and the law of double negation.

We now turn to infinite calculi, and in order to measure and control the complexities of infinite derivations we need control over the ranks and ordinal parameters of formulas occurring in infinite derivations.

Definition 7.1 (Rank and parameter set).

1. *The* rank, $rk(A)$, *of a closed* $\mathcal{L}^{\infty}(n)$ *formula* A *in negation-normal form is defined inductively as follows:*
   (a) $rk(A) := rk(\neg A) := 0$ *for closed atomic* $\mathcal{L}_1$ *formulas* A,
   (b) $rk(P_{\mathfrak{A}}^{<\alpha}(t)) := rk(\neg P_{\mathfrak{A}}^{<\alpha}(t)) := \omega\alpha$ *for* $\alpha < \Omega$,
   (c) $rk(P_{\mathfrak{A}}(t)) := rk(\neg P_{\mathfrak{A}}(t)) := \Omega$,
   (d) $rk(P_{\mathfrak{A}}^{(m)}(t)) := rk(\neg P_{\mathfrak{A}}^{(m)}(t)) := \Omega$,
   (e) $rk(A \wedge B) := rk(A \vee B) := \max\{rk(A), rk(B)\} + 1$,
   (f) $rk(\exists x A[x]) := rk(\forall x A[x]) := rk(A[0]) + 1$,
2. *The* parameter set, $|A|$, *of a closed* $\mathcal{L}^{\infty}(n)$ *formula* A *in negation-normal form is defined to be the set of the ordinals* α *occurring in subformulas* $P_{\mathfrak{A}}^{<\alpha}(t)$ *in* A.

Note that the definition of rank ensures that $rk(\mathfrak{A}[P^{<\beta}, s]) < rk(P_{\mathfrak{A}}^{<\alpha}(s))$ for $\beta < \alpha$.

Given any natural number n, we now introduce an infinitary system $\widehat{\mathsf{ID}}_n^{\infty}$ in Tait-style, and use the capital Greek letters Γ, Θ, Λ, possibly with subscripts, for finite sets of closed $\mathcal{L}^{\infty}(n)$ formulas in negation-normal form. Also, we write (for example) Γ, Θ, A, B for $\Gamma \cup \Theta \cup \{A, B\}$. If Γ is the set $\{A_1, \dots, A_n\}$ of closed $\mathcal{L}^{\infty}(n)$ formulas in negation-normal form, then $|\Gamma| := |A_1| \cup \dots \cup |A_n|$ is the parameter set of Γ.

Axioms of $\widehat{\mathsf{ID}}_n^{\infty}$:

1. Γ, A whenever A is a true atomic $\mathcal{L}_1$ formula.
2. $\Gamma, \neg B$ whenever B is a false atomic $\mathcal{L}_1$ formula.
3. $\Gamma, \neg \mathsf{R}(s), \mathsf{R}(t)$ for numerically equivalent closed terms s and t.
4. $\Gamma, \neg P_{\mathfrak{A}}^{(m)}(s), P_{\mathfrak{A}}^{(m)}(t)$ for numerically equivalent closed terms s and t.

Basic rules of inference of $\widehat{\mathsf{ID}}_n^{\infty}$:

$$(\vee)\ \frac{\Gamma, A, B}{\Gamma, A \vee B} \qquad\qquad (\wedge)\ \frac{\Gamma, A \qquad \Gamma, B}{\Gamma, A \wedge B}$$

$$(\exists)\ \frac{\Gamma, A[s]}{\Gamma, \exists x A[x]} \qquad\qquad (\forall)\ \frac{\Gamma, A[s] \text{ for all closed } s}{\Gamma, \forall x A[x]}$$

$$(P_{\mathfrak{A}}^{(m)})\ \frac{\Gamma, \mathfrak{A}[P_{\mathfrak{A}}^{(m)}, s]}{\Gamma, P_{\mathfrak{A}}^{(m)}(s)} \text{ if } m \leq n \qquad (\neg P_{\mathfrak{A}}^{(m)})\ \frac{\Gamma, \neg\mathfrak{A}[P_{\mathfrak{A}}^{(m)}, s]}{\Gamma, \neg P_{\mathfrak{A}}^{(m)}(s)} \text{ if } m \leq n$$

Cuts of $\widehat{\mathsf{ID}}_n^\infty$:

$$(\text{cut})\ \frac{\Gamma, A \qquad \Gamma, \neg A}{\Gamma}$$

Closure rules of $\widehat{\mathsf{ID}}_n^\infty$:

$$(\text{Cl-}P_{\mathfrak{A}})\ \frac{\Gamma, \mathfrak{A}[P_{\mathfrak{A}}, s]}{\Gamma, P_{\mathfrak{A}}(s)}$$

Ordinal rules of inference of $\widehat{\mathsf{ID}}_n^\infty$:

$$(P_{\mathfrak{A}}^{<\alpha})\ \frac{\Gamma, \mathfrak{A}[P_{\mathfrak{A}}^{<\beta}, s]}{\Gamma, P_{\mathfrak{A}}^{<\alpha}(s)}\ \text{if}\ \beta < \alpha \qquad (\neg P_{\mathfrak{A}}^{<\alpha})\ \frac{\Gamma, \neg\mathfrak{A}[P_{\mathfrak{A}}^{<\xi}, s]\ \text{for all}\ \xi < \alpha}{\Gamma, \neg P_{\mathfrak{A}}^{<\alpha}(s)}$$

$$(P_{\mathfrak{A}})\ \frac{\Gamma, \mathfrak{A}[P_{\mathfrak{A}}^{<\beta}, s]}{\Gamma, P_{\mathfrak{A}}(s)}\ \text{if}\ \beta < \Omega \qquad (\neg P_{\mathfrak{A}})\ \frac{\Gamma, \neg\mathfrak{A}[P_{\mathfrak{A}}^{<\xi}, s]\ \text{for all}\ \xi < \Omega}{\Gamma, \neg P_{\mathfrak{A}}(s)}$$

The formulas A and $\neg A$ in the premises of (cut) are called the *cut formulas* of this cut. The *rank* of a cut is the rank of its cut formulas.

For the ordinal assignment to proofs and the subsequent cut elimination and collapsing we follow [Buchholz, 1992] and make use of his approach to operator controlled derivations.

Definition 7.2. *Let $\mathcal{H}$ be a derivation operator and let Γ be a finite set of closed $\mathcal{L}^\infty(n)$ formulas in negation-normal form. Then $\widehat{\mathsf{ID}}_n^\infty, \mathcal{H} \vdash_\rho^\alpha \Gamma$ is defined for all ordinals α and ρ by induction on α.*

1. *If Γ is an axiom of $\widehat{\mathsf{ID}}_n^\infty$ and $|\Gamma| \cup \{\alpha\} \subseteq \mathcal{H}(\emptyset)$, then $\widehat{\mathsf{ID}}_n^\infty, \mathcal{H} \vdash_\rho^\alpha \Gamma$ for all ordinals ρ.*
2. *If $\widehat{\mathsf{ID}}_n^\infty, \mathcal{H} \vdash_\rho^{\alpha_0} \Gamma_i$ and $\alpha_0 < \alpha$ for every premise of a basic inference of $\widehat{\mathsf{ID}}_n^\infty$ or a cut of rank less than ρ and if $|\Gamma| \cup \{\alpha\} \subseteq \mathcal{H}(\emptyset)$ for the conclusion Γ of this rule, then $\widehat{\mathsf{ID}}_n^\infty, \mathcal{H} \vdash_\rho^\alpha \Gamma$.*
3. *If $\widehat{\mathsf{ID}}_n^\infty, \mathcal{H} \vdash_\rho^{\alpha_0} \Gamma, \mathfrak{A}[P_{\mathfrak{A}}^{<\sigma}, s]$ for some $\sigma < \tau$ and $\sigma, \alpha_0 < \alpha$ and if $|\Gamma, P_{\mathfrak{A}}^{<\tau}(s)| \cup \{\alpha\} \subseteq \mathcal{H}(\emptyset)$, then $\widehat{\mathsf{ID}}_n^\infty, \mathcal{H} \vdash_\rho^\alpha \Gamma, P_{\mathfrak{A}}^{<\tau}(s)$.*
4. *If $\widehat{\mathsf{ID}}_n^\infty, \mathcal{H}[\sigma] \vdash_\rho^{\alpha_\sigma} \Gamma, \neg\mathfrak{A}[P_{\mathfrak{A}}^{<\sigma}, s]$ and $\alpha_\sigma < \alpha$ for all $\sigma < \tau$ and if $|\Gamma, \neg P_{\mathfrak{A}}^{<\tau}(s)| \cup \{\alpha\} \subseteq \mathcal{H}(\emptyset)$, then $\widehat{\mathsf{ID}}_n^\infty, \mathcal{H} \vdash_\rho^\alpha \Gamma, \neg P_{\mathfrak{A}}^{<\tau}(s)$.*
5. *If $\widehat{\mathsf{ID}}_n^\infty, \mathcal{H} \vdash_\rho^{\alpha_0} \Gamma, \mathfrak{A}[P_{\mathfrak{A}}^{<\sigma}, s]$ for some $\sigma < \Omega$ and $\sigma, \alpha_0 < \alpha$ and if $|\Gamma, P_{\mathfrak{A}}(s)| \cup \{\alpha\} \subseteq \mathcal{H}(\emptyset)$, then $\widehat{\mathsf{ID}}_n^\infty, \mathcal{H} \vdash_\rho^\alpha \Gamma, P_{\mathfrak{A}}(s)$.*
6. *If $\widehat{\mathsf{ID}}_n^\infty, \mathcal{H}[\sigma] \vdash_\rho^{\alpha_\sigma} \Gamma, \neg\mathfrak{A}[P_{\mathfrak{A}}^{<\sigma}, s]$ and $\alpha_\sigma < \alpha$ for all $\sigma < \Omega$ and if $|\Gamma, \neg P_{\mathfrak{A}}(s)| \cup \{\alpha\} \subseteq \mathcal{H}(\emptyset)$, then $\widehat{\mathsf{ID}}_n^\infty, \mathcal{H} \vdash_\rho^\alpha \Gamma, \neg P_{\mathfrak{A}}(s)$.*
7. *If $\widehat{\mathsf{ID}}_n^\infty, \mathcal{H}[\sigma] \vdash_\rho^{\alpha_0} \Gamma, \mathfrak{A}[P_{\mathfrak{A}}, s]$ and $\alpha_0 + 1 < \alpha$ and if $|\Gamma, P_{\mathfrak{A}}(s)| \cup \{\alpha\} \subseteq \mathcal{H}(\emptyset)$, then $\widehat{\mathsf{ID}}_n^\infty, \mathcal{H} \vdash_\rho^\alpha \Gamma, P_{\mathfrak{A}}(s)$.*

We write $\widehat{\mathsf{ID}}_n^\infty, \mathcal{H} \frac{|{<}\alpha}{\;\rho\;} \Gamma$ to mean there exists an $\alpha_0 < \alpha$ with $\widehat{\mathsf{ID}}_n^\infty, \mathcal{H} \frac{|\alpha_0}{\;\rho\;} \Gamma$.

In the remaining part of this article we describe (or sketch) how to employ the infinitary systems for establishing the upper proof-theoretic bound of the theories $\widehat{\mathsf{ID}}_n$. Our main reference is again [Buchholz, 1992] where an analogous analysis has been carried out in full details for a theory that is similar to (or even more complicated than) our systems as far as impredicative methods of proof theory are concerned. We state the main results and believe that it should not be too complicated for the reader to fill in the missing details.

First observe that the infinitary systems $\widehat{\mathsf{ID}}_n^\infty$ have the property that all instances of complete induction on the natural numbers and all instances of least fixed point induction are provable, in particular by making use of the infinitary rules $(\forall)$ and $(\neg P_{\mathfrak{A}})$. However, the price is that we have to deal with complex derivations of infinitary depths. As a consequence we obtain a canonical embedding theorem.

Theorem 7.3 (Embedding). *Let A be a closed $\mathcal{L}^\bullet(n)$ formula A in negation-normal form. If $\widehat{\mathsf{ID}}_n^\bullet \vdash A$, then there exists a natural number k such that for all derivation operators $\mathcal{H}$,*

$$\widehat{\mathsf{ID}}_n^\infty, \mathcal{H} \frac{|{<}\varepsilon_{\Omega+1}}{\;\Omega+k\;} A.$$

Now we move on to cut elimination. It is easy to convince oneself that the axioms and rules of inference of the infinitary systems $\widehat{\mathsf{ID}}_n^\infty$ and the definitions of the ranks of closed $\mathcal{L}^\infty(n)$ formulas in negation-normal form are so that cuts of ranks greater than Ω can be eliminated without any problems. As usual, $\omega_0(\alpha) := \alpha$ and $\omega_{k+1}(\alpha) := \omega^{\omega_k(\alpha)}$ for all ordinals α and natural numbers k.

Lemma 7.4 (Partial cut elimination). *If Γ is a finite set of closed $\mathcal{L}^\infty(n)$ formulas in negation-normal form, then we have for all derivation operators $\mathcal{H}$, all ordinals α, and all natural numbers k:*

$$\widehat{\mathsf{ID}}_n^\infty, \mathcal{H} \frac{|\alpha}{\;\Omega+1+k\;} \Gamma \quad \Longrightarrow \quad \widehat{\mathsf{ID}}_n^\infty, \mathcal{H} \frac{|\omega_k(\alpha)}{\;\Omega+1\;} \Gamma.$$

The next step is to eliminate the fixed points $P_{\mathfrak{A}}^{(m)}$. To achieve this, we can make use of standard elimination procedures for finitely many fixed points by asymmetric interpretations as, for example, in [Cantini, 1985], [Jäger and Strahm, 1996], or [Marzetta and Strahm, 1997].

Lemma 7.5 (Elimination of fixed points). *If Γ is a finite set of closed $\mathcal{L}^\infty(n)$ formulas in negation-normal form, then we have for all derivation operators $\mathcal{H}$ and all ordinals α:*

$$\widehat{\mathsf{ID}}_{n+1}^\infty, \mathcal{H} \frac{|{<}\alpha}{\;\Omega+1\;} \Gamma \quad \Longrightarrow \quad \widehat{\mathsf{ID}}_n^\infty, \mathcal{H} \frac{|{<}\varphi\alpha 0}{\;\Omega+1\;} \Gamma.$$

Cut formulas of rank less than Ω are eliminated by methods of predicative cut elimination as presented in [Schütte, 1977]; for all details concerning predicative cut elimination in the presence of derivation functions see [Buchholz, 1992].

Lemma 7.6 (Predicative cut elimination). *If Γ is a finite set of closed $\mathcal{L}^{\infty}(0)$ formulas in negation-normal form, then we have for all derivation operators $\mathcal{H}$, all ordinals α, and all ordinals β and ρ with $\beta, \rho < \Omega$ and $\rho \in \mathcal{H}(\emptyset)$:*

$$\widehat{\mathsf{ID}}_n^{\infty}, \mathcal{H} \left|\frac{\alpha}{\beta+\omega^{\rho}}\right. \Gamma \quad \Longrightarrow \quad \widehat{\mathsf{ID}}_n^{\infty}, \mathcal{H} \left|\frac{\varphi\rho\alpha}{\beta}\right. \Gamma.$$

So it only remains to deal with cut formulas of the form $P_{\mathfrak{A}}(t)$ and $\neg P_{\mathfrak{A}}(t)$, and here the boundedness and collapsing techniques enter the picture. Let *POS* be the collection of all closed $\mathcal{L}^{\infty}(0)$ formulas in negation-normal form that do not contain subformulas of the form $\neg P_{\mathfrak{A}}(t)$; i.e., *POS* is the collection of all closed $\mathcal{L}^{\infty}(0)$ formulas in negation-normal form that are positive in the least fixed point relations $P_{\mathfrak{A}}$. In addition, if A belongs to *POS* and α is an ordinal less than Ω, then $A^{(<\alpha)}$ is the formula obtained from A if all occurrences of $P_{\mathfrak{A}}(t)$ are replaced by $P_{\mathfrak{A}}^{<\alpha}(t)$. For the proof of this boundedness and collapsing lemma consult again [Buchholz, 1992].

Lemma 7.7 (Boundedness and Collapsing).

1. *For all finite sets Γ of closed $\mathcal{L}^{\infty}(n)$ formulas in negation-normal form, all and elements A of POS, all derivation operators $\mathcal{H}$, and all ordinals α, β, ρ such that $\alpha \le \beta < \Omega$ and $\beta \in \mathcal{H}(\emptyset)$ we have:*

$$\widehat{\mathsf{ID}}_0^{\infty}, \mathcal{H} \left|\frac{\alpha}{\rho}\right. \Gamma, A \quad \Longrightarrow \quad \widehat{\mathsf{ID}}_0^{\infty}, \mathcal{H} \left|\frac{\alpha}{\rho}\right. \Gamma, A^{(<\beta)}.$$

2. *Suppose that Γ is a finite subset of POS and σ an ordinal such that $|\Gamma| \subseteq \mathbf{C}(\sigma + 1, \psi(\sigma+1))$ and $\sigma \in \mathcal{H}_{\sigma}[|\Gamma|](\emptyset)$. Then we have for all ordinals α and $\beta := \sigma + \omega^{\Omega+\alpha}$:*

$$\widehat{\mathsf{ID}}_0^{\infty}, \mathcal{H}_{\sigma}[|\Gamma|] \left|\frac{\alpha}{\Omega+1}\right. \Gamma \quad \Longrightarrow \quad \widehat{\mathsf{ID}}_0^{\infty}, \mathcal{H}_{\beta}[|\Gamma|] \left|\frac{\psi\beta}{\psi\beta}\right. \Gamma.$$

Combining Theorem 7.3 with the series of Lemmas 7.4 to 7.7 and carrying through some ordinal calculations, we obtain complete cut elimination for the closed $\mathcal{L}_1$ formulas provable in one of the theories $\widehat{\mathsf{ID}}_n$.

Theorem 7.8 (Complete cut elimination). *Let A be a closed formula of the language $\mathcal{L}_1$ and suppose that A is provable in $\widehat{\mathsf{ID}}_n$ for some natural number $n \ge 1$. Then there exist a derivation operator $\mathcal{H}$ and an ordinal $\alpha < \psi(\Gamma_{\Omega+1})$ such that $\widehat{\mathsf{ID}}_0^{\infty}, \mathcal{H} \left|\frac{\alpha}{0}\right. A$.*

By standard proof theory this result immediately gives us the upper bound result for the theory $\widehat{\mathsf{ID}}^{\bullet}_{<\omega}$: If $\widehat{\mathsf{ID}}^{\bullet}_{<\omega}$ proves $TI[\lhd, \mathsf{R}]$ for some primitive recursive well ordering $\lhd$, then there exists a natural number n large enough such that $\widehat{\mathsf{ID}}_n$ proves

$TI[\lhd, \mathsf{R}]$. In view of the previous theorem this implies

$$\widehat{\mathsf{ID}}_0^\infty, \mathcal{H} \left|\frac{\alpha}{0}\right. TI[\lhd, \mathsf{R}]$$

for some derivation operator $\mathcal{H}$ and some ordinal $\alpha < \psi(\Gamma_{\Omega+1})$. Finally, as shown in [Schütte, 1977], we can conclude that the depth of a cut free derivation of $TI[\lhd, \mathsf{R}]$ essentially bounds the order type of this well ordering.

Corollary 7.9. $|\widehat{\mathsf{ID}}^\bullet_{<\omega}| \leq \psi(\Gamma_{\Omega+1})$.

It only remains to see this upper bound result in the context of Theorem 3.2, Corollary 3.6, Theorem 5.2, Corollary 5.3, and Theorem 6.1 in order to conclude the ordinal analysis of the main theories of this article.

Corollary 7.10 (Proof-theoretic ordinal).

$$\begin{aligned}\psi(\Gamma_{\Omega+1}) &= |\widehat{\mathsf{ID}}^\bullet_{<\omega}| = |\mathsf{FP}^\bullet_0| = |\mathsf{ATR}^\bullet_0| \\ &= |\Sigma^1_1\text{-}\mathsf{DC}^\bullet_0 + (\mathsf{SUB}^\bullet)| = |\Sigma^1_1\text{-}\mathsf{AC}^\bullet_0 + (\mathsf{SUB}^\bullet)| = |\mathcal{U}(\mathsf{ID}_1)|.\end{aligned}$$

8 Discussion

We have identified several systems in classical logic of strength $\psi(\Gamma_{\Omega+1})$ (cf. Corollary 7.10). Our results are a parallel to those characterizing classical systems of strength Γ_0,

$$\begin{aligned}\Gamma_0 &= |\widehat{\mathsf{ID}}_{<\omega}| = |\mathsf{FP}_0| = |\mathsf{ATR}_0| \\ &= |\Sigma^1_1\text{-}\mathsf{DC}_0 + (\mathsf{SUB})| = |\Sigma^1_1\text{-}\mathsf{AC}_0 + (\mathsf{SUB})| = |\mathcal{U}(\mathsf{NFA})|.\end{aligned}$$

(See [Feferman, 1982] for $\widehat{\mathsf{ID}}_{<\omega}$, [Avigad, 1996] for FP_0, [Simpson, 2009] for the subsystems of second order arithmetic, and Feferman and [Feferman and Strahm, 2000] for $\mathcal{U}(\mathsf{NFA})$.)

A companion article in preparation shall establish similar results for constructive systems. In particular, we shall verify the conjecture of Hancock [Hancock, 2000] by studying a predicative type theory in the style of [Martin-Löf, 1975] (i.e., a dependent type theory with an externally indexed hierarchy of predicative universes $(U_n)_{n<\omega}$) extended with a single well-ordering type (belonging to all universes) corresponding to the constructive tree ordinals (a type *Ord* with constructors zero of type *Ord*, successor of type *Ord* → *Ord* and limit of type (*Nat* → *Ord*) → *Ord*). This will parallel the result of Feferman [Feferman, 1982] that the strength of the predicative type theory itself is Γ_0. Also of interest is an analogous system of explicit mathematics, similarly containing a hierarchy of universes (as in [Feferman, 1982]) and a type of constructive tree ordinals.

References

J. Avigad. On the relationship between ATR_0 and $\widehat{ID}_{<\omega}$. *J. Symb. Logic*, 61(3):768–779, 1996.

H. Bachmann. Die Normalfunktionen und das Problem der ausgezeichneten Folgen von Ordnungszahlen. *Vierteljahresschr. Naturforsch. Ges. Zürich*, 95:115–147, 1950.

K. J. Barwise and J. Schlipf. On recursively saturated models of arithmetic. In D. H. Saracino and V. B. Weispfenning, editors, *Model Theory and Algebra*, volume 498 of *Lecture Notes in Mathematics*, pages 42–55. Springer, 1975.

M. J. Beeson. *Foundations of Constructive Mathematics: Metamathematical Studies*, volume 6 of *Ergebnisse der Mathematik und ihrer Grenzgebiete (3)*. Springer, 1985.

U. Buchholtz. *Unfolding of Systems of Inductive Definitions*. PhD thesis, Department of Mathematics, Stanford University, 2014.

W. Buchholz. A simplified version of local predicativity. In P. Aczel, H. Simmons, and S. S. Wainer, editors, *Proof theory: A Selection of Papers from the Leeds Proof Theory Programme 1990*, pages 115–147. Cambridge Univ. Press, 1992.

W. Buchholz and K. Schütte. *Proof Theory and Impredicative Subsystems of Analysis*, volume 2 of *Studies in Proof Theory, Monographs*. Bibliopolis, 1988.

W. Buchholz, S. Feferman, W. Pohlers, and W. Sieg. *Iterated Inductive Definitions and Subsystems of Analysis: Recent Proof-Theoretical Studies*, volume 897 of *Lecture Notes in Mathematics*. Springer, 1981.

A. Cantini. A note on a predicatively reducible theory of iterated elementary induction. *Boll. Unione Mat. Ital. (B)*, 6:413–430, 1985.

A. Cantini. On the relation between choice and comprehension principles in second order arithmetic. *J. Symb. Logic*, 51(2):360–373, 1986.

J. N. Crossley and J. Bridge Kister. Natural well-orderings. *Arch. Math. Logik Grundlagenforsch.*, 26(1–2):57–76, 1986/87.

S. Feferman. Iterated inductive fixed-point theories: application to Hancock's conjecture. In G. Metakides, editor, *The Patras Symposion*, pages 171–196. North-Holland, 1982.

S. Feferman. Gödel's program for new axioms: Why, where, how and what? In P. Hájek, editor, *Gödel '96*, volume 6 of *Lecture Notes in Logic*, pages 3–22. Springer, 1996.

S. Feferman and G. Jäger. Choice principles, the bar rule and autonomously iterated comprehension schemes in analysis. *J. Symb. Logic*, 48(1):63–70, 1983.

S. Feferman and T. Strahm. The unfolding of non-finitist arithmetic. *Ann. Pure Appl. Logic*, 104 (1-3):75–96, 2000.

P. G. Hancock. *Ordinals and Interactive Programs*. PhD thesis, Laboratory for Foundations of Computer Science, University of Edinburgh, 2000.

W. A. Howard. A system of abstract constructive ordinals. *J. Symb. Logic*, 37:355–374, 1972.

G. Jäger and T. Strahm. Some theories with positive induction of ordinal strength $\varphi\omega 0$. *J. Symb. Logic*, 61(3):818–842, 1996.

G. Jäger, R. Kahle, A. Setzer, and T. Strahm. The proof-theoretic analysis of transfinitely iterated fixed point theories. *J. Symb. Logic*, 64(1):53–67, 1999.

P. Martin-Löf. An intuitionistic theory of types: predicative part. In *Logic Colloquium '73 (Bristol, 1973)*, volume 80 of *Studies in Logic and the Foundations of Mathematics*, pages 73–118. North-Holland, 1975.

M. Marzetta and T. Strahm. The μ quantification operator in explicit mathematics with universes and iterated fixed point theories with ordinals. *Arch. Math. Logic*, 37(5+6):391–413, 1997.

L. W. Miller. Normal functions and constructive ordinal notations. *J. Symb. Logic*, 41(2):439–459, 1976.

W. Pohlers. *Proof Theory. An Introduction*, volume 1407 of *Lecture Notes in Mathematics*. Springer, 1989.

W. Pohlers. *Proof Theory. The First Step into Impredicativity*. Universitext. Springer, 2009.

K. Schütte. *Proof Theory*, volume 225 of *Grundlehren der mathematischen Wissenschaften*. Springer, 1977.

S. G. Simpson. *Subsystems of Second Order Arithmetic*. Perspectives in Logic. Cambridge Univ. Press, 2nd edition, 2009.

Thierry Coquand and Henri Lombardi

Some Remarks about Normal Rings

Abstract: We give a constructive proof that $R[X]$ is normal when R is normal. We apply this result to an operation needed for studying the henselization of a local ring. Our proof is based on the case where R is without zero divisors, which is more involved than the case where R is an integral domain. We have to use a constructive deciphering technique that replaces the use of minimal primes (in classical mathematics) by suitable explicit localizations in a suitable tree.

Keywords: Normal ring, Pf-ring, Constructive mathematics, Gcd tree.

Mathematics Subject Classification 2010: 13B22, 13B30, 13B40, 03F65

1 Introduction

An *integrally closed domain* R is an integral domain whose integral closure in its field of fraction is R itself. An element b is *integral* over an ideal I iff b satisfies an integral relation

$$b^n + u_1 b^{n-1} + \cdots + u_n = 0$$

with u_l in I^l. We can reformulate the definition of being integrally closed by stating that whenever b is integral over $\langle a\rangle$ then b belongs to $\langle a\rangle$. In this form, this definition makes sense even if R is an arbitrary ring (not necessarily a domain) and this characterizes the notion of *normal* ring. It can be checked that this is equivalent to the following: any localization of R at a prime ideal is an integrally closed integral domain ([Ducos et al., 2004, Proposition 6.4]).

This paper is mainly concerned with the analysis of the following classical result: if R is an integrally closed domain then so is $R[X]$. We first recall a proof which reduces this result to Kronecker's Theorem ([Lombardi and Quitté, 2015, Theorem 3.3]). Interestingly, the argument depends in a crucial how we interpret constructively the notion of "integral domain". Logically, to be an integral domain can be stated as

$$\forall x.\forall y.\ xy = 0 \rightarrow [x = 0 \vee y = 0] \tag{1}$$

which is classically, but not constructively, equivalent to

$$\forall x.\ x = 0 \vee [\forall y.xy = 0 \rightarrow y = 0]. \tag{2}$$

Thierry Coquand: Department of Computer Science, Chalmers, University of Göteborg, 41296 Göteborg, Sweden, e-mail: thierry.coquand@cse.gu.se
Henri Lombardi: Laboratoire de Mathématiques. Université de Franche-Comté. F-25030 Besançon Cedex, e-mail: henri.lombardi@univ-fcomte.fr

On this form, this means that any element is 0 or is regular. This Definition (2) is actually the usual definition of integral domain in constructive mathematics (Lombardi and Quitté [2015], Mines et al. [1988]). With this definition the argument using Kronecker's Theorem makes sense constructively.

The definition (1) also has been considered in constructive algebra: a ring satisfying this condition is called a ring *without zero divisors* (Lombardi and Quitté [2015]). The main part of this paper presents a proof that if R is a normal ring without zero divisors than so is $R[X]$. What is surprising is that this proof seems to require a technique which is used for analyzing argument involving *minimal* prime ideal ([Lombardi and Quitté, 2015, Section XV-7]). Furthermore, the proof involves the introduction of the notion of *gcd tree* of two polynomials, which is important in other context (Alonso et al. [2014]). Going from Definition (2) to Definition (1), classically equivalent, requires a much more complex argument.

The advantage of Definition (1) is that it is now relatively easy to conclude from this that, more generally, if R is normal (without any integrality condition) then so is $R[X]$. The last section analyzes a connected operation useful for studying the henselization of a local ring.

2 Constructible and Gcd trees

Given a reduced ring R we define the notion of *constructible tree* for R. This is a binary tree. To each node of this tree is associated a reduced ring, and R is associated to the root of the tree. Such a tree can only grow in the following way: we choose a leaf, and an element a of the ring S associated to this leaf. We add then two sons to this node: to the left branch we associate the ring $S[1/a]$ and to the right branch the ring $S/\sqrt{\langle a\rangle}$. Any such tree defines a partition of the constructible spectrum of R (Johnstone [1986]).

If we look at the leftmost branch of this tree, we see that the leaf is of the form $R[1/(a_1 \cdots a_n)]$ and so is a localization of the ring R.

The main proofs in this note will be by induction on the size of a given constructible tree.

If we have two polynomials P and Q in $R[X]$ we can associate a constructible tree which corresponds to the formal computation of the gcd of P and Q. To each leaf S are also associated polynomials A, B, G, P_1, Q_1 in $S[X]$, with G monic, which witness the computation of the gcd of P and Q

$$AP_1 + BQ_1 = 1 \qquad P = GP_1 \qquad Q = GQ_1.$$

Notice that for building this tree, R does not need to be discrete (i.e. to have a decidable equality). Here is a simple example: $P = X^2$ and $Q = aX + b$. We start by the two branches $S_0 = R[1/a]$ and $S_1 = R/\sqrt{\langle a \rangle}$. Over S_0 we have the two branches $S_{00} = S_0[1/b]$ and $S_{01} = S_0/\sqrt{\langle b \rangle}$. Over S_1 we have the two branches $S_{10} = S_1[1/b]$ and $S_{11} = S_1/\sqrt{\langle b \rangle}$. The gcd is 1 over S_{00} and S_{10}, and is X over S_{01} and is X^2 over S_{11}.

This tree is called the *gcd tree* of P and Q.

If one of the polynomial is monic, one can reduce the size of this tree by using subresultants (Apéry and Jouanoulou [2006]).

3 Kronecker's Theorem

We shall only need a simple case of Kronecker's Theorem ([Lombardi and Quitté, 2015, Theorem 3.3]).

Theorem 3.1. *Let R be a ring, if* $X^m + a_1X^{m-1} + \cdots + a_m$ *divides a polynomial of the form* $X^n + b_1X^{n-1} + \cdots + b_n$ *in* $R[X]$ *then* $a_1, \ldots, a_m$ *are integral over the subring of R generated by* $b_1, \ldots, b_n$.

Proof: We introduce the splitting algebra[1] T of $X^n + b_1X^{n-1} + \cdots + b_n$ (Lombardi and Quitté [2015]) so that $X^n + b_1X^{n-1} + \cdots + b_n = (X - t_1) \cdots (X - t_n)$ in $T[X]$. The ring R embeds in T and a_i is a polynomial in $t_1, \ldots, t_n$. □

Corollary 3.2. *Let R be a ring, if* $Y + a_0X^m + \cdots + a_m$ *divides a polynominal of the form* $Y^n + b_1Y^{n-1} + \cdots + b_n$ *in* $R[X, Y]$ *then all coefficients* $a_0, \ldots, a_m$ *are roots of polynomials of the form* $Z^l + p_1Z^{l-1} + \cdots + p_l$ *where* p_i *is a homogeneous polynomial of degree i in* $b_1, \ldots, b_n$ *where* b_j *has weight j.*

Proof: It is enough to look at the case where $a_0, \ldots, a_m$ are indeterminates, and R is a polynomial ring on $a_0, \ldots, a_m$ and some other indeterminates. By replacing Y by X^N for N big enough, we get that each $a_0, \ldots, a_m$ is integral over $\mathbb{Z}[b_1, \ldots, b_n]$ and hence each a_k is root of a polynomial of the form $Z^l + p_1Z^{l-1} + \cdots + p_l$ where p_i is a polynomial in $b_1, \ldots, b_n$. By replacing Y by Y/c where c is another indeterminate, we get that p_i is homogeneous of degree i in $b_1, \ldots, b_n$ where b_j has weight j. □

1 $T = R[X_1, \ldots, X_n]/J(f) = R[x_1, \ldots, x_n]$ where $J(f)$ is the ideal of symmetric relators necessary to identify $\prod_{j=1}^n (X - x_j)$ with $f(X)$ in $T[X]$. We let $x = x_1$ and the quotient ring $R[x] = R[X]/\langle f \rangle$ is identified with a subring of T. If $g(X, Y) = \frac{f(X)-f(Y)}{X-Y}$ then $g(x_1, X) = \prod_{i=2}^n (X - x_i)$ in $T[X]$ and $g(x_1, x_j) = 0$ for $j \geq 2$.

Corollary 3.3. *Let R be a normal integral domain, then $R[X]$ is a normal integral domain.*

Proof: We assume given P and Q in $R[X]$ such that P is integral over $\langle Q\rangle$ and we want to show that P is in $\langle Q\rangle$ in $R[X]$. Let K be the total fraction field of R. Since R is an integral domain, we can consider R to be a subring of K. Since $K[X]$ is euclidean, we know that P is in $\langle Q\rangle$ in $K[X]$ and we have $cP = HQ$ for some regular element c. Since P is integral over $\langle Q\rangle$ we have a relation

$$P^n + A_1 Q P^{n-1} + \cdots + A_n Q^n = 0$$

with $A_1, \dots, A_n$ in $R[X]$ and so we can write

$$Y^n + A_1 c Y^{n-1} + \cdots + A_n c^n = (Y - H)S(X, Y)$$

where $S(X, Y)$ is a monic polynomial in $R[X][Y]$. Using Corollary 3.2, it follows that all coefficients of H are integral over $\langle c\rangle$ and hence are in $\langle c\rangle$ since R is normal. We can then write $H = cH_1$ and so $c(P - QH_1) = 0$ in $R[X]$. It follows that we have $P = QH_1$ and hence P is in $\langle Q\rangle$ in $R[X]$. □

This is the argument we are going to adapt in the case where R is normal and without zero divisors.

4 Polynomial ring

We assume that R is normal without zero divisors and we show that $R[X]$ is normal. We assume given P and Q in $R[X]$ such that P is integral over $\langle Q\rangle$ and we want to show that P is in $\langle Q\rangle$ in $R[X]$.

Lemma 4.1. *If R is normal then R is reduced.*

Proof: If $b^2 = 0$ then b is integral over $\langle 0\rangle$ and so is in $\langle 0\rangle$. □

Lemma 4.2. *If R is normal then so is $R[1/a]$.*

Proof: For c and b in R, if c is integral over $\langle b\rangle$ in $R[1/a]$ we have a relation $(a^N c)^n + u_1 b (a^N c)^{n-1} + \cdots + u_n b^n = 0$ with $u_1, \dots, u_n$ in R. Since R is normal we have $a^N c$ in $\langle b\rangle$. □

Lemma 4.3. *If R is without zero divisors then so is $R[1/a]$.*

Proof: We take two elements $v = b/a^n$ and $w = c/a^m$ of $R[1/a]$ with b and c in R. If we have $vw = 0$ in $R[1/a]$ we have $a^p bc = 0$ in R for some $p \geqslant 0$. We have then $a^p b = 0$ in R or $a^p c = 0$ in R, which implies that $v = 0$ or $w = 0$ in $R[1/a]$. □

From now on in this section, we assume R to be a normal ring without zero divisors.

Lemma 4.4. *If P is integral over $\langle Q\rangle$ and is in $\langle Q\rangle$ in $R[1/a][X]$ then $a = 0$ or P is in $\langle Q\rangle$ in $R[X]$.*

Proof: We have H in $R[X]$ such that $a^N P = QH$ for some N. We write $c = a^N$. Since P is integral over $\langle Q\rangle$ we have a relation

$$P^n + A_1 QP^{n-1} + \cdots + A_n Q^n = 0$$

with $A_1, \ldots, A_n$ in $R[X]$ and so

$$Q^n(H^n + A_1 cH^{n-1} + \cdots + A_n c^n) = 0$$

in $R[X]$. Hence either $Q = 0$ in which case $P = 0$ is in $\langle Q\rangle$ or we can write

$$Y^n + A_1 cY^{n-1} + \cdots + A_n c^n = (Y - H)S(X, Y)$$

where $S(X, Y)$ is a monic polynomial in $R[X][Y]$. Using the corollary of Kronecker's Theorem 3.2, it follows that all coefficients of H are integral over $\langle c\rangle$ and hence are in $\langle c\rangle$ since R is normal. We can then write $H = cH_1$ and so $c(P - QH_1) = 0$ in $R[X]$. It follows that we have $c = 0$, that is equivalent to $a = 0$, or $P = QH_1$ and hence P is in $\langle Q\rangle$ in $R[X]$. □

Lemma 4.5. *If we have P and Q in $R[X]$ and a constructible tree for R such as, at all leaves S of this tree, we have P in $\langle Q\rangle$ in $S[X]$. Then P is in $\langle Q\rangle$ in $R[X]$.*

Proof: Look at the leftmost branch of this tree, indexed by elements $a_1, \ldots, a_l$, so that $S = S'[1/a_l]$ where $S' = R[1/(a_1 \cdots a_{l-1})]$ is without zero divisors by Lemma 4.3 and is normal by Lemma 4.2. Using Lemma 4.4 we get that $a_l = 0$ in S' or P is in $\langle Q\rangle$ in $S'[X]$. In the second case, we can shorten the leftmost branch to $a_1, \ldots, a_l$ and get a smaller tree. In the first case where $a_l = 0$ in S', this means that the right son $S'/\langle a\rangle$ of S' is equal to S' and we also can shorten the tree. We conclude by tree induction. □

Theorem 4.6. *If R is normal and without zero divisors then so is $R[X]$.*

Proof: We take P and Q in $R[X]$ and we assume that we have a relation

$$P^n + A_1 QP^{n-1} + \cdots + A_n Q^n = 0$$

with $n \geqslant 1$ and $A_1, \cdots, A_n$ in $R[X]$. We have to show that P is in $\langle Q\rangle$ in $R[X]$.

We look now at the gcd tree of P and Q as defined in the first section. At all leaves S of this tree, we have P_1, Q_1, G, A, B in $S[X]$ satisfying

$$P = GP_1, \quad Q = GQ_1, \quad AP_1 + BQ_1 = 1$$

in $S[X]$ and G is monic. Since G is monic and

$$P^n + A_1QP^{n-1} + \cdots + A_nQ^n = G^n(P_1^n + A_1Q_1P_1^{n-1} + \cdots + A_nQ_1^n) = 0$$

we have

$$P_1^n + A_1Q_1P_1^{n-1} + \cdots + A_nQ_1^n = 0$$

and Q_1 divides P_1^n. With $AP_1 + BQ_1 = 1$ this implies that Q_1 is a unit and so P is in $\langle Q\rangle$ in $S[X]$. We can now apply Lemma 4.5. □

5 Normal ring

We say that the ring is *locally* without zero divisors ([Lombardi and Quitté, 2015, Lemma VIII-3.2]) if, and only if, whenever $ab = 0$ then there exists u such that $ua = 0$ and $(1-u)b = 0$. These rings are often called pf-rings. In this note, only the notion of rings without zero divisors and locally without zero divisors will play a role.

Lemma 5.1. *If R is normal then R is locally without zero divisors.*

Proof: If $ab = 0$ then $b^2 - (a+b)b = 0$ so b is integral over $\langle a+b\rangle$ and so is in $\langle a+b\rangle$. We can write $b = (a+b)u$ and so $ua = (1-u)b$. This implies $ua^2 = (1-u)ba = 0$ and so $ua = (1-u)b = 0$ since R is reduced. □

Theorem 5.2. *If R is normal then so is* $R[X]$.

Proof: By Lemma 5.1, R is locally without zero divisors. Assume then that a polynomial $P \in R[X]$ is integral over $\langle Q\rangle$ in $R[X]$. Following the proof of Theorem 4.6, each time we use $ab = 0 \to a = 0$ or $b = 0$, we split the "current ring $R[1/v]$" in two rings $R[1/vu]$ and $R[1/v(1-u)]$ by using an u such that $ua = 0$ and $(1-u)b = 0$. We find finally $u_1, \ldots, u_m$ in R such that $\langle u_1, \ldots, u_m\rangle = 1$ and P belongs to $\langle Q\rangle$ in each $R[1/u_j][X]$. It follows that P is in $\langle Q\rangle$ as required. □

6 The ring $R_{\{f\}}$

Let f be a monic polynomial in $R[X]$. We can consider the extension $S = R[X]/\langle f\rangle$ where f has a root x. We let f_X be the formal derivative of f w.r.t. X, and we define $R_{\{f\}}$ to be the localization $S[1/f_X(x)]$. This construction is important to study the properties of henselization of a local ring (Raynaud [1970]).

The goal of this section is to show that $R_{\{f\}}$ is normal whenever R is normal. As in the previous section, we can first assume that R is without zero divisors, and then use that a normal ring is locally without zero divisors to conclude. So in the rest of the section, we assume that R is a normal ring without zero divisors.

If $f = gh$ is the product of two monic polynomials g and h we have $R_{\{f\}}$ isomorphic to $R_{\{g\}}[1/h(x)] \times R_{\{h\}}[1/g(x)]$. This remark is important since by using Lemma 4.2 we can reason by induction on the degree of f to show that $R_{\{f\}}$ is normal if R is normal.

Lemma 6.1. *If R is normal without zero divisors, and a in R and $T = R[1/a]$ and $f = gf_1$ with g and f_1 monic in $T[X]$ then we have g and f_1 in $R[X]$ or $a = 0$.*

Proof: Using Kronecker's Theorem 3.1, each coefficient of g and f_1 is integral over R. Since R is normal and without zero divisors, this implies that $a = 0$ or g and f_1 are in $R[X]$. □

We have the trace map $\mathrm{tr} : S \to R$. If we introduce the splitting algebra ([Lombardi and Quitté, 2015, Definition III-4.1]) of f and write $f = (X - x_1) \cdots (X - x_n)$ with $x = x_1$ then the trace of $h(x) \in S$ is $h(x_1) + \cdots + h(x_n)$. If $v = h(x)$ in S is integral over $\langle a \rangle_S$ with a in R then all elements $h(x_1), \ldots, h(x_n)$ are integral over $\langle a \rangle_R$ and so $\mathrm{tr}(v)$ is also integral over $\langle a \rangle_X$ and so is in $\langle a \rangle_X$ since R is normal. Also if we write $f(X) - f(Y) = (X - Y)g(X, Y) = \sum_i g_i(Y)X^i$, we have $f_X(T) = g(T, T)$ and $g(x_1, x_j) = 0$ for $j \neq 1$. So we get for all $v = h(x) = h(x_1)$ in S

$$f_X(x)h(x) = g(x_1, x_1)h(x_1) = \sum_i g_i(x_1)h(x_1)\, x_1^i$$

and for $j \neq 1$

$$0 = g(x_1, x_j)h(x_j) = \sum_i g_i(x_j)h(x_j)\, x_1^i$$

so that, by summation, we get Tate's formula ([Raynaud, 1970, Chapter VII, 1])

$$f_X(x)v = \sum_i \mathrm{tr}(g_i(x)v)\, x^i.$$

Since each $g_i(x)$ is integral over R, we can state the following lemma.

Lemma 6.2. *If R is normal, if a in R and if v in S is integral over $\langle a \rangle_S$ then $f_X(x)v$ is in $\langle a \rangle_S$.*

Theorem 6.3. *If R is normal then $R_{\{f\}}$ is normal.*

Proof: We assume given p, q in $R[X]$ such that $p(x)$ is integral over $\langle q(x) \rangle$ in S so that we have a relation $p(x)^n + u_1(x)q(x)p(x)^{n-1} + \cdots + u_n(x)q(x)^n = 0$. The goal is to show that $p(x)$ is in $q(x)S[f_X(x)^{-1}]$.

We look at the gcd tree of q and f, and the leftmost branch of this tree. At the leaf of this branch we have a list of elements that we force to be invertible $a_1, \ldots, a_n$ and q_1, f_1, g, A, B in $R[a^{-1}]$ with $a = a_1 \ldots a_n$ such that

$$f = gf_1 \quad q = gq_1 \quad 1 = Af_1 + Bq_1.$$

Furthermore g and hence f_1 are monic since f is monic.

If $f = f_1$ we have $g = 1$ and $q = q_1$. In this case we have $c = Af + Bq$ where $c = (a_1 \dots a_n)^m$ for some m and so we have $c = B(x)q(x)$ in $S = R[X]/\langle f \rangle$. We then have a relation

$$(p(x)B(x))^n + u_1(x)c(p(x)B(x))^{n-1} + \dots + u_n(x)c^n = 0$$

and hence, by Lemma 6.2, we get that $p(x)B(x)$ is in $\langle c \rangle$ in $S[f_X(x)^{-1}]$. Hence we have $l(x)$ in S and N such that

$$f_X(x)^N p(x)B(x) = cl(x) = q(x)B(x)l(x)$$

and so

$$c(p(x)f_X(x)^N - l(x)q(x)) = 0.$$

We have then $c = 0$ or $p(x)$ is in $\langle q(x) \rangle$ in $R[f_X(x)^{-1}]$. So either we have the desired conclusion that $p(x)$ is in $\langle q(x) \rangle$ or we have $a_n = 0$ in $R[1/(a_1 \cdots a_{n-1})]$ and we can shorten the computation tree of the gcd of f and q.

If f and f_1 have not the same degree, we have found a proper decomposition $f = gf_1$ of f in $R[1/a]$ with $a = a_1 \cdots a_n$. In this case, since R is normal, by Lemma 6.1, we have two subcases

- either g and f_1 are in $R[X]$ and we can conclude by induction on the degree of f, using that $R_{\{f\}}$ isomorphic to $R_{\{g\}}[1/f_1] \times R_{\{f_1\}}[1/g]$,
- or $a = 0$ and as in the previous case, we can shorten the computation tree of the gcd of f and q.

□

As in ([Lombardi and Quitté, 2015, VIII-4.4]), we say that a ring is a Prüfer ring if it is arithmetic and reduced. A coherent Prüfer ring is an arithmetic *pp*-ring.

Corollary 6.4. *If R is a Prüfer ring of Krull dimension $\leqslant 1$ then so is $R_{\{f\}}$.*

Proof: We use the fact that a ring is normal coherent ring of Krull dimension $\leqslant 1$ if, and only if, it is Prüfer and of Krull dimension $\leqslant 1$ (Ducos et al. [2004]). We have shown that $R_{\{f\}}$ is normal. Since $S = R[X]/\langle f \rangle$ is an integral extension of R it is also of Krull dimension $\leqslant 1$ (Coquand et al. [2009]) and so is its localization $R_{\{f\}}$. Finally, S is a finite free R-module, and so it is coherent if R is coherent and so is its localization $R_{\{f\}}$. □

This gives an alternative proof to the main result of (Coquand et al. [2010]), that $R_{\{f\}}$ is Prüfer when $R = k[X]$, in the case where f is monic in Y. It is possible however to reduce the general case to this case, by a change of variables.

References

M. E. Alonso, T. Coquand, and H. Lombardi. Revisiting Zariski main theorem from a constructive point of view. *Journal of Algebra*, 406:46–68, 2014. 10.1016/j.jalgebra.2014.02.003. URL http://dx.doi/10.1016/j.jalgebra.2014.02.003.

François Apéry and Jean-Pierre Jouanoulou. *Élimination: le cas d'une variable.* Hermann, Paris, 2006.

Thierry Coquand, Lionel Ducos, Henri Lombardi, and Claude Quitté. Constructive Krull dimension. I. Integral extensions. *Journal of Algebra and its Applications*, 8(1):129–138, 2009. 10.1142/S0219498809003199. URL http://dx.doi/10.1142/S0219498809003199.

Thierry Coquand, Henri Lombardi, and Claude Quitté. Curves and coherent Prüfer rings. *Journal of Symbolic Computation*, 45(12):1378–1390, 2010. 10.1016/j.jsc.2010.06.016. URL http://dx.doi/10.1016/j.jsc.2010.06.016.

L. Ducos, H. Lombardi, C. Quitté, and M. Salou. Théorie algorithmique des anneaux arithmétiques, des anneaux de Prüfer et des anneaux de Dedekind. *Journal of Algebra*, 281(2): 604–650, 2004. 10.1016/j.jalgebra.2004.06.030. URL http://dx.doi/10.1016/j.jalgebra.2004.06.030.

Peter T. Johnstone. *Stone spaces*, volume 3 of *Cambridge Studies in Advanced Mathematics*. Cambridge University Press, Cambridge, 1986. Reprint of the 1982 edition.

Henri Lombardi and Claude Quitté. *Commutative algebra: constructive methods*, volume 20 of *Algebra and Applications*. Springer, Dordrecht, revised edition, 2015. 10.1007/978-94-017-9944-7. URL http://dx.doi/10.1007/978-94-017-9944-7. Finite projective modules, Translated from the French by Tania K. Roblot.

Ray Mines, Fred Richman, and Wim Ruitenburg. *A course in constructive algebra.* Universitext. Springer, New York, 1988. 10.1007/978-1-4419-8640-5. URL http://dx.doi/10.1007/978-1-4419-8640-5.

Michel Raynaud. *Anneaux locaux henséliens.* Lecture Notes in Mathematics, Vol. 169. Springer, Berlin-New York, 1970.

Kosta Došen

On Sets of Premises

Abstract: Conceiving of premises as collected into sets or multisets, instead of sequences, may lead to triviality for classical and intuitionistic logic in general proof theory, where we investigate identity of deductions. Any two deductions with the same premises and the same conclusions become equal. In terms of categorial proof theory, this is a consequence of a simple fact concerning adjunction with a full and faithful functor applied to the adjunction between the diagonal functor and the product biendofunctor, which corresponds to the conjunction connective.

Keywords: Sequent, Deduction, Identity of deductions, Contraction, Isomorphism of formulae, Categories, Adjunction, Diagonal functor, Product, Conjunction

Mathematics Subject Classification 2010: 03F03, 03F07

1 Introduction

General proof theory addresses the question "What is a proof?", or rather "What is a deduction?" – a deduction being a hypothetical proof, i.e. a proof with hypotheses – by dealing with questions related to normal forms for deductions, and in particular with the question of identity criteria for deductions. It deals with the structure of deductions, as exhibited, for example, with the help of the typed lambda calculus in the Curry-Howard correspondence, and not with their strength measured by ordinals, which is what one finds in proof theory that arose out of Hilbert's programme.

Much of general proof theory is the field of categorial proof theory. Fundamental notions of category theory like the notion of adjoint functor, and very important structures like cartesian closed categories, came to be of central concern for logic in that field. Through results of categorial proof theory called coherence results, which provide a model theory for equality of deductions, logic finds new ties with geometry, topology and algebra (see the books [Došen, 1999], [Došen and Petrić, 2004] and [Došen and Petrić, 2007a], the more recent introductory survey [Došen, 2014], and references therein).

Kosta Došen: Faculty of Philosophy, University of Belgrade, and Mathematical Institute, Serbian Academy of Sciences and Arts, Knez Mihailova 36, p.f. 367, 11001 Belgrade, e-mail: kosta@mi.sanu.ac.rs

In general proof theory, and in particular in categorial proof theory, one looks for an algebra of deductions, and for that, one concentrates on the operations of this algebra, which come with the inference rules. As an equational theory, the algebra of deductions involves the question of identity criteria for deductions, the central question of general proof theory. (This question may be found, at least implicitly, in Hilbert's 24th problem; see [Došen, 2015].)

In categorial proof theory one usually studies a freely generated category of a certain kind equationally presented. This freely generated category is constructed out of syntactical material, as in universal algebra one constructs a freely generated algebra of a certain kind equationally presented by factoring through an equivalence relation on terms. In categories we have partial algebras—the arrow terms out of which the equivalence classes are built have types, their sources and targets—but there is no significant mathematical difference in the construction when compared with what one has in universal algebra (see [Došen and Petrić, 2004], Chapter 2, in particular in Section 2.3). The objects of this freely constructed categories are propositions, i.e. formulae, and the arrows, i.e. the equivalence classes of arrow terms, are deductions, i.e. equivalence classes of particular derivations, whose sources are premises and whose targets are conclusions. For deductions we have the partial operation of composition and identity deductions (this is essential for them; see [Došen, 2016]). The categories in question are interesting if they are not preorders, i.e., not all arrows with the same source and the same target are equal. Otherwise, the proof theory is trivial: any deductions with the same premises and the same conclusions become equal.

In terms of categorial proof theory, assuming that premises are collected into sets leads to assuming that for every proposition A we have that A and $A \wedge A$ are isomorphic, where $\wedge$ is the conjunction connective. Isomorphism is understood here as in category theory: there are arrows, i.e. deductions, from A to $A \wedge A$ and back, which composed give identity arrows, i.e. identity deductions (see [Došen and Petrić, 2012] and references therein). We are led to assume moreover that the associativity and commutativity of conjunction give isomorphisms.

It is not difficult to establish that a category $\mathcal{K}$ with binary product $\times$ is a preorder (i.e., any two arrows with the same source and the same target are equal) iff for every object B of $\mathcal{K}$ the diagonal arrow from B to $B \times B$ is an isomorphism. It is also not difficult to establish the related fact that $\mathcal{K}$ is a preorder iff for every object B of $\mathcal{K}$ the first-projection arrow from $B \times B$ to B is equal to the second-projection arrow with the same source and the same target. We will put these facts within a more general categorial context involving adjunction, which should shed light upon them. This is the main goal of this paper, and achieving this goal, together with some related matters (like those in the last section), is the novelty it should bring.

In categorial proof theory the binary connective of conjunction, both classical and intuitionistic, is identified with binary product. So what we mentioned above indicates that assuming that premises are collected into sets leads to preordered categories where objects are propositions, i.e., formulae, and arrows are deductions. This makes the proof theory trivial.

Assuming that premises are collected into multisets leads to assuming that for every proposition A the permutation deduction from $A \wedge A$ to $A \wedge A$, which permutes the two occurrences of A in $A \wedge A$, is equal to the identity deduction from $A \wedge A$ to $A \wedge A$. Although that assumption leads to the same absolute trivialization of the proof theory of classical and intuitionistic logic, the trivialization obtained with it for linear logic is less absolute. The same less absolute trivialization is obtained for the proof theory of relevant logic by assuming moreover that A and $A \wedge A$ are isomorphic. (What is this less absolute trivialization will be explained in the last section, at the end of the paper.)

2 Sequents

Gentzen's sequents are expressions of the form $A_1, \ldots, A_n \vdash B_1, \ldots, B_m$ where $A_1, \ldots, A_n, B_1, \ldots, B_m$ are formulae of an object language, like a language of propositional logic or a first-order language. Instead of the turnstile $\vdash$ Gentzen writes $\rightarrow$ (which is more commonly used nowadays for the binary connective of implication; we use it below, as usual, for separating the sources and targets of arrows in categories), for A and B he uses Gothic letters, and for n and m Greek letters (see [Gentzen, 1935], Section I.2.3). The natural numbers n and m may also be zero; when n is zero $A_1, \ldots, A_n$ is the empty word, and analogously for m and $B_1, \ldots, B_m$. For what we have to say in this paper we could restrict ourselves to sequents where m is one.

The comma in sequents is an auxiliary symbol that serves to separate formulae in sequences, which however is not essential. A sequent could as well be $A_1 \ldots A_n \vdash B_1 \ldots B_m$, but it could be difficult, though not impossible, to see where A_i ends and A_{i+1} begins in the sequence $A_i A_{i+1}$. Instead of $p, p, p \wedge q \vdash r, r$ we would have the less perspicuous $ppp \wedge q \vdash rr$, which however is not ambiguous. (It becomes more perspicuous when we do not omit the outermost parentheses of formulae, as in $pp(p \wedge q) \vdash rr$.)

So Gentzen's sequents may be conceived as expressions of the form $\Gamma \vdash \Delta$ where Γ and Δ are finite, possibly empty, sequences of formulae. (Capital Greek letters as schemata in sequents originate from Gentzen; see op. cit.) To economize upon considerations involving structural rules, a number of authors seem to think they are improving upon Gentzen if they take in $\Gamma \vdash \Delta$ that Γ and Δ are not *sequences* of formulae, but other sorts of finite collections of formulae, *multisets* (i.e. sets of occurrences) of formulae or *sets* of formulae. One should note immediately that with that $\Gamma \vdash \Delta$ seizes to be a word of a formal language, as usually conceived. If Γ and Δ are multisets or sets, then $\Gamma \vdash \Delta$ is not a *sequence* of symbols. It could be conceived as a triple $(\Gamma, \vdash, \Delta)$, in which case $\vdash$ is not essential. A sequent could be identified with the ordered pair (Γ, Δ).

Such a move is not without its dangers. We are not usually interested in particular sequents, but in sequent schemata. In other words, we want our sequents to be closed under substitution. If we continue using the notation $A_1, \ldots, A_n \vdash B_1, \ldots, B_m$ as it is usually done, and if $A_1, \ldots, A_n$ and $B_1, \ldots, B_m$ are conceived as sets of formulae, then by substituting p for q from the sequent $p, q \vdash p$ we obtain as a substitution instance $p \vdash p$, and the application of the structural rule of thinning on the left is transformed into the figure on the right:

$$\frac{p \vdash p}{p, q \vdash p} \qquad\qquad \frac{p \vdash p}{p \vdash p}$$

which does not look like an application of thinning.

This may be embarrassing, but need not be calamitous. It becomes really dangerous when we are interested not only in provability, but in proofs, i.e. not only in deducibility, but in deductions, and try to characterize identity of deductions. If we do that in category theory, i.e. in categorial proof theory, making the move that corresponds to switching from sequences to multisets or sets is dangerous, and may result in collapse. It may trivialize matters: any two deductions with the same premises and the same conclusions will be equal. The corresponding categories will be preorders.

3 Sequents in categorial proof theory

In categorial proof theory sequents of the simple kind $A \vdash B$, where A and B are single formulae, give the types of arrows $f: A \to B$, with A being the source and B the target. In the presence of conjunction $\wedge$, which serves to replace the comma on the left-hand side of Gentzen's sequents, and the constant $\top$ which replaces the empty sequence, together with disjunction $\vee$ and the constant $\bot$ for the right-hand side, we can mimic Gentzen's sequents, as Gentzen himself envisaged in [Gentzen, 1935] (Section I.2.4).

Conjunction, classical or intuitionistic (they are the same), corresponds in this perspective to binary product, and $\top$ corresponds to a terminal object (which may be conceived as nullary product). The associativity natural isomorphism for binary product justifies Gentzen's switch to sequences. The matter with the commutativity, i.e. symmetry, natural isomorphism for binary product is more tricky. (We deal with it in the last section.) We may identify $A \wedge B$ and $B \wedge A$, but we should nevertheless distinguish the identity deduction for $A \wedge A$ to $A \wedge A$ from the deduction that permutes the two occurrences of A.

To the principle of contraction there does not however correspond an isomorphism: A and $A \wedge A$ are not isomorphic, as the sets A and $A \times A$ are not isomorphic if A is finite with more than one member. To assume that contraction is tied to an isomorphism leads to triviality. Matters are analogous with binary coproduct and an initial object, which correspond respectively $\vee$ and $\bot$, but we will concentrate on conjunction and product in this paper.

By conjunction we understand here a binary connective to which a meet operation of a semilattice corresponds algebraically. In the terminology of substructural logics, this is an additive conjunction, for which we have the two natural-deduction rules of conjunction elimination, the first-projection and second-projection rules that correspond to Gentzen's structural rule of thinning. (We will consider on another occasion what happens if we assume that contraction is an isomorphism for a multiplicative conjunction that does not involve thinning, like a conjunction we find in relevant logic.)

4 Adjunction with a full and faithful functor

We will appeal to a result concerning adjoint functors dual to Theorem 1 of Section IV.3 of [Mac Lane, 1971]. (The proof of this result below will be more direct and simpler than the proof in loc. cit.; it will not appeal to the Yoneda lemma.)

Let F be a functor from a category $\mathcal{K}$ to a category $\mathcal{A}$, and G a functor from $\mathcal{A}$ to $\mathcal{K}$, such that G is right-adjoint to F, with members of the counit of the adjunction being $\varphi_A : FGA \to A$, for A an object of the category $\mathcal{A}$, and members of the unit of the adjunction being $\gamma_B : B \to GFB$, for B an object of the category $\mathcal{K}$ (we follow the notational conventions of [Došen, 1999], Chapter 4, save that we write $\mathcal{K}$ instead of $\mathcal{B}$, to distinguish this category clearly from its object B). Then we have the following.

Proposition 4.1. *The functor F is faithful iff for every object B of $\mathcal{K}$ the arrow γ_B is monic.*

Proof: From left to right we have:

$$\begin{aligned}\gamma_B \cdot g_1 = \gamma_B \cdot g_2 &\Rightarrow F\gamma_B \cdot Fg_1 = F\gamma_B \cdot Fg_2, \text{ since } F \text{ is a functor}\\ &\Rightarrow \varphi_{FB} \cdot F\gamma_B \cdot Fg_1 = \varphi_{FB} \cdot F\gamma_B \cdot Fg_2\\ &\Rightarrow Fg_1 = Fg_2, \text{ by a triangular equation of adjunction}\\ &\Rightarrow g_1 = g_2, \text{ since } F \text{ is faithful.}\end{aligned}$$

From right to left we have:

$$\begin{aligned}Fg_1 = Fg_2 &\Rightarrow GFg_1 \cdot \gamma_C = GFg_2 \cdot \gamma_C\\ &\Rightarrow \gamma_B \cdot g_1 = \gamma_B \cdot g_2, \text{ by the naturality of } \gamma\\ &\Rightarrow g_1 = g_2, \text{ since } \gamma_B \text{ is monic.}\end{aligned}$$

□

For $\mathbf{1}_B : B \to B$ being an identity arrow we have the following.

Proposition 4.2. *The functor F is full iff for every object B of $\mathcal{K}$ there is an arrow $h_B : GFB \to B$ of $\mathcal{K}$ such that $\gamma_B \cdot h_B = \mathbf{1}_{GFB}$.*

Proof: From left to right, the fullness of F implies that for every object B of $\mathcal{K}$ there is an arrow $h_B : GFB \to B$ such that $Fh_B = \varphi_{FB}$. We have:

$$\begin{aligned}\gamma_B \cdot h_B &= GFh_B \cdot \gamma_{GFB}, \text{ by the naturality of } \gamma\\ &= G\varphi_{FB} \cdot \gamma_{GFB}, \text{ since } Fh_B = \varphi_{FB}\\ &= \mathbf{1}_{GFB}, \text{ by a triangular equation of adjunction.}\end{aligned}$$

From right to left, take $f : FB_1 \to FB_2$ and $j = h_{B_2} \cdot Gf \cdot \gamma_{B_1} : B_1 \to B_2$. We have:

$$\begin{aligned}Fj &= Fh_{B_2} \cdot FGf \cdot F\gamma_{B_1}, \text{ since } F \text{ is a functor}\\ &= \varphi_{FB_2} \cdot F\gamma_{B_2} \cdot Fh_{B_2} \cdot FGf \cdot F\gamma_{B_1}, \text{ by a triangular equation of adjunction}\\ &= \varphi_{FB_2} \cdot FGf \cdot F\gamma_{B_1}, \text{ since } \gamma_{B_2} \cdot h_{B_2} = \mathbf{1}_{GFB_2} \text{ and } F \text{ is a functor}\\ &= f \cdot \varphi_{FB_1} \cdot F\gamma_{B_1}, \text{ by the naturality of } \varphi\\ &= f, \text{ by a triangular equation of adjunction.}\end{aligned}$$

□

Corollary 4.3. *The functor F is full and faithful iff for every object B of $\mathcal{K}$ the arrow γ_B is an isomorphism.*

Proof: From left to right, we have h_B by the fullness of F and Proposition 4.2. Then we have:

$$\begin{aligned}F(h_B \cdot \gamma_B) &= Fh_B \cdot F\gamma_B, \text{ since } F \text{ is a functor}\\ &= \varphi_{FB} \cdot F\gamma_B, \text{ since } Fh_B = \varphi_{FB}\\ &= F\mathbf{1}_B,\end{aligned}$$

by a triangular equation of adjunction and the functoriality of F. By the faithfulness of F we obtain that $h_B \circ \gamma_B = \mathbf{1}_B$.

From right to left, we use Propositions 4.1 and 4.2 from right to left. □

As another corollary we have that if F is full and faithful, then the adjunction is trivial, in the sense that every two canonical arrows of adjunction of the same type, i.e. with the same source and the same target, are equal (for trivial adjunctions see [Došen, 1999], Sections 4.6.2 and 4.11).

5 Adjunction with the diagonal functor

Consider now the diagonal functor D from a category $\mathcal{K}$ to the product category $\mathcal{K} \times \mathcal{K}$. This functor is always faithful. We have on the other hand the following.

Proposition 5.1. *The functor D is full iff $\mathcal{K}$ is a preorder.*

Proof: From left to right, take the arrows $g, g' : B_1 \to B_2$ of $\mathcal{K}$. By the fullness of D, for the arrow $(g, g') : (B_1, B_1) \to (B_2, B_2)$ of $\mathcal{K} \times \mathcal{K}$ we have an arrow h of $\mathcal{K}$ such that $(g, g') = Dh = (h, h)$. Hence $g = g'$.

From right to left, we have for every arrow $(g, g') : (B_1, B_1) \to (B_2, B_2)$ of $\mathcal{K} \times \mathcal{K}$ that $g = g'$. Hence $(g, g') = (g, g) = Dg$. □

The category $\mathcal{K}$ has a product biendofunctor $\times$ iff this functor is right adjoint to the diagonal functor D. The members $\gamma_B : B \to B \times B$ of the unit of this adjunction, which can be called diagonal arrows, correspond to the contraction arrows of conjunction $w_B : B \vdash B \wedge B$. If these arrows are isomorphisms, then, by Proposition 4.2 (and Corollary 4.3 from right to left), the diagonal functor D is full, and, by Proposition 5.1 from left to right, $\mathcal{K}$ is a preorder. In proof-theoretical terms, any two deductions with the same premises and the same conclusions are equal. So if we want our proof theory to be nontrivial, we should not permit the contraction arrows w_B to be isomorphisms.

Conversely, if $\mathcal{K}$ is a preorder, by Proposition 5.1 from right to left, the diagonal functor is full, and, by Corollary 4.3 from left to right, the arrows $\gamma_B : B \to B \times B$ are isomorphisms. So we have established the following for categories $\mathcal{K}$ with binary product $\times$.

Proposition 5.2. *The category $\mathcal{K}$ is a preorder iff for every object B of $\mathcal{K}$ the diagonal arrow $\gamma_B : B \to B \times B$ is an isomorphism.*

Note that it is possible in a category $\mathcal{K}$ with binary product $\times$ to have for every object B that B and $B \times B$ are isomorphic without the diagonal map being an isomorphism. We will consider that matter in the next section.

We have established Proposition 5.2 without mentioning explicitly the projection arrows, but they are in the background. It is easy to see that $\mathcal{K}$ is a preorder iff for every object B of $\mathcal{K}$ we have that the first-projection arrow $k^1_{B,B} : B \times B \to B$ is equal to the second-projection arrow $k^2_{B,B} : B \times B \to B$. From left to right this is trivial, while for the other direction we have that if $k^1_{B,B} = k^2_{B,B}$, then for any two arrows $g, g' : C \to B$ of $\mathcal{K}$ we have for $\langle g, g' \rangle : C \to B \times B$ that $k^1_{B,B} \circ \langle g, g' \rangle = k^2_{B,B} \circ \langle g, g' \rangle$, and hence $g = g'$.

If the diagonal functor D is full, then for the members

$$(k^1_{B,B}, k^2_{B,B}) : (B \times B, B \times B) \to (B, B)$$

of the counit of the adjunction of D with the biendofunctor $\times$ we have that $k^1_{B,B} = k^2_{B,B}$. So if the projection arrows $k^1_{B,B}$ and $k^2_{B,B}$ are not equal, then D is not full, and, by Proposition 5.1, the category $\mathcal{K}$ is not a preorder. That $\mathcal{K}$ is not a preorder follows of course immediately from the inequality of $k^1_{B,B}$ and $k^2_{B,B}$, but it is worth seeing how Proposition 5.1 and with it the wider perspective of the preceding section are involved.

This shows how the non-triviality of $\mathcal{K}$ hinges on distinguishing the two projection arrows from $B \times B$ to B, i.e. the two deductions from $B \wedge B$ to B based on the two natural-deduction rules of conjunction elimination (see the end of Section 3).

In a category with binary product we derive $k^1_{B,B} = k^2_{B,B}$ from the assumption that the diagonal arrow $w_B : B \to B \times B$, called γ_B above, is an isomorphism in the following manner. We have in any such category that $k^1_{B,B} \circ w_B = k^2_{B,B} \circ w_B = \mathbf{1}_B$, and then, by the isomorphism of w_B, we obtain $k^1_{B,B} = k^2_{B,B}$.

When $k^1_{B,B} = k^2_{B,B}$, then $k^1_{B,B}$, i.e. $k^2_{B,B}$, is the arrow inverse to $w_B : B \to B \times B$, which makes of the diagonal arrow w_B an isomorphism. We have $k^1_{B,B} \circ w_B = \mathbf{1}_B$ anyway. We also have:

$$\begin{aligned} w_B \circ k^1_{B,B} &= \langle k^1_{B,B}, k^1_{B,B} \rangle, \text{ by the naturality of } w_B \\ &= \langle k^1_{B,B}, k^2_{B,B} \rangle, \text{ since } k^1_{B,B} = k^2_{B,B} \\ &= \mathbf{1}_{B \times B}. \end{aligned}$$

The reduction to triviality brought by assuming that the diagonal arrows w_B are isomorphisms could have been shown by appealing only to these comments about $k^1_{B,B} = k^2_{B,B}$. We preferred however to put the matter in a wider perspective given by the preceding section. This shows that the matters we deal with are not peculiar to the adjunction with the diagonal functor.

The triviality of the adjunctions involving the quantifiers that are considered in [Došen and Petrić, 2009] (Section 1.4) has the same roots involving sets. The universally quantified formula $\forall xA$ interpreted over a domain with two objects a and b stands either for the conjunction $A^x_a \wedge A^x_b$ or the conjunction $A^x_b \wedge A^x_a$. Moreover, $A^x_a \wedge A^x_a$ and A^x_a are not distinguished. Quantifiers involve sets of objects, and neither multisets nor sequences. So from the point of view of general proof theory the assertion that quantifiers are like conjunctions and disjunctions should be taken with a grain of salt.

6 Collapse with sets and multisets of premises

We have noted after Proposition 5.2 that it is possible in a category with binary product $\times$ to have for every object B that B and $B \times B$ are isomorphic without the diagonal arrow $w_B : B \to B \times B$ being an isomorphism. The isomorphism may be produced by something else. To see that, take the category $\mathcal{C}$ with binary product freely generated out of a nonempty set P of generating objects, which are usually taken to be syntactical objects, letters, like propositional variables. (The construction of such equationally presented syntactical freely generated categories is described in detail in [Došen and Petrić, 2004], Chapter 2, in particular in Section 2.3.) Out of $\mathcal{C}$ we can build another category with binary product, which we call $\mathcal{C}'$, by replacing the objects of $\mathcal{C}$, which are like propositional conjunctive formulae, by sets of letters, free generators, occurring in these formulae. If P is the singleton $\{p\}$, then we have in $\mathcal{C}'$ a single object $\{p\}$ (i.e. P itself). The category $\mathcal{C}'$ is not a preorder if we keep in it the structure of arrows of $\mathcal{C}$. (That this is possible is shown by the one-object category that is a skeleton of the category of denumerably infinite sets with functions as arrows.[1]) So the diagonal arrow $w_{\{p\}} : \{p\} \to \{p\}$, though its source and target are both $\{p\}$, need not be equal to the identity arrow $\mathbf{1}_{\{p\}} : \{p\} \to \{p\}$. Analogously, the projection arrows $k^1_{\{p\},\{p\}} : \{p\} \to \{p\}$ and $k^2_{\{p\},\{p\}} : \{p\} \to \{p\}$ need not be equal to $\mathbf{1}_{\{p\}}$, and they need not be mutually equal.

Once we have passed to $\mathcal{C}'$ and understand this category proof-theoretically, the temptation is however big to take the diagonal arrow $w_{\{p\}}$, which corresponds to contraction, to be the identity arrow $\mathbf{1}_{\{p\}}$. In proof theory the switch to sets of premises is usually made in order not to be obliged to keep an account of structural rules. If we have to continue keeping this account, what is the advantage of the switch? We are moreover required to keep an account of contraction when it is invisible, when it consists in passing from $\{p\}$ to $\{p\}$.

When $w_{\{p\}}$ is identified with $\mathbf{1}_{\{p\}}$ we have collapse, as we have seen in this paper (and in particular in Proposition 5.2). Collapse also ensues when $k^1_{\{p\},\{p\}} =$

1 I am grateful to Zoran Petrić for suggesting this example.

$k^2_{\{p\},\{p\}} = \mathbf{1}_{\{p\}} : \{p\} \to \{p\}$, which seems tempting and natural to assume when the structural rule of thinning is invisible, when it adds a premise we already have, and premises are collected into sets (see the trivial thinning figure mentioned in Section 2).

Suppose now $\mathcal{C}$ is a freely generated category with a symmetric monoidal multiplication biendofunctor $\otimes$ (see [Mac Lane, 1971], Section VII.7; these categories are called symmetric associative in [Došen and Petrić, 2004], Chapter 5), which corresponds to the multiplicative conjunction of linear logic, and is tied to collecting premises into multisets. In $\mathcal{C}$ we have a symmetry natural isomorphism $c_{B_1,B_2} : B_1 \otimes B_2 \to B_2 \otimes B_1$. Now $\mathcal{C}'$ is obtained from $\mathcal{C}$ by replacing the objects of $\mathcal{C}$, i.e. propositional formulae, by sets of occurrences of letters in these formulae, i.e. multisets of letters occurring in these formulae. As before, $\mathcal{C}'$ is not a preorder if we keep in it the structure of arrows of $\mathcal{C}$, and in particular continue distinguishing the symmetry isomorphisms $c_{p,p} : p \otimes p \to p \otimes p$ from the identity arrows $\mathbf{1}_{p\otimes p} : p \otimes p \to p \otimes p$. The corresponding arrows in $\mathcal{C}'$ go from the multiset $\{p, p\}$ to the multiset $\{p, p\}$.

The assumption $c_{B,B} = \mathbf{1}_{B\otimes B}$ is however dangerous. If $\otimes$ happens to be binary product $\times$, this assumption leads to collapse, as did the isomorphism of the diagonal arrows and the equality of the projection arrows. This is because we have $k^1_{B,B} \circ c_{B,B} = k^2_{B,B}$ in categories with binary product, where $c_{B,B}$ is defined as $\langle k^2_{B,B}, k^1_{B,B}\rangle$. So collecting premises into multisets in classical and intuitionistic logic is as dangerous as collecting them into sets.

The situation is somewhat different in linear logic. We have the assumption $c_{B,B} = \mathbf{1}_{B\otimes B}$ in a category called $\mathbf{S}'$ in [Došen and Petrić, 2004] (Section 6.5; this category is obtained out of the symmetric associative category freely generated by a nonempty set of generating objects), which is a preorder, and this shows that the assumption $c_{B,B} = \mathbf{1}_{B\otimes B}$ makes that all the canonical arrows tied to multiplicative conjunction in linear logic that have the same source and the same target are equal. This does not mean however that the assumption will lead in general to preorder.

If among the free generators we do not have only objects, but also different arrows with the same source and the same target, they will not become equal because of $c_{B,B} = \mathbf{1}_{B\otimes B}$. This is shown by an inductive argument establishing that for every equation derivable for such a freely generated category, for every generating arrow f, we cannot have f on one side of the equation without having it on the other.[2] The same holds for relevant logic and categories related to it (see [Petrić, 2002] and [Došen and Petrić, 2007b]).

---

2 I am grateful to Zoran Petrić for suggesting this simple argument.

In these categories we have, as members of a natural transformation corresponding to contraction, diagonal arrows from B to $B \otimes B$, with $\otimes$ corresponding to multiplicative conjunction as above. In addition to $c_{B,B} = \mathbf{1}_{B\otimes B}$, we are led to assume that these diagonal arrows are isomorphisms if we assume that premises are collected into sets, and we will show in another place how this makes equal all the canonical arrows tied to multiplicative conjunction that have the same source and the same target. This assumption of isomorphism will however not lead in general to preorder, for reasons analogous to those in the preceding paragraph. So, in linear and relevant logic, with premises collected into multisets or sets the danger is still there, but not as big as in classical and intuitionistic logic.

Acknowledgment: Work on this paper was supported by the Ministry of Education, Science and Technological Development of Serbia, while the Alexander von Humboldt Foundation has, within the frame of Humboldt-Kollegs, supported the presentation of a talk partly related to the paper at the conference *Proof* in Bern in September 2013. I am grateful to the organizers of that conference for their hospitality, and to Miloš Adžić and Zoran Petrić for discussing the ideas of the paper, reading a draft of it and making useful comments.

References

K. Došen. *Cut Elimination in Categories*. Kluwer, Dordrecht, 1999.

K. Došen. Algebras of deductions in category theory. In D. Jokanović et al., editor, *Third Mathematical Conference of the Republic of Srpska, Proceedings, Trebinje 2013, Zbornik radova*, volume I, pages 11–18, Trebinje, 2014. Univerzitet u Istočnom Sarajevu, Fakultet za proizvodnju i menadžment. (available at: http://www.mi.sanu.ac.rs/~kosta/Dosen%20Algebras%20of%20Deductions.pdf).

K. Došen. General proof theory. In P. Schroeder-Heister et al., editor, *Logic, Methodology and Philosophy of Science – Proceedings of the 14th International Congress (Nancy 2011): Logic and Science Facing the New Technologies*, pages 149–151, London, 2015. College Publications. (preprint available at: http://www.mi.sanu.ac.rs/~kosta/Dosen_General%20Proof%20Theory%20nov2011.pdf).

K. Došen. On the paths of categories: An introduction to deduction. In T. Piecha and P. Schroeder-Heister et al., editors, *Advances in Proof-Theoretic Semantics*, pages 65–77, Cham, 2016. Springer. (preprint available at: http://www.mi.sanu.ac.rs/~kosta/Dosen%20On%20the%20Paths.pdf).

K. Došen and Z. Petrić. *Proof-Theoretical Coherence*. KCL Publications (College Publications), London, 2004. (revised version of 2007 available at: http://www.mi.sanu.ac.rs/~kosta/publications.htm).

K. Došen and Z. Petrić. *Proof-Net Categories*. Polimetrica, Monza, 2007a. (preprint of 2005, inessentially different from the printed text, available at: http://www.mi.sanu.ac.rs/~kosta/publications.htm).

K. Došen and Z. Petrić. Relevant categories and partial functions. *Publications de l'Institut Mathématique, N.S.*, 82 (96):17–23, 2007b. (available at: http://arXiv.org).

K. Došen and Z. Petrić. Coherence in linear predicate logic. *Annals of Pure and Applied Logic*, 158:125–153, 2009. (available at: http://arXiv.org).

K. Došen and Z. Petrić. Isomorphic formulae in classical propositional logic. *Mathematical Logic Quarterly*, 58:5–17, 2012. (available at: http://arXiv.org).

G. Gentzen. Untersuchungen über das logische Schließen. *Mathematische Zeitschrift*, 39: 176–210, 405–431, 1935. (English translation: *Investigations into logical deduction*, in *The Collected Papers of Gerhard Gentzen*, M.E. Szabo, editor, pages 68-131, 312-317, Amsterdam, 1969. North-Holland).

S. Mac Lane. *Categories for the Working Mathematician*. Springer, Berlin, 1971. Second edition, 1998.

Z. Petrić. Coherence in substructural categories. *Studia Logica*, 70:271–296, 2002. Available at: http://arXiv.org.

Hajime Ishihara and Takako Nemoto

Non-Deterministic Inductive Definitions and Fullness

Abstract: In this paper, we deal with the non-deterministic inductive definition principle NID with the weak notion of a set-generated class introduced by van den Berg and with the strong notion of a set-generated class adopted by Aczel et al. We introduce a principle, called nullary NID, and prove that nullary NID is equivalent to Fullness in a subsystem of the constructive Zermelo-Fraenkel set theory.

Keywords: Constructive set theory, Non-deterministic inductive definition, Fullness

Mathematics Subject Classification 2010: 03E70, 03F65

1 Introduction

The notion of a *set-generated class* was introduced in [Aczel, 2006] for dcpos using some terminology from domain theory. A partially ordered class is a *directed complete partial order* (dcpo) if each directed subset has a least upper bound, where a subset is *directed* if any pair of elements of the subset has an upper bound in the subset. A dcpo X is *set-generated* if there is a subset G of X such that, for each $a \in X$, $\{x \in G \mid x \leq a\}$ is a directed subset whose least upper bound is a. If we restrict our attention to a class X of subsets of a set with the inclusion $\subseteq$ as a partial order, then we may say that X is *set-generated* if there exists a subset G of X such that

$$\forall \alpha \in X \forall \tau \in \mathrm{Fin}(\alpha) \exists \beta \in G[\tau \subseteq \beta \subseteq \alpha],$$

where $\mathrm{Fin}(\alpha)$ is the set of finitely enumerable subsets of α.

This definition was adopted in [Aczel et al., 2015], and it was shown that the notion of a set-generated class plays crucial roles in predicative constructive mathematics. However, in an early draft of [Aczel et al., 2015] and in [van den Berg, 2013], a weaker notion of a set-generated class was employed: a class X of subsets of a set is set-generated if there exists a subset G of X such that

$$\forall \alpha \in X \forall x \in \alpha \exists \beta \in G[x \in \beta \subseteq \alpha].$$

Hajime Ishihara and Takako Nemoto: School of Information Science, Japan Advanced Institute of Science and Technology, Nomi, Ishikawa 923-1292, Japan, e-mail: ishihara@jaist.ac.jp, t-nemoto@jaist.ac.jp

Note that the strong notion of a set-generated class is crucial in some applications, for example, [Ishihara and Kawai, 2015].

In [van den Berg, 2013], the principle NID on non-deterministic inductive definitions and set-generated classes were introduced, using the weaker notion of a set-generated class in the constructive set theory CZF. A *rule* on a set S is a pair (a, b) of subsets a and b of S, and a rule is called *elementary* if a is a singleton and *finitary* if a is finitely enumerable. A subset α of S is *closed under* the rule (a, b) if

$$a \subseteq \alpha \Rightarrow b \between \alpha,$$

where $b \between \alpha \Leftrightarrow \exists x \in b(x \in \alpha)$, that is, the intersection of b and α is inhabited. For a set R of rules on S, we call a subset α of S *R-closed* if it is closed under each rule in R. The NID principle is that for each set S and set R of rules on S, the class of R-closed subsets of S is set-generated. If we restrict rules in NID to elementary and finitary rules, we call them elementary NID and finitary NID, respectively.

On the other hand, [Aczel et al., 2015] characterized set-generated classes with the strong notion using generalized geometric theories and the set generation axiom (SGA) in CZF. In [van den Berg, 2013], it was discussed on a relation between finitary NID and SGA, and revealed some aspect between the weak notion and the strong notion of a set-generated class. He also showed that elementary NID implies Fullness, a theorem in CZF, which is an important axiom in a subsystem of CZF that implies Exponentiation (the class of functions between sets is a set).

In this paper, we introduce another NID principle, called *nullary* NID, which is weaker than finitary NID, and prove that nullary NID is equivalent to Fullness in a subsystem of CZF, that is, the elementary constructive set theory **ECST**. We also show that elementary NID implies nullary NID, and that nullary, elementary and finitary NID with the weak notion of a set-generated class are equivalent to respectively nullary, elementary and finitary NID with the strong notion.

2 The elementary constructive set theory

The constructive Zermelo-Fraenkel set theory CZF, founded by [Aczel, 1978, 1982, 1986], grew out of Myhill's constructive set theory [Myhill, 1975] as a formal system for Bishop's constructive mathematics, and permits a quite natural interpretation in Martin-Löf type theory [Martin-Löf, 1984]. Aczel and Rathjen introduced the elementary constructive set theory **ECST** which is a subsystem of CZF in their book draft [Aczel and Rathjen, 2010] written by extending their research report [Aczel and Rathjen, 2001].

Definition 1. *The language of a constructive set theory contains variables for sets and the binary predicates* $=$ *and* $\in$. *The axioms and rules are those of intuitionistic predicate logic with equality. In addition,* ECST *has the following set theoretic axioms:*

- ***Extensionality:*** $\forall a \forall b(\forall x(x \in a \leftrightarrow x \in b) \to a = b)$.
- ***Pairing:*** $\forall a \forall b \exists c \forall x(x \in c \leftrightarrow x = a \vee x = b)$.
- ***Union:*** $\forall a \exists b \forall x(x \in b \leftrightarrow \exists y \in a(x \in y))$.
- ***Restricted Separation:***

$$\forall a \exists b \forall x(x \in b \leftrightarrow x \in a \wedge \varphi(x))$$

  for every restricted formula $\varphi(x)$. *Here a formula* $\varphi(x)$ *is restricted, or* Δ_0, *if all the quantifiers occurring in it are bounded, i.e. of the form* $\forall x \in c$ *or* $\exists x \in c$.
- ***Replacement:***

$$\forall a(\forall x \in a \exists! y \varphi(x, y) \to \exists b \forall y(y \in b \leftrightarrow \exists x \in a \varphi(x, y)))$$

  for every formula $\varphi(x, y)$.
- ***Strong Infinity:***

$$\exists a[0 \in a \wedge \forall x(x \in a \to x + 1 \in a) \\ \wedge \forall y(0 \in y \wedge \forall x(x \in y \to x + 1 \in y) \to a \subseteq y)],$$

  where $x + 1$ *is* $x \cup \{x\}$, *and* 0 *is the empty set* $\emptyset$.

Let a and b be sets. Using Replacement and Union, the *cartesian product* $a \times b$ of a and b consisting of the ordered pairs $(x, y) = \{\{x\}, \{x, y\}\}$ with $x \in a$ and $y \in b$ can be introduced in ECST. Similarly, we can introduce the *disjoint union* $a + b = \{(0, x) \mid x \in a\} \cup \{(1, y) \mid y \in b\}$, where $1 = \{0\}$. A *relation* r between a and b is a subset of $a \times b$. A relation $r \subseteq a \times b$ is *total* (or is a *multivalued function*) if for every $x \in a$ there exists $y \in b$ such that $(x, y) \in r$. The class of total relations between a and b is denoted by $\mathrm{mv}(a, b)$, or more formally

$$r \in \mathrm{mv}(a, b) \Leftrightarrow r \subseteq a \times b \wedge \forall x \in a \exists y \in b((x, y) \in r).$$

A *function* from a to b is a total relation $f \subseteq a \times b$ such that for every $x \in a$ there is exactly one $y \in b$ with $(x, y) \in f$. The class of functions from a to b is denoted by b^a, or more formally

$$f \in b^a \Leftrightarrow f \in \mathrm{mv}(a, b) \wedge \forall x \in a \forall y, z \in b((x, y) \in f \wedge (x, z) \in f \to y = z).$$

We use ω for the unique set a such that $0 \in a \wedge \forall x(x \in a \to x + 1 \in a)$, ensured by Strong Infinity. For a set S, let $\mathrm{Fin}(S)$ denote the class $\{\mathrm{ran}(f) \mid f \in S^n, n \in \omega\}$

of finitely enumerable subsets of S, and let $\mathrm{Fin}^+(S)$ denote the class $\{\sigma \in \mathrm{Fin}(S) \mid \exists x \in S(x \in \sigma)\}$ of finitely enumerable inhabited subsets of S.

The constructive set theory CZF is obtained from ECST by replacing Replacement by

Strong Collection:

$$\forall a(\forall x \in a\exists y\varphi(x, y) \to \exists b(\forall x \in a\exists y \in b\varphi(x, y) \wedge \forall y \in b\exists x \in a\varphi(x, y)))$$

for every formula $\varphi(x, y)$,

and adding

- **Subset Collection:**

$$\forall a\forall b\exists c\forall u(\forall x \in a\exists y \in b\varphi(x, y, u) \to \exists d \in c(\forall x \in a\exists y \in d\varphi(x, y, u) \wedge \forall y \in d\exists x \in a\varphi(x, y, u)))$$

  for every formula $\varphi(x, y, u)$, and
- **$\in$-Induction:** $\forall a(\forall x \in a\varphi(x) \to \varphi(a)) \to \forall a\varphi(a)$, for every formula $\varphi(a)$.

In ECST, Subset Collection implies

Fullness: $\forall a\forall b\exists c(c \subseteq \mathrm{mv}(a, b) \wedge \forall r \in \mathrm{mv}(a, b)\exists s \in c(s \subseteq r))$,

and Fullness and Strong Collection imply Subset Collection. The notable consequence of Fullness is that b^a forms a set, that is

Exponentiation: $\forall a\forall b\exists c\forall f(f \in c \leftrightarrow f \in b^a)$.

Note that, in the presence of Exponentiation (and hence Fullness), the classes $\mathrm{Fin}(S)$ and $\mathrm{Fin}^+(S)$ are sets for each set S, by Replacement, Union and Restricted Separation.

For a set S, we write $\mathrm{Pow}(S)$ for the power class of S which is not a set in ECST nor in CZF:

$$a \in \mathrm{Pow}(S) \Leftrightarrow a \subseteq S.$$

3 Non-deterministic inductive definitions and Fullness

In this section, we work within the subsystem ECST of CZF.

Definition 2. *Let S be a set. Then a rule on S is a pair (a, b) of subsets a and b of S. A rule is called nullary if a is empty, elementary if a is a singleton and finitary if* *a is finitely enumerable. A subset α of S is closed under the rule* (a, b) *if*

$$a \subseteq \alpha \Rightarrow b \between \alpha.$$

For a set R of rules on S, we call a subset α of S R-closed if it is closed under each rule in R.

Remark 3. *Note that if a rule is nullary or elementary, then it is finitary.*

Definition 4. *Let S be a set and let X be a subclass of* $\mathrm{Pow}(S)$. *Then X is set-generated if there exists a subset G of X such that*

$$\forall \alpha \in X \forall x \in \alpha \exists \beta \in G(x \in \beta \subseteq \alpha),$$

and strongly set-generated if there exists a subset G of X such that

$$\forall \alpha \in X \forall \sigma \in \mathrm{Fin}(\alpha) \exists \beta \in G(\sigma \subseteq \beta \subseteq \alpha).$$

Definition 5. *Let* NID *and* NID* *denote the principles that for each set S and set R of rules on S, the class of R-closed subsets of S is set-generated and strongly set-generated, respectively. The principles obtained by restricting R in* $\mathrm{NID}^{(*)}$ *to a set of nullary, elementary and finitary rules are called nullary* $\mathrm{NID}^{(*)}$, *elementary* $\mathrm{NID}^{(*)}$ *and finitary* $\mathrm{NID}^{(*)}$, *respectively.*

Remark 6. *Trivially,* NID* *implies* NID, *and note that finitary* $\mathrm{NID}^{(*)}$ *implies nullary* $\mathrm{NID}^{(*)}$ *and elementary* $\mathrm{NID}^{(*)}$.

For a set S, let $*_S$ denote the set $\{x \in S \mid x \notin x\}$ which is not in S.

Theorem 7. *Nullary* NID *is equivalent to Fullness.*

Proof: Suppose nullary NID. Let A and B be sets, and define a set R of nullary rules on $(A \times B) \cup \{*_{A\times B}\}$ by

$$R = \{(\emptyset, \{x\} \times B) \mid x \in A\} \cup \{(\emptyset, \{*_{A\times B}\})\}.$$

Then, by nullary NID, there exists a subset G of the class X of R-closed subsets of $(A \times B) \cup \{*_{A\times B}\}$ such that

$$\forall \alpha \in X \forall z \in \alpha \exists \beta \in G(z \in \beta \subseteq \alpha).$$

Let $C = \{\beta \cap (A \times B) \mid \beta \in G\}$. Then for each $\beta \in G$ and $x \in A$, since $\beta \between \{x\} \times B$, there exists $y \in B$ such that $(x, y) \in \beta$, and hence $(x, y) \in \beta \cap (A \times B)$. Therefore $C \subseteq \mathrm{mv}(A, B)$. For each $r \in \mathrm{mv}(A, B)$, since $\{x\} \times B \between r \cup \{*_{A\times B}\}$ for each $x \in A$ and $*_{A\times B} \in r \cup \{*_{A\times B}\}$, we have $r \cup \{*_{A\times B}\} \in X$, and hence there exists $\beta \in G$ such that $*_{A\times B} \in \beta \subseteq r \cup \{*_{A\times B}\}$. Thus $\beta \cap (A \times B) \subseteq (r \cup \{*_{A\times B}\}) \cap (A \times B) = r$.

Conversely, suppose Fullness. Let R be a set of nullary rules on a set S. Then, by Fullness, there exists a set $C \subseteq \mathrm{mv}(R, S)$ such that

$$\forall r \in \mathrm{mv}(R, S) \exists s \in C(s \subseteq r).$$

Let

$$G = \{\{x\} \cup \mathrm{ran}(r) \mid x \in S, r \in C, \forall((\emptyset, b), y) \in r(y \in b)\}.$$

If $r \in C$ and $\forall((\emptyset, b), y) \in r(y \in b)$, then for each $(\emptyset, b) \in R$, we have $b \between \mathrm{ran}(r)$, and hence $b \between \{x\} \cup \mathrm{ran}(r)$ for each $x \in S$. Therefore G is a subset of the class of R-closed subsets of S. For each R-closed subset α of S, since $r = \{((\emptyset, b), y) \mid (\emptyset, b) \in R, y \in b \cap \alpha\} \in \mathrm{mv}(R, S)$, there exists $s \in C$ such that $s \subseteq r$. Note that, since $\forall((\emptyset, b), y) \in r(y \in b)$ and $\mathrm{ran}(r) \subseteq \alpha$, we have $\forall((\emptyset, b), y) \in s(y \in b)$ and $\mathrm{ran}(s) \subseteq \alpha$. Therefore if $x \in \alpha$, then $\{x\} \cup \mathrm{ran}(s) \in G$ and $x \in \{x\} \cup \mathrm{ran}(s) \subseteq \{x\} \cup \alpha = \alpha$. □

Proposition 8. *Elementary* NID *implies nullary* NID.

Proof: Let R be a set of nullary rules on a set S, and define a set R' of elementary rules on $S \cup \{*_S\}$ by

$$R' = \{(\{*_S\}, b) \mid (\emptyset, b) \in R\} \cup \{(\{x\}, \{*_S\})\}) \mid x \in S\}.$$

Then, by elementary NID, there exists a subset G' of the class X' of R'-closed subsets of $S \cup \{*_S\}$ such that

$$\forall \alpha' \in X' \forall z \in \alpha' \exists \beta' \in G'(z \in \beta' \subseteq \alpha').$$

Let $G = \{\beta' \cap S \mid \beta' \in G', *_S \in \beta'\}$. Then for each $\beta' \cap S \in G$ and $(\emptyset, b) \in R$, since $*_S \in \beta'$, we have $b \between \beta'$, and hence $b \between \beta' \cap S$. Therefore G is a set of R-closed subsets of S. Let α be an R-closed subset of S. Then for each $(\emptyset, b) \in R$, since $b \between \alpha$, we have $b \between \alpha \cup \{*_S\}$, and $*_S \in \alpha \cup \{*_S\}$. Hence $\alpha \cup \{*_S\} \in X'$. Therefore for each $x \in \alpha$, since $x \in \alpha \cup \{*_S\}$, there exists $\beta' \in G'$ such that $x \in \beta' \subseteq \alpha \cup \{*_S\}$, and so $*_S \in \beta'$. Thus $\beta' \cap S \in G$ and $x \in \beta' \cap S \subseteq (\alpha \cup \{*_S\}) \cap S = \alpha$. □

Remark 9. *Note that, assuming nullary, elementary or finitary* NID*, we have Fullness, and hence the classes* $\mathrm{Fin}(S)$ *and* $\mathrm{Fin}^+(S)$ *are sets for each set* S.

For a subset α of the disjoint union $A + B$, let $(\alpha)_0$ and $(\alpha)_1$ denote the sets $\{x \in A \mid (0, x) \in \alpha\}$ and $\{y \in B \mid (1, y) \in \alpha\}$, respectively.

Proposition 10. *Nullary, elementary and finitary* NID *imply nullary, elementary and finitary* NID**, respectively.*

Proof: Let R be a set of nullary rules on a set S, and define a set R' of nullary rules on $\mathrm{Fin}(S)$ by

$$R' = \{(\emptyset, \mathrm{Fin}^+(b)) \mid (\emptyset, b) \in R\}.$$

Then, by nullary NID, there exists a subset G' of the class X' of R'-closed subsets of $\mathrm{Fin}(S)$ such that

$$\forall \alpha' \in X' \forall \sigma \in \alpha' \exists \beta' \in G'(\sigma \in \beta' \subseteq \alpha').$$

Let $G = \{\bigcup \beta' \mid \beta' \in G'\}$. Then for each $\beta' \in G'$ and $(\emptyset, b) \in R$, since $\mathrm{Fin}^+(b) \between \beta'$, there exists $\sigma \in \mathrm{Fin}^+(b) \cap \beta'$, hence there exists $x \in b$ such that $x \in \sigma \in \beta'$, and therefore $b \between \bigcup \beta'$. Thus G is a set of R-closed subsets of S. Let α be an R-closed subset of S. Then for each $(\emptyset, \mathrm{Fin}^+(b)) \in R'$, since $(\emptyset, b) \in R$, we have $b \between \alpha$, and hence $\mathrm{Fin}^+(b) \between \mathrm{Fin}(\alpha)$. Therefore $\mathrm{Fin}(\alpha) \in X'$. Thus for each $\sigma \in \mathrm{Fin}(\alpha)$ there exists $\beta' \in G'$ such that $\sigma \in \beta' \subseteq \mathrm{Fin}(\alpha)$, and so $\sigma \subseteq \bigcup \beta' \subseteq \bigcup \mathrm{Fin}(\alpha) = \alpha$.

Let R be a set of elementary or finitary rules on a set S, and define a set R' of elementary and finitary rules on $S + \mathrm{Fin}(S)$ by

$$\begin{aligned} R' = {} & \{(1 \times a, 1 \times b) \mid (a, b) \in R\} \\ & \cup \{(1 \times a, \{(1, \sigma) \mid \sigma \in \mathrm{Fin}(b)\}) \mid (a, b) \in R\} \\ & \cup \{(\{(1, \sigma)\}, \{(0, x)\}) \mid \sigma \in \mathrm{Fin}(S), x \in \sigma\}. \end{aligned}$$

Note that R' is elementary and finitary if R is elementary and finitary, respectively. Then, by elementary or finitary NID, there exists a subset G' of the class X' of R'-closed subsets of $S + \mathrm{Fin}(S)$ such that

$$\forall \alpha' \in X' \forall z \in \alpha' \exists \beta' \in G'(z \in \beta' \subseteq \alpha').$$

Let $G = \{(\beta')_0 \mid \beta' \in G'\}$. Then for each $\beta' \in G'$ and $(a, b) \in R$ with $a \subseteq (\beta')_0$, since $1 \times a \subseteq \beta'$, we have $1 \times b \between \beta'$, and hence $b \between (\beta')_0$. Therefore G is a set of R-closed subsets of S. Let α be an R-closed subset of S. Then $\alpha + \mathrm{Fin}(\alpha)$ is an R'-closed subset of $S + \mathrm{Fin}(S)$. In fact, for each $(a, b) \in R$, if $1 \times a \subseteq \alpha + \mathrm{Fin}(\alpha)$, then, since $a \subseteq \alpha$, we have $b \between \alpha$, and hence $1 \times b \between \alpha + \mathrm{Fin}(\alpha)$ and $(1, \sigma) \in \alpha + \mathrm{Fin}(\alpha)$ for some $\sigma \in \mathrm{Fin}(b)$. For $(\{(1, \sigma)\}, \{(0, x)\}) \in R'$ where $\sigma \in \mathrm{Fin}(S)$ and $x \in \sigma$, if $\{(1, \sigma)\} \subseteq \alpha + \mathrm{Fin}(\alpha)$, then, since $\sigma \in \mathrm{Fin}(\alpha)$, we have $x \in \sigma \subseteq \alpha$, and hence $(0, x) \in \alpha + \mathrm{Fin}(\alpha)$. Therefore $\alpha + \mathrm{Fin}(\alpha) \in X'$. Thus for each $\sigma \in \mathrm{Fin}(\alpha)$ there exists $\beta' \in G'$ such that $(1, \sigma) \in \beta' \subseteq \alpha + \mathrm{Fin}(\alpha)$, and since β' is R'-closed and $(0, x) \in \beta'$ for each $x \in \sigma$, we have $\sigma \subseteq (\beta')_0 \subseteq (\alpha + \mathrm{Fin}(\alpha))_0 = \alpha$. □

Corollary 11. *Nullary, elementary and finitary* NID *are equivalent to respectively nullary, elementary and finitary* NID*.

Acknowledgment: The authors thank the anonymous referee for helpful comments and suggestions, and the Japan Society for the Promotion of Science (JSPS), Core-to-Core Program (A. Advanced Research Networks) for supporting the research.

References

Peter Aczel. The type theoretic interpretation of constructive set theory. In *Logic Colloquium '77 (Proc. Conf., Wrocław, 1977)*, volume 96 of *Stud. Logic Foundations Math.*, pages 55–66. North-Holland, Amsterdam-New York, 1978.

Peter Aczel. The type theoretic interpretation of constructive set theory: choice principles. In *The L. E. J. Brouwer Centenary Symposium (Noordwijkerhout, 1981)*, volume 110 of *Stud. Logic Found. Math.*, pages 1–40. North-Holland, Amsterdam, 1982.

Peter Aczel. The type theoretic interpretation of constructive set theory: inductive definitions. In *Logic, methodology and philosophy of science, VII (Salzburg, 1983)*, volume 114 of *Stud. Logic Found. Math.*, pages 17–49. North-Holland, Amsterdam, 1986.

Peter Aczel. Aspects of general topology in constructive set theory. *Ann. Pure Appl. Logic*, 137 (1-3):3–29, 2006.

Peter Aczel and Michael Rathjen. Notes on constructive set theory. Technical Report No. 40, Institut Mittag-Leffler, The Royal Swedish Academy of Sciences, 2001.

Peter Aczel and Michael Rathjen. CST. Book draft, 2010. URL https://www1.maths.leeds.ac.uk/~rathjen/book.pdf.

Peter Aczel, Hajime Ishihara, Takako Nemoto, and Yasushi Sangu. Generalized geometric theories and set-generated classes. *Math. Structures Comput. Sci.*, 25(7):1466–1483, 2015.

Hajime Ishihara and Tatsuji Kawai. Completeness and cocompleteness of the categories of basic pairs and concrete spaces. *Math. Structures Comput. Sci.*, 25(8):1626–1648, 2015.

Per Martin-Löf. *Intuitionistic type theory*, volume 1 of *Studies in Proof Theory. Lecture Notes.* Bibliopolis, Naples, 1984. Notes by Giovanni Sambin.

John Myhill. Constructive set theory. *J. Symbolic Logic*, 40(3):347–382, 1975.

Benno van den Berg. Non-deterministic inductive definitions. *Arch. Math. Logic*, 52(1-2):113–135, 2013.

Ioannis Kokkinis and Thomas Studer

Cyclic Proofs for Linear Temporal Logic

Abstract: Annotated sequents provide an elegant approach for the design of deductive systems for temporal logics. Their proof theory, however, is notoriously difficult. It is not even clear how to syntactically show the admissibility of weakening. In this paper, we establish weakening by purely proof-theoretic methods, thus solving an open problem by Brünnler and Lange. We also investigate the role of cut in annotated sequent systems.

Keywords: Proof theory, Temporal logic

Mathematics Subject Classification 2010: 03B44, 03F03

1 Introduction

The proof theory of temporal logics, and of modal fixed point logics in general, is notoriously difficult. It is not even clear how to design a finitary deductive system for linear time temporal logic LTL with nice proof-theoretic properties. In this context, deductive systems featuring infinite long proof branches (together with a global soundness condition) and their cyclic variants have recently obtained much attention, see, for instance, [Brotherston and Simpson, 2011, Brünnler and Lange, 2008, Bucheli et al., 2010, Niwinski and Walukiewicz, 1996, Shamkanov, 2014, Studer, 2008].

Brünnler and Lange [Brünnler and Lange, 2008] proposed an elegant formalism with cyclic proofs for LTL using focus games from Lange and Stirling [Lange and Stirling, 2001] as an inspiration. The main technical feature of their system are annotated sequents, which are employed to derive greatest fixed points. However, some very basic proof-theoretic problems turn out to be surprisingly hard in this setting. For instance, despite the admissibility of several structural rules, including cut, being proved semantically in [Brünnler and Lange, 2008], it remains to prove the same facts by proper proof-theoretic methods. Even the admissibility of weakening, which is quite trivial for most types of sequent calculi, is far from being simple for annotated sequents due to the presence of sequent contexts in the annotations.

Ioannis Kokkinis, Thomas Studer: Institut für Informatik, Universität Bern, Neubrückstrasse 10, 3012 Bern, Switzerland, e-mail: kokkinis@inf.unibe.ch, tstuder@inf.unibe.ch

As Brünnler and Lange point out [Brünnler and Lange, 2008], the problem with weakening is to be expected since the fact that a certain statement is provable by induction does not imply that a weaker statement is also provable by induction.

In this paper we provide a solution to this open problem and establish the admissibility of weakening by proof-theoretic means.

Moreover, we present a series of examples that explain the design of Brünnler and Lange's system. Their system is based on annotations that are sets of sets of formulas and it uses several rules to unfold greatest fixed points. Our examples show that these features are necessary in order to have completeness, that means their system is as simple as a cut-free system can be.

However, we also show that if we add a cut-rule, then the system can be made much simpler. That is, we can have fewer rules and annotations of a simpler form. This provides a very nice and instructive example on the role of cut in proofs of induction statements.

2 Sequent Systems for LTL

In this section we see an approach for defining a sound and complete sequent calculus for LTL. Note that we only study the unary fragment of LTL. This is enough to discuss the proof-theoretic problems but simplifies the presentation.

We start with defining the syntax and semantics for LTL-sequents. Then we recall a naive way of giving an LTL-sequent calculus and explain its shortcomings. We finish this section with introducing the sequent calculus LT1, which is based on so-called annotations.

2.1 Syntax and Semantics

We start with a countable set of atomic propositions, which we call Prop. The language of the sequent calculus, $\mathcal{L}_{\mathbb{S}}$, is then described by the following grammar: (gesetzt mit in Latex wie oben angegeben)

$$A ::= P \mid \overline{P} \mid A \wedge A \mid A \vee A \mid \Box A \mid \diamond A \mid \bigcirc A$$

where $P \in \text{Prop}$ and $\overline{P}$ denotes the negation of P. In this paper, we assume right associativity for all binary connectives.

We define the set of sequents, Seq, by:

$$\text{Seq} := \{\Gamma \mid \Gamma \text{ is a finite subset of } \mathcal{L}_{\mathbb{S}}\}$$

We will use capital greek letters like $\Gamma, \Delta, \Sigma, \ldots$ for sequents, capital latin letters like $A, B, C, D, \ldots$ for $\mathcal{L}_\mathcal{S}$-formulas and the letters P, Q for elements of Prop, all of them possibly primed or with subscripts. As usual union is represented by comma, i.e.:

$$\Gamma, \Delta \text{ stands for } \Gamma \cup \Delta$$
$$\Gamma, A \text{ stands for } \Gamma \cup \{A\}$$

We define the negation of an $\mathcal{L}_\mathcal{S}$-formula as usual by:

$$\overline{\overline{P}} := P \qquad \overline{\bigcirc A} := \bigcirc \overline{A}$$
$$\overline{A \vee B} := \overline{A} \wedge \overline{B} \qquad \overline{\Box A} := \Diamond \overline{A}$$
$$\overline{A \wedge B} := \overline{A} \vee \overline{B} \qquad \overline{\Diamond A} := \Box \overline{A}$$

Let $\Gamma \in$ Seq. We define the following sequent:

$$\bigcirc \Gamma := \{\bigcirc A \mid A \in \Gamma\}$$

Now we can define the notion of LTL-model and validity.

Definition 1 (LTL-model). *An* LTL-*model or simply a model is a function μ that maps natural numbers to sets of atomic propositions, i.e.:*[1]

$$\mu : \mathbb{N} \to \mathcal{P}(\mathsf{Prop})$$

Let μ be a model. Every natural number i represents a point in time and the set $\mu(i)$ represents the facts that hold at the time-point i. The expression $\mu, i \models A$ stands for *model μ satisfies formula A at time-point i*. The relation $\models$ is defined as follows:

Definition 2 (Satisfiability of $\mathcal{L}_\mathcal{S}$-formulas). *Let μ be a model and let $i \in \mathbb{N}$. We have:*

$$\begin{aligned}
\mu, i \models P &\iff P \in \mu(i)\\
\mu, i \models \overline{P} &\iff P \notin \mu(i)\\
\mu, i \models A \wedge B &\iff \mu, i \models A \text{ and } \mu, i \models B\\
\mu, i \models A \vee B &\iff \mu, i \models A \text{ or } \mu, i \models B\\
\mu, i \models \Box A &\iff \forall j \geq i(\mu, j \models A)\\
\mu, i \models \Diamond A &\iff \exists j \geq i(\mu, j \models A)\\
\mu, i \models \bigcirc A &\iff \mu, i+1 \models A
\end{aligned}$$

1 $\mathcal{P}$ stands for powerset

A formula A is *valid*, denoted by $\models A$, iff $(\forall\mu)(\forall i)[\mu, i \models A]$, i.e. iff A holds in all models at all time-points.

Let $\Gamma \in$ Seq. We assume that Γ is semantically equivalent to the disjunction of its elements, i.e. we set

$$\mu, i \models \Gamma \iff \mu, i \models \bigvee_{A \in \Gamma} A$$

2.2 A Naive Approach

We begin with a "naive" sequent calculus for LTL, called $\mathsf{LT}_{\mathsf{naive}}$, which is given in Fig. 1. This system contains the usual propositional axioms and rules (aid, $\vee$, $\wedge$), the rule for $\bigcirc$ and the rules for unfolding $\Box$ and $\Diamond$.

As usual $\mathsf{L} \vdash A$ means that the logic L proves the formula A. When L is clear from the context we may simply write $\vdash A$. By $\mathsf{L} \overset{n}{\vdash} A$ we mean that there is a derivation of A in L with depth at most n.

$$\text{aid}\ \frac{}{\Gamma, P, \overline{P}} \qquad \vee\ \frac{\Gamma, A, B}{\Gamma, A \vee B} \qquad \wedge\ \frac{\Gamma, A \qquad \Gamma, B}{\Gamma, A \wedge B}$$

$$\Box\ \frac{\Gamma, A \qquad \Gamma, \bigcirc\Box A}{\Gamma, \Box A} \qquad \Diamond\ \frac{\Gamma, A, \bigcirc\Diamond A}{\Gamma, \Diamond A} \qquad \bigcirc\ \frac{\Gamma}{\Sigma, \bigcirc\Gamma}$$

Fig. 1. System $\mathsf{LT}_{\mathsf{naive}}$

It is easy to see that system $\mathsf{LT}_{\mathsf{naive}}$ is sound with respect to LTL-models. But what about completeness? The first observation, see [Brünnler and Lange, 2008], is that system $\mathsf{LT}_{\mathsf{naive}}$ almost works: the only thing that goes wrong is that we cannot derive induction principles as is shown in the following example.

Example 3. *The valid sequent*

$$\Gamma = \Diamond(P \wedge \bigcirc\overline{P}), \overline{P}, \Box P$$

cannot be proved in $\mathsf{LT}_{\mathsf{naive}}$. *This sequent is semantically equivalent to the temporal induction axiom*

$$P \wedge \Box(P \to \bigcirc P) \to \Box P$$

where the connective → is interpreted in the standard way.

An attempt to prove Γ in $\mathsf{LT}_{\mathsf{naive}}$ *will lead to a derivation like the following:*

$$
\dfrac{\dfrac{}{\Diamond(P \wedge \bigcirc\overline{P}), \overline{P}, P}\,\text{aid} \qquad \dfrac{\dfrac{\dfrac{}{\bigcirc\Diamond(P \wedge \bigcirc\overline{P}), P, \overline{P}, \bigcirc\Box P}\,\text{aid} \qquad \dfrac{\Diamond(P \wedge \bigcirc\overline{P}), \overline{P}, \Box P}{\bigcirc\Diamond(P \wedge \bigcirc\overline{P}), \bigcirc\overline{P}, \overline{P}, \bigcirc\Box P}\,\bigcirc}{\bigcirc\Diamond(P \wedge \bigcirc\overline{P}), P \wedge \bigcirc\overline{P}, \overline{P}, \bigcirc\Box P}\,\wedge}{\Diamond(P \wedge \bigcirc\overline{P}), \overline{P}, \bigcirc\Box P}\,\Diamond}{\Diamond(P \wedge \bigcirc\overline{P}), \overline{P}, \Box P}\,\Box
$$

Observe that the endsequent reappears in the top right of the derivation. Hence there is no proof of Γ in $\mathsf{LT}_{\mathsf{naive}}$.

When we try to prove a formula that contains the operator $\Box$, the proof-search will fail like it did in the above example for sequent Γ. However, the obvious idea of just closing a cyclic branch as axiomatic will lead to an unsound system as is illustrated in the next example, see [Brünnler and Lange, 2008].

Example 4. *Consider the non-valid sequent $\Delta = \Box P, \bigcirc\Diamond\Box P$. If we could close all the cyclic branches then we would have the following proof for Δ in* $\mathsf{LT}_{\mathsf{naive}}$*:*

$$
\dfrac{\dfrac{\dfrac{\Box P, \bigcirc\Diamond\Box P}{\Diamond\Box P}\,\Diamond}{P, \bigcirc\Diamond\Box P}\,\bigcirc \qquad \dfrac{\dfrac{\Box P, \bigcirc\Diamond\Box P}{\Box P, \Diamond\Box P}\,\Diamond}{\bigcirc\Box P, \bigcirc\Diamond\Box P}\,\bigcirc}{\Box P, \bigcirc\Diamond\Box P}\,\Box
$$

Hence, a better idea, than simply closing every cyclic branch, is required. Brünnler and Lange's idea ([Brünnler and Lange, 2008]) is to close a cyclic branch if there is a formula such that whenever the $\Box$-rule is applied to it between the two occurrences of the cyclic sequent, the branch is along the right premise. Thus, in Example 3, the rightmost branch of the $\mathsf{LT}_{\mathsf{naive}}$-derivation for Γ would be closed, thus yielding a correct proof. In Example 4, however, the left branch in the derivation for Δ would not be closed and hence this would not be a proof for the non-valid sequent Δ. In order to implement this idea, we have to enrich our syntax with the so-called *annotations*.

2.3 Annotations – System LT1

We define the set of *annotated* formulas, $\mathcal{L}_{\mathsf{ann}}$:

$$\mathcal{L}_{\mathsf{ann}} := \{\Box_\Gamma A, \bigcirc\Box_\Gamma A \mid A \in \mathcal{L}_{\mathcal{S}}, \Gamma \in \mathsf{Seq}\}$$

In $\Box_\Gamma A$, the sequent Γ is called an annotation. We define the set of annotated sequents:

$$\mathsf{Seq}_{\mathsf{ann}} := \{\, \Gamma \mid \Gamma \text{ is a finite subset of } \mathcal{L}_\mathcal{S} \cup \mathcal{L}_{\mathsf{ann}} \text{ that contains at most one annotated formula} \,\}$$

The semantics of $\Box_\Gamma A$ is defined as follows. Let μ be an LTL-model and let $i \in \mathbb{N}$. We have:

$$\mu, i \models \Box_\Gamma A \quad \Longleftrightarrow \quad \forall j \geq i\Big((\forall i \leq k \leq j(\mu, k \models \Gamma)) \Longrightarrow \mu, j \models A\Big)$$

System LT1 is given by the axioms and rules in Fig. 2. For the $\bigcirc$-rule, we assume $\Sigma \in \mathsf{Seq}$, i.e. Σ does not contain annotated formulas. For rep and foc, we also assume $\Gamma \in \mathsf{Seq}$. System LT1 contains all the rules of the system $\mathsf{LT}_{\mathsf{naive}}$ plus the

$$\mathsf{aid}\ \frac{}{\Gamma, P, \overline{P}} \qquad \vee\ \frac{\Gamma, A, B}{\Gamma, A \vee B} \qquad \wedge\ \frac{\Gamma, A \qquad \Gamma, B}{\Gamma, A \wedge B}$$

$$\mathsf{rep}\ \frac{}{\Gamma, \Box_\Gamma A} \qquad \mathsf{foc}\ \frac{\Gamma, A \qquad \Gamma, \bigcirc\Box_\Gamma A}{\Gamma, \Box A}$$

$$\Box\ \frac{\Gamma, A \qquad \Gamma, \bigcirc\Box A}{\Gamma, \Box A} \qquad \diamond\ \frac{\Gamma, A, \bigcirc \diamond A}{\Gamma, \diamond A} \qquad \bigcirc\ \frac{\Gamma}{\Sigma, \bigcirc\Gamma}$$

Fig. 2. System LT1

rule foc and the axiom rep. The name of the rule foc stands for focus and implies that we focus on a specific $\Box$-formula, i.e. the formula $\Box A$ in the conclusion of the foc-rule. We focus on this formula by annotating it with its context, i.e. Γ. When the annotated formula appears in exactly the same context (i.e. Γ), then we can close the branch as axiomatic using the axiom rep (the name of this axiom stands for repetition).

The sequent

$$\diamond(P \wedge \bigcirc\overline{P}), \overline{P}, \Box P$$

(which could not be proved in system $\mathsf{LT}_{\mathsf{naive}}$) can now be derived in LT1 as shown below:

$$
\text{foc}\,\dfrac{\text{aid}\,\dfrac{}{\diamond(P\wedge\bigcirc\overline{P}),\overline{P},P}\qquad \diamond\,\dfrac{\wedge\,\dfrac{\text{aid}\,\dfrac{}{\bigcirc\diamond(P\wedge\bigcirc\overline{P}),P,\overline{P},\bigcirc\Box_\Gamma P}\qquad \bigcirc\,\dfrac{\text{rep}\,\dfrac{}{\diamond(P\wedge\bigcirc\overline{P}),\overline{P},\Box_\Gamma P}}{\bigcirc\diamond(P\wedge\bigcirc\overline{P}),\bigcirc\overline{P},\overline{P},\bigcirc\Box_\Gamma P}}{\bigcirc\diamond(P\wedge\bigcirc\overline{P}),P\wedge\bigcirc\overline{P},\overline{P},\bigcirc\Box_\Gamma P}}{\diamond(P\wedge\bigcirc\overline{P}),\overline{P},\bigcirc\Box_\Gamma P}}{\underbrace{\diamond(P\wedge\bigcirc\overline{P}),\overline{P},\Box P}_{\Gamma}}
$$

Brünnler and Lange [Brünnler and Lange, 2008] present a soundness proof for a system very similar to LT1. However, as we will see later, system LT1 only is complete if we add a cut rule. We define system $\mathsf{LT1}^{\mathsf{cut}}$ to be LT1 enriched with the rule cut:

$$
\text{cut}\,\dfrac{\Gamma, A \qquad \overline{A}, \Delta}{\Gamma, \Delta}
$$

In this rule, A is an $\mathcal{L}_{\mathbb{S}}$-formula.

Before showing the completeness and incompleteness results we need to show that weakening is syntactically admissible in system $\mathsf{LT1}^{\mathsf{cut}}$.

Definition 5. *A derivation satisfies the* next-property *iff any branch from the end-sequent to any instance of* foc *goes through at least one* $\bigcirc$*-rule.*

Lemma 6. *Let* $\Gamma \in \mathsf{Seq}$. *If* $\mathsf{LT1}^{\mathsf{cut}} \overset{n}{\vdash} \Gamma$, *then there is an* $\mathsf{LT1}^{\mathsf{cut}}$*-proof of* Γ *satisfying the next-property.*

Proof: By induction on n and a case distinction on the last rule.

1. Γ is the conclusion of aid. Then the claim holds trivially.
2. Γ is the conclusion of rep. This case is not possible because of our assumption $\Gamma \in \mathsf{Seq}$.
3. Γ is the conclusion of foc. Then $\Gamma = \Delta, \Box A$ and the given proof $\mathcal{E}$ of Γ has the following form:

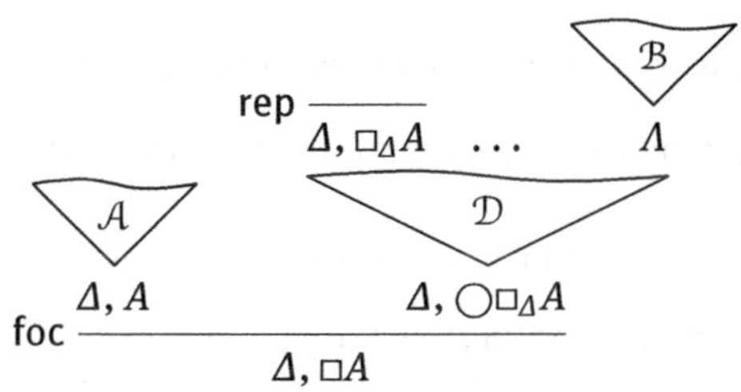

In the above derivation the sequent Λ contains no annotations, i.e. $\Lambda \in \mathsf{Seq}$. All the other sequents in derivation $\mathcal{D}$ contain annotations. Obviously Λ is obtained by an application of the rule cut.

First we observe that by the induction hypothesis, there are proofs $\mathcal{A}'$ and $\mathcal{B}'$ of Δ, A and Λ respectively that satisfy the next-property.
Further we have that in the derivation $\mathcal{D}$

$$\begin{aligned}&\text{any branch from } \Delta, \bigcirc\Box_{\Delta}A \text{ to some } \Delta, \Box_{\Delta}A\\ &\text{goes through a } \bigcirc\text{-rule}\end{aligned} \tag{1}$$

and

$$\text{there are no instances of foc.} \tag{2}$$

Let $\mathcal{D}'$ be the derivation that is obtained from $\mathcal{D}$ by dropping all annotations occurring in threads starting from $\bigcirc\Box_{\Delta}A$, i.e. by replacing sequents of the form $\Sigma, \Box_{\Delta}A$ and $\Sigma, \bigcirc\Box_{\Delta}A$ by $\Sigma, \Box A$ and $\Sigma, \bigcirc\Box A$, rescpectively.
Hence we have the following proof of $\Delta, \Box A$:

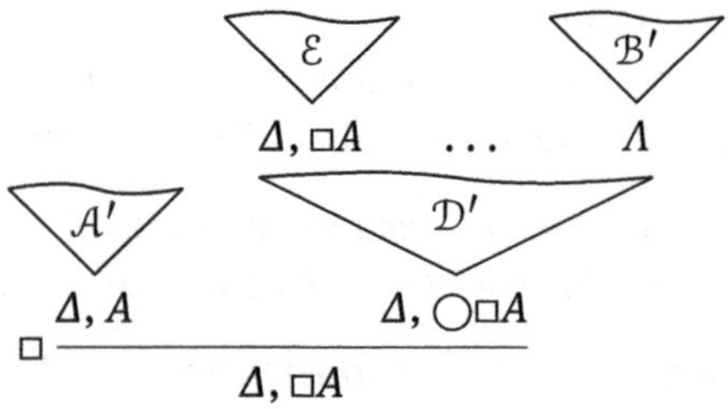

This proof satisfies the next-property. Indeed:
(a) The proofs $\mathcal{A}'$ and $\mathcal{B}'$ satisfy the next-property.
(b) In the derivation $\mathcal{D}'$ every branch from $\Delta, \bigcirc\Box A$ to some $\Delta, \Box A$ goes through at least one $\bigcirc$-rule (because of (1)).
(c) The derivation $\mathcal{D}'$ contains no foc-rules (because of (2)).

Hence any branch from the conclusion $\Delta, \Box A$ to an instance of foc goes through a $\bigcirc$-rule.

4. In all other cases, the claim follows easily by the induction hypothesis. □

By Lemma 6 we immediately get the following weakening result.

Corollary 7 (Weakening for non-annotated sequents). *For any* $\Gamma, \Delta \in \mathsf{Seq}$ *we have:*

$$\mathsf{LT1}^{\mathsf{cut}} \vdash \Gamma \Longrightarrow \mathsf{LT1}^{\mathsf{cut}} \vdash \Gamma, \Delta$$

Proof: Let $\mathsf{LT1}^{\mathsf{cut}} \vdash \Gamma$. By Lemma 6 we have a proof $\mathcal{D}$ for Γ that satisfies the next-property. We prove the claim by induction on the length of $\mathcal{D}$. Let α be the lowermost rule in $\mathcal{D}$. Since $\mathcal{D}$ satisfies the next-property and $\Gamma \in \mathsf{Seq}$, the rule α can be aid, $\lor$, $\land$, $\Box$, $\Diamond$, cut or $\bigcirc$. If α is aid or $\bigcirc$, then claim follows by built-in weaking. Otherwise it follows by the induction hypothesis. □

By a similar proof, we can show invertibiltiy of $\vee$.

Corollary 8 (Invertibility of the $\vee$-rule). *For any $\Gamma \in$ Seq and $A, B \in \mathcal{L}_{\mathcal{S}}$ we have:*

$$\mathsf{LT1}^{\mathsf{cut}} \vdash \Gamma, A \vee B \Longrightarrow \mathsf{LT1}^{\mathsf{cut}} \vdash \Gamma, A, B$$

3 Completeness

In this section, we show that system $\mathsf{LT1}^{\mathsf{cut}}$ is complete by embedding a complete Hilbert system for LTL in $\mathsf{LT1}^{\mathsf{cut}}$. We will also show that the cut-free system LT1 is not complete and that we need more complex annotations to obtain a complete cut-free system.

3.1 A Hilbert System for LTL

The language $\mathcal{L}_{\mathcal{H}}$ is described by the following grammar:

$$\phi := P \mid \neg\phi \mid \phi \wedge \phi \mid \Box\phi \mid \bigcirc\phi$$

where $P \in$ Prop. Additionally we will use the following abbreviations:

$$\phi \vee \psi := \neg(\neg\phi \wedge \neg\psi) \qquad \Diamond\phi := \neg\Box\neg\phi$$
$$\phi \rightarrow \psi := \neg(\phi \wedge \neg\psi) \qquad \phi \leftrightarrow \psi := (\phi \rightarrow \psi) \wedge (\psi \rightarrow \phi)$$

We will use the Greek letters $\phi, \psi, \rho, \ldots$ for $\mathcal{L}_{\mathcal{H}}$-formulas possibly primed or with subscripts.

Table 1 shows the Hilbert system Σ_{LTL} for LTL.

Table 1. Axioms and Rules of the system Σ_{LTL}

Axioms		**Rules**	
(P)	$\vdash \phi$, ϕ is a propositional tautology	(MP)	$\phi, \phi \rightarrow \psi \vdash \psi$
(Fun)	$\vdash \bigcirc\neg\phi \leftrightarrow \neg\bigcirc\phi$	($\mathsf{N}_{\bigcirc}$)	$\phi \vdash \bigcirc\phi$
($\mathsf{K}_{\bigcirc}$)	$\vdash \bigcirc(\phi \rightarrow \psi) \rightarrow (\bigcirc\phi \rightarrow \bigcirc\psi)$	(Ind)	$\phi \rightarrow \psi, \phi \rightarrow \bigcirc\phi \vdash \phi \rightarrow \Box\psi$
(Alw)	$\vdash \Box\phi \rightarrow \phi \wedge \bigcirc\Box\phi$		

Satisfiability of $\mathcal{L}_{\mathcal{H}}$-formulas in an LTL-model is defined as follows:

Definition 9 (Satisfiability of $\mathcal{L}_{\mathcal{H}}$-formulas). *Let μ be a model and $i \in \mathbb{N}$. We have:*

$$
\begin{array}{rcl}
\mu, i \models P & \Longleftrightarrow & P \in \mu(i) \\
\mu, i \models \neg\phi & \Longleftrightarrow & \mu, i \not\models \phi \\
\mu, i \models \phi \wedge \psi & \Longleftrightarrow & \mu, i \models \phi \text{ and } \mu, i \models \psi \\
\mu, i \models \square\phi & \Longleftrightarrow & \forall j \geq i(\mu, j \models \phi) \\
\mu, i \models \bigcirc\phi & \Longleftrightarrow & \mu, i+1 \models \phi
\end{array}
$$

Validity for $\mathcal{L}_{\mathcal{H}}$-formulas is defined in the same way as for $\mathcal{L}_{\mathcal{S}}$-formulas. The following theorem is proved in Chapter 2 of [Kröger and Merz, 2008].

Theorem 10 (Soundness and Completeness). *The system Σ_{LTL} is sound and complete with respect to* LTL*-models, i.e. we have for all $\mathcal{L}_{\mathcal{H}}$-formulas ϕ:*

$$\models \phi \quad \Longleftrightarrow \quad \Sigma_{\mathsf{LTL}} \vdash \phi$$

3.2 System LT1 + cut is Complete

In this subsection we will show that the system $\mathsf{LT1}^{\mathsf{cut}}$ is complete by embedding system Σ_{LTL} to system $\mathsf{LT1}^{\mathsf{cut}}$. Before proving the embedding we need some auxiliary definitions and lemmata.

As a first step we extend axiom aid to all $\mathcal{L}_{\mathcal{S}}$-formulas.

Lemma 11. *[Extension of axiom* aid *to $\mathcal{L}_{\mathcal{S}}$-formulas] Let $\Gamma \in$* Seq *and let $A \in \mathcal{L}_{\mathcal{S}}$. Then we have the following in system $\mathsf{LT1}^{\mathsf{cut}}$:*

$$\vdash \Gamma, A, \overline{A}$$

Proof: The proof is by induction on the structure of A. The only interesting cases are $A = \square B$ and $A = \diamond B$. So assume $A = \square B$. It holds $\overline{A} = \diamond\overline{B}$. Then we have the following derivation in $\mathsf{LT1}^{\mathsf{cut}}$:

$$
\dfrac{\dfrac{\dfrac{\dfrac{}{B, \bigcirc\diamond\overline{B}, \overline{B}}\,\text{i.h.}}{B, \diamond\overline{B}}\,\diamond
\qquad
\dfrac{\dfrac{\dfrac{}{\square_{\diamond\overline{B}}B, \diamond\overline{B}}\,\mathsf{rep}}{\bigcirc\square_{\diamond\overline{B}}B, \bigcirc\diamond\overline{B}, \overline{B}}\,\bigcirc}{\bigcirc\square_{\diamond\overline{B}}B, \diamond\overline{B}}\,\diamond}{\square B, \diamond\overline{B}}\,\mathsf{foc}}{\Gamma, \square B, \diamond\overline{B}}\,\text{Corollary 7}
$$

The case $A = \diamond B$ simply is dual to the shown case. □

An easy induction on the structure of the formula A also yields the following lemma.

Lemma 12. *Let $A \in \mathcal{L}_{\mathcal{S}}$. It holds that:*

$$\overline{\overline{A}} = A$$

Now we define two translation functions between the languages $\mathcal{L}_{\mathcal{H}}$ and $\mathcal{L}_{\mathcal{S}}$.

We define the function $\sigma : \mathcal{L}_{\mathcal{S}} \to \mathcal{L}_{\mathcal{H}}$ inductively:

$$\begin{aligned} \sigma(P) &= P & \sigma(\overline{P}) &= \neg P \\ \sigma(A \wedge B) &= \sigma(A) \wedge \sigma(B) & \sigma(A \vee B) &= \sigma(A) \vee \sigma(B) \\ \sigma(\Box A) &= \Box\sigma(A) & \sigma(\diamond A) &= \diamond\sigma(A) \\ \sigma(\bigcirc A) &= \bigcirc\sigma(A) \end{aligned}$$

We define the function $\tau : \mathcal{L}_{\mathcal{H}} \to \mathcal{L}_{\mathcal{S}}$ inductively:

$$\begin{aligned} \tau(P) &= P & \tau(\Box\phi) &= \Box\tau(\phi) \\ \tau(\neg\phi) &= \overline{\tau(\phi)} & \tau(\bigcirc\phi) &= \bigcirc\tau(\phi) \\ \tau(\phi \wedge \psi) &= \tau(\phi) \wedge \tau(\psi) \end{aligned}$$

Some simple calculations show that the function τ behaves as expected with respect to the propositional connectives.

Lemma 13. *Let $\phi, \psi \in \mathcal{L}_{\mathcal{H}}$. It holds:*
1. $\tau(\phi \to \psi) = \overline{\tau(\phi)} \vee \tau(\psi)$
2. $\tau(\phi \vee \psi) = \tau(\phi) \vee \tau(\psi)$
3. $\tau(\phi \leftrightarrow \psi) = (\overline{\tau(\phi)} \vee \tau(\psi)) \wedge (\overline{\tau(\psi)} \vee \tau(\phi))$
4. $\tau(\diamond\phi) = \diamond\tau(\phi)$

It is straightforward to show that τ is the inverse of σ.

Lemma 14. *Let $A \in \mathcal{L}_{\mathcal{S}}$. It holds:*

$$\tau(\sigma(A)) = A$$

Now we can prove the embedding Lemma:

Lemma 15 (Embedding of Σ_{LTL} into $\mathsf{LT1}^{\mathsf{cut}}$).

$$(\forall\phi \in \mathcal{L}_{\mathcal{H}})[\Sigma_{\mathsf{LTL}} \vdash \phi \Longrightarrow \mathsf{LT1}^{\mathsf{cut}} \vdash \tau(\phi)]$$

Proof: By induction on the length of the derivation $\Sigma_{\mathsf{LTL}} \vdash \phi$. We distinguish the following cases:

1. ϕ is a propositional tautology. Then $\tau(\phi)$ is also a propositional tautology. Hence, clearly $\mathsf{LT1}^{\mathsf{cut}} \vdash \tau(\phi)$.
2. ϕ is an instance of the axiom (Fun). That means there is a $\psi \in \mathcal{L}_{\mathcal{H}}$ such that $\phi = \neg\bigcirc\psi \leftrightarrow \bigcirc\neg\psi$. From Lemma 12 and Lemma 13 we get:

$$\tau(\phi) = (\bigcirc\tau(\psi) \vee \bigcirc\overline{\tau(\psi)}) \wedge (\bigcirc\tau(\psi) \vee \bigcirc\overline{\tau(\psi)})$$

So, we have the following derivation in $\mathsf{LT1}^{\mathsf{cut}}$:

$$\wedge\,\dfrac{\vee\,\dfrac{\text{Lemma 11}\,\dfrac{}{\bigcirc\tau(\psi), \bigcirc\overline{\tau(\psi)}}}{\bigcirc\tau(\psi) \vee \bigcirc\overline{\tau(\psi)}} \qquad \vee\,\dfrac{\text{Lemma 11}\,\dfrac{}{\bigcirc\tau(\psi), \bigcirc\overline{\tau(\psi)}}}{\bigcirc\tau(\psi) \vee \bigcirc\overline{\tau(\psi)}}}{(\bigcirc\tau(\psi) \vee \bigcirc\overline{\tau(\psi)}) \wedge (\bigcirc\tau(\psi) \vee \bigcirc\overline{\tau(\psi)})}$$

3. ϕ is an instance of axiom ($\mathsf{K}_{\bigcirc}$). That means there are $\psi, \rho \in \mathcal{L}_{\mathcal{H}}$ such that $\phi = \bigcirc(\psi \to \rho) \to \bigcirc\psi \to \bigcirc\rho$. By Lemma 12 and Lemma 13 we get:

$$\tau(\phi) = \bigcirc\left(\tau(\psi) \wedge \overline{\tau(\rho)}\right) \vee \bigcirc\overline{\tau(\psi)} \vee \bigcirc\tau(\rho)$$

So, we have the following derivation:

$$\vee\,\dfrac{\vee\,\dfrac{\bigcirc\,\dfrac{\wedge\,\dfrac{\text{Lemma 11}\,\dfrac{}{\tau(\psi), \overline{\tau(\psi)}, \tau(\rho)} \qquad \text{Lemma 11}\,\dfrac{}{\overline{\tau(\rho)}, \overline{\tau(\psi)}, \tau(\rho)}}{\tau(\psi) \wedge \overline{\tau(\rho)}, \overline{\tau(\psi)}, \tau(\rho)}}{\bigcirc\left(\tau(\psi) \wedge \overline{\tau(\rho)}\right), \bigcirc\overline{\tau(\psi)}, \bigcirc\tau(\rho)}}{\bigcirc\left(\tau(\psi) \wedge \overline{\tau(\rho)}\right) \vee \bigcirc\overline{\tau(\psi)}, \bigcirc\tau(\rho)}}{\bigcirc\left(\tau(\psi) \wedge \overline{\tau(\rho)}\right) \vee \bigcirc\overline{\tau(\psi)} \vee \bigcirc\tau(\rho)}$$

4. ϕ is an instance of the axiom (Alw). That means there is a $\psi \in \mathcal{L}_{\mathcal{H}}$ such that $\phi = \Box\psi \to \psi \wedge \bigcirc\Box\psi$. From Lemma 12 and Lemma 13 we get:

$$\tau(\phi) = \diamond\overline{\tau(\psi)} \vee (\tau(\psi) \wedge \bigcirc\Box\tau(\psi))$$

So, we have the following derivation in $\mathsf{LT1}^{\mathsf{cut}}$:

$$\vee\,\dfrac{\diamond\,\dfrac{\wedge\,\dfrac{\text{Lemma 11}\,\dfrac{}{\overline{\tau(\psi)}, \bigcirc\diamond\overline{\tau(\psi)}, \tau(\psi)} \qquad \text{Lemma 11}\,\dfrac{}{\overline{\tau(\psi)}, \bigcirc\diamond\overline{\tau(\psi)}, \bigcirc\Box\tau(\psi)}}{\overline{\tau(\psi)}, \bigcirc\diamond\overline{\tau(\psi)}, \tau(\psi) \wedge \bigcirc\Box\tau(\psi)}}{\diamond\overline{\tau(\psi)}, \tau(\psi) \wedge \bigcirc\Box\tau(\psi)}}{\diamond\overline{\tau(\psi)} \vee (\tau(\psi) \wedge \bigcirc\Box\tau(\psi))}$$

5. ϕ is the conclusion of an application of (MP). That means there is a $\psi \in \mathcal{L}_{\mathcal{H}}$ such that $\Sigma_{\mathsf{LTL}} \vdash \psi$ and $\Sigma_{\mathsf{LTL}} \vdash \psi \to \phi$. Then we find the following derivation in $\mathsf{LT1}^{\mathsf{cut}}$:

$$\text{cut}\ \dfrac{\text{i.h.}\ \dfrac{}{\tau(\psi)} \qquad \text{Corollary 8}\ \dfrac{\text{Lemma 13 1}\ \dfrac{\text{i.h.}\ \dfrac{}{\tau(\psi \to \phi)}}{\overline{\tau(\psi)} \lor \tau(\phi)}}{\overline{\tau(\psi)}, \tau(\phi)}}{\tau(\phi)}$$

6. ϕ is the conclusion of an application of ($\mathsf{N}_{\bigcirc}$). That means that there is a $\psi \in \mathcal{L}_{\mathcal{H}}$ such that $\phi = \bigcirc\psi$ and that $\Sigma_{\mathsf{LTL}} \vdash \psi$. In $\mathsf{LT1}^{\mathsf{cut}}$ we have the following derivation:

$$\bigcirc\ \dfrac{\text{i.h.}\ \dfrac{}{\tau(\psi)}}{\bigcirc\tau(\psi)}$$

And since $\bigcirc\tau(\psi) = \tau(\bigcirc\psi) = \tau(\phi)$ we have that $\mathsf{LT1}^{\mathsf{cut}} \vdash \tau(\phi)$.

7. ϕ is the conclusion of an application of (Ind). That means there are $\psi, \rho \in \mathcal{L}_{\mathcal{H}}$ such that $\phi = \psi \to \Box\rho$, and that $\Sigma_{\mathsf{LTL}} \vdash \psi \to \rho$ and $\Sigma_{\mathsf{LTL}} \vdash \psi \to \bigcirc\psi$. By Lemma 13 1 we have:

$$\tau(\phi) = \tau(\psi \to \Box\rho) = \overline{\tau(\psi)} \lor \Box\tau(\rho)$$
$$\tau(\psi \to \rho) = \overline{\tau(\psi)} \lor \tau(\rho)$$
$$\tau(\psi \to \bigcirc\psi) = \overline{\tau(\psi)} \lor \bigcirc\tau(\psi)$$

We find the following derivation in $\mathsf{LT1}^{\mathsf{cut}}$:

$$\lor\ \dfrac{\dfrac{\text{Corollary 8}\ \dfrac{\text{i.h.}\ \dfrac{}{\overline{\tau(\psi)} \lor \tau(\rho)}}{\overline{\tau(\psi)}, \tau(\rho)} \qquad \text{cut}\ \dfrac{\text{Corollary 8}\ \dfrac{\text{i.h.}\ \dfrac{}{\overline{\tau(\psi)} \lor \bigcirc\tau(\psi)}}{\overline{\tau(\psi)}, \bigcirc\tau(\psi)} \qquad \dfrac{\text{rep}\ \dfrac{}{\tau(\psi), \Box_{\overline{\tau(\psi)}}\tau(\rho)}}{\bigcirc\tau(\psi), \bigcirc\Box_{\overline{\tau(\psi)}}\tau(\rho)}}{\overline{\tau(\psi)}, \bigcirc\Box_{\overline{\tau(\psi)}}\tau(\rho)}}{\overline{\tau(\psi)}, \Box\tau(\rho)}\ \text{foc}}{\overline{\tau(\psi)} \lor \Box\tau(\rho)}$$

□

To establish completeness of $\mathsf{LT1}^{\mathsf{cut}}$, we need the following lemma, which can easily be shown by induction on A.

Lemma 16. *Let $A \in \mathcal{L}_{\mathcal{S}}$, let μ be a model and let $i \in \mathbb{N}$. It holds:*

$$\mu, i \models A \quad \Longrightarrow \quad \mu, i \models \sigma(A)$$

Finally we can prove completeness of system $\mathsf{LT1}^{\mathsf{cut}}$.

Theorem 17. *System* $\mathsf{LT1}^{\mathsf{cut}}$ *is complete for* $\mathcal{L}_{\mathcal{S}}$*-formulas, i.e. for all* $\mathcal{L}_{\mathcal{S}}$*-formulas* A

$$\models A \quad \Longrightarrow \quad \mathsf{LT1}^{\mathsf{cut}} \vdash A$$

Proof: Let $A \in \mathcal{L}_{\mathcal{S}}$. We have:

$$\models A \Longrightarrow (\forall\mu)(\forall i \in \mathbb{N})[\mu, i \models A] \overset{\text{Lemma 16}}{\Longrightarrow}$$

$$(\forall\mu)(\forall i \in \mathbb{N})[\mu, i \models \sigma(A)] \Longrightarrow \models \sigma(A) \overset{\text{Theorem 10}}{\Longrightarrow}$$

$$\Sigma_{\mathsf{LTL}} \vdash \sigma(A) \overset{\text{Lemma 15}}{\Longrightarrow} \mathsf{LT1}^{\mathsf{cut}} \vdash \tau(\sigma(A)) \overset{\text{Lemma 14}}{\Longrightarrow} \mathsf{LT1}^{\mathsf{cut}} \vdash A \qquad \square$$

3.3 System LT1 is not Complete

Now we show that if we remove the rule cut from system $\mathsf{LT1}^{\mathsf{cut}}$, the resulting system (i.e. LT1) is not complete. Let Γ be the following valid sequent:

$$\overline{P}, \diamond(P \wedge \bigcirc\bigcirc\overline{P}), \Box(P \vee \bigcirc P)$$

Γ is semantically equivalent to the following $\mathcal{L}_{\mathcal{H}}$-formula:

$$P \wedge \Box(P \to \bigcirc\bigcirc P) \to \Box(P \vee \bigcirc P)$$

which expresses a valid induction statement in LTL.

The following derivation is an attempt to prove Γ in LT1.

$$\text{foc}\;\dfrac{\vee\;\dfrac{\text{aid}\;\dfrac{}{\overline{P}, \diamond(P \wedge \bigcirc\bigcirc\overline{P}), P, \bigcirc P}}{\overline{P}, \diamond(P \wedge \bigcirc\bigcirc\overline{P}), P \vee \bigcirc P} \qquad \diamond\;\dfrac{\wedge\;\dfrac{\text{aid}\;\dfrac{}{\overline{P}, P, \bigcirc\diamond(P \wedge \bigcirc\bigcirc\overline{P}), \bigcirc\Box_{\Delta}P} \qquad \mathcal{D}}{\overline{P}, P \wedge \bigcirc\bigcirc\overline{P}, \bigcirc\diamond(P \wedge \bigcirc\bigcirc\overline{P}), \bigcirc\Box_{\Delta}(P \vee \bigcirc P)}}{\overline{P}, \diamond(P \wedge \bigcirc\bigcirc\overline{P}), \bigcirc\Box_{\Delta}(P \vee \bigcirc P)}}{\underbrace{\overline{P}, \diamond(P \wedge \bigcirc\bigcirc\overline{P})}_{\Delta}, \Box(P \vee \bigcirc P)}$$

$$\mathcal{D}\left\{\;\bigcirc\;\dfrac{\bigcirc\overline{P}, \diamond(P \wedge \bigcirc\bigcirc\overline{P}), \Box_{\Delta}(P \vee \bigcirc P)}{\overline{P}, \bigcirc\bigcirc\overline{P}, \bigcirc\diamond(P \wedge \bigcirc\bigcirc\overline{P}), \bigcirc\Box_{\Delta}(P \vee \bigcirc P)}\right.$$

The reason we cannot prove sequent Γ in system LT1 is that in system LT1 it is impossible to "get rid of an annotation". So, in the above proof-attempt for Γ there is no way we can drop Δ from sequent $\bigcirc\overline{P}, \diamond(P \wedge \bigcirc\bigcirc\overline{P}), \Box_{\Delta}(P \vee \bigcirc P)$, which is the sequent on the top of the derivation $\mathcal{D}$. So, the only way to prove an annotated sequent is to reach either axiom rep or axiom aid from it. In our case it is impossible to reach rep from $\bigcirc\overline{P}, \diamond(P \wedge \bigcirc\bigcirc\overline{P}), \Box_{\Delta}(P \vee \bigcirc P)$, since it would require the application of a $\bigcirc$-rule, which is not possible (the conclusion of a $\bigcirc$-rule cannot contain a formula of the form $\Box_\Delta A$). It is also impossible to reach aid from $\bigcirc\overline{P}, \diamond(P \wedge \bigcirc\bigcirc\overline{P}), \Box_\Delta(P \vee \bigcirc P)$ since the sequent $\bigcirc\overline{P}, \diamond(P \wedge \bigcirc\bigcirc\overline{P})$ is not valid and our system is sound. Thus, proof-search for Γ in LT1 fails and, therefore, system LT1 is not complete.

We can tackle the problem of "being unable to get rid of annotations" by introducing a new rule. For any $\Delta \in \mathsf{Seq}$ we define the following rule:

$$\Box_\Delta \; \frac{\Gamma, A \qquad \Gamma, \bigcirc\Box_\Delta A}{\Gamma, \Box_\Delta A}$$

Rule $\Box_\Delta$ is sound with respect to LTL-models for any Δ. As we can see, rule $\Box_\Delta$ allows us to drop the annotation from an annotated sequent (the left premise of the rule $\Box_\Delta$ is an unannotated sequent). We define LT1$^+$ to be system LT1 plus the rule $\Box_\Delta$ for any $\Delta \in \mathsf{Seq}$.

In system LT1$^+$ we can prove the sequent $\overline{P}, \diamond(P \wedge \bigcirc\bigcirc\overline{P}), \Box(P \vee \bigcirc P)$ as follows:

$$\text{foc}\;\dfrac{\vee\;\dfrac{\text{aid}\;\dfrac{}{\overline{P}, \diamond(P \wedge \bigcirc\bigcirc P), P, \bigcirc P}}{\overline{P}, \diamond(P \wedge \bigcirc\bigcirc\overline{P}), P \vee \bigcirc P} \qquad \diamond\;\dfrac{\wedge\;\dfrac{\text{aid}\;\dfrac{}{\overline{P}, P, \bigcirc \diamond (P \wedge \bigcirc\bigcirc\overline{P}), \bigcirc\Box_\Delta(P \vee \bigcirc P)} \qquad \mathcal{D}_1}{P, P \wedge \bigcirc\bigcirc P, \bigcirc \diamond (P \wedge \bigcirc\bigcirc\overline{P}), \bigcirc\Box_\Delta(P \vee \bigcirc P)}}{\overline{P}, \diamond(P \wedge \bigcirc\bigcirc\overline{P}), \bigcirc\Box_\Delta(P \vee \bigcirc P)}}{\underbrace{\overline{P}, \diamond(P \wedge \bigcirc\bigcirc\overline{P})}_{\Delta}, \Box(P \vee \bigcirc P)}$$

$$\mathcal{D}_1 \left\{ \bigcirc\;\dfrac{\Box_\Delta\;\dfrac{\vee\;\dfrac{\bigcirc\;\dfrac{\text{aid}\;\dfrac{}{\overline{P}, P}}{\bigcirc\overline{P}, \diamond(P \wedge \bigcirc\bigcirc\overline{P}), P, \bigcirc P}}{\bigcirc\overline{P}, \diamond(P \wedge \bigcirc\bigcirc\overline{P}), P \vee \bigcirc P} \qquad \diamond\;\dfrac{\mathcal{D}_2}{\bigcirc\overline{P}, \diamond(P \wedge \bigcirc\bigcirc\overline{P}), \bigcirc\Box_\Delta(P \vee \bigcirc P)}}{\bigcirc\overline{P}, \diamond(P \wedge \bigcirc\bigcirc\overline{P}), \Box_\Delta(P \vee \bigcirc P)}}{\overline{P}, \bigcirc\bigcirc\overline{P}, \bigcirc \diamond (P \wedge \bigcirc\bigcirc\overline{P}), \bigcirc\Box_\Delta(P \vee \bigcirc P)} \right.$$

$$\mathcal{D}_2 \left\{ \wedge\;\dfrac{\bigcirc\;\dfrac{\text{rep}\;\dfrac{}{\overline{P}, \diamond(P \wedge \bigcirc\bigcirc\overline{P}), \Box_\Delta(P \vee \bigcirc P)}}{\bigcirc\overline{P}, P, \bigcirc \diamond (P \wedge \bigcirc\bigcirc\overline{P}), \bigcirc\Box_\Delta(P \vee \bigcirc P)} \qquad \mathcal{D}_3}{\bigcirc\overline{P}, P \wedge \bigcirc\bigcirc\overline{P}, \bigcirc \diamond (P \wedge \bigcirc\bigcirc\overline{P}), \bigcirc\Box_\Delta(P \vee \bigcirc P)} \right.$$

$$\mathcal{D}_3 \left\{ \bigcirc\;\dfrac{\text{rep}\;\dfrac{}{\overline{P}, \diamond(P \wedge \bigcirc\bigcirc\overline{P}), \Box_\Delta(P \vee \bigcirc P)}}{\bigcirc\overline{P}, \bigcirc\bigcirc\overline{P}, \bigcirc \diamond (P \wedge \bigcirc\bigcirc\overline{P}), \bigcirc\Box_\Delta(P \vee \bigcirc P)} \right.$$

However, LT1$^+$ is still too simple: it fails to prove all valid sequents. Take for example the following valid sequent, which we call Σ:

$$\bigcirc\Box(\overline{P} \vee \overline{Q}), \diamond(\bigcirc P \wedge \bigcirc Q)$$

where P, Q are different elements of Prop. We show that Σ cannot be derived in $\mathsf{LT1}^+$.

We set $C = \overline{P} \vee \overline{Q}$ and $D = \bigcirc P \wedge \bigcirc Q$. A proof-attempt for Σ in $\mathsf{LT1}^+$ is as follows:

$$\diamond\,\dfrac{\wedge\,\dfrac{\bigcirc\,\dfrac{\mathcal{D}_1(P,Q)}{\bigcirc\Box C, \bigcirc\diamond D, \bigcirc P} \qquad \bigcirc\,\dfrac{\mathcal{D}_1(Q,P)}{\bigcirc\Box C, \bigcirc\diamond D, \bigcirc Q}}{\bigcirc\Box C, \bigcirc\diamond D, D}}{\bigcirc\Box C, \diamond D}$$

where $\mathcal{D}_1(P,Q)$ is

$$\text{foc}\,\dfrac{\vee\,\dfrac{\text{aid}\,\dfrac{}{\overline{P}, \overline{Q}, \diamond D, P}}{C, \diamond D, P} \qquad \diamond\,\dfrac{\wedge\,\dfrac{\bigcirc\,\dfrac{\text{rep}\,\dfrac{}{\diamond D, P, \Box_{\diamond D,P} C}}{\bigcirc\diamond D, \bigcirc P, P, \bigcirc\Box_{\diamond D,P} C} \qquad \bigcirc\,\dfrac{\mathcal{D}_2(P,Q)}{\bigcirc\diamond D, \bigcirc Q, P, \bigcirc\Box_{\diamond D,P} C}}{\bigcirc\diamond D, D, P, \bigcirc\Box_{\diamond D,P} C}}{\diamond D, P, \bigcirc\Box_{\diamond D,P} C}}{\Box C, \diamond D, P}$$

and $\mathcal{D}_2(P,Q)$ is

$$\Box_{\diamond D,P}\,\dfrac{\vee\,\dfrac{\text{aid}\,\dfrac{}{\diamond D, Q, \overline{P}, \overline{Q}}}{\diamond D, Q, C} \qquad \diamond\,\dfrac{\wedge\,\dfrac{\bigcirc\,\dfrac{\text{rep}\,\dfrac{}{\diamond D, P, \Box_{\diamond D,P} C}}{\bigcirc\diamond D, \bigcirc P, Q, \bigcirc\Box_{\diamond D,P} C} \qquad \bigcirc\,\dfrac{\diamond D, Q, \Box_{\diamond D,P} C}{\bigcirc\diamond D, \bigcirc Q, Q, \bigcirc\Box_{\diamond D,P} C}}{\bigcirc\diamond D, D, Q, \bigcirc\Box_{\diamond D,P} C}}{\diamond D, Q, \bigcirc\Box_{\diamond D,P} C}}{\diamond D, Q, \Box_{\diamond D,P} C}$$

The reason we cannot prove Σ in $\mathsf{LT1}^+$ is that it is impossible to prove the sequent $\diamond D, Q, \Box_{\diamond D,P} C$ in $\mathsf{LT1}^+$. Since $\diamond D, Q, \Box_{\diamond D,P} C$ is no instance of axiom rep, it is natural to try to prove it by applying the $\Box_{\diamond,P}$ rule first. However, as we can see in derivation $\mathcal{D}_2(P,Q)$ this leads again to sequent $\diamond D, Q, \Box_{\diamond D,P} C$. Hence proof-search for Σ fails, i.e. $\mathsf{LT1}^+$ cannot be complete. However, the fact that an annotated sequent, i.e. $\diamond D, Q, \Box_{\diamond D,P} C$, is cyclic in $\mathcal{D}_2(P,Q)$ gives us a hint for how we should improve the principle for closing cyclic branches. What if we keep a set of sequents rather than a single sequent in the annotation? This is the approach of Brünnler and Lange [Brünnler and Lange, 2008], which we study in the next section.

4 Histories

So far, all our systems could only store one sequent in the annotation. In the previous section we have seen that this not enough to define a complete cut-free sequent system for LTL. Brünnler and Lange [Brünnler and Lange, 2008] present a system that is very similar to LT1. The only difference is that their annotations contain sets of sequents rather than single sequents. Based on their approach, we present a system LT2, for which we syntactically establish weakening. This solves an open problem of Brünnler and Lange.

4.1 System LT2

We start with defining the set of histories (finite sets of sequents), which we call His:

$$\mathsf{His} = \{H \mid H \text{ is a finite subset of } \mathsf{Seq}\}$$

We will use the letters H and G for histories, possibly primed or with subscripts. We also define the set $\mathcal{L}_{\mathsf{his}}$:

$$\mathcal{L}_{\mathsf{his}} := \{\sqcup_H A, \bigcirc\sqcup_H A \mid A \in \mathcal{L}_{\mathbb{S}}, H \in \mathsf{His}\}$$

We will refer to $\mathcal{L}_{\mathsf{his}}$-formulas as formulas with histories, or when there is no danger of confusion as annotated formulas. We also define the set of sequents with histories:

$$\mathsf{Seq}_{\mathsf{his}} := \{\, \Gamma \mid \Gamma \text{ is a finite subset of } \mathcal{L}_{\mathbb{S}} \cup \mathcal{L}_{\mathsf{his}} \text{ that contains at most one } \mathcal{L}_{\mathsf{his}}\text{-formula} \,\}$$

We assume that a history is semantically equivalent to the conjunction of its elements, i.e. for a model μ and $i \in \mathbb{N}$:

$$\mu, i \models H \iff \mu, i \models \bigwedge_{\Gamma \in H} \Gamma \iff \mu, i \models \bigwedge_{\Gamma \in H} \bigvee_{A \in \Gamma} A$$

The semantics of $\Box_H A$ is defined like that of $\Box_\Gamma A$. We have:

$$\mu, i \models \Box_H A \iff \forall j \geq i\Big((\forall i \leq k \leq j(\mu, k \models H)) \Longrightarrow \mu, j \models A\Big)$$

In Figure 3 we present system LT2. Again we assume $\Sigma \in \mathsf{Seq}$ in the $\bigcirc$-rule. For rep, $\Box_H$ and foc, we also assume $\Gamma \in \mathsf{Seq}$.

Whenever rule $\Box_\Delta$ is applied in an LT1$^+$-proof, the sequent in the annotation does not change. In system LT2, an application of rule $\Box_H$ leads to a new sequent being added to the annotation (history). Thus, rule $\Box_H$ is a generalization of rule $\Box_\Delta$, which implies that system LT1$^+$ is a subsystem of LT2.

System LT2 proves the sequent:

$$\Sigma = \bigcirc\Box(\overline{P} \vee \overline{Q}), \diamond(\bigcirc P \wedge \bigcirc Q)$$

$$\text{aid}\ \frac{}{\Gamma, P, \overline{P}} \qquad \vee\ \frac{\Gamma, A, B}{\Gamma, A \vee B} \qquad \wedge\ \frac{\Gamma, A \qquad \Gamma, B}{\Gamma, A \wedge B}$$

$$\text{rep}\ \frac{}{\Gamma, \Box_{H,\Gamma} A} \qquad \text{foc}\ \frac{\Gamma, A \qquad \Gamma, \bigcirc\Box_{\{\Gamma\}} A}{\Gamma, \Box A} \qquad \Box_H\ \frac{\Gamma, A \qquad \Gamma, \bigcirc\Box_{H,\Gamma} A}{\Gamma, \Box_H A}$$

$$\Box\ \frac{\Gamma, A \qquad \Gamma, \bigcirc\Box A}{\Gamma, \Box A} \qquad \diamond\ \frac{\Gamma, A, \bigcirc\diamond A}{\Gamma, \diamond A} \qquad \bigcirc\ \frac{\Gamma}{\Sigma, \bigcirc\Gamma}$$

Fig. 3. System LT2

which as we saw in section 3 cannot be proved in LT1$^+$. Hence, system LT1$^+$ is proper subsystem of system LT2. We now present the proof of Σ in LT2. As before we set $D = \bigcirc P \wedge \bigcirc Q$ and $C = \overline{P} \vee \overline{Q}$. Moreover, we sometimes write, e.g., $\Box_{\Delta,\Gamma}$ for $\Box_{\{\Delta,\Gamma\}}$.

$$\diamond\ \dfrac{\wedge\ \dfrac{\bigcirc\ \dfrac{\mathcal{D}_1(P, Q)}{\bigcirc\Box C, \bigcirc\diamond D, \bigcirc P} \qquad \bigcirc\ \dfrac{\mathcal{D}_1(Q, P)}{\bigcirc\Box C, \bigcirc\diamond D, \bigcirc Q}}{\bigcirc\Box C, \bigcirc\diamond D, D}}{\bigcirc\Box C, \diamond D}$$

where $\mathcal{D}_1(P, Q)$ is

$$\text{foc}\ \dfrac{\vee\ \dfrac{\text{aid}\ \dfrac{}{\overline{P}, \overline{Q}, \diamond D, P}}{C, \diamond D, P} \qquad \diamond\ \dfrac{\wedge\ \dfrac{\bigcirc\ \dfrac{\text{rep}\ \dfrac{}{\diamond D, P, \Box_{\{\diamond D, P\}} C}}{\bigcirc\diamond D, \bigcirc P, P, \bigcirc\Box_{\{\diamond D, P\}} C} \qquad \bigcirc\ \dfrac{\mathcal{D}_2(P, Q)}{\bigcirc\diamond D, \bigcirc Q, P, \bigcirc\Box_{\{\diamond D, P\}} C}}{\bigcirc\diamond D, D, P, \bigcirc\Box_{\{\diamond D, P\}} C}}{\diamond D, P, \bigcirc\Box_{\{\diamond D, P\}} C}}{\Box C, \diamond D, P}$$

$\mathcal{D}_2(P, Q)$ is

$$\Box_{\{\diamond D, P\}}\ \dfrac{\vee\ \dfrac{\text{aid}\ \dfrac{}{\diamond D, Q, \overline{P}, \overline{Q}}}{\diamond D, Q, C} \qquad \diamond\ \dfrac{\mathcal{D}_3(P, Q)}{\diamond D, Q, \bigcirc\Box_{\{\diamond D, P\},\{\diamond D, Q\}} C}}{\diamond D, Q, \Box_{\{\diamond D, P\}} C}$$

and $\mathcal{D}_3(P, Q)$ is

$$\wedge\ \dfrac{\bigcirc\ \dfrac{\text{rep}\ \dfrac{}{\diamond D, P, \Box_{\{\diamond D, P\},\{\diamond D, Q\}} C}}{\bigcirc\diamond D, \bigcirc P, Q, \bigcirc\Box_{\{\diamond D, P\},\{\diamond D, Q\}} C} \qquad \bigcirc\ \dfrac{\text{rep}\ \dfrac{}{\diamond D, Q, \Box_{\{\diamond D, P\},\{\diamond D, Q\}} C}}{\bigcirc\diamond D, \bigcirc Q, Q, \bigcirc\Box_{\{\diamond D, P\},\{\diamond D, Q\}} C}}{\bigcirc\diamond D, D, Q, \bigcirc\Box_{\{\diamond D, P\},\{\diamond D, Q\}} C}$$

As we mentioned before, Brünnler and Lange's system with histories is very similar to our LT2, so using the same techniques as in [Brünnler and Lange, 2008] we can show soundness and completeness. Actually, we can prove completeness for a system LT2′, in which the $\bigcirc$-rule allows only weakinging with atomic formulas. This system can then be embedded in LT2, thus showing the completeness of LT2.

Theorem 18. *System* LT2 *is sound and complete with respect to* LTL*-models, i.e. for all $\mathcal{L}_{\mathcal{S}}$-formulas A we have*

$$\models A \iff \mathsf{LT2} \vdash A$$

4.2 Weakening

Brünnler and Lange [Brünnler and Lange, 2008] formulated the problem of syntactically showing the admissibility of weakening in annotated systems. Here we provide a solution to this problem for LT2.

We use the same approach as for LT1. That is we show weakening as a corollary of the next-property. However, the presence of histories in LT2 requires some care. We will need the following two auxiliary statements, which can be shown simultaneously by induction on n.

Lemma 19. *Let H and $G \in$ His. It holds:*

1. *If* LT2 $\overset{n}{\vdash} \Gamma, \Box_H A$, *then* LT2 $\overset{n}{\vdash} \Gamma, \Box_{H,G} A$.
2. *If* LT2 $\overset{n}{\vdash} \Gamma, \bigcirc\Box_H A$, *then* LT2 $\overset{n}{\vdash} \Gamma, \bigcirc\Box_{H,G} A$.

The analogue of Lemma 6 for system LT2 is the following lemma:

Lemma 20. *Let $\Gamma \in$ Seq. If* LT2 $\overset{n}{\vdash} \Gamma$, *then there is an* LT2*-proof of Γ satisfying the next-property.*

Proof: Again the proof is by induction on n and a case distinction on the last rule. We only show the case for foc. Then $\Gamma = \Delta, \Box A$ and the given proof $\mathcal{E}$ of Γ has the following form (where the derivation $\mathcal{H}$ is shown in the next page):

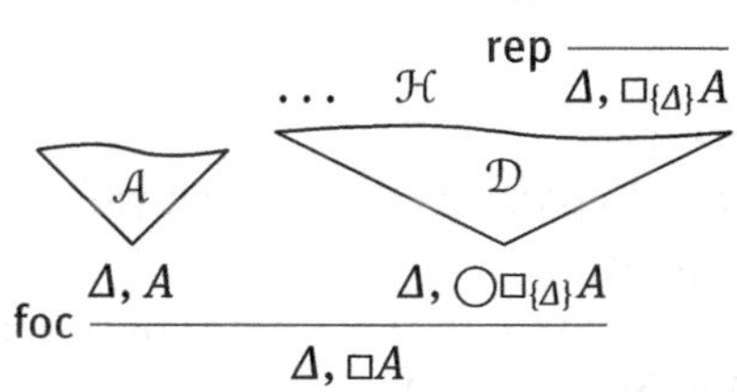

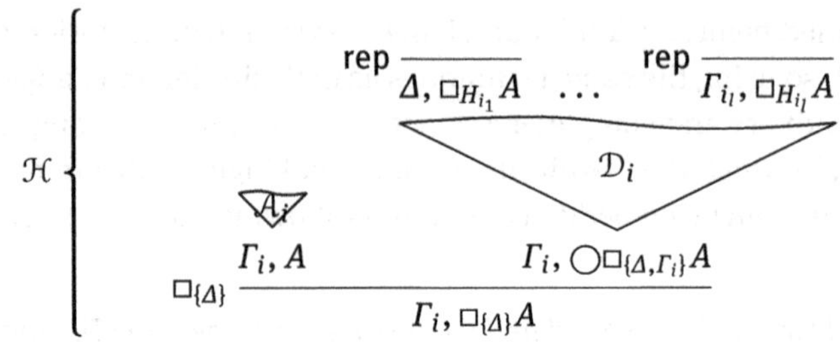

We have that in the derivation $\mathcal{D}$:

$$\text{any branch from } \Delta, \bigcirc\Box A \text{ to some } \Gamma_i, \Box A \text{ or some } \Delta, \Box A \text{ goes through a } \bigcirc\text{-rule} \tag{3}$$

and

$$\text{there are no instances of foc} \tag{4}$$

Furthermore, we observe that if $\Delta \in H_{i_k}$ (that is when $\Delta, \Box_{H_{i_k}} A$ is an instance of rep), then from $\mathsf{LT2} \vdash \Delta, \bigcirc\Box_{\{\Delta\}} A$ and Lemma 19 we get a proof $\mathcal{B}_{i_k}$ for $\Delta, \bigcirc\Box_{H_{i_k}} A$.

We let $\mathcal{D}_i'$ be the the derivation that results from $\mathcal{D}_i$ by deleting Δ from all histories occurring in threads starting from $\bigcirc\Box_{\{\Delta,\Gamma_i\}} A$.

Hence we obtain the following proofs of $\Gamma_i, \Box A$, which we denote by $\mathcal{C}_i$.

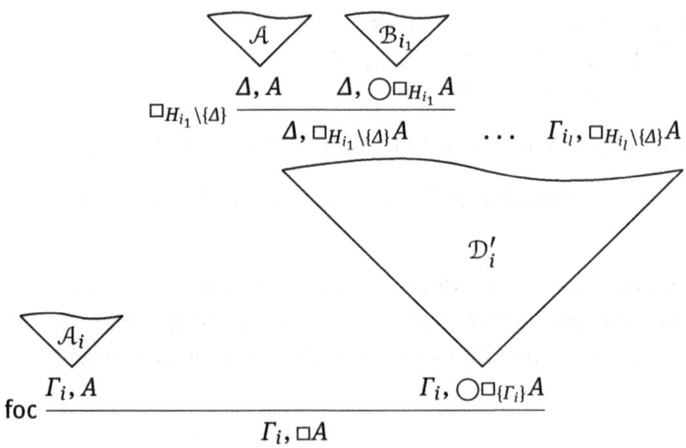

Now we proceed as follows:

1. We apply the induction hypothesis to $\mathcal{A}$, which yields a proof $\mathcal{A}'$ of Δ, A that satisfies the next-property.
2. We let $\mathcal{D}'$ be the derivation that results from $\mathcal{D}$ by dropping the annotation Δ in the threads starting from $\bigcirc\Box_{\{\Delta\}} A$.

We find that in the derivation $\mathcal{D}'$,

any branch from $\Delta, \bigcirc\Box A$ to some $\Gamma_i, \Box A$ or some $\Delta, \Box A$ goes through a $\bigcirc$-rule

because of (3) and

there are no instances of foc

because of (4).

Finally we obtain the following proof of $\Delta, \Box A$.

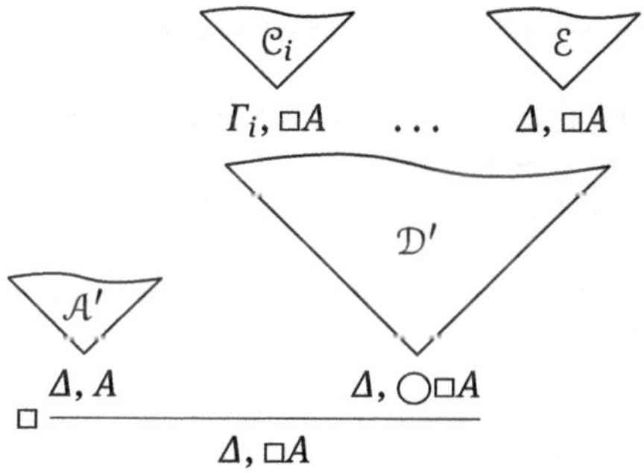

This proof satisfies the next-property. Indeed, we have:

1. the proof $\mathcal{A}'$ satisfies the next-property;
2. any branch from $\Delta, \bigcirc\Box A$ to some $\Gamma_i, \Box A$ or to some $\Delta, \Box A$ goes through a $\bigcirc$-rule;
3. the derivation $\mathcal{D}'$ does not contain instances of foc.

Hence any branch from the conclusion $\Delta, \Box A$ to an instance of foc goes through a $\bigcirc$-rule. □

We get weakening for LT2 as a corollary of Lemma 20. The proof is the same as for Corollary 7.

Corollary 21 (Weakening for non-annotated sequents). *For any $\Gamma, \Delta \in$ Seq we have:*

$$\mathsf{LT2} \vdash \Gamma \quad \Longrightarrow \quad \mathsf{LT2} \vdash \Gamma, \Delta$$

Acknowledgment: We would like to thank the anonymous referee for valuable comments that helped us improve the paper substantially.

References

James Brotherston and Alex Simpson. Sequent calculi for induction and infinite descent. *Journal of Logic and Computation*, 21(6):1177–1216, December 2011.

Kai Brünnler and Martin Lange. Cut-free sequent systems for temporal logic. *The Journal of Logic and Algebraic Programming*, 76(2):216–225, 2008.

Samuel Bucheli, Roman Kuznets, and Thomas Studer. Two ways to common knowledge. In Thomas Bolander and Torben Braüner, editors, *Proceedings of the 6th Workshop on Methods for Modalities (M4M–6 2009), Copenhagen, Denmark, 12–14 November 2009*, Electronic Notes in Theoretical Computer Science, pages 83–98. Elsevier, 2010. 10.1016/j.entcs.2010.04.007. URL http://www.iam.unibe.ch/ltgpub/2010/bks10a.pdf.

Fred Kröger and Stephan Merz. *Temporal Logic and State Systems*. Springer, Berlin, 2008.

Martin Lange and Colin Stirling. Focus games for satisfiability and completeness of temporal logic. In *LICS*, 2001.

Damian Niwinski and Igor Walukiewicz. Games for the mu-calculus. *Theoretical Computer Science*, 163(1&2):99–116, 1996.

Daniyar S. Shamkanov. Circular proofs for Gödel-Löb logic. *CoRR*, abs/1401.4002, 2014. URL http://arxiv.org/abs/1401.4002.

Thomas Studer. On the proof theory of the modal mu-calculus. *Studia Logica*, 89:343–363, 2008. 10.1007/s11225-008-9133-6.

Roman Kuznets

Craig Interpolation via Hypersequents

Abstract: In this paper, we describe a novel constructive method of proving the Craig interpolation property (CIP) based on cut-free hypersequent calculi and apply the method to prove the CIP for the modal logic S5.

Keywords: Craig interpolation, Hypersequent, Structural proof theory, Modal logic

Mathematics Subject Classification 2010: 03B45, 03C40, 03F07, 03B70

1 Introduction

Interpolation is one of the standard properties of a logical system, studied alongside decidability, complexity, and semantic completeness. Interpolation is closely related to algebraic properties such as amalgamation (see [Gabbay and Maksimova, 2005] for an overview and the history of the subject). The logic L is said to have the *Craig interpolation property* (CIP) if, whenever $\mathsf{L} \vdash A \to B$, there exists a formula C "in the common language" of A and B such that $\mathsf{L} \vdash A \to C$ and $\mathsf{L} \vdash C \to B$. The formula C is then called the *interpolant* of A and B. In this paper, we consider modal logics, hence, the "common language" simply means having the same propositional variables.

One of the methods for proving the CIP constructively and efficiently[1], is by employing a cut-free (or, more generally, an analytic) proof system and by constructing an interpolant by induction on the derivation of $A \to B$ (properly represented in this proof system). Such a method based on sequent calculi is well known and had been used for many a system, e.g., for classical and intuitionistic propositional logics and for many modal logics. However, sequent calculi do not seem to be expressive enough to capture many interesting modal logics. In fact, one of the first modal logics to be considered in modern times, S5, has so far resisted all attempts at being captured by a cut-free sequent calculus. Moreover, it was shown by Lellmann and Pattinson [2013] that such a calculus does not exist under reasonable restrictions on the type of rules used.

1 Rather than by the exhaustive search of all potential interpolants.

Roman Kuznets: Institut für Computersprachen, Technische Universität Wien,
e-mail: roman@logic.at

The simplest extension of the sequent formalism that is known to be capable of capturing S5 is the formalism of hypersequents, which was first introduced by Minc [1971]. (Mints[2] used the term "cortege" rather than "hypersequent" and employed a slightly more cumbersome notation than we are used to now.[3]) Hypersequents under their proper name were later independently rediscovered by Pottinger [1983] and Avron [1987].

It is, thus, natural to generalize the constructive method of showing the CIP from sequents to hypersequents. In his seminal survey, Avron [1996] writes:

> The only rule [...] which brings moments of synchronization into proofs is external contraction. [I]ts presence is the explanation why in hypersequential calculi cut-elimination usually does not imply the Craig interpolation theorem.

According to Avron, the rule of external contraction

$$\mathsf{EC}\ \frac{\mathcal{G} \mid \Gamma \Rightarrow \Delta \mid \Gamma \Rightarrow \Delta}{\mathcal{G} \mid \Gamma \Rightarrow \Delta}$$

is the main obstacle to such a generalization. However, the modal logic S5, captured by a hypersequent calculus with the external contraction rule EC, is known to enjoy the CIP. Thus, EC does not, in general, prevent the CIP from holding.

In a joint work [Fitting and Kuznets, 2015], we developed a method of generalizing the syntactic proof of the CIP from sequents to nested sequents. While hypersequents are not exactly subsystems of nested sequents because several of hypersequent rules are not present in nested calculi, it is reasonable to view hypersequents as substructures of nested sequents. Indeed, a hypersequent is a set (multiset, sequence) of ordinary sequents, whereas a nested sequent is a tree of sequents. Both hypersequents and nested sequents are *internal calculi*, meaning that each hypersequent and each nested sequent can be translated into a formula. Based on the commonly used translations, components of a hypersequent correspond to children of the root of a nested sequent. Thus, it seems reasonable to expect that the method for proving the CIP via nested sequents from [Fitting and Kuznets, 2015] can be adapted to hypersequents. In this paper, we show that this is indeed the case and that the external contraction does not present any difficulties for our method. We develop the method using a hypersequent calculus for S5 from [Avron, 1996] as an example.

2 Spelled "Minc" in the translation.

3 The full text of the original Russian version of 1968 is available at http://mi.mathnet.ru/eng/tm/v98/p88.

2 The Modal Logic S5

Being the fifth of Lewis's systems, the modal logic S5 is one of the oldest modern modal logics in existence and is often used, for instance, as a logic of knowledge [Fagin et al., 1995]. Its language, which we will call the *modal language*, is defined by the grammar

$$A ::= P \mid \bot \mid (A \to A) \mid \Box A,$$

where P is taken from a countably infinite set Prop of propositional variables. The Boolean constant $\top$, the Boolean connectives $\neg$, $\wedge$, and $\vee$, and the modality $\Diamond$ are defined in the standard way. We use the standard conventions regarding the omission of redundant parentheses.

The modal logic S5 is axiomatized by all instances of propositional tautologies in the modal language above and the following modal axioms:

- t: $\Box A \to A$;
- 4: $\Box A \to \Box\Box A$;
- 5: $\neg\Box A \to \Box\neg\Box A$.

The epistemic reading of these axioms, when $\Box A$ is understood as "A is known," states the factivity of knowledge and the positive and negative introspection of the knowledge agent. The inference rules are *modus ponens* and *modal necessitation*:

$$\mathsf{MP}\ \frac{A \to B \quad A}{B} \qquad \text{and} \qquad \mathsf{Nec}\ \frac{A}{\Box A}.$$

The logic S5 has a particularly simple Kripke semantics:

Definition 2.1 (Kripke frames and models). *A* Kripke frame *(for* S5*) is* $(W, \sim)$, *a pair of a non-empty set* W *and of an equivalence relation* $\sim$ *on* W. *Elements of* W *are called* worlds. *If* $u \sim v$, *we say that the world* u *is* indistinguishable *from the world* v. *A* Kripke model *(for* S5*) is a triple* $(W, \sim, V)$ *where* $(W, \sim)$ *is a Kripke frame (for* S5*) and* $V \colon W \to 2^{\mathrm{Prop}}$ *is a valuation that assigns to each world from* W *the set of propositional variables true at this world.*

Definition 2.2 (Truth and validity in Kripke models). *Given a Kripke model* $\mathcal{M} = (W, \sim, V)$, *the relation of truth between modal formulas and worlds* w *in this model is defined as follows:*

for each propositional variable $P \in \mathrm{Prop}$: $\quad \mathcal{M}, w \Vdash P \quad$ *iff* $\quad P \in V(w)$;

$\mathcal{M}, w \nVdash \bot$;

$\mathcal{M}, w \Vdash A \to B \quad$ *iff* $\quad \mathcal{M}, w \nVdash A$ *or* $\mathcal{M}, w \Vdash B$;

$\mathcal{M}, w \Vdash \Box A \quad$ *iff* $\quad \mathcal{M}, v \Vdash A$ *for all* $v \in W$ *such that* $v \sim w$.

We say that A is valid in $\mathcal{M}$, *written* $\mathcal{M} \Vdash A$, *if* $\mathcal{M}, w \Vdash A$ *for all* $w \in W$.

Theorem 2.3 (Completeness of Kripke semantics for S5, [Fagin et al., 1995]). *For any modal formula A,*

$$\mathsf{S5} \vdash A \qquad \textit{iff} \qquad A \textit{ is valid in all Kripke models for } \mathsf{S5}.$$

3 Hypersequents for S5

Definition 3.1 (Sequents and hypersequents). *A* sequent *is a figure $\Gamma \Rightarrow \Delta$ where Γ and Δ are finite (possibly empty) multisets of modal formulas. A* hypersequent $\mathcal{G}$ *is a finite (possibly empty) sequence*

$$\Gamma_1 \Rightarrow \Delta_1 \mid \ldots \mid \Gamma_n \Rightarrow \Delta_n$$

of sequents $\Gamma_i \Rightarrow \Delta_i$, called (sequent) components *of $\mathcal{G}$. We define the length $\|\mathcal{G}\|$ of a hypersequent $\mathcal{G}$ to be the number of sequent components it contains.*

We denote finite multisets of formulas by $\Gamma, \Delta, \Pi, \Sigma$, etc. and denote hypersequents by $\mathcal{G}, \mathcal{H}$, etc. For a multiset of formulas Γ, we define $\Box\Gamma := \{\Box C \mid C \in \Gamma\}$.

Definition 3.2 (Hypersequent system for S5). *The hypersequent system* HS5 *for the logic* S5 *is presented in Figure 1. It is essentially the one presented in [Avron, 1996]. The modifications are slight and clearly do not affect the cut-free completeness of the system. We list the differences between our presentation compared to that by Avron [1996]:*

- *sequent components consist of pairs of multisets of formulas rather than pairs of sequences of formulas (a hypersequent, however, remains a sequence of components);*
- *Boolean connectives are restricted to $\bot$ and $\rightarrow$;*
- *in [Avron, 1996] rules can be applied to any component of a hypersequent, whereas we restrict the applications to the last component of the hypersequent, except for the external exchange rule;*
- *it is not specified what type of propositional rules, additive or multiplicative, is used in [Avron, 1996].*

To state the completeness of the hypersequent calculus HS5 with respect to the logic S5, we use a translation from hypersequents to formulas:

$$\text{id}\ \frac{}{A \Rightarrow A} \qquad \text{id}_{\bot}\ \frac{}{\bot \Rightarrow}$$

$$\rightarrow\Rightarrow\ \frac{\mathcal{G} \mid \Gamma \Rightarrow \Delta, A \quad \mathcal{G} \mid \Gamma, B \Rightarrow \Delta}{\mathcal{G} \mid \Gamma, A \rightarrow B \Rightarrow \Delta} \qquad \Rightarrow\rightarrow\ \frac{\mathcal{G} \mid \Gamma, A \Rightarrow \Delta, B}{\mathcal{G} \mid \Gamma \Rightarrow \Delta, A \rightarrow B}$$

$$\text{W}\Rightarrow\ \frac{\mathcal{G} \mid \Gamma \Rightarrow \Delta}{\mathcal{G} \mid \Gamma, A \Rightarrow \Delta} \qquad \Rightarrow\text{W}\ \frac{\mathcal{G} \mid \Gamma \Rightarrow \Delta}{\mathcal{G} \mid \Gamma \Rightarrow \Delta, A} \qquad \text{EW}\ \frac{\mathcal{G}}{\mathcal{G} \mid \Gamma \Rightarrow \Delta}$$

$$\text{C}\Rightarrow\ \frac{\mathcal{G} \mid \Gamma, A, A \Rightarrow \Delta}{\mathcal{G} \mid \Gamma, A \Rightarrow \Delta} \qquad \Rightarrow\text{C}\ \frac{\mathcal{G} \mid \Gamma \Rightarrow \Delta, A, A}{\mathcal{G} \mid \Gamma \Rightarrow \Delta, A} \qquad \text{EC}\ \frac{\mathcal{G} \mid \Gamma \Rightarrow \Delta \mid \Gamma \Rightarrow \Delta}{\mathcal{G} \mid \Gamma \Rightarrow \Delta}$$

$$\text{Ex}\ \frac{\mathcal{G} \mid \Gamma \Rightarrow \Delta \mid \Lambda \Rightarrow \Theta \mid \mathcal{H}}{\mathcal{G} \mid \Lambda \Rightarrow \Theta \mid \Gamma \Rightarrow \Delta \mid \mathcal{H}}$$

$$\Box\Rightarrow\ \frac{\mathcal{G} \mid \Gamma, A \Rightarrow \Delta}{\mathcal{G} \mid \Gamma, \Box A \Rightarrow \Delta} \qquad \Rightarrow\Box\ \frac{\mathcal{G} \mid \Box\Gamma \Rightarrow A}{\mathcal{G} \mid \Box\Gamma \Rightarrow \Box A} \qquad \text{MS}\ \frac{\mathcal{G} \mid \Box\Lambda, \Gamma \Rightarrow \Box\Phi, \Delta}{\mathcal{G} \mid \Box\Lambda \Rightarrow \Box\Phi \mid \Gamma \Rightarrow \Delta}$$

Fig. 1. Cut-free hypersequent system HS5 for the modal logic S5 (following [Avron, 1996]).

Definition 3.3 (Formula translation). *The formula translation of a hypersequent is defined as*

$$\underline{\Gamma_1 \Rightarrow \Delta_1 \mid \ldots \mid \Gamma_n \Rightarrow \Delta_n} := \bigvee_{i=1}^{n} \Box\left(\bigwedge \Gamma_i \rightarrow \bigvee \Delta_i\right).$$

As usual, the empty disjunction is understood to be $\bot$ and the empty conjunction is $\top$. (Strictly speaking, the order of formulas in $\bigwedge \Gamma_i$ and $\bigvee \Delta_i$ has to be specified even though all possible end results are pairwise provably equivalent in S5. *To be formally correct, we assume an arbitrary but fixed order on modal formulas and use it in the conjunctions/disjunctions of multisets Γ_i and Δ_i.)*

Theorem 3.4 (Completeness of HS5, [Avron, 1996]). *For any hypersequent $\mathcal{G}$,*

$$\mathsf{HS5} \vdash \mathcal{G} \qquad \Longleftrightarrow \qquad \mathsf{S5} \vdash \underline{\mathcal{G}}.$$

Corollary 3.5. *For arbitrary modal formulas A and B,*

$$\mathsf{HS5} \vdash A \Rightarrow B \qquad \Longleftrightarrow \qquad \mathsf{S5} \vdash A \to B.$$

4 Preparing for Interpolation

Interpolation is always performed between two entities, e.g., between formulas A and B. The last corollary of the preceding section shows that we can equivalently interpolate between the antecedent and consequent of a single-component hypersequent. However, as in the case of two-sided sequents, such a division does not remain stable along a hypersequent derivation because the $\to$-introducing rules move formulas from one side of $\Rightarrow$ to the other one, which affects the set of variables common between the antecedent(s) and the consequent(s) in unpredictable ways. Instead, we must supply a hypersequent with an extra layer of structure, splitting all formulas, antecedent and consequent alike, into *left* formulas, i.e., eventually contributing to A, and *right* formulas, eventually contributing to B in the endsequent $A \Rightarrow B$:

Definition 4.1 (Split hypersequents). *A* split hypersequent $\widetilde{\mathcal{G}}$ *is a hypersequent where each antecedent and each consequent is partitioned into two multisets by a semicolon:*

$$\Gamma_1; \Pi_1 \Rightarrow \Delta_1; \Sigma_1 \mid \dots \mid \Gamma_n; \Pi_n \Rightarrow \Delta_n; \Sigma_n$$

(the semicolon can be omitted if an antecedent or a consequent is empty, i.e., we write $\Gamma; \Pi \Rightarrow$ instead of $\Gamma; \Pi \Rightarrow\ ;$). For the split hypersequent $\widetilde{\mathcal{G}}$ above, its left (right) sides are obtained by dropping all right (left) formulas:

$$L\widetilde{\mathcal{G}} := \Gamma_1 \Rightarrow \Delta_1 \mid \dots \mid \Gamma_n \Rightarrow \Delta_n,$$
$$R\widetilde{\mathcal{G}} := \Pi_1 \Rightarrow \Sigma_1 \mid \dots \mid \Pi_n \Rightarrow \Sigma_n.$$

As before, the length of a split hypersequent is the number of its components, i.e., for the split hypersequent above $\|\widetilde{\mathcal{G}}\| := n$. It is obvious that $\|\widetilde{\mathcal{G}}\| = \|L\widetilde{\mathcal{G}}\| = \|R\widetilde{\mathcal{G}}\|$.

Given that we plan to find interpolants between the left and right formulas of a given split hypersequent, we need to split all the rules of the calculus HS5.

Definition 4.2 (Split hypersequent calculus SHS5). *The calculus* SHS5 *is presented in Figures 2 and 3.*

$$\mathsf{EW}\ \frac{\mathfrak{G}}{\mathfrak{G} \mid \Gamma; \Pi \Rightarrow \Delta; \Sigma} \qquad \mathsf{EC}\ \frac{\mathfrak{G} \mid \Gamma; \Pi \Rightarrow \Delta; \Sigma \mid \Gamma; \Pi \Rightarrow \Delta; \Sigma}{\mathfrak{G} \mid \Gamma; \Pi \Rightarrow \Delta; \Sigma}$$

$$\mathsf{Ex}\ \frac{\mathfrak{G} \mid \Gamma; \Pi \Rightarrow \Delta; \Sigma \mid \Lambda; \Phi \Rightarrow \Theta; \Psi \mid \mathfrak{H}}{\mathfrak{G} \mid \Lambda; \Phi \Rightarrow \Theta; \Psi \mid \Gamma; \Pi \Rightarrow \Delta; \Sigma \mid \mathfrak{H}}$$

$$\Box^{\mathsf{l}} \Rightarrow \frac{\mathfrak{G} \mid \Gamma, A; \Pi \Rightarrow \Delta; \Sigma}{\mathfrak{G} \mid \Gamma, \Box A; \Pi \Rightarrow \Delta; \Sigma} \qquad \Rightarrow \Box^{\mathsf{l}}\ \frac{\mathfrak{G} \mid \Box\Gamma; \Box\Pi \Rightarrow A;}{\mathfrak{G} \mid \Box\Gamma; \Box\Pi \Rightarrow \Box A;}$$

$$\Box^{\mathsf{r}} \Rightarrow \frac{\mathfrak{G} \mid \Gamma; \Pi, A \Rightarrow \Delta; \Sigma}{\mathfrak{G} \mid \Gamma; \Pi, \Box A \Rightarrow \Delta; \Sigma} \qquad \Rightarrow \Box^{\mathsf{r}}\ \frac{\mathfrak{G} \mid \Box\Gamma; \Box\Pi \Rightarrow ; A}{\mathfrak{G} \mid \Box\Gamma; \Box\Pi \Rightarrow ; \Box A}$$

$$\mathsf{MS}\ \frac{\mathfrak{G} \mid \Box\Lambda, \Gamma; \Box\Theta, \Pi \Rightarrow \Box\Phi, \Delta; \Box\Psi, \Sigma}{\mathfrak{G} \mid \Box\Lambda; \Box\Theta \Rightarrow \Box\Phi; \Box\Psi \mid \Gamma; \Pi \Rightarrow \Delta; \Sigma}$$

Fig. 2. Cut-free split hypersequent system SHS5 for the modal logic S5: modal and external structural rules.

It is easy to see that the split system is nothing but the split of the original system:

Theorem 4.3 (Equivalence of HS5 and SHS5). *For arbitrary finite multisets Γ, Π, Δ, and Σ of modal formulas,*

$$\mathsf{SHS5} \vdash \Gamma; \Pi \Rightarrow \Delta; \Sigma \qquad \Longleftrightarrow \qquad \mathsf{HS5} \vdash \Gamma \cup \Pi \Rightarrow \Delta \cup \Sigma.$$

Proof: Both directions are proved by induction on the depth of the derivation. The crucial observation is that each split rule of SHS5 becomes an ordinary rule of HS5 if one takes the union of left and right formulas separately in each antecedent and each consequent. Vice versa, for each split of the conclusion of any rule of HS5, there is a split of the premise(s) that turns this rule into a rule of SHS5. □

Corollary 4.4. *For arbitrary modal formulas A and B,*

$$\mathsf{SHS5} \vdash A; \Rightarrow ; B \qquad \Longleftrightarrow \qquad \mathsf{S5} \vdash A \to B.$$

$$\mathsf{id}^{\mathsf{ll}}\ \frac{}{A;\ \Rightarrow A;}\qquad \mathsf{id}^{\mathsf{rl}}\ \frac{}{;A \Rightarrow A;}\qquad \mathsf{id}^{\mathsf{lr}}\ \frac{}{A;\ \Rightarrow\ ;A}\qquad \mathsf{id}^{\mathsf{rr}}\ \frac{}{;A \Rightarrow\ ;A}$$

$$\mathsf{id}^{\mathsf{l}}_{\bot}\ \frac{}{\bot;\ \Rightarrow}\qquad \mathsf{id}^{\mathsf{r}}_{\bot}\ \frac{}{;\bot \Rightarrow}$$

$$\Rightarrow\to^{\mathsf{l}}\ \frac{\widetilde{\mathcal{G}} \mid \Gamma, A; \Pi \Rightarrow \Delta, B; \Sigma}{\widetilde{\mathcal{G}} \mid \Gamma; \Pi \Rightarrow \Delta, A \to B; \Sigma}\qquad \Rightarrow\to^{\mathsf{r}}\ \frac{\widetilde{\mathcal{G}} \mid \Gamma; \Pi, A \Rightarrow \Delta; \Sigma, B}{\widetilde{\mathcal{G}} \mid \Gamma; \Pi \Rightarrow \Delta; \Sigma, A \to B}$$

$$\to^{\mathsf{l}}\Rightarrow\ \frac{\widetilde{\mathcal{G}} \mid \Gamma; \Pi \Rightarrow \Delta, A; \Sigma \quad \widetilde{\mathcal{G}} \mid \Gamma, B; \Pi \Rightarrow \Delta; \Sigma}{\widetilde{\mathcal{G}} \mid \Gamma, A \to B; \Pi \Rightarrow \Delta; \Sigma}$$

$$\to^{\mathsf{r}}\Rightarrow\ \frac{\widetilde{\mathcal{G}} \mid \Gamma; \Pi \Rightarrow \Delta; \Sigma, A \quad \widetilde{\mathcal{G}} \mid \Gamma; \Pi, B \Rightarrow \Delta; \Sigma}{\widetilde{\mathcal{G}} \mid \Gamma; \Pi, A \to B \Rightarrow \Delta; \Sigma}$$

$$\mathsf{W}^{\mathsf{l}}\Rightarrow\ \frac{\widetilde{\mathcal{G}} \mid \Gamma; \Pi \Rightarrow \Delta; \Sigma}{\widetilde{\mathcal{G}} \mid \Gamma, A; \Pi \Rightarrow \Delta; \Sigma}\qquad \Rightarrow\mathsf{W}^{\mathsf{l}}\ \frac{\widetilde{\mathcal{G}} \mid \Gamma; \Pi \Rightarrow \Delta; \Sigma}{\widetilde{\mathcal{G}} \mid \Gamma; \Pi \Rightarrow \Delta, A; \Sigma}$$

$$\mathsf{W}^{\mathsf{r}}\Rightarrow\ \frac{\widetilde{\mathcal{G}} \mid \Gamma; \Pi \Rightarrow \Delta; \Sigma}{\widetilde{\mathcal{G}} \mid \Gamma; \Pi, A \Rightarrow \Delta; \Sigma}\qquad \Rightarrow\mathsf{W}^{\mathsf{r}}\ \frac{\widetilde{\mathcal{G}} \mid \Gamma; \Pi \Rightarrow \Delta; \Sigma}{\widetilde{\mathcal{G}} \mid \Gamma; \Pi \Rightarrow \Delta; \Sigma, A}$$

$$\mathsf{C}^{\mathsf{l}}\Rightarrow\ \frac{\widetilde{\mathcal{G}} \mid \Gamma, A, A; \Pi \Rightarrow \Delta; \Sigma}{\widetilde{\mathcal{G}} \mid \Gamma, A; \Pi \Rightarrow \Delta; \Sigma}\qquad \Rightarrow\mathsf{C}^{\mathsf{l}}\ \frac{\widetilde{\mathcal{G}} \mid \Gamma; \Pi \Rightarrow \Delta, A, A; \Sigma}{\widetilde{\mathcal{G}} \mid \Gamma; \Pi \Rightarrow \Delta, A; \Sigma}$$

$$\mathsf{C}^{\mathsf{r}}\Rightarrow\ \frac{\widetilde{\mathcal{G}} \mid \Gamma; \Pi, A, A \Rightarrow \Delta; \Sigma}{\widetilde{\mathcal{G}} \mid \Gamma; \Pi, A \Rightarrow \Delta; \Sigma}\qquad \Rightarrow\mathsf{C}^{\mathsf{r}}\ \frac{\widetilde{\mathcal{G}} \mid \Gamma; \Pi \Rightarrow \Delta; \Sigma, A, A}{\widetilde{\mathcal{G}} \mid \Gamma; \Pi \Rightarrow \Delta; \Sigma, A}$$

Fig. 3. Cut-free split hypersequent system SHS5 for the modal logic S5: propositional and internal structural rules

Proof: The statement follows from Corollary 3.5 and Theorem 4.3. □

Remark 4.5. *The same reasons that compelled us to split each component of a hypersequent explain why the cut rule cannot be processed by a constructive proof of the CIP based on the induction on a sequent-like derivation. The cut rule violates the subformula property by removing a formula completely from each of the two premises. This removal can shrink the set of common variables in unpredictable ways, thus, rendering the interpolants from the induction hypothesis suddenly unusable.*

Splitting a hypersequent is no different from splitting a sequent and is standard for constructive proofs of the CIP. Thus, by itself, it does not yet help to extend the method to hypersequents. The crucial idea behind our method of proving the CIP is that interpolation should be done on the component level, i.e., instead of having a formula interpolant for the whole hypersequent, responsible, in particular, for encoding the hypersequent structure and its transformations, we allow each sequent component to have its own formula interpolant and combine these componentwise interpolants by explicit operations creating structures that parallel the structure of the hypersequent being interpolated. This view of interpolation signifies a departure from the very definition of interpolation. Thus, we must both present the intuition behind our view and demonstrate that the two definitions coincide for the final result of our interpolation procedure (but not during the intermediate stages, where our interpolation statements cannot be translated to the standard ones). To this end, we present an alternative semantics for hypersequents that provides the intuition for the structure of our interpolants.

Definition 4.6 (Connected worlds). *Let $\mathcal{M} = (W, \sim, V)$ be a Kripke model. A sequence of worlds $\vec{w} = w_1, \ldots, w_n$ from W is called $\mathcal{M}$-connected if $w_1 \sim w_i$ for each $2 \leq i \leq n$. It immediately follows that $w_i \sim w_j$ for all $1 \leq i \leq j \leq n$.*

Definition 4.7 (Componentwise semantics). *Let $\mathcal{M} = (W, \sim, V)$ be a Kripke model. A sequent $\Gamma \Rightarrow \Delta$ holding at a world $w \in W$ of $\mathcal{M}$ is defined as follows:*

$$\mathcal{M}, w \vDash \Gamma \Rightarrow \Delta \quad \Longleftrightarrow \quad \begin{array}{l} \mathcal{M}, w \nVdash A \text{ for some } A \in \Gamma \quad \text{or} \\ \mathcal{M}, w \Vdash B \text{ for some } B \in \Delta. \end{array}$$

Let $\mathcal{G}$ be a hypersequent $\Gamma_1 \Rightarrow \Delta_1 \mid \ldots \mid \Gamma_n \Rightarrow \Delta_n$ and $\vec{w} = w_1, \ldots, w_n$ be a sequence of length $\|\mathcal{G}\|$ of worlds from W. The hypersequent $\mathcal{G}$ holding on the sequence $\vec{w}$ in $\mathcal{M}$ is defined as follows:

$$\mathcal{M}, \vec{w} \vDash \Gamma_1 \Rightarrow \Delta_1 \mid \ldots \mid \Gamma_n \Rightarrow \Delta_n \quad \Longleftrightarrow \quad \mathcal{M}, w_i \vDash \Gamma_i \Rightarrow \Delta_i \text{ for some } 1 \leq i \leq n.$$

(In particular, we define $\mathcal{M}, \varepsilon \nvDash \varepsilon$, i.e., the empty hypersequent does not hold on the empty sequence of worlds.)

Definition 4.8 (Componentwise validity). *A hypersequent* $\mathcal{G}$ *is* componentwise valid *if* $\mathcal{M}, \vec{w} \models \mathcal{G}$ *for any Kripke model* $\mathcal{M}$ *and any* $\mathcal{M}$*-connected sequence* $\vec{w}$ *of length* $\|\mathcal{G}\|$.

Lemma 4.9 (Equivalence of two semantics). *A hypersequent* $\mathcal{G}$ *is componentwise valid iff its formula interpretation is valid in all Kripke models.*

Proof: Let $\mathcal{G}$ be a hypersequent $\Gamma_1 \Rightarrow \Delta_1 \mid \ldots \mid \Gamma_n \Rightarrow \Delta_n$. We prove that it is componentwise invalid iff its formula interpretation is invalid. Our first observation is that for any component $\Gamma \Rightarrow \Delta$, any Kripke model $\mathcal{M} = (W, \sim, V)$, and any world $w \in W$, we have

$$\mathcal{M}, w \not\models \Gamma \Rightarrow \Delta \qquad \Longleftrightarrow \qquad \mathcal{M}, w \not\Vdash \bigwedge \Gamma \to \bigvee \Delta. \tag{1}$$

In the following sequence of statements, each statement is equivalent to the previous one.

1. $\underline{\mathcal{G}}$ is invalid;
2. there is a Kripke model $\mathcal{M} = (W, \sim, V)$ and there is a world $v \in W$ such that $\mathcal{M}, v \not\Vdash \bigvee_{i=1}^{n} \Box(\bigwedge \Gamma_i \to \bigvee \Delta_i)$; (by definitions of $\underline{\mathcal{G}}$ and of validity for formulas)
3. there exists a Kripke model $\mathcal{M} = (W, \sim, V)$, a world $v \in W$, and worlds $w_1, \ldots, w_n \in W$ such that $v \sim w_i$ and $\mathcal{M}, w_i \not\Vdash \bigwedge \Gamma_i \to \bigvee \Delta_i$ for each $i = 1, \ldots, n$; (by definition of $\Vdash$)
4. there exists a Kripke model $\mathcal{M} = (W, \sim, V)$ and an $\mathcal{M}$-connected sequence $\vec{w} = w_1, \ldots, w_n$ such that $\mathcal{M}, w_i \not\Vdash \bigwedge \Gamma_i \to \bigvee \Delta_i$ for each $i = 1, \ldots, n$; (v can be set to w_1 because $\sim$ is an equivalence relation)
5. there exists a Kripke model $\mathcal{M} = (W, \sim, V)$ and an $\mathcal{M}$-connected sequence $\vec{w} = w_1, \ldots, w_n$ such that $\mathcal{M}, w_i \not\models \Gamma_i \Rightarrow \Delta_i$ for each $i = 1, \ldots, n$; (by (1))
6. there exists a Kripke model $\mathcal{M} = (W, \sim, V)$ and an $\mathcal{M}$-connected sequence $\vec{w}$ of length n such that $\mathcal{M}, \vec{w} \not\models \mathcal{G}$; (by definition of $\models$)
7. $\mathcal{G}$ is componentwise invalid. (by definition of componentwise validity)

□

Having transferred validity to the level of sequent components, we now define the objects that will serve as interpolants.

Definition 4.10 (Hyperformulas and g-hyperformulas). *An* hyperformula *is a sequence of modal formulas, written in the hypersequent notation:* $C_1 \mid \ldots \mid C_n$. *By analogy with sequence of worlds, we use vector notation* $\vec{C}$ *for hyperformulas and define the length of a hyperformula to be the number of formulas in it, i.e.,* $\|C_1 \mid \ldots \mid C_n\| := n$.

Each hyperformula is a generalized hyperformula *of the same length. If* $\mho_1$ *and* $\mho_2$ *are generalized hyperformulas of length n each, then* $(\mho_1 \ovee \mho_2)$ *and* $(\mho_1 \owedge \mho_2)$ *are also generalized hyperformulas of length n. For brevity's sake, we sometimes call generalized hyperformulas simply* g-hyperformulas.

Definition 4.11 (Componentwise semantics for (g-)hyperformulas). *Let a triple* $\mathcal{M} = (W, \sim, V)$ *be a Kripke model. Let* $\vec{C} = C_1 \mid \ldots \mid C_n$ *be a hyperformula and* $\vec{w} = w_1, \ldots, w_n$ *be a sequence of length* $\|\vec{C}\|$ *of worlds from* W*. We define*

$$\mathcal{M}, \vec{w} \models C_1 \mid \ldots \mid C_n \qquad \textit{iff} \qquad \mathcal{M}, w_i \models C_i \textit{ for some } 1 \leq i \leq n.$$

For arbitrary g-hyperformulas $\mho_1$ *and* $\mho_2$ *and arbitrary sequence* $\vec{w}$ *of worlds from* W *such that* $\|\mho_1\| = \|\mho_2\| = \|\vec{w}\|$*, we define*

$$\mathcal{M}, \vec{w} \models \mho_1 \owedge \mho_2 \quad \textit{iff} \quad \mathcal{M}, \vec{w} \models \mho_1 \textit{ and } \mathcal{M}, \vec{w} \models \mho_2;$$
$$\mathcal{M}, \vec{w} \models \mho_1 \ovee \mho_2 \quad \textit{iff} \quad \mathcal{M}, \vec{w} \models \mho_1 \textit{ or } \mathcal{M}, \vec{w} \models \mho_2.$$

Thus, with respect to the componentwise semantics, $\owedge$ *and* $\ovee$ *on hyperformulas correspond to* $\wedge$ *and* $\vee$ *on formulas respectively.*

Two g-hyperformulas $\mho_1$ *and* $\mho_2$ *are called* componentwise equivalent, *written* $\mho_1 \mathrel{=\!\!|\!\models} \mho_2$*, iff* $\|\mho_1\| = \|\mho_2\|$ *and, for any Kripke model* $\mathcal{M}$ *and any* $\mathcal{M}$*-connected sequence* $\vec{w}$ *of length* $\|\mho_1\|$*, we have*

$$\mathcal{M}, \vec{w} \models \mho_1 \qquad \Longleftrightarrow \qquad \mathcal{M}, \vec{w} \models \mho_2.$$

Definition 4.12 (Componentwise interpolant). *A* componentwise interpolant *of a split hypersequent* $\widetilde{\mathcal{G}}$ *is a g-hyperformula* $\mho$ *such that* $\|\mho\| = \|\widetilde{\mathcal{G}}\|$ *and for each Kripke model* $\mathcal{M}$ *and each* $\mathcal{M}$*-connected sequence* $\vec{w}$ *with* $\|\vec{w}\| = \|\mho\|$*,*

- *if* $\mathcal{M}, \vec{w} \not\models \mho$*, then* $\mathcal{M}, \vec{w} \models L\widetilde{\mathcal{G}}$*;*
- *if* $\mathcal{M}, \vec{w} \models \mho$*, then* $\mathcal{M}, \vec{w} \models R\widetilde{\mathcal{G}}$*;*
- $\mho$ *contains only propositional variables common to* $L\widetilde{\mathcal{G}}$ *and* $R\widetilde{\mathcal{G}}$*.*

If $\mho$ *is a componentwise interpolant of* $\widetilde{\mathcal{G}}$*, we write* $\widetilde{\mathcal{G}} \longleftarrow \mho$*.*

Lemma 4.13 (Translation from componentwise to formula interpolants). *If a g-hyperformula* $C^1 \owedge \cdots \owedge C^m$ *of length 1 is a componentwise interpolant of a split hypersequent* $A; \Rightarrow ; B$*, then the formula* $C^1 \wedge \cdots \wedge C^m$ *is an interpolant of the formulas* A *and* B*.*

Proof: Indeed, $L(A; \Rightarrow ; B)$ is $A \Rightarrow$ and $R(A; \Rightarrow ; B)$ is $\Rightarrow B$. Thus, the formulas C^j contain only propositional variables common to A and B.

Let $\mathcal{M} = (W, \sim, V)$ be a Kripke model and $w \in W$ be a world in it. We have

$$\mathcal{M}, w \not\Vdash \bigwedge_{i=1}^{m} C^i \implies \mathcal{M}, w \not\models \bigotimes_{i=1}^{m} C^i \implies \mathcal{M}, w \models A \Rightarrow \implies \mathcal{M}, w \not\Vdash A,$$

$$\mathcal{M}, w \Vdash \bigwedge_{i=1}^{m} C^i \implies \mathcal{M}, w \models \bigotimes_{i=1}^{m} C^i \implies \mathcal{M}, w \models \Rightarrow B \implies \mathcal{M}, w \Vdash B.$$

Hence, by the completeness of S5, we have

$$\mathsf{S5} \vdash A \to C^1 \wedge \dots \wedge C^m \qquad \text{and} \qquad \mathsf{S5} \vdash C^1 \wedge \dots \wedge C^m \to B. \qquad \square$$

Lemma 4.14 (External disjunction elimination).

1. *For arbitrary modal formulas $A_1, \dots, A_n, B_1, \dots, B_n$,*

$$(A_1 \mid \dots \mid A_n) \ovee (B_1 \mid \dots \mid B_n) \dashv\models A_1 \vee B_1 \mid \dots \mid A_n \vee B_n.$$

2. *Any g-hyperformula can be transformed to a componentwise equivalent external conjunction of hyperformulas without changing the set of propositional variables occurring in it.*

In principle, any g-hyperformula of length 1 can be translated into a formula in this way, but translating $\owedge$ is sufficient because with respect to the componentwise semantics the external conjunction and disjunction $\owedge$ and $\ovee$ behave like $\wedge$ and $\vee$ with respect to Boolean semantics. In particular, any g-hyperformula can be transformed to a componentwise equivalent DNF or CNF using the standard algorithm. The simple proof of the following lemma is left to the reader.

Lemma 4.15 (External disjunction elimination).

1. *For arbitrary modal formulas $A_1, \dots, A_n, B_1, \dots, B_n$,*

$$(A_1 \mid \dots \mid A_n) \ovee (B_1 \mid \dots \mid B_n) \dashv\models A_1 \vee B_1 \mid \dots \mid A_n \vee B_n.$$

2. *Any g-hyperformula can be transformed to a componentwise equivalent external conjunction of hyperformulas without changing the set of propositional variables occurring in it.*

5 The Interpolation Algorithm

We divide our description of the algorithm for constructing interpolants into three parts. First, we present the propositional and structural rules in Figures 4 and 5 respectively and prove correctness for the most interesting cases. The modal rules (presented later in Figure 6) require additional auxiliary lemmas.

$$\text{id}^{ll}\ \frac{}{A;\Rightarrow A;\ \longleftarrow\ \bot} \qquad \text{id}^{rl}\ \frac{}{;A \Rightarrow A;\ \longleftarrow\ \neg A}$$

$$\text{id}^{lr}\ \frac{}{A;\Rightarrow ;A\ \longleftarrow\ A} \qquad \text{id}^{rr}\ \frac{}{;A \Rightarrow ;A\ \longleftarrow\ \top}$$

$$\text{id}^{l}_{\bot}\ \frac{}{\bot;\Rightarrow\ \longleftarrow\ \bot} \qquad \text{id}^{r}_{\bot}\ \frac{}{;\bot\Rightarrow\ \longleftarrow\ \top}$$

$$\rightarrow^{l}\Rightarrow\ \frac{\widetilde{\mathcal{G}} \mid \Gamma;\Pi \Rightarrow \Delta, A;\Sigma\ \longleftarrow\ \mho_1 \quad \widetilde{\mathcal{G}} \mid \Gamma, B;\Pi \Rightarrow \Delta;\Sigma\ \longleftarrow\ \mho_2}{\widetilde{\mathcal{G}} \mid \Gamma, A \rightarrow B;\Pi \Rightarrow \Delta;\Sigma\ \longleftarrow\ \mho_1 \circledvee \mho_2}$$

$$\rightarrow^{r}\Rightarrow\ \frac{\widetilde{\mathcal{G}} \mid \Gamma;\Pi \Rightarrow \Delta;\Sigma, A\ \longleftarrow\ \mho_1 \quad \widetilde{\mathcal{G}} \mid \Gamma;\Pi, B \Rightarrow \Delta;\Sigma\ \longleftarrow\ \mho_2}{\widetilde{\mathcal{G}} \mid \Gamma;\Pi, A \rightarrow B \Rightarrow \Delta;\Sigma\ \longleftarrow\ \mho_1 \circledwedge \mho_2}$$

For rules $\rightarrow^{l}\Rightarrow$ and $\rightarrow^{r}\Rightarrow$, it is required that $\|\mho_1\| = \|\mho_2\| = \|\widetilde{\mathcal{G}}\| + 1$.

Fig. 4. Interpolation algorithm: Propositional rules. For the propositional rules not depicted above, the given interpolant for the premise is to be used as an interpolant for the conclusion.

Remark 5.1. *Note that the empty (split) hypersequent can never occur in a derivation in* HS5 *(or in* SHS5*), thus, we need not care how to properly define transformations to CNF and DNF for g-hyperformulas of length* 0.

Lemma 5.2 (Algorithm correctness I: propositional and structural rules). *All the rules depicted in Figure 4 produce a componentwise interpolant for the conclusion of the rule whenever componentwise interpolants (transformed to a proper form) are given for all the premises. Further, any componentwise interpolant of the premise of a propositional or structural rule not depicted in Figure 4 is also a componentwise interpolant for the conclusion of the same rule.*

Proof: We consider several representative cases, leaving the rest to the reader. Throughout the proof we assume $\mathcal{M} = (W, \sim, V)$ to be an arbitrary Kripke model, $w \in W$ to be an arbitrary world from it, and $\vec{w}$ to be an arbitrary $\mathcal{M}$-connected sequence of worlds of appropriate length. We also omit the model from the $\Vdash$ statements about formulas and from $\vDash$ statements about hyperformulas and hypersequents.

Rule id^{rl} $\dfrac{}{;A \Rightarrow A;\ \longleftarrow\ \neg A}$. It is clear that all propositional variables in $\neg A$ are common between $\ \Rightarrow A$ and $A \Rightarrow\ $. We need to consider arbitrary $\mathcal{M}$-connected sequences of length 1, i.e., arbitrary worlds. We have

$$w \nvDash \neg A \quad\Longrightarrow\quad w \nVdash \neg A \quad\Longrightarrow\quad w \vDash\ \Rightarrow A \quad\Longrightarrow\quad w \vDash L(\ ;A \Rightarrow A;\ ),$$

$$w \vDash \neg A \quad\Longrightarrow\quad w \Vdash \neg A \quad\Longrightarrow\quad w \vDash A \Rightarrow\ \quad\Longrightarrow\quad w \vDash R(\ ;A \Rightarrow A;\ ).$$

Rules not depicted in Figure 4. They all work in the same way. It is sufficient to note that, for all of them, if one side, left or right, of the premise hypersequent holds on $\vec{w}$ then the same side of the conclusion hypersequent also holds on $\vec{w}$ (and, of course, to verify the common-variable condition).

Rule $\rightarrow^{l} \Rightarrow$ $\dfrac{\widetilde{\mathcal{G}} \mid \Gamma; \Pi \Rightarrow \Delta, A; \Sigma \longleftarrow \mho_1 \quad \widetilde{\mathcal{G}} \mid \Gamma, B; \Pi \Rightarrow \Delta; \Sigma \longleftarrow \mho_2}{\widetilde{\mathcal{G}} \mid \Gamma, A \rightarrow B; \Pi \Rightarrow \Delta; \Sigma \longleftarrow \mho_1 \oslash \mho_2}$.

Assume that $\mho_1$ and $\mho_2$ are interpolants of the premises, in particular, $\|\mho_1\| = \|\mho_2\| = \|\widetilde{\mathcal{G}}\| + 1$, making $\mho_1 \oslash \mho_2$ well defined. It is easy to see that the common-variable condition for the conclusion is fulfilled. Let $\vec{w} = \vec{v}, u$. For the left side,

$$\begin{aligned}
&\vec{w} \nvDash \mho_1 \oslash \mho_2 \quad\Longrightarrow\quad \vec{w} \nvDash \mho_1 \ \text{ and }\ \vec{w} \nvDash \mho_2 \quad\Longrightarrow\\
&\quad \vec{w} \vDash L\widetilde{\mathcal{G}} \mid \Gamma \Rightarrow \Delta, A \ \text{ and }\ \vec{w} \vDash L\widetilde{\mathcal{G}} \mid \Gamma, B \Rightarrow \Delta \quad\Longrightarrow\\
&\qquad \big(\vec{v} \vDash L\widetilde{\mathcal{G}} \ \text{ or }\ u \vDash \Gamma \Rightarrow \Delta \ \text{ or }\ u \Vdash A\big) \text{ and}\\
&\qquad \big(\vec{v} \vDash L\widetilde{\mathcal{G}} \ \text{ or }\ u \vDash \Gamma \Rightarrow \Delta \ \text{ or }\ u \nVdash B\big) \quad\Longrightarrow\\
&\quad \big(\vec{v} \vDash L\widetilde{\mathcal{G}} \ \text{ or }\ u \vDash \Gamma \Rightarrow \Delta \ \text{ or }\ u \nVdash A \rightarrow B\big) \quad\Longrightarrow\\
&\qquad\qquad \vec{w} \vDash L\widetilde{\mathcal{G}} \mid \Gamma, A \rightarrow B \Rightarrow \Delta.
\end{aligned}$$

The argument for the right side is even simpler. If $\vec{w} \vDash \mho_1 \oslash \mho_2$, then at least one of $\mho_1$ and $\mho_2$ holds on $\vec{w}$ making the right side of the corresponding premise hypersequent true on $\vec{w}$. But the right side of the conclusion hypersequent is the same as that of both premise hypersequents.

Rule EC $\dfrac{\widetilde{\mathcal{G}} \mid \Gamma; \Pi \Rightarrow \Delta; \Sigma \mid \Gamma; \Pi \Rightarrow \Delta; \Sigma \longleftarrow \bigotimes_{j=1}^{m} (\vec{C}^j \mid A^j \mid B^j)}{\widetilde{\mathcal{G}} \mid \Gamma; \Pi \Rightarrow \Delta; \Sigma \longleftarrow \bigotimes_{j=1}^{m} (\vec{C}^j \mid A^j \vee B^j)}$.

Recall that it was this rule that was identified as the main obstacle to using hypersequents for proving the CIP in [Avron, 1996]. Once again, the common-variable condition presents no difficulties. Let $\vec{w} = \vec{v}, u$ be a sequence of length $\|\widetilde{G}\| + 1$. The crucial observations for this case are that

$$\vec{v}, u \vDash \vec{C}^j \mid A^j \vee B^j \qquad \Longleftrightarrow \qquad \vec{v}, u, u \vDash \vec{C}^j \mid A^j \mid B^j$$

for each $j = 1, \ldots, m$ and that

$$\vec{v}, u, u \vDash S(\widetilde{\mathcal{G}} \mid \Gamma; \Pi \Rightarrow \Delta; \Sigma \mid \Gamma; \Pi \Rightarrow \Delta; \Sigma) \quad \Longrightarrow \quad \vec{v}, u \vDash S(\widetilde{\mathcal{G}} \mid \Gamma; \Pi \Rightarrow \Delta; \Sigma)$$

for S being either L or R. □

$$\text{EW}\ \frac{\widetilde{\mathcal{G}} \longleftarrow \bigotimes_{j=1}^{m} \vec{C}^j}{\widetilde{\mathcal{G}} \mid \Gamma; \Pi \Rightarrow \Delta; \Sigma \longleftarrow \bigotimes_{j=1}^{m} (\vec{C}^j \mid \bot)}$$

$$\text{EC}\ \frac{\widetilde{\mathcal{G}} \mid \Gamma; \Pi \Rightarrow \Delta; \Sigma \mid \Gamma; \Pi \Rightarrow \Delta; \Sigma \longleftarrow \bigotimes_{j=1}^{m} (\vec{C}^j \mid A^j \mid B^j)}{\widetilde{\mathcal{G}} \mid \Gamma; \Pi \Rightarrow \Delta; \Sigma \longleftarrow \bigotimes_{j=1}^{m} (\vec{C}^j \mid A^j \vee B^j)}$$

$$\text{Ex}\ \frac{\widetilde{\mathcal{G}} \mid \Gamma; \Pi \Rightarrow \Delta; \Sigma \mid \Lambda; \Phi \Rightarrow \Theta; \Psi \mid \widetilde{\mathcal{H}} \longleftarrow \bigotimes_{j=1}^{m} (\vec{C}^j \mid A^j \mid B^j \mid \vec{D}^j)}{\widetilde{\mathcal{G}} \mid \Lambda; \Phi \Rightarrow \Theta; \Psi \mid \Gamma; \Pi \Rightarrow \Delta; \Sigma \mid \widetilde{\mathcal{H}} \longleftarrow \bigotimes_{j=1}^{m} (\vec{C}^j \mid B^j \mid A^j \mid \vec{D}^j)}$$

In order to apply the rules EW, EC, and Ex, the interpolant of the premise must be first transformed to an external conjunction of hyperformulas by Lemma 4.15. In addition, it is required that

$$\|\widetilde{G}\| = \|\vec{C}^j\| \text{ and } \|\widetilde{\mathcal{H}}\| = \|\vec{D}^j\| \text{ for each } j = 1, \ldots, m.$$

Fig. 5. Interpolation algorithm: Structural rules. For the structural rules not depicted above, the given interpolant for the only premise is to be used as an interpolant for the conclusion.

Restricting some interpolants to be external conjunctions of hyperformulas in Figure 4 is for convenience more than out of necessity. The same structural transformations applied to arbitrary g-hyperformulas would have worked equally well, but would have been more cumbersome to describe. The modal rules, in contrast, require logical transformations that affect the structure of the interpolant, making it necessary to impose an even more restrictive interpolant format to process the $\Rightarrow \Box^{\mathrm{l}}$ rule. Our immediate goal is to show that this special format can always be achieved. We formulate this as a lemma but leave its simple proof to the reader.

Lemma 5.3 (Separation of a g-hyperformula component). *Let $\bot^n$ be an abbreviation for $\underbrace{\bot \mid \ldots \mid \bot}_{n}$.*

1. *For any hyperformula $\vec{C}$ of length n and any formula A,*
$$\vec{C} \mid A \quad \dashv\vDash \quad (\bot^n \mid A) \otimes (\vec{C} \mid \bot).$$
2. *For arbitrary formulas $A^1, \ldots, A^m$ with $m \geq 2$ and any $n \geq 0$,*
$$\bigotimes_{j=1}^{m}(\bot^n \mid A^j) \quad \dashv\vDash \quad \bot^n \mid \bigwedge_{j=1}^{m} A^j.$$
3. *For any hyperformula $\vec{C}$ of length n,*
$$\vec{C} \mid \bot \quad \dashv\vDash \quad (\bot^n \mid \top) \otimes (\vec{C} \mid \bot).$$
4. *For any g-hyperformula $\mho$ of length $n + 1$, there is a componentwise equivalent g-hyperformula of the form*
$$\bigovee_{j=1}^{m}\left((\bot^n \mid B^j) \otimes \bigotimes_{k_j=1}^{l_j}(\vec{C}^{j,k_j} \mid \bot)\right)$$
with the same set of propositional variables.

Finally, before getting our hands dirty with the modal rules, we formulate auxiliary statements that will make the main arguments more transparent by separating tedious technical details to stand-alone lemmas.

Lemma 5.4 (Boxed formulas invariant within a connected component). *Let $\mathcal{M} = (W, \sim, V)$ be an arbitrary Kripke model and u and z be worlds from W.*
$$u \sim z \quad \Longrightarrow \quad (u \Vdash \Box G \quad \Longleftrightarrow \quad z \Vdash \Box G). \tag{2}$$

Proof: By a standard semantic argument. □

Lemma 5.5 (Hypersequent necessitation in one component). *Let $\mathcal{M} = (W, \sim, V)$ be an arbitrary Kripke model. For any hypersequent $\mathcal{H}$, any finite multiset Ξ of modal formulas, any modal formula F, any sequence $\vec{v}$ of worlds from W, and any world $u \in W$ such that $\|\mathcal{H}\| = \|\vec{v}\|$, we have*

1. $\mathcal{M}, \vec{v}, u' \vDash \mathcal{H} \mid \Box\Xi \Rightarrow$ *for some* $u' \sim u$ $\Longrightarrow$ $\mathcal{M}, \vec{v}, u \vDash \mathcal{H} \mid \Box\Xi \Rightarrow$;
2. $\mathcal{M}, \vec{v}, u' \vDash \mathcal{H} \mid \Box\Xi \Rightarrow F$ *for all* $u' \sim u$ $\Longrightarrow$ $\mathcal{M}, \vec{v}, u \vDash \mathcal{H} \mid \Box\Xi \Rightarrow \Box F$.

Proof: We distinguish three possibilities:

1. $\mathcal{M}, \vec{v} \vDash \mathcal{H}$. It follows that $\mathcal{M}, \vec{v}, u \vDash \mathcal{H} \mid \Box\Xi \Rightarrow$ and $\mathcal{M}, \vec{v}, u \vDash \mathcal{H} \mid \Box\Xi \Rightarrow \Box F$.
2. $\mathcal{M}, u' \nVdash \Box G$ for some $G \in \Xi$ and some $u' \sim u$. By Lemma 5.4, $\mathcal{M}, u \nVdash \Box G$. It follows immediately that $\mathcal{M}, \vec{v}, u \vDash \mathcal{H} \mid \Box\Xi \Rightarrow$ and $\mathcal{M}, \vec{v}, u \vDash \mathcal{H} \mid \Box\Xi \Rightarrow \Box F$.
3. If $\mathcal{M}, \vec{v} \nvDash \mathcal{H}$ and $\mathcal{M}, u' \Vdash \Box G$ for all $G \in \Xi$ and all $u' \sim u$, then assumption of part 1 of the lemma is not satisfied. However, whenever the assumption of part 2 is satisfied, we have $\mathcal{M}, u' \Vdash F$ for all $u' \sim u$. Thus, $\mathcal{M}, u \Vdash \Box F$ and $\mathcal{M}, \vec{v}, u \vDash \mathcal{H} \mid \Box\Xi \Rightarrow \Box F$. □

Lemma 5.6 (Algorithm correctness II: modal rules). *All the rules depicted in Figure 6 produce a componentwise interpolant for the conclusion of the rule whenever componentwise interpolants (in a proper form) are given for the premise. Further, any componentwise interpolant of the premise of a modal rule not depicted in Figure 6 is also a componentwise interpolant for the conclusion of the same rule.*

Proof: We omit the rule $\Box^r \Rightarrow$ because it is very similar to the case of $\Box^l \Rightarrow$ we demonstrate. Throughout the proof we assume $\mathcal{M} = (W, \sim, V)$ to be an arbitrary Kripke model, w to be an arbitrary world from W, and $\vec{w}$ to be an arbitrary connected sequence of worlds from W of appropriate length. We also omit the model from the $\Vdash$ statements about formulas and from $\vDash$ statements about hyperformulas and hypersequents. Finally, we omit the trivial proofs that the common-variables condition is satisfied for all the rules.

Rule $\Box^l \Rightarrow$.

We need to show that any interpolant of the premise hypersequent is also an interpolant of the conclusion hypersequent. The only change from the premise to the conclusion is that A is replaced with $\Box A$ in the antecedent. Thus, it is sufficient to observe that, due to the reflexivity of $\sim$,

$$w \nVdash A \quad \Longrightarrow \quad w \nVdash \Box A.$$

Rule MS

$$\frac{\widetilde{\mathcal{G}} \mid \Box\Lambda, \Gamma; \Box\Theta, \Pi \Rightarrow \Box\Phi, \Delta; \Box\Psi, \Sigma \longleftarrow \bigotimes_{j=1}^{m}(\vec{C}^j \mid A^j)}{\widetilde{\mathcal{G}} \mid \Box\Lambda; \Box\Theta \Rightarrow \Box\Phi; \Box\Psi \mid \Gamma; \Pi \Rightarrow \Delta; \Sigma \longleftarrow \bigotimes_{j=1}^{m}(\vec{C}^j \mid \bot \mid A^j)}.$$

The key observations are (2) and that, for any sequence $\vec{v}$ of worlds of length $\|\widetilde{\mathcal{G}}\|$, for arbitrary worlds u and z, and for each $j = 1, \ldots, m$,

$$\vec{v}, u \vDash \vec{C}^j \mid A^j \quad \Longleftrightarrow \quad \vec{v}, z, u \vDash \vec{C}^j \mid \bot \mid A^j$$

The requirement for the sequence of worlds to be connected is used here for the first time to allow for the application of (2).

$$\textbf{Rule} \Rightarrow \Box^{\mathsf{r}} \quad \frac{\widetilde{\mathcal{G}} \mid \Box\Gamma; \Box\Pi \Rightarrow ; A \longleftarrow \bigotimes_{j=1}^{m}(\vec{C}^j \mid B^j)}{\widetilde{\mathcal{G}} \mid \Box\Gamma; \Box\Pi \Rightarrow ; \Box A \longleftarrow \bigotimes_{j=1}^{m}(\vec{C}^j \mid \Box B^j)}.$$

Let $\vec{w} = \vec{v}, u$. For the left side, let us assume that $\vec{w} \nvDash \bigotimes_{j=1}^{m}(\vec{C}^j \mid \Box B^j)$. Then there is $1 \le J \le m$ such that $\vec{v} \nvDash \vec{C}^J$ and $u \nVdash \Box B^J$. It follows that there exists a world $u' \in W$ such that $u \sim u'$ and $u' \nVdash B^J$. Thus, on the $\mathcal{M}$-connected sequence $\vec{v}, u'$, the given interpolant $\bigotimes_{j=1}^{m}(\vec{C}^j \mid B^j)$ of the premise hypersequent does not hold, making the left side of the premise hypersequent true on the same sequence: $\vec{v}, u' \vDash L\widetilde{\mathcal{G}} \mid \Box\Gamma \Rightarrow$. By Lemma 5.5.1, we have $\vec{w} \vDash L\widetilde{\mathcal{G}} \mid \Box\Gamma \Rightarrow$ for the left side of the conclusion hypersequent.

For the right side, assume that $\vec{w} \vDash \bigotimes_{j=1}^{m}(\vec{C}^j \mid \Box B^j)$. Then for each $1 \le j \le m$ either $\vec{v} \vDash \vec{C}^j$ or $u \Vdash \Box B^j$. Since $u \Vdash \Box B^j$ implies $u' \Vdash B^j$ for all $u' \sim u$, it follows that $\vec{v}, u' \vDash \bigotimes_{j=1}^{m}(\vec{C}^j \mid B^j)$ for all $u' \sim u$, making the right side of the premise hypersequent true on all these sequences, which are $\mathcal{M}$-connected: $\vec{v}, u' \vDash R\widetilde{\mathcal{G}} \mid \Box\Pi \Rightarrow A$ for all $u' \sim u$. By Lemma 5.5.2, $\vec{w} \vDash R\widetilde{\mathcal{G}} \mid \Box\Pi \Rightarrow \Box A$ for the right side of the conclusion hypersequent.

$$\textbf{Rule} \Rightarrow \Box^{\mathsf{l}} \quad \frac{\widetilde{\mathcal{G}} \mid \Box\Gamma; \Box\Pi \Rightarrow A; \longleftarrow \bigoplus_{j=1}^{m}\left((\bot^n \mid B^j) \oslash \bigotimes_{k_j=1}^{l_j}(\vec{C}^{j,k_j} \mid \bot)\right)}{\widetilde{\mathcal{G}} \mid \Box\Gamma; \Box\Pi \Rightarrow \Box A; \longleftarrow \bigoplus_{j=1}^{m}\left((\bot^n \mid \Diamond B^j) \oslash \bigotimes_{k_j=1}^{l_j}(\vec{C}^{j,k_j} \mid \bot)\right)}.$$

Let us consider $\vec{w} = \vec{v}, u$. For the left side, assume that

$$\vec{w} \nvDash \bigoplus_{j=1}^{m}\left((\bot^n \mid \Diamond B^j) \oslash \bigotimes_{k_j=1}^{l_j}(\vec{C}^{j,k_j} \mid \bot)\right).$$

In other words, for each $j = 1, \ldots, m$ there exists $1 \le K_j \le l_j$ such that

$$u \nVdash \Diamond B^j \qquad \text{or} \qquad \vec{v} \nVdash \vec{C}^{j,K_j}.$$

It follows that for each $u' \sim u$ and for each $j = 1, \ldots, m$,

$$u' \not\Vdash B^j \qquad \text{or} \qquad \vec{v} \not\Vdash \vec{C}^{j,K_j}.$$

In other words, for each $u' \sim u$,

$$\vec{v}, u' \not\models \bigovee_{j=1}^{m} \left((\bot^n \mid B^j) \oslash \bigowedge_{k_j=1}^{l_j} (\vec{C}^{j,k_j} \mid \bot) \right),$$

making the left side of the premise hypersequent true on all such sequences, which are $\mathcal{M}$-connected: for all $u' \sim u$, we have $\vec{v}, u' \models L\widetilde{\mathcal{G}} \mid \Box\Gamma \Rightarrow A$. By Lemma 5.5.2, we have $\vec{w} \models L\widetilde{\mathcal{G}} \mid \Box\Gamma \Rightarrow \Box A$, i.e., the left side of the conclusion hypersequent holds on $\vec{w}$.

For the right side, assume that

$$\vec{w} \models \bigovee_{j=1}^{m} \left((\bot^n \mid \diamond B^j) \oslash \bigowedge_{k_j=1}^{l_j} (\vec{C}^{j,k_j} \mid \bot) \right).$$

In other words, for some $1 \le J \le m$,

$$u \Vdash \diamond B^J \qquad \text{and} \qquad \text{for all } 1 \le k_J \le l_J \quad \vec{v} \Vdash \vec{C}^{J,k_J}.$$

It follows that, for some $u' \sim u$,

$$u' \Vdash B^J \qquad \text{and} \qquad \text{for all } 1 \le k_J \le l_J \quad \vec{v} \Vdash \vec{C}^{J,k_J}.$$

In other words, for some $u' \sim u$,

$$\vec{v}, u' \models \bigovee_{j=1}^{m} \left((\bot^n \mid B^j) \oslash \bigowedge_{k_j=1}^{l_j} (\vec{C}^{j,k_j} \mid \bot) \right),$$

making the right side of the premise hypersequent true on the sequence $\vec{v}, u'$, which is $\mathcal{M}$-connected: we have

$$\vec{v}, u' \models R\widetilde{\mathcal{G}} \mid \Box\Pi \Rightarrow$$

for this u'. By Lemma 5.5.1, we have

$$\vec{w} \models L\widetilde{\mathcal{G}} \mid \Box\Pi \Rightarrow$$

i.e., the right side of the conclusion hypersequent holds on $\vec{w}$. □

Putting all together, we conclude that

Theorem 5.7. S5 *enjoys the CIP.*

Proof: Let S5 $\vdash A \to B$. By Corollary 4.4, we have SHS5 $\vdash A; \Rightarrow ; B$. By Lemmas 5.2 and 5.6, we can construct a componentwise interpolant $\mho$ of $A; \Rightarrow ; B$. By Lemma 4.15, this $\mho$ can be efficiently transformed to another componentwise interpolant $\mho'$, which is an external conjunction of hyperformulas of length 1. By Lemma 4.13, this $\mho'$ can be efficiently transformed to a formula interpolant C of A and B. □

$$\Rightarrow \Box^{\mathsf{l}} \frac{\mathcal{G} \mid \Box\Gamma; \Box\Pi \Rightarrow A; \quad \longleftarrow \quad \bigovee_{j=1}^{m} \left((\bot^n \mid B^j) \oslash \bigowedge_{k_j=1}^{l_j} (\vec{C}^{j,k_j} \mid \bot) \right)}{\mathcal{G} \mid \Box\Gamma; \Box\Pi \Rightarrow \Box A; \quad \longleftarrow \quad \bigovee_{j=1}^{m} \left((\bot^n \mid \Diamond B^j) \oslash \bigowedge_{k_j=1}^{l_j} (\vec{C}^{j,k_j} \mid \bot) \right)}$$

$$\Rightarrow \Box^{\mathsf{r}} \frac{\mathcal{G} \mid \Box\Gamma; \Box\Pi \Rightarrow ; A \quad \longleftarrow \quad \bigowedge_{j=1}^{m} (\vec{C}^j \mid B^j)}{\mathcal{G} \mid \Box\Gamma; \Box\Pi \Rightarrow ; \Box A \quad \longleftarrow \quad \bigowedge_{j=1}^{m} (\vec{C}^j \mid \Box B^j)}$$

$$\mathsf{MS} \frac{\mathcal{G} \mid \Box\Lambda, \Gamma; \Box\Theta, \Pi \Rightarrow \Box\Phi, \Delta; \Box\Psi, \Sigma \quad \longleftarrow \quad \bigowedge_{j=1}^{m} (\vec{C}^j \mid A^j)}{\mathcal{G} \mid \Box\Lambda; \Box\Theta \Rightarrow \Box\Phi; \Box\Psi \mid \Gamma; \Pi \Rightarrow \Delta; \Sigma \quad \longleftarrow \quad \bigowedge_{j=1}^{m} (\vec{C}^j \mid \bot \mid A^j)}$$

For $\Rightarrow \Box^{\mathsf{r}}$ and MS, the interpolant of the premise must be transformed to an external conjunction of hyperformulas. The rule $\Rightarrow \Box^{\mathsf{l}}$ requires a more elaborate form guaranteed by Lemma 5.3. It is required that

$$\|\mathcal{G}\| = \|\vec{C}^j\| \text{ for each } j = 1, \ldots, m, \tag{MS}$$

and that

$$\|\mathcal{G}\| = \|\vec{C}^{j,k_j}\| = n \text{ for each } j = 1, \ldots, m \text{ and } k_j = 1, \ldots, l_j, \tag{$\Rightarrow \Box^{\mathsf{l}}$}$$

where $\bot^n = \underbrace{\bot \mid \ldots \mid \bot}_{n}$.

Fig. 6. Interpolation algorithm: Modal rules. For the modal rules $\Box^{\mathsf{l}} \Rightarrow$ and $\Box^{\mathsf{r}} \Rightarrow$ not depicted above, the given interpolant for the premise is to be used as an interpolant for the conclusion.

Example 5.8. *To illustrate our method, we apply it to a properly adapted sample derivation from [Avron, 1996] in Figure 7. The rules for handling* $\neg$ *are easily derivable from those we presented for* $\rightarrow$*. Note that the interpolant* $P \vee \Box\bot$ *of* P *and* $\Box\neg\Box\neg P$ *constructed by the algorithm contains redundancies: the formula* P *alone could serve as an interpolant.*

$$
\begin{array}{ll}
\mathsf{id}^{lr} & \overline{P;\ \Rightarrow\ ;P \longleftarrow P} \\
\neg^{r} \Rightarrow & \overline{P;\neg P \Rightarrow\ \longleftarrow P} \\
\Box^{r} \Rightarrow & \overline{P;\Box\neg P \Rightarrow\ \longleftarrow P} \\
\mathsf{MS} & \overline{;\Box\neg P \Rightarrow\ \mid P;\ \Rightarrow\ \longleftarrow \bot \mid P} \\
\mathsf{Ex} & \overline{P;\ \Rightarrow\ \mid ;\Box\neg P \Rightarrow\ \longleftarrow P \mid \bot} \\
\Rightarrow \neg^{r} & \overline{P;\ \Rightarrow\ \mid\ \Rightarrow ;\neg\Box\neg P \longleftarrow P \mid \bot} \\
\Rightarrow \Box^{r} & \overline{P;\ \Rightarrow\ \mid\ \Rightarrow ;\Box\neg\Box\neg P \longleftarrow P \mid \Box\bot} \\
\mathsf{Ex} & \overline{\Rightarrow ;\Box\neg\Box\neg P \mid P;\ \Rightarrow\ \longleftarrow \Box\bot \mid P} \\
\Rightarrow \mathsf{W}^{r} & \overline{\Rightarrow ;\Box\neg\Box\neg P \mid P;\ \Rightarrow ;\Box\neg\Box\neg P \longleftarrow \Box\bot \mid P} \\
\mathsf{Ex} & \overline{P;\ \Rightarrow ;\Box\neg\Box\neg P \mid\ \Rightarrow ;\Box\neg\Box\neg P \longleftarrow P \mid \Box\bot} \\
\mathsf{W}^{l} \Rightarrow & \overline{P;\ \Rightarrow ;\Box\neg\Box\neg P \mid P;\ \Rightarrow ;\Box\neg\Box\neg P \longleftarrow P \mid \Box\bot} \\
\mathsf{EC} & \overline{P;\ \Rightarrow ;\Box\neg\Box\neg P \longleftarrow P \vee \Box\bot}
\end{array}
$$

Fig. 7. Application of our algorithm to the derivation of $P \Rightarrow \Box\neg\Box\neg P$ from [Avron, 1996].

6 Conclusion and Future Work

To the best of our knowledge, we have presented the first method of proving interpolation constructively by induction on a hypersequent derivation. The method was developed for the classical modal logic S5. We plan to extend this method to

- various other classical hypersequent systems,
- grafted sequent calculi, recently developed in [Kuznets and Lellmann, 2016], which combine hypersequent and nested sequent calculi,
- non-classical hypersequent systems, especially ones for intermediate logics.

Acknowledgment: This work is funded by the Austrian Science Fund (FWF) grant M 1770-N25.

References

Arnon Avron. A constructive analysis of **RM**. *Journal of Symbolic Logic*, 52(4):939–951, December 1987. 10.2307/2273828.

Arnon Avron. The method of hypersequents in the proof theory of propositional non-classical logics. In Wilfrid Hodges, Martin Hyland, Charles Steinhorn, and John Truss, editors, *Logic: From Foundations to Applications: European Logic Colloquium*, pages 1–32. Clarendon Press, 1996.

Ronald Fagin, Joseph Y. Halpern, Yoram Moses, and Moshe Y. Vardi. *Reasoning about Knowledge*. MIT Press, 1995.

Melvin Fitting and Roman Kuznets. Modal interpolation via nested sequents. *Annals of Pure and Applied Logic*, 166(3):274–305, March 2015. 10.1016/j.apal.2014.11.002.

Dov M. Gabbay and Larisa Maksimova. *Interpolation and Definability: Modal and Intuitionistic Logic*, volume 46 of *Oxford Logic Guides*. Clarendon Press, 2005.

Roman Kuznets and Björn Lellmann. Grafting hypersequents onto nested sequents. *Logic Journal of the IGPL*, In Press, 2016. 10.1093/jigpal/jzw005.

Björn Lellmann and Dirk Pattinson. Correspondence between modal Hilbert axioms and sequent rules with an application to S5. In Didier Galmiche and Dominique Larchey-Wendling, editors, *Automated Reasoning with Analytic Tableaux and Related Methods, 22nd International Conference, TABLEAUX 2013, Nancy, France, September 16–19, 2013, Proceedings*, volume 8123 of *Lecture Notes in Computer Science*, pages 219–233. Springer, 2013. 10.1007/978-3-642-40537-2_19.

G. E. Minc. On some calculi of modal logic. In V. P. Orevkov, editor, *The Calculi of Symbolic Logic. I*, volume 98 of *Proceedings of the Steklov Institute of Mathematics*, pages 97–124. AMS, 1971. Originally published in Russian in 1968.

Garrel Pottinger. Uniform, cut-free formulations of T, S_4, and S_5. *Journal of Symbolic Logic*, 48 (3):900, September 1983. 10.2307/2273495. Abstract.

Serafina Lapenta and Ioana Leuştean

A General View on Normal Form Theorems for Łukasiewicz Logic with Product

Abstract: In this survey paper we explore the connection between the Pierce-Birkhoff conjecture and Łukasiewicz logic with product. Conservative extensions of Łukasiewicz logic can be defined by adding an internal product or a multiplication with scalars from $[0, 1]$. The corresponding models reflect an algebraic hierarchy of lattice-ordered structures, from groups to algebras. We prove a general version of the normal form theorem and we state a local version of the Pierce-Birkhoff conjecture.

Keywords: Łukasiewicz logic, MV-algebras, Normal form theorem, Pierce-Birkhoff conjecture.

Mathematics Subject Classification 2010: 06D35, 06F25, 03B50

1 Introduction

The modern evolution of Łukasiewicz logic, defined in [Łukasiewicz and Tarski, 1930], is strongly connected with its algebraic counterpart: the theory of MV-algebras. Introduced in [Chang, 1958], MV-algebras stand to Łukasiewicz propositional logic as boolean algebras stand to classical logic. We refer to [Cignoli et al., 2000] for a comprehensive study of their general theory and to [Mundici, 2011] for advanced topics. Łukasiewicz logic and MV-algebras are also studied in the general context of t-norm based logics [Hájek, 1998].

MV-algebras are structures $(A, \oplus, ^*, 0)$ of type $(2, 1, 0)$, satisfying some specific identities. The theory of MV-algebras was highlighted by Mundici's categorical equivalence between MV-algebras and Abelian lattice-ordered groups with strong unit (ℓu-groups) [Mundici, 1986]. As a consequence, for any MV-algebra A there exists a unique, up to isomorphism, ℓu-group (G, u) such that A is isomorphic with the unit interval $[0, u]$ of G, endowed with an MV-algebra structure by $x^* = u - x$ and $x \oplus y = (x + y) \wedge u$ for any $x, y \in [0, u]$. Note that the MV-algebraic sum $\oplus$, which can be seen as a non-idempotent disjunction, is the group addition

Serafina Lapenta: Department of Mathematics, University of Salerno, Via Giovanni Paolo II, 132 Fisciano (SA), Italy, e-mail: slapenta@unisa.it

Ioana Leuştean: Department of Computer Science, Faculty of Mathematics and Computer Science, University of Bucharest, Academiei nr.14, sector 1, C.P. 010014, Bucharest, Romania, e-mail: ioana@fmi.unibuc.ro

$+$ truncated to the unit interval. Further operations are defined as follows: 1 is 0^*, the Łukasiewicz implication is $x \to y = x^* \oplus y = (u - x + y) \wedge u$ and the Łukasiewicz conjunction is $x \odot y = (x^* \oplus y^*)^* = (x + y - u) \vee 0$ for any $x, y \in [0, u]$. The standard MV-algebra is the real unit interval $[0, 1]$ equipped with the above defined operations. Chang's completeness theorem states that an equation is satisfied in all MV-algebras if and only if it is satisfied in the MV-algebra $[0, 1]$.

Since the real interval $[0, 1]$ is closed to the product operation, a natural problem was to find a complete axiomatization for the variety generated by $([0, 1], \oplus, \cdot, ^*, 0)$, where $([0, 1], \oplus, ^*, 0)$ is the standard MV-algebra and $\cdot : [0, 1] \times [0, 1] \to [0, 1]$ is the product of real numbers. Enriching the structure of MV-algebras, the product was defined as an internal operation for PMV-algebras [Di Nola and Dvurečenskij, 2001] and as a scalar multiplication with scalars from $[0, 1]$ for Riesz MV-algebras [Di Nola and Leuştean, 2014]. In [Lapenta and Leuştean, 2015] the authors defined and studied MV-algebras with both internal product and scalar multiplication under the name of fMV-algebras. The equivalence between MV-algebras and ℓu-groups is generalized for each case, leading to equivalences with particular classes of f-rings, Riesz spaces and f-algebras. Moreover, connections between these structures are proved in [Lapenta and Leuştean, 2016] using the MV-algebraic tensor product defined in [Mundici, 1999].

The logical systems developed for PMV-algebras [Horčík and Cintula, 2004] and Riesz MV-algebras [Di Nola and Leuştean, 2014] are conservative extensions of Łukasiewicz logic. One of the main theorems of Łukasiewicz logic states that, for $n \geq 1$, the term functions with n variables are exactly the continuous $[0, 1]$-valued piecewise linear functions with integer coefficients defined on $[0, 1]^n$ [McNaughton, 1951]. This can be seen as a normal form theorem for Łukasiewicz logic. A similar result was proved in [Di Nola and Leuştean, 2014] for the logical system that has Riesz MV-algebras as models; in this case the piecewise linear functions have real coefficients. In [Montagna and Panti, 2001, Introduction] it is stated that a similar result for PMV-algebras is related to the Pierce-Birkhoff conjecture [Birkhoff and Pierce, 1956] and our aim was to make a deeper investigation of this connection. Consequently, we characterized a subclass of fMV-algebras such that the normal form theorem of the corresponding logical system is a local version of the Pierce-Birkhoff conjecture [Lapenta and Leuştean, 2015].

We overview four equational theories, whose underlying models are MV-algebras, Riesz MV-algebras, PMV-algebras and fMV-algebras. At the core of our presentation lies Theorem 4.9, a general normal form result. The already known normal form theorems for MV-algebras, Riesz MV-algebras and fMV-algebras are straightforward consequences, as well as the normal form theorem for PMV-algebras, which is a new result. In the last section of our paper we emphasize the relation with the Pierce-Birkhoff conjecture.

2 The algebras of Łukasiewicz logic with product

In this section we present the structures of Łukasiewicz logic extended with a product operation that can be either an internal binary one, or a scalar multiplication with scalars from $[0,1]$.

An *MV-algebra* is a structure $(A, \oplus, ^*, 0)$ of type (2,1,0) which satisfies the following properties for any $x, y \in A$:

$$\begin{array}{ll} (A, \oplus, 0) \text{ is an Abelian monoid}, & (x^*)^* = x \\ (x^* \oplus y)^* \oplus y = (y^* \oplus x)^* \oplus x, & 0^* \oplus x = 0^* \end{array}$$

We refer to [Cignoli et al., 2000] for all the unexplained notions related to MV-algebras. In any MV-algebra A we can define the following: $1 \stackrel{def}{=} 0^*$, $x \odot y \stackrel{def}{=} (x^* \oplus y^*)^*$, $x \vee y \stackrel{def}{=} (x \odot y^*) \oplus y$ and $x \wedge y \stackrel{def}{=} (x \oplus y^*) \odot y$ for any $x, y \in A$. Hence $(A, \vee, \wedge, 0, 1)$ is a bounded distributive lattice such that $x \leq y$ if and only if $x \odot y^* = 0$.

A *Riesz MV-algebra* [Di Nola and Leuştean, 2014] is a structure

$$(R, \oplus, ^*, 0, \{r \mid r \in [0,1]\})$$

such that $(R, \oplus, ^*, 0)$ is an MV-algebra and $\{r \mid r \in [0,1]\}$ is a family of unary operation such that the following properties hold for any $x, y \in A$ and $r, q \in [0,1]$:

$$\begin{array}{ll} r(x \odot y^*) = (rx) \odot (ry)^*, & (r \odot q^*) \cdot x = (rx) \odot (qx)^*, \\ r(qx) = (rq)x, & 1x = x. \end{array}$$

A *PMV-algebra*[1] [Di Nola and Dvurečenskij, 2001, Montagna, 2000] is a structure $(P, \oplus, ^*, \cdot, 0)$ such that $(P, \oplus, ^*, 0)$ is an MV-algebra, the operation $\cdot : P \times P \to P$ is associative and commutative, and the following identities hold for any $x, y, z \in P$:

$$\begin{array}{l} z \cdot (x \odot y^*) = (z \cdot x) \odot (z \cdot y)^*, \\ x \cdot 1 = x. \end{array}$$

An *fMV-algebra*[2] [Lapenta and Leuştean, 2015] is a structure

$$(A, \oplus, \cdot, ^*, \{r \mid r \in [0,1]\}, 0)$$

1 We assume that PMV-algebras are commutative and unital, while the definition from [Di Nola and Dvurečenskij, 2001] is more general.

2 As for PMV-algebras, we assume that *f*MV-algebras are commutative and unital, while the definition from [Lapenta and Leuştean, 2015] is more general.

which satisfies the following properties:

$$(A, \oplus, \cdot, ^*, 0) \text{ is a PMV-algebra,}$$
$$(A, \oplus, ^*, \{r \mid r \in [0,1]\}, 0) \text{ is a Riesz MV-algebra,}$$
$$r(x \cdot y) = (rx) \cdot y = x \cdot (ry) \text{ for any } x, y, z \in A \text{ and } \alpha \in [0,1].$$

For $x, y, r \in [0,1]$ we define $x \oplus y = \min\{x+y, 1\}$, $x^* = 1 - x$, while $x \cdot y = xy$ and rx coincide with the product of real numbers. In consequence the real interval $[0,1]$ naturally becomes an MV-algebra, a PMV-algebra, a Riesz MV-algebra and an fMV-algebra. While $[0,1]$, endowed with the appropriate structure, generates the variety of MV-algebras and Riesz MV-algebras, this is no longer true for PMV-algebras and fMV-algebras. This issue will be further discussed in Section 5.

We mention in the sequel one of the most relevant results in the theory of MV-algebras: its connection with the theory of Abelian lattice-ordered groups.

An *ℓu-group* is a pair (G, u), where G is an Abelian lattice-ordered group [Bigard et al., 1977] and u is a strong unit. If (G, u) is an ℓu-group, then $[0,u]_G = ([0,u], \oplus, ^*, 0)$ is an MV-algebra, where $[0,u] = \{x \in G \mid 0 \leq x \leq u\}$ and $x \oplus y = u \wedge (x+y)$, $x^* = u - x$ for any $x \in [0,u]$.

If **MV** is the category of MV-algebras with MV-algebra homomorphisms and **uAG** is the category of ℓu-groups equipped with lattice-ordered group homomorphisms that preserve the strong unit, then one defines a functor $\Gamma : \mathbf{uAG} \to \mathbf{MV}$ by $\Gamma(G,u) = [0,u]_G$ and $\Gamma(h) = h|_{[0,u_1]_{G_1}}$, where (G,u) is an ℓu-group and $h : G_1 \to G_2$ is a morphism in **uAG** between (G_1, u_1) and (G_2, u_2). In [Mundici, 1986], Mundici proved that the functor Γ establishes a categorical equivalence between **uAG** and **MV**.

It is clear that all mentioned algebraic structures are deeply related to each other. In particular, all of them have an MV-algebra reduct. Therefore we can define forgetful functors from the categories **PMV** of PMV-algebras, **RMV** of Riesz MV-algebras and **fMV** of fMV-algebras to **MV**. The categorical equivalence between MV-algebras and ℓu-groups can be generalized for each of this structures to an equivalence with an appropriate class of unital lattice-ordered structures having a lattice-ordered group reduct with a strong unit [Di Nola and Dvurečenskij, 2001, Di Nola and Leuştean, 2014, Lapenta and Leuştean, 2015]. Denoted by **uR** the category of unital f-rings with strong unit, by **uRS** the category of Riesz spaces with strong unit and by **ufAlg** the category of unital f-algebras with strong unit (*fu*-algebras), we have the following commutative diagram,

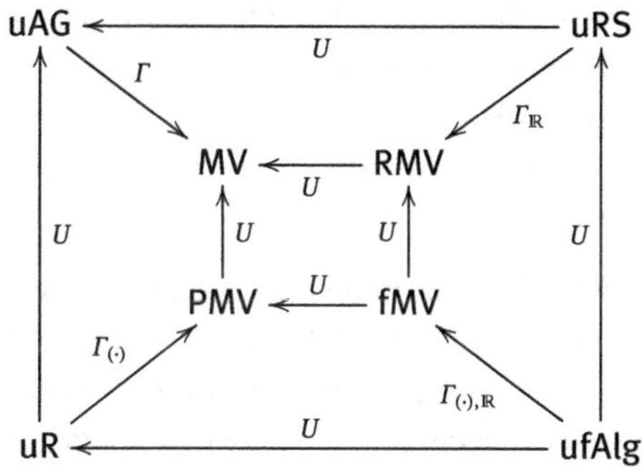

3 On the Pierce-Birkhoff conjecture

At the end of the paper [Birkhoff and Pierce, 1956], the authors asked for a characterization of the "*free, commutative, real ℓ-algebra (ℓ-group) with n generators*" and they conjectured that "*it is isomorphic with the l-group of real functions which are continuous and piecewise polynomial of degree at most n over a finite number of pieces*". They asked "*the same problem for the free (commutative) ℓ-rings, for free f-rings*", saying that: "*The former is probably very difficult*".

Definition 3.1. *Let $n \geq 1$ be a natural number.*

- *A function $f : \mathbb{R}^n \to \mathbb{R}$ is a* piecewise polynomial *(PWP) function if it is continuous and there is a finite set of polynomials $\{p_1, \dots, p_k\} \subseteq \mathbb{R}[x_1, \cdots, x_n]$ such that for any $(a_1, \dots, a_n) \in \mathbb{R}^n$ there exists $i \in \{1, \dots, k\}$ with $f(a_1, \dots, a_n) = p_i(a_1, \dots, a_n)$.*
- *A continuous function $f : \mathbb{R}^n \to \mathbb{R}$ is a* inf-sup-polynomial-definable *(ISD) function if there is a finite set of polynomials $\{q_{ij} | 1 \leq i \leq m, 1 \leq j \leq k\} \subseteq \mathbb{R}[x_1, \cdots, x_n]$ such that $f = \bigvee_{i=1}^m \bigwedge_{j=1}^k q_{ij}$.*

We denote by PWP(n) the set of all PWP-functions and by ISD(n) the set of all ISD-functions defined as above. Our notations are inspired by [Delzell, 1989].

The Pierce-Birkhoff conjecture states that

$$PWP(n) = ISD(n) \text{ for any } n \geq 2.$$

In this form, it was formulated by Henriksen and Isbell. The proof for $n \leq 2$ is made in [Mahé, 1984, 2007], where an unpublished proof of Gus Efroymson is also quoted. In [Mahé, 2007] the author proves that any PWP-function has a representation by inf and sup of polynomials in the whole $\mathbb{R}^3$, except for the union of arbitrary small balls, depending on the PWP-function. We refer to [Madden, 2011] for a comprehensive survey on the subject.

Nowadays the conjecture is still inspiring. In [Lucas et al., 2015] the authors say: *"This paper represents a step in our program towards the proof of the Pierce–Birkhoff conjecture."*

Remark 3.2. *In Definition 3.1 one may consider piecewise linear functions instead of piecewise polynomial functions. Such functions are called* piecewise homogeneous linear *in [Beynon, 1974], where the following result is proved:*

the vector lattice of all piecewise homogeneous linear functions $f : \mathbb{R}^n \to \mathbb{R}$ coincides with the vector lattice of all functions $f : \mathbb{R}^n \to \mathbb{R}$ which can be expressed in the form $\bigvee_{i=1}^{m} \bigwedge_{j=1}^{k} q_{ij}$, where $q_{ij} : \mathbb{R}^n \to \mathbb{R}$ are linear functionals for any i, j.

4 Term functions and piecewise polynomial functions

Let S be a subring of $\mathbb{R}$. Let $\mathcal{L}_0 = \{\oplus, ^*, 0\}$ be the language of MV-algebras, i.e. $\oplus$ is a binary operation, * is unary and 0 is a constant. Let $\mathcal{L}_1 = \mathcal{L}_0 \cup \{\cdot\}$, $\mathcal{L}_{0,S} = \mathcal{L}_0 \cup \{\delta_r \mid r \in [0,1] \cap S\}$, $\mathcal{L}_{1,S} = \mathcal{L}_0 \cup \{\cdot\} \cup \{\delta_r \mid r \in [0,1] \cap S\}$, where $\cdot$ is a binary operation and δ_r is unary operation for any $r \in S \cap [0,1]$. We assume that $\mathcal{L} \in \{\mathcal{L}_0, \mathcal{L}_1, \mathcal{L}_{0,S}, \mathcal{L}_{1,S}\}$ and that $[0,1]$ is an $\mathcal{L}$-algebra as follows: $x \oplus y = \min\{x+y, 1\}$, $x^* = 1-x$, $x \cdot y = xy$ is the product of real numbers and $\delta_r x = rx$ for any $r \in S$, $x, y \in [0,1]$.

For $n \geq 1$ we define in the usual way the set $Term(n, \mathcal{L})$ of $\mathcal{L}$-terms in n variables (denoted by $X_1, \cdots, X_n$). If $t \in Term(n, \mathcal{L})$ then $\widetilde{t} : [0,1]^n \to [0,1]$ is the term function determined by t, when $[0,1]$ is assumed to have the corresponding $\mathcal{L}$-structure. We further set

$$TF(n, \mathcal{L}) = \{\widetilde{t} : [0,1]^n \to [0,1] \mid t \in Term(n, \mathcal{L})\}.$$

Our aim is to characterize the elements of $TF(n, \mathcal{L})$, by means of piecewise polynomial functions.

Remark 4.1. *In order to express better our general results we introduce the notion of $\mathcal{L}$-*polynomial function *as follows:*

- *an $\mathcal{L}_0$-*polynomial function *is an affine linear function $p : \mathbb{R}^n \to \mathbb{R}$ with integer coefficients;*
- *an $\mathcal{L}_1$-*polynomial function *is a polynomial function $p : \mathbb{R}^n \to \mathbb{R}$ with integer coefficients;*
- *an $\mathcal{L}_{0,S}$-*polynomial function *is an affine linear function $p : \mathbb{R}^n \to \mathbb{R}$ with coefficients from S;*

– *an* $\mathcal{L}_{1,S}$-polynomial function *is a polynomial function* $p : \mathbb{R}^n \to \mathbb{R}$ *with coefficients from S.*

Notation 4.2. *In the sequel, the map* $\varrho : \mathbb{R} \to [0,1]$ *is defined by* $\varrho(x) = x \wedge 1 \vee 0$, *for any* $x \in \mathbb{R}$.

Definition 4.3. *Let* $n \geq 1$ *be a natural number and let S be a subring of* $\mathbb{R}$.

– *A continuous function* $f : [0,1]^n \to [0,1]$ *is a* unital piecewise polynomial function with coefficients from S *if there exists a finite set of polynomials, called components,* $\{p_i : \mathbb{R}^n \to \mathbb{R} \mid 1 \leq i \leq k\} \subseteq S[x_1, \ldots, x_n]$ *such that for any* $(a_1, \ldots, a_n) \in \mathbb{R}^n$ *there exists* $i \in \{1, \ldots, k\}$ *with* $f(a_1, \ldots, a_n) = p_i(a_1, \ldots, a_n)$. *We denote by* $PWP(n, \mathcal{L})$ *the set of all such functions whose components are* $\mathcal{L}$*-polynomial functions.*
– *A continuous function* $f : [0,1]^n \to [0,1]$ *is a* unital inf-sup-definable function with coefficients from S *if there exists a finite set of polynomials, called components,* $\{q_{ij} : \mathbb{R}^n \to \mathbb{R} | 1 \leq i \leq m, 1 \leq j \leq k\} \subseteq S[x_1, \ldots, x_n]$ *such that* $f = \bigvee_{i=1}^{m} \bigwedge_{j=1}^{k} (\varrho \circ q_{ij})$. *We denote by* $ISD(n, \mathcal{L})$ *the set of all such functions whose components are* $\mathcal{L}$*-polynomial functions.*

The following result generalizes [McNaughton, 1951, Theorem 2], [Cignoli et al., 2000, Proposition 3.1.8], [Di Nola and Leuştean, 2014, Theorem 10], [Lapenta and Leuştean, 2015, Proposition 3.5]. We give the proof for the sake of completeness.

Proposition 4.4. $TF(n, \mathcal{L}) \subseteq PWP(n, \mathcal{L})$.

Proof: Let t be a term in $Term(n, \mathcal{L})$. The result will be proved by structural induction on t.

– If $t = X_i$ for some $i \leq n$, then $\tilde{t} = \pi_i^n$ and it trivially belongs to $PWP(n, \mathcal{L})$.
– If $t = t_1^*$, then $\tilde{t} = (\tilde{t_1})^*$. By induction hypothesis there exists an integer h and some polynomials $q_1, \ldots, q_h \in S[x_1, \ldots, x_n]$ such that for any point in the n-cube, $\tilde{t_1}$ coincide with one of them. Then $1 - q_1, \ldots, 1 - q_h$ are the components of $\tilde{t}$.
– If $t = t_1 \oplus t_2$, let $q_1, \ldots, q_m$ be the components of $\tilde{t_1}$ and $p_1, \ldots, p_k$ be the components of $\tilde{t_2}$. Then $\tilde{t}$ is defined by the polynomials $\{1\} \cup \{s_{ij}\}_{i,j}$, where $s_{ij} = 1 - q_i + p_j$ for any $i \in \{1, \ldots, s\}$ and $j \in \{1, \ldots, r\}$.
– If t is $\delta_r(t_1)$ for some $r \in [0,1] \cap S$ and $q_1, \ldots, q_s$ are the components of $\tilde{t_1}$, then $rq_1, \ldots, rq_s$ are the components of $\tilde{t}$.
– If $t = t_1 \cdot t_2$, let $q_1, \ldots, q_m$ be the components of $\tilde{t_1}$ and $p_1, \ldots, p_k$ be the components of $\tilde{t_2}$. Then $\tilde{t}$ is defined by the polynomials $q_i \cdot p_j$, for any $i \in \{1, \ldots, s\}$ and $j \in \{1, \ldots, r\}$. □

Lemma 4.5. *[Di Nola and Leuştean, 2014, Lemma 9] For any $x, y \in \mathbb{R}$ the following properties hold:*
(a) $(x \vee 0) + (y \vee 0) \geq (x + y) \vee 0$,
(b) $x \geq 0$ *iff* $\varrho(-x) = 0$,
(c) $\varrho(x) = \varrho(x \vee 0)$.

Lemma 4.6. *[Di Nola and Leuştean, 2014, Lemma 10] If $g : [0, 1]^n \to \mathbb{R}$ and $h : [0, 1]^n \to [0, 1]$ then the following properties hold.*
(a) $\varrho \circ (g + h) = ((\varrho \circ g) \oplus h) \odot (\varrho \circ (g + 1))$.
(b) $\varrho \circ (1 - g) = 1 - (\varrho \circ g)$.

The following result is a generalization of [McNaughton, 1951, Theorem 1], [Cignoli et al., 2000, Lemma 3.1.9], [Di Nola and Leuştean, 2014, Proposition 6], [Lapenta and Leuştean, 2015, Proposition 4.3]. In spite of the fact that the proof is very similar to the one in [Lapenta and Leuştean, 2015], it covers at least one additional relevant case: for $\mathcal{L} = \mathcal{L}_1$, this result and Theorem 4.9 lead to a better understanding of the free PMV-algebra with n free generators.

We note that, following the proof, one can extract an algorithm that generates the term associated to a given $\mathcal{L}$-polynomial function (see Figure 1).

Proposition 4.7. *Let S be a subring of $\mathbb{R}$.*

(a) *For any $\mathcal{L}$-polynomial function $p : [0, 1]^n \to \mathbb{R}$ there exists $t \in Term(n, \mathcal{L})$ such that $\varrho \circ p = \widetilde{t}$.*

(b) *For any $g \in ISD(n, \mathcal{L})$ there exists $t \in Term(n, \mathcal{L})$ such that $g = \widetilde{t}$.*

Proof: We give the proof for $\mathcal{L} = \mathcal{L}_{1,S}$, which is the most general one.

(a) Let $p : [0, 1]^n \to \mathbb{R}$ be an polynomial function. Let k be the degree of p. Then

$$p(x_1, \ldots, x_n) = \sum_{i_1+\ldots+i_n \leq k} c_{i_1,\ldots,i_n} x_1^{i_1} \cdots x_n^{i_n},$$

where $c_{i_1,\ldots,i_n} \in S$ for any choice of the indexes. Since any $c_{i_1,\ldots,i_n}$ can be written as a sum of a finite number of elements in $[-1, 1] \cap S$, we assume that

$$p(x_1, \ldots, x_n) = r_m y_m + \cdots + r_{p+1} y_{p+1} + r_p + \cdots + r_1$$

where $m \geq 1$ and $p \geq 0$ are natural numbers, $p \leq m$, $r_j \in ([-1, 1] \cap S) \setminus \{0\}$ for any $j \in \{1, \cdots, m\}$ and $y_j \in \{x_1^{i_1} \cdots x_n^{i_n} \mid i_1 + \cdots + i_n \leq k\}$ for any $j \in \{p + 1, \cdots, m\}$.
We prove the theorem by induction on $m \geq 1$. In the sequel we denote by x an element $(x_1, \ldots, x_n)$ from $[0, 1]^n$.
Initial step $m = 1$. We have $p(\mathrm{x}) = r$ for any $\mathrm{x} \in [0, 1]^n$ or $p(\mathrm{x}) = r x_1^{i_1} \cdots x_n^{i_n}$ for any $\mathrm{x} \in [0, 1]^n$ where $r \in ([-1, 1] \cap S) \setminus \{0\}$ and $\{i_1, \ldots, i_n\}$ is a suitable set of indexes.

If $r \in [-1, 0) \cap S$ then $\varrho \circ p = 0$ so $\varrho \circ p = \tilde{t}$ for $t = 0$.

If $r \in (0, 1] \cap S$ then $p = \varrho \circ p$. It follows that $p = \tilde{t}$ where $t = \delta_r(0^*)$ if $p(\mathrm{x}) = r$ for any $\mathrm{x} \in [0, 1]^n$ and $t = \delta_r(X_1^{i_1} \cdots X_n^{i_n})$ if $p(\mathrm{x}) = rx_1^{i_1} \cdots x_n^{i_n}$ for any $\mathrm{x} \in [0, 1]^n$.
Induction step. We take $p = g + h$ where $\varrho \circ g = \widetilde{t_1}$ for some term t_1 and there is $r \in ([-1, 1] \cap S) \setminus \{0\}$ and a suitable choice of indexes for $i_1, \ldots, i_n$ such that $h(\mathrm{x}) = r$ for any $\mathrm{x} \in [0, 1]^n$, or $h(\mathrm{x}) = rx_1^{i_1} \cdots x_n^{i_n}$ for any $\mathrm{x} \in [0, 1]^n$. We consider two cases.
Case 1. If $r \in (0, 1]$ then $h : [0, 1]^n \to [0, 1]$ so $\varrho \circ p = ((\varrho \circ g) \oplus h) \odot (\varrho \circ (1 + g))$ by Lemma 4.6 (a). Following the initial step, there is a term t_2 such that $h = \widetilde{t_2}$. We notice that $1 + g = 1 - (-g)$ and the induction hypothesis holds for $(-g)$, then there is a term t_3 such that $\varrho \circ (-g) = \widetilde{t_3}$. It follows by Lemma 4.6 (b) that $\varrho \circ (1 + g) = 1 - \widetilde{t_3} = \widetilde{t_3}^*$, and $\varrho \circ p = \tilde{t}$ where $t = (t_1 \oplus t_2) \odot t_3^*$.
Case 2. If $r \in [-1, 0)$, then $g + h = (g - 1) + (1 + h)$ and $1 + h : [0, 1]^n \to [0, 1]$. By Lemma 4.6 (a) we get

$$\varrho \circ p = ((\varrho \circ (g - 1)) \oplus (1 + h)) \odot (\varrho \circ g).$$

Following the initial step, there is a term t_2 such that $-h = \widetilde{t_2}$, so $1 + h = 1 - (-h) = \widetilde{t_2}^*$. In the sequel we have to find a term t_3 that corresponds to $\varrho \circ (g - 1)$, where

$$g(\mathrm{x}) = r_m y_m + \cdots + r_{p+1} y_{p+1} + r_p + \cdots + r_1$$

with $r_j \in ([-1, 1] \cap S) \setminus \{0\}$ for any $j \in \{1, \cdots, m\}$ and y_j in $\{x_1^{i_1} \cdots x_n^{i_n} \mid i_1 + \cdots + i_n \leq k\}$ for any $j \in \{p + 1, \cdots, m\}$.
Case 2.1. If $r_j \leq 0$ for any $j \in \{1, \cdots, m\}$ then $g - 1 \leq 0$, so $\varrho \circ (g - 1) = 0 = \widetilde{t_3}$ with $t_3 = 0$.
Case 2.2. If there is $j_0 \in \{1, \cdots, p\}$ such that $r_{j_0} > 0$, then

$$(g - 1)(\mathrm{x}) = r_m y_m + \cdots + r_{p+1} y_{p+1} + r_p + \cdots + (r_{j_0} - 1) + \cdots + r_1$$

and $r_{j_0} - 1 \in [-1, 0)$, so the induction hypothesis applies to $g - 1$. Then there exists a term t_3 such that $\varrho \circ (g - 1) = \widetilde{t_3}$.
Case 2.3. If there is $j_0 \in \{p + 1, \cdots, m\}$ such that $r_{j_0} > 0$, then we set $h_0(\mathrm{x}) = r_{j_0} y_{j_0}$ and

$$g_0(\mathrm{x}) = g(\mathrm{x}) - r_{j_0} y_{j_0} - 1.$$

It follows that $g - 1 = g_0 + h_0$ such that g_0 satisfies the induction hypothesis and $h_0 : [0, 1]^n \to [0, 1]$. We are in the hypothesis of *Case 1*, so there exists a term t_3 such that $\varrho \circ (g - 1) = \widetilde{t_3}$.
Summing up, we get $\varrho \circ (g + h) = \tilde{t}$ with $t = ((t_2 \oplus t_3^*) \odot t_1)$.

(b) This is straightforward by (a): for any g_{ij} there exist a term $t_{ij} \in Term_n(S)$ such that $\varrho \circ g_{ij} = \widetilde{t_{ij}}$.

Then

$$g = \varrho \circ g = \bigvee_{i\in I} \bigwedge_{j\in J} \varrho \circ g_{ij} = \bigvee_{i\in I} \bigwedge_{j\in J} \widetilde{t_{ij}}.$$

If $t = \bigvee_{i\in I} \bigwedge_{j\in J} t_{ij}$, we get $g = \widetilde{t}$. □

Remark 4.8. *If $f : [0,1]^n \to \mathbb{R}$ is a polynomial function with real coefficients then we represent $f : [0,1]^n \to \mathbb{R}$ as*

$$f(x_1, \ldots, x_n) = r_m y_m + \cdots + r_{p+1} y_{p+1} + r_p + \cdots + r_1$$

where $p \leq m$, $r_j \in [-1,1] \setminus \{0\}$, and $y_j = x_1^{i_j^1} \cdots x_n^{i_j^n}$ is a monomial for any $j > p$. Using this representation one can easily extract an algorithm from the proof of Theorem 4.9, which returns an $\mathcal{L}$-term corresponding to f. In Figure 1 we describe the algorithm for $\mathcal{L} = \mathcal{L}_{1,S}$.

We note that the polynomial f is represented by

$$(p, r_1, \ldots, r_p, (r_{p+1}, \mathrm{i}_{p+1}), \ldots, (r_m, \mathrm{i}_m)),$$

where $\mathrm{i}_j = (i_j^1, \ldots, i_j^n)$ for any $j \in \{p+1, \ldots, m\}$. In particular cases the representation can be simplified as follows:

- *for $\mathcal{L} \in \{\mathcal{L}_0, \mathcal{L}_{0,S}\}$ the monomials y_j satisfy the condition $i_j^1 + \cdots + i_j^n = 1$, so y_j are of the form x_k;*
- *for $\mathcal{L} \in \{\mathcal{L}_0, \mathcal{L}_1\}$ and for any j we have $r_j \in \{-1, 1\}$.*

The algorithm for $\mathcal{L} = \mathcal{L}_{0,\mathbb{R}}$ is described in [Gerla et al., 2013].

The following theorem summarizes our results.

Theorem 4.9. *$ISD(n, \mathcal{L}) \subseteq TF(n, \mathcal{L}) \subseteq PWP(n, \mathcal{L})$ for $\mathcal{L} \in \{\mathcal{L}_0, \mathcal{L}_1, \mathcal{L}_{0,S}, \mathcal{L}_{1,S}\}$.*

5 Normal form theorems and the Pierce-Birkhoff conjecture

Let S be a subring of $\mathbb{R}$ and $\mathcal{L}$ as above. In the light of Theorem 4.9, one may ask if the following equalities hold:

$$(*) \qquad ISD(n, \mathcal{L}) = TF(n, \mathcal{L}) = PWP(n, \mathcal{L}).$$

This problem has the same flavor as the Pierce-Birkhoff conjecture. Moreover, for $\mathcal{L} = \mathcal{L}_0$ this is the McNaughton theorem [McNaughton, 1951, Mundici, 1994]. In the following we discuss the property $(*)$ from logical perspective.

```
// we use the notation i_k = (i_k^1, ..., i_k^n)
function Term(p, r_1, ..., r_p, (r_{p+1}, i_{p+1}), ...(r_m, i_m))
{
(F1) if r_k ≤ 0 for any k ∈ {1, ..., m} then return(0);
(F2) find k ∈ {1, ..., m} such that r_k > 0;
     if k ≤ p then ψ := δ_{r_k}(0*) else ψ := δ_{r_k}(X_1^{i_k^1} ··· X_n^{i_k^n});
(F3) if m = 1 then return(ψ);
(F4) if k ≤ p then
{φ = Term(p − 1, r_1, ..., r_{k−1}, r_{k+1}, ...r_p, (r_{p+1}, i_{p+1}), ..., (r_m, i_m));
χ =Term(p − 1,−r_1, ...,−r_{k−1}, −r_{k+1}, ...− r_p, (−r_{p+1}, i_{p+1}), ..., (−r_m, i_m))
} else
{φ = Term(p, r_1, ..., r_p, ..., (r_{k−1}, i_{k−1})(r_{k+1}, i_{k+1}), ..., (r_m, i_m));
χ = Term(p, −r_1, ..., −r_p, ..., (−r_{k−1}, i_{k−1}), (−r_{k+1}, i_{k+1}), ..., (−r_m, i_m))
}; return((φ ⊕ ψ) ⊙ ¬χ)
}
```

Fig. 1. The term corresponding to a polynomial function

If $\mathcal{L} \in \{\mathcal{L}_1, \mathcal{L}_{0,S}, \mathcal{L}_{1,S}\}$ and $[0,1]_{\mathcal{L}}$ is an $\mathcal{L}$-algebra defined in the previous section, then $TF(n, \mathcal{L})$ is an $\mathcal{L}$-algebra with pointwise operations. By general results in universal algebra $TF(n, \mathcal{L})$ is the free $\mathcal{L}$-structure in the variety generated by $[0,1]_{\mathcal{L}}$.

We already mentioned that $\mathcal{L}_0$ is the language of Łukasiewicz logic. One can easily see that $\mathcal{L}_1$, $\mathcal{L}_{0,\mathbb{R}}$, $\mathcal{L}_{1,\mathbb{R}}$ are, respectively, the language of PMV-algebras, Riesz MV-algebras and *f*MV-algebras.

We further note that $[0,1]_{\mathcal{L}_0}$ generates the variety of MV-algebras [Chang, 1959, Cignoli et al., 2000] and $[0,1]_{\mathcal{L}_{0,\mathbb{R}}}$ generates the variety of Riesz MV-algebras [Di Nola and Leuştean, 2014]. Let $\mathcal{L}\mathbb{R}$ be the propositional calculus which has Riesz MV-algebras as models. In this context $(*)$ holds [Di Nola and Leuştean, 2014] and it can be seen as a local version of the result from Remark 3.2.

The situation is different for PMV-algebras, since the standard model $[0,1]_{\mathcal{L}_1}$ generates only a proper subvariety [Horčík and Cintula, 2004]. Moreover, due to a result of Isbell [Isbell, 1972] this variety is not finitely axiomatizable. Montagna proved that the proper quasi-variety of PMV$^+$-algebras defined by the quasi-identity:

$$x \cdot x = 0 \Rightarrow x = 0 \text{ for any } x$$

is generated, as a quasi-variety, by $[0,1]_{\mathcal{L}_1}$ [Montagna, 2005]. Consequently, the equality $(*)$ for $\mathcal{L}_1$ is a normal form theorem for the system PL', described in [Horčík and Cintula, 2004], that has PMV$^+$-algebras as models. A similar analysis is made in [Lapenta and Leuştean, 2015] for fMV-algebras; in this case the quasi-variety generated by the standard model $[0, 1]_{\mathcal{L}_{1,\mathbb{R}}}$ is the class of FR$^+$-algebras and the corresponding propositional calculus is denoted $\mathcal{FMVL}^+$.

Finally we note that all these logical systems are conservative extensions of Łukasiewicz ∞-valued logic and the $\mathcal{L}$-structure $TF(n, \mathcal{L})$ is, up to isomorphism, the Lindenbaum-Tarski algebra of the corresponding propositional calculus. The property $(*)$ is a normal form theorem and we summarize below the results known so far.

Table 1. Normal form theorem : summary of known results

$\mathcal{L}$	Logic	Algebra	$ISD(n, \mathcal{L}) = PWP(n, \mathcal{L})$
$\mathcal{L}_0$	Łuk	MV-algebras	true [McNaughton, 1951]
$\mathcal{L}_{0,\mathbb{R}}$	$\mathcal{L}\mathbb{R}$	Riesz MV-algebras	true [Di Nola and Leuştean, 2014]
$\mathcal{L}_1$	PL'	PMV$^+$-algebras	open problem
$\mathcal{L}_{1,\mathbb{R}}$	$\mathcal{FMVL}^+$	FR$^+$-algebras	true if $n \leq 2$ open problem if $n > 2$ [Lapenta and Leuştean, 2015]

In the following we emphasize the results for the system $\mathcal{FMVL}^+$. Note that in this case the unital piecewise polynomial functions from Definition 4.3 have the components as the continuous piecewise polynomial functions from the Pierce-Birkhoff conjecture.

If $f : \mathbb{R}^n \to [0, 1]$ is a continuous piecewise polynomial function then $f|_{[0,1]^n}$ is a continuous piecewise polynomial function defined on the unit cube. In general it is not known if any piecewise polynomial function defined on $[0, 1]^n$ is the restriction of a continuous piecewise polynomial function defined on $\mathbb{R}^n$, but for $n = 2$ a positive answer is given in [Fischer and Marshall, 2013]. As a direct consequence and using Mahé's proof for the Pierce-Birkhoff conjecture [Mahé, 1984], for $n \leq 2$ we get

$$ISD(n, \mathcal{L}_{1,\mathbb{R}}) = TF(n, \mathcal{L}_{1,\mathbb{R}}) = PWP(n, \mathcal{L}_{1,\mathbb{R}}).$$

The above result is a normal form theorem for the propositional calculus $\mathcal{FMVL}^+$ and describes the functions corresponding to the formulas in two variables. The following result can be interpreted as a local version of the Pierce-Birkhoff theorem.

Conjecture 5.1. *[Lapenta and Leuştean, 2015] For $n > 2$,*

$$ISD(n, \mathcal{L}_{1,\mathbb{R}}) = TF(n, \mathcal{L}_{1,\mathbb{R}}) = PWP(n, \mathcal{L}_{1,\mathbb{R}}).$$

We note that the above result does not immediately imply and it is not immediately implied by the original Pierce-Birkhoff conjecture and it might be equally hard to prove.

Acknowledgment: Ioana Leuştean was partially supported by a grant of the Romanian National Authority for Scientific Research and Innovation, CNCS-UEFISCDI, project number PN-II-RU-TE-2014-4-0730.

References

W. M. Beynon. Combinatorial aspects of piecewise polynomial functions. *Journal . Math. Soc.*, 2:719–727, 1974.

A. Bigard, K. Keimel, and S. Wolfstein. *Groupes et anneuax reticules*. Springer-Verlag, Lecture Notes in Mathematics 608, 1977.

G. Birkhoff and R.S. Pierce. Lattice-ordered rings. *An. Acad. Brasil. Cienc.*, 28:41–69, 1956.

C.C. Chang. Algebraic analysis of many-valued logics. *Trans. Amer. Math. Soc.*, 88:467–490, 1958.

C.C. Chang. A new proof of the completeness of the Łukasiewicz axioms. *Transactions of the American Mathematical Society*, 93:74–80, 1959.

R. Cignoli, I. M. L. D'Ottaviano, and D. Mundici. *Algebraic foundation of many-valued Reasoning*. Kluver Academic Publ. Dordrecht, Trends in Logic 7, 2000.

C.N. Delzell. On the Pierce-Birkhoff conjecture over ordered fields. *Rocky Mountains Journal of Mathematics*, 19(3):651–668, 1989.

A. Di Nola and A. Dvurečenskij. Product MV-algebras. *Multiple-Valued Logics*, 6:193–215, 2001.

A. Di Nola and I. Leuştean. Łukasiewicz logic and riesz spaces. *Soft Computing*, 18(12):2349–2363, 2014.

A. Fischer and M. Marshall. Extending piecewise polynomial functions in two variables. *Annales de la faculté des Sciences de Toulouse Mathématiques*, 22:253–268, 2013.

B. Gerla, A. Di Nola, and I. Leuştean. Adding Real Coefficients to Łukasiewicz logic: An Application to Neural Networks. In Masulli F., Pasi G., and Yager R., editors, *Lecture notes in Computer Science 8256, WILF 2013, Proceedings*, pages 77–85. 2013.

P. Hájek. *Metamathematics of Fuzzy Logic*. Kluver Dordrecht, Trends in Logic 4, 1998.

R. Horčík and P. Cintula. Product Łukasiewicz logic. *Archive for Mathematical Logic*, 43(4): 477–503, 2004.

J R. Isbell. Notes on Ordered Rings. *Algebra Univ.*, 1:393–399, 1972.

S. Lapenta and I. Leuştean. Towards understanding the Pierce-Birkhoff conjecture via MV-algebras. *Fuzzy sets and systems*, 276:114–130, 2015.

S. Lapenta and I. Leuştean. Scalar extensions for the algebraic structures of Łukasiewicz logic. *Journal of Pure and Applied Algebra*, 220:1538–1553, 2016.

F. Lucas, D. Schaub, and M. Spivakovsky. On the pierce–birkhoff conjecture. *Journal of Algebra*, 435:124 – 158, 2015.

J. Łukasiewicz and A. Tarski. Untersuchungen über den aussagenkalkül. *Comptes Rendus de la Société des Sciences et des Lettres de Varsovie*, 23:30–50, 1930.

J. Madden. Henriksen and Isbell on f-rings. *Topology and Its Applications*, 158:1768–1773, 2011.

L. Mahé. On the Birkhoff-Pierce conjecture. *Rocky Mountains Journal*, 14(4):983–985, 1984.

L. Mahé. On the Birkhoff-Pierce conjecture in three variables. *J. Pure Appl. Algebra*, 211:459–470, 2007.

R. McNaughton. A theorem about infinite-valued sentential logic. *Journal of Symbolic Logic*, 16: 1–13, 1951.

F. Montagna. An algebraic approach to Propositional Fuzzy Logic. *Journal of Logic, Language and Information*, 9:91–124, 2000.

F. Montagna. Subreducts of MV-algebras with product and product residuation. *Algebra Universalis*, 53:109–137, 2005.

F. Montagna and G. Panti. Adding structure to MV-algebras. *J. Pure and Applied Algebra*, 164: 365–387, 2001.

D. Mundici. Interpretation of ACF*-algebras in Łukasiewicz sentential calculus. *J. Funct. Anal.*, 65:15–63, 1986.

D. Mundici. A constructive proof of McNaughton Theorem in infinite-valued logic. *J. Symb. Log.*, 59:596–602, 1994.

D. Mundici. Tensor Products and the Loomis–Sikorski Theorem for MV-Algebras. *Advances in Applied Mathematics*, 22(2):227 – 248, 1999.

D. Mundici. *Advanced Łukasiewicz calculus and MV-algebras*. Springer, Trends in Logic 35, 2011.

Maria Emilia Maietti and Giuseppe Rosolini

Relating Quotient Completions via Categorical Logic

Abstract: In previous work we introduced the notion of elementary quotient completion with respect to an elementary doctrine. We also generalized the notion of exact completion of a regular category as an exact completion of an existential elementary doctrine. Here we characterize when the elementary quotient completion of an elementary existential doctrine coincides with an exact completion. We do this by employing the categorical logic of the various notions of doctrines involved in our analysis. The outcome is that the two completions coincide when a choice rule holds in the starting existential elementary doctrine.

Keywords: Quotient completion, Split fibration, Universal construction

Mathematics Subject Classification 2010: 03G30, 03D15, 18C50, 03B20, 03F55

1 Introduction

The need of completing a structure with quotients arises in various fields in mathematics and in computer science, including foundation of mathematics, type theory and category theory, in order to get models closed under quotients though based on much weaker structures.

In previous work [Maietti and Rosolini, 2013b,a] we used the language of category theory to introduce a notion of quotient completion for Lawvere's elementary doctrines, see [Lawvere, 1969a, 1970]. The notion includes well-known instances of quotient completion in category theory and in type theory: for examples, the exact completion of a lex category as in [Carboni and Celia Magno, 1982], Joyal's arithmetic universes (see [Maietti, 2010]) and setoid models of type theory are all instances of elementary quotient completions.

In another work [Maietti and Rosolini, 2015] we reviewed and generalized the notion of exact completion of a regular category as a free construction from an existential elementary doctrine. Examples of this construction are toposes obtained from triposes, see [Hyland et al., 1980], in particular the effective topos of [Hyland, 1982].

Maria Emilia Maietti: Dipartimento di Matematica Pura ed Applicata, Università di Padova, via Trieste 63, 35121 Padova, Italy, e-mail: maietti@math.unipd.it
Giuseppe Rosolini: DIMA, Università di Genova, via Dodecaneso 35, 16146 Genova, Italy, e-mail: rosolini@unige.it

The aim of this paper is to present a logical characterization of when the elementary quotient completion of an existential elementary doctrine happens to coincide with its exact completion.

To this purpose in the first part of the paper we review the mentioned notions of quotient completion related to an elementary doctrine. Differently from the previous work, here we employ the internal logic for elementary doctrine as presented in [Jacobs, 1999]. Finally we show that the two completions coincide when the starting elementary existential doctrine satisfies a rule of choice.

2 Notions of elementary doctrine

In previous work [Maietti and Rosolini, 2013b,a] we employed the notion of elementary doctrine to generalize the completion of a categorical structure with quotients. The main idea was to relativize the concept of quotient completion to a many sorted logic, represented categorically by a doctrine, validating the logical structure needed to speak of equivalence relations. In *loc.cit.* we found appropriate to use Lawvere's concept of elementary doctrine, which represents a many sorted logic with conjunctions and propositional equality, and related to it we defined the notion of *elementary quotient completion*.

F.W. Lawvere introduced the notion of elementary doctrine in a series of seminal papers, see [Lawvere, 1969a,b, 1970], to synthetize the structural properties of logical systems, see also [Lawvere and Rosebrugh, 2003] for a unified survey. Lawvere's crucial intuition was to consider logical languages and theories as hyperdoctrines to study their 2-categorical properties, *e.g.* connectives and quantifiers are determined by structural adjunctions. That approach proved extremely fruitful, see [Makkai and Reyes, 1977, Carboni, 1982, Lambek and Scott, 1986, Jacobs, 1999, Taylor, 1999, van Oosten, 2008] and references therein.

Taking advantage of the category-theoretical presentation of logic by doctrines, we now first introduce a general notion of elementary doctrine which we found appropriate to study the notion of quotient of an equivalence relation, see [Maietti and Rosolini, 2013b, 2015].

Since doctrines can be presented equivalently in the form of fibrations, see *e.g.* [Jacobs, 1999], we may use the names interchangeably.

Denote by **InfSL** the 2-category of inf-semilattice: objects are posets with finite meets, 1-morphisms are functions between them which preserves finite meets, and 2-morphisms are given by the pointwise order on the homsets.

Definition 2.1. *Let* **C** *be a category with finite products. An* ***elementary doctrine on*** **C** *is a functor* $P\colon \mathbf{C}^{op} \longrightarrow \mathbf{InfSL}$[1]*—the value* $P(A)$ *is the* ***fibre*** *on* A*—such that, for every object* A *in* **C***, there is an object* δ_A*, which we shall call* ***fibered equality****, in* $P(A \times A)$ *such that*

1. *for every object* X *and morphism* $e := \langle \mathrm{pr}_1, \mathrm{pr}_2, \mathrm{pr}_2\rangle\colon X \times A \to X \times A \times A$ *in* **C***, the assignment that sends* α *in* $P(X \times A)$ *to the meet*

$$\exists_e(\alpha) := P_{\mathrm{id}_X \times \mathrm{pr}_1}(\alpha) \wedge P_{\langle \mathrm{pr}_2, \mathrm{pr}_3\rangle}(\delta_A),$$

computed in $P(X{\times}A{\times}A)$*, determines a left adjoint to* $P_e\colon P(X{\times}A{\times}A) \to P(X{\times}A)$ *where* pr_i*,* $i = 1, 2$*, are the projections from* $X \times A$ *to each of the two factors.*

Remark 2.2.

1. *Since* T *is terminal in* **C***, condition (1.) for* $X = T$ *yields that the assignment that sends* α *in* $P(A)$ *to the meet*

$$\exists_{\langle \mathrm{id}_A, \mathrm{id}_A\rangle}(\alpha) := P_{\mathrm{pr}_1}(\alpha) \wedge \delta_A,$$

computed in $P(A \times A)$*, determines a left adjoint to* $P_{\langle \mathrm{id}_A, \mathrm{id}_A\rangle}\colon P(A \times A) \to P(A)$*.*
2. *The previous remark yields that* δ_A *is unique for each object* A *in* **C***.*
3. *Since* $\langle \mathrm{pr}_2, \mathrm{pr}_1\rangle \circ \langle \mathrm{id}_A, \mathrm{id}_A\rangle = \langle \mathrm{id}_A, \mathrm{id}_A\rangle$*, one has that*

$$\exists_{\langle \mathrm{id}_A, \mathrm{id}_A\rangle}(\alpha) = P_{\mathrm{pr}_2}(\alpha) \wedge \delta_A$$

for every α *in* $P(A)$*.*
4. *In* $P(A \times A)$ *it is* $\top \le P_{\langle \mathrm{id}_A, \mathrm{id}_A\rangle}(\delta_A)$ *and, in* $P(A \times A)$*, it is* $\delta_A \le P_{f \times f}(\delta_B)$ *when* $f\colon A \to B$*.*
5. *For* α_1 *in* $P(X_1 \times Y_1)$ *and* α_2 *in* $P(X_2 \times Y_2)$*, write* $\alpha_1 \boxtimes \alpha_2$ *for the object*

$$P_{\langle \mathrm{pr}_1, \mathrm{pr}_3\rangle}(\alpha_1) \wedge P_{\langle \mathrm{pr}_2, \mathrm{pr}_4\rangle}(\alpha_2)$$

in $P(X_1 \times X_2 \times Y_1 \times Y_2)$ *where* pr_i*,* $i = 1, 2, 3, 4$*, are the projections from* $X_1 \times X_2 \times Y_1 \times Y_2$ *to each of the four factors.*
Condition 2.1.1. is to request that $\delta_{A \times B} = \delta_A \boxtimes \delta_B$ *for every pair of objects* A *and* B *in* **C***.*

Examples 2.3.

1. *The first example of elementary doctrine is that of powersets. The domain category is the category* **Set** *of sets and functions. The terminal object can be fixed*

---

1 We may refer to such a functor as an indexed inf-semilattice; we always write P_f for the value of the indexing functor P on a morphism f.

as $\{\emptyset\}$, *and the product of two sets is given by the set of ordered pairs encoded as usual* $(x, y) := \{\{x\}, \{x, y\}\}$ *with obvious projections.*

The elementary doctrine $\mathbf{P}\colon \mathbf{Set}^{op} \longrightarrow$ **InfSL** *on* **Set** *is given as the powerset* $\mathbf{P}(A)$. *The action on functions* $f\colon A \to B$ *is by inverse image, i.e.* $\mathbf{P}_f := f^{-1}\colon \mathbf{P}(B) \to \mathbf{P}(A)$.

The elementary structure is given by the diagonal subset of $A \times A$.

2. *The standard categorical example of an indexed poset is the fibration of subobjects, a generalization of the previous example to an arbitrary category. Consider a category* $\mathbf{C}$ *with binary products, a terminal object and pullbacks of monomorphisms—for convenience, we shall assume that* $\mathbf{C}$ *is also well-powered, here as well as any other time when size issues become relevant. The functor* $S\colon \mathbf{C}^{op} \longrightarrow$ **InfSL** *assigns to any object* A *in* $\mathbf{C}$ *the poset* $S(A)$ *of subobjects of* A *in* $\mathbf{C}$ *and it is easy to check that the poset is an inf-semilattice. For a morphism* $f\colon B \to A$, *the assignment that maps a subobject in* $S(A)$ *to that represented by the left-hand morphism in any pullback of it along* f *produces a functor* $S_f\colon S(A) \to S(B)$ *that preserves products.*

   The elementary structure is provided by the diagonal morphisms.
3. *Another categorical example is given by a category* $\mathbf{S}$ *with binary products and weak pullbacks, by defining the doctrine functor of weak subobjects* $\Psi\colon \mathbf{S}^{op} \longrightarrow$ **InfSL** *which evaluates as the poset reflection of each comma category* $\mathbf{S}/A$ *at each object* A *of* $\mathbf{S}$, *introduced in [Grandis, 2000].*
4. *An example of elementary doctrine from first order logic is the Lindenbaum-Tarski algebras of well-formed formulas of a theory* $\mathcal{T}$ *with equality in the first order language* $\mathcal{L}$. *The domain category is the category* $\mathbf{V}$ *of lists of variables and term substitutions:*
   - ***object of*** $\mathbf{V}$ *are lists of distinct variables* $(x_{j_1}, \ldots, x_{j_n})$;
   - ***morphisms*** $(t_1/x_{k_1}, \ldots, t_m/x_{k_m})\colon (x_{j_1}, \ldots, x_{j_n}) \to (x_{k_1}, \ldots, x_{k_m})$ *are lists of substitutions for variables where each term* t_j, $i = 1, \ldots, m$, *is built in* $\mathcal{L}$ *on the variables* $x_{j_1}, \ldots, x_{j_n}$;
   - ***composition*** $(x_{j_1}, \ldots, x_{j_n}) \xrightarrow{(t_i/x_{k_i})_{i=1}^m} (x_{k_1}, \ldots, x_{k_m}) \xrightarrow{(s_i/x_{h_i})_{i=1}^{\ell}} (x_{h_1}, \ldots, x_{h_\ell})$ *is given by simultaneous substitutions*

$$(x_{j_1}, \ldots, x_{j_n}) \xrightarrow{\quad (s_i(t_1/x_{k_1}, \ldots, t_m/x_{k_m})/x_{h_i})_{i=1}^{\ell}) \quad} (x_{h_1}, \ldots, x_{h_\ell}).$$

The terminal object is the empty list $()$; *the product of two objects* $(x_{j_1}, \ldots, x_{j_n})$ *and* $(x_{k_1}, \ldots, x_{k_m})$ *is given by the list* $(x_{j_1}, \ldots, x_{j_n}, x_{J+k_1}, \ldots, x_{J+k_m})$, *where* $J = \max\{j_1, \ldots, j_n\}$. *Projections are given by substitution of the variables in* $(x_{j_1}, \ldots, x_{j_n})$ *by the first* n *variables in* $(x_{j_1}, \ldots, x_{j_n}, x_{J+k_1}, \ldots, x_{J+k_m})$, *and of the variables in* $(x_{k_1}, \ldots, x_{k_m})$ *with the last* m *variables occurring in* $(x_{j_1}, \ldots, x_{j_n}, x_{J+k_1}, \ldots, x_{J+k_m})$.

*The elementary doctrine $LT\colon \mathbf{V}^{\mathrm{op}} \longrightarrow$ **InfSL** on $\mathbf{V}$ is given as follows: for a list of distinct variables $\vec{x}$, the inf-semilattice $LT(\vec{x})$ has*

- ***objects** are equivalence classes of well-formed formulas of $\mathcal{L}$ with no more free variables than $x_{j_1},\ldots,x_{j_n}$ with respect to provable reciprocal consequence $W \dashv\vdash_{\mathcal{T}} W'$ in $\mathcal{T}$;*

morphisms$[W] \to [V]$ are the provable consequences $W \vdash_{\mathcal{T}} V$ in $\mathcal{T}$ for some pair of representatives (hence for any pair);

- ***composition** is given by the cut rule in the logical calculus;*
- ***identities** $[W] \to [W]$ are given by the logical rules $W \vdash_{\mathcal{T}} W$.*

For a list of distinct variables $\vec{x} := (x_{j_1}, \ldots, x_{j_n})$, the poset $LT(\vec{x})$ has finite meets: the top element is any true formula, for example $x = x$ with x variable in $\vec{x}$ and the meet of a pair of formulas is obtained by conjunction.

Elementary doctrines are the cloven Eq-fibrations of [Jacobs, 1999] and, as explained in *loc.cit.*, there is a deductive calculus associated to those which is that of the $\wedge$=-fragment over type theory with just a unit type and a binary product type constructor. From now on, we shall employ the logical language introduced in *loc.cit.* and write

$$a_1\colon A_1, \ldots, a_k\colon A_k \mid \phi_1(a_1, \ldots, a_k), \ldots, \phi_n(a_1, \ldots, a_k) \vdash \psi(a_1, \ldots, a_k)$$

in place of

$$\phi_1 \wedge \ldots \wedge \phi_n \le \psi$$

in $P(A_1 \times \ldots \times A_k)$. Note that, in line with *loc.cit.*, $\delta_A(a, a')$ will be written as $a\colon A, a'\colon A \mid a =_A a'$. Also we write $a\colon A \mid \alpha \dashv\vdash \beta$ to abbreviate $a\colon A \mid \alpha \vdash \beta$ and $a\colon A \mid \beta \vdash \alpha$.

It is possible to express precisely relationships between the examples once one considers the 2-category **ED** of elementary doctrines:

- **the 1-morphisms** are pairs $(F, \mathfrak{b})$ where $F\colon \mathbf{C} \to \mathbf{D}$ is a functor and $\mathfrak{b}\colon P \stackrel{.}{\to} R\circ F^{\mathrm{op}}$ is a natural transformation as in the diagram

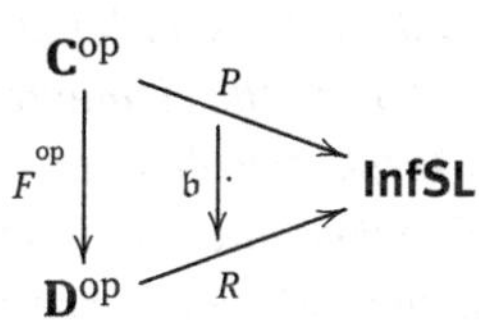

where the functor F preserves products and, for every object A in $\mathbf{C}$, the functor $\mathfrak{b}_A\colon P(A) \to R(F(A))$ preserves finite meets and

$$\mathfrak{b}_{A\times A}(\delta_A) = R_{\langle F(\mathrm{pr}_1), F(\mathrm{pr}_2)\rangle}(\delta_{F(A)});$$

- **the 2-morphisms** are natural transformations $\theta\colon F \stackrel{\cdot}{\to} G$ such that

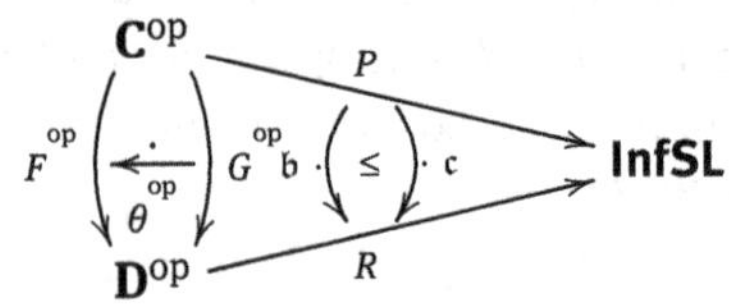

so that, for every A in $\mathbf{C}$ and every α in $P(A)$, one has $\mathfrak{b}_A(\alpha) \leq_{F(A)} R_{\theta_A}(\mathfrak{c}_A(\alpha))$.

Remark 2.4. *Expressing in the internal logic the action of a 1-morphism $(F, \mathfrak{b})$ we have the following translation table—simply writing $(-)^F$ and $(-)^{\mathfrak{b}}$ for the action of F and $\mathfrak{b}$ respectively.*

$$\begin{array}{ccc}
P & \xrightarrow{(F,\mathfrak{b})} & R \\
A \times B & \longmapsto & A^F \times B^F \\
a\colon A & \longmapsto & a\colon A^F \\
(s,t) & \longmapsto & (s^F, t^F) \\
f(s) & \longmapsto & f^F(s^F) \\
\top & \longmapsto & \top \\
\phi \wedge \psi & \longmapsto & \phi^{\mathfrak{b}} \wedge \psi^{\mathfrak{b}} \\
s =_A t & \longmapsto & s^F =_{A^F} t^F
\end{array}$$

In the same vein, a 2-morphism $\theta\colon (F, \mathfrak{b}) \to (G, \mathfrak{c})$ determines a family of terms $(x\colon X^F \mid \theta_X(x)\colon X^G)$, as X varies in $\mathbf{C}$, such that, for every term $(a\colon A \mid t\colon B)$ in $\mathbf{C}$, the terms $(a\colon A^F \mid \theta_B(t^F)\colon B^G)$ and $(a\colon A^F \mid t^G(\theta_A(a))\colon B^G)$ are equal and

$$a\colon A^F \mid \alpha^{\mathfrak{b}}(a) \vdash \alpha^{\mathfrak{c}}(\theta(a))$$

for every A in $\mathbf{C}$ and α in $P(A)$.

Examples 2.5.

1. *Given a theory $\mathcal{T}$ with equality in a first order language $\mathcal{L}$ (say with a single sort), a 1-morphism $(F, \mathfrak{b})\colon LT \to S$ from the elementary doctrine $LT\colon \mathbf{V}^{\mathrm{op}} \longrightarrow$ **InfSL** as in 2.3.4. into $S\colon \mathbf{Set}^{\mathrm{op}} \longrightarrow$ **InfSL**, the elementary doctrine in 2.3.1. with $\mathbf{C} = \mathbf{Set}$, determines a model $\mathfrak{M}$ of $\mathcal{T}$ where the set underlying the interpretation is $F(x_1)$. In fact, there is an equivalence between the category $\mathbf{ED}(LT, S)$ and the category of models of $\mathcal{T}$ and $\mathcal{L}$-homomorphisms.*
2. *Given a category $\mathbf{C}$ with products and pullbacks, one can consider the two indexed inf-semilattices: that of subobjects $S\colon \mathbf{C}^{\mathrm{op}} \longrightarrow$ **InfSL**, and the other $\Psi\colon \mathbf{C}^{\mathrm{op}} \longrightarrow$ **InfSL**, obtained by the poset reflection of each comma category $\mathbf{C}/A$, for A in $\mathbf{C}$. The inclusion of the poset $S(A)$ of subobjects over A into the*

poset reflection of $\mathbf{C}/A$ extend to a 1-morphism from S to Ψ which is an equivalence exactly when every morphism in $\mathbf{C}$ can be factored as a retraction followed by a monic.

The notion of comprehension is related to the characterization of doctrines of subobjects. Though very general, we shall present that notion in the particular case of an elementary doctrine.

Definition 2.6. *Given any elementary doctrine $P\colon \mathbf{C}^{op} \longrightarrow$ **InfSL**, and an object α in some $P(A)$, a **comprehension** of α is a morphism $\{a\colon A \mid \alpha(a)\}\colon X \to A$ in $\mathbf{C}$ such that*

$$x\colon X \mid \top \vdash \alpha(\{a\colon A \mid \alpha(a)\}(x))$$

and, for every $f\colon Z \to A$ such that

$$z\colon Z \mid \top \vdash \alpha(f(z))$$

there is a unique morphism $f'\colon Z \to X$ such that $f = \{a\colon A \mid \alpha(a)\} \circ f'$.[2]

Intuitively, the comprehension morphism represents the subsets of elements in the object A obtained by comprehension with the predicate α.

*In previous works [Maietti and Rosolini, 2013b,a] we said that an elementary doctrine P **has comprehensions** if every α has a comprehension, and that P **has full comprehensions** if, moreover, $\alpha \leq \beta$ in $P(A)$ whenever $\{\alpha\}$ factors through $\{\beta\}$.*

However to view comprehensions as logical constructors following Jacobs [1999], we need to assume that a choice of comprehensions is available in the doctrine as follows.

*An elementary doctrine P **has a choice of comprehensions** if there is a function $\{-\}$ that assigns a comprehension to every object α in the fibre $P(A)$*[3]*, and that $P\colon \mathbf{C}^{op} \longrightarrow$ **InfSL** **has a choice of full comprehensions** if it has a choice comprehensions and, for any $\mathbf{C}$-object A, one has $a\colon A \mid \alpha \vdash \beta$ holds in case there is a factorization*

$$\begin{array}{ccc} X & \longrightarrow & Y \\ & {\scriptstyle \{a:A|\alpha(a)\}} \searrow \quad \swarrow {\scriptstyle \{a:A|\beta(a)\}} & \\ & A & \end{array}$$

Remark 2.7. *The notion of comprehension connects an abstract indexed poset with one of subobjects of the base when this has pullbacks of monomorphisms. We now recall a standard result in the case of interest for us, see e.g. [Jacobs, 1999] where*

2 Hence a comprehension is necessarily monic.

3 It is a function $\{-\} \in \prod_{A \in \mathbf{C}_0} \mathbf{C}_1^{P(A)_0}$.

also a very elegant, more abstract view of comprehensions as right adjoint is considered.

*Let $P\colon \mathbf{C}^{op} \longrightarrow$ **InfSL** be an elementary doctrine, and suppose that P has a choice of comprehensions. Consider the indexed inf-semilattice $S\colon \mathbf{C}^{op} \longrightarrow$ **InfSL** of subobjects of the base category **C**. The function computing comprehensions can be extended to a natural transformation*

$$\mathbf{C}^{op} \underset{S}{\overset{P}{\rightrightarrows}} \mathbf{InfSL} \qquad \{\!|-|\!\}\colon P \Rightarrow S$$

in a unique way. All functors $\{\!|-|\!\}\colon P(A) \to S(A)$ are faithful, and are fully faithful exactly when P has full comprehensions.

From now on when *we simply say that an elementary doctrine P has comprehensions, or full comprehensions, we assume that it has a choice of comprehensions, or full comprehensions just to be able to use the internal categorical logic.*

A special case of comprehensions are the diagonal morphisms and the following definition considers just that possibility.

Definition 2.8. *An elementary doctrine $P\colon \mathbf{C}^{op} \longrightarrow$ **InfSL** **has comprehensive diagonals** if every diagonal morphism $\langle \mathrm{id}_A, \mathrm{id}_A\rangle\colon A \to A \times A$ is a comprehension.*

Lemma 2.9. *Let $P\colon \mathbf{C}^{op} \longrightarrow$ **InfSL** be an elementary doctrine. The following are equivalent:*
1. *P has comprehensive diagonals.*
2. *For every object A in **C**, the diagonal $\langle \mathrm{id}_A, \mathrm{id}_A\rangle\colon A \to A \times A$ is a comprehension of the corresponding fibered equality δ_A.*
3. *For any two morphisms $f, g\colon A \to B$ in **C**, it is*

$$f = g \text{ in } \mathbf{C} \text{ iff } a\colon A \mid \top \vdash f(a) =_B g(a).$$

It is easy to determine a 2-reflection of elementary doctrines into the full 2-subcategory of elementary doctrines with comprehensive diagonals once one notices that the condition

$$a\colon A \mid \top \vdash f(a) =_B g(a)$$

ensures that $P_f = P_g$, *i.e.* that, for every β in $P(B)$, one has that

$$a\colon A \mid \beta(f(a)) \vdash \beta(g(a)) \qquad \text{and} \qquad a\colon A \mid \beta(g(a)) \vdash \beta(f(a)).$$

So the reflection takes an elementary doctrine $P\colon \mathbf{C}^{op} \longrightarrow$ **InfSL** to the elementary doctrine $P_{\mathrm{x}}\colon \mathbf{C}_P^{op} \longrightarrow$ **InfSL**, induced by P on the quotient category $\mathbf{C}_P$ of **C** with respect to the equivalence relation where $f \sim g$ when

$$a\colon A \mid \top \vdash f(a) =_B g(a).$$

We may refer to the doctrine P_{x} as the ***extensional reflection of*** P.

Remark 2.10. *When an elementary doctrine $P\colon \mathbf{C}^{op} \longrightarrow$ **InfSL** has full comprehensions and comprehensive diagonals, the base category $\mathbf{C}$ has equalizers. The equalizer of $A \underset{g}{\overset{f}{\rightrightarrows}} B$ is computed as $\{a\colon A \mid f(a) =_B g(a)\}\colon E \to A$.*

Remark 2.11. *In the situation considered in 2.7, the pair $(\mathrm{id}_{\mathbf{C}}, \{-\})$ is a 1-morphism in* **ED**

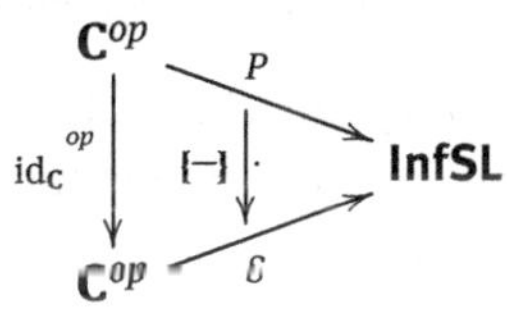

exactly when

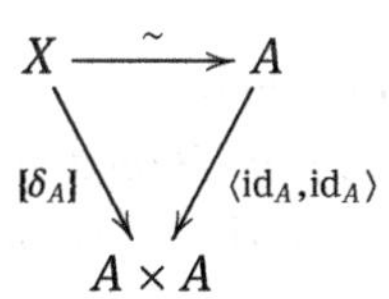

for any objects A in $\mathbf{C}$, i.e. when P has comprehensive diagonals.

Indeed, if P is an elementary doctrine with comprehensions and comprehensive diagonals, then the base category has pullbacks. So P embeds into the indexed poset S of subobjects of $\mathbf{C}$. One can think that comprehensions and comprehsive diagonals force P to "look like" a poset of subobjects of $\mathbf{C}$. Actually, this will be the main situation we shall be interested in the following sections.

With that perspective, we also introduce the notion of existential elementary doctrine from [Lawvere, 1969a], which presents existential quantification as a left adjoint to the substitution functor, see also [Jacobs, 1999].

Definition 2.12. *An elementary doctrine $P\colon \mathbf{C}^{op} \longrightarrow$ **InfSL** is **existential** if, for A_1 and A_2 in $\mathbf{C}$, for a(ny) projection $\mathrm{pr}_i\colon A_1 \times A_2 \to A_i$, $i = 1, 2$, the functor $P_{\mathrm{pr}_i}\colon P(A_i) \to P(A_1 \times A_2)$ has a left adjoint $\exists_{\mathrm{pr}_i}$, to which we shall unimaginatively refer as **existential**, and these left adjoints satisfy the*

- ***Frobenius reciprocity:*** *for $\mathrm{pr}\colon X \to A$ a projection, α in $P(A)$, β in $P(X)$, the canonical morphism $\exists_{\mathrm{pr}}(P_{\mathrm{pr}}(\alpha) \wedge_X \beta) \le \alpha \wedge_A \exists_{\mathrm{pr}}(\beta)$ in $P(A)$ is iso;*

– ***Beck-Chevalley condition:*** *for any pullback diagram*

$$\begin{array}{ccc} X' & \xrightarrow{\mathrm{pr}'} & A' \\ {\scriptstyle f'}\big\downarrow & & \big\downarrow{\scriptstyle f} \\ X & \xrightarrow{\mathrm{pr}} & A \end{array}$$

with pr *a projection (hence also* pr′ *a projection), for any* β *in* $P(X)$*, the canonical morphism* $\exists_{\mathrm{pr}'} P_{f'}(\beta) \leq P_f \exists_{\mathrm{pr}}(\beta)$ *in* $P(A')$ *is iso.*

Definition 2.13. *Let* **CEED** *be the 2-full 2-subcategory of* **ED** *consisting of existential elementary doctrines with comprehensions whose first components F of 1-morphisms* $(F, \mathfrak{b})$ *preserve comprehensions and existential adjoints.*

Similarly to the case of elementary doctrines, [Jacobs, 1999] shows that the deductive calculus associated to existential elementary doctrines is that of the $\wedge$=$\exists$-fragment over type theory with a unit type and a binary product type constructor.

Remark 2.14. *We recall a result of G.M. Kelly in [Kelly, 1992] which is related to the situation considered in 2.7 when the elementary doctrine* $P\colon \mathbf{C}^{\mathrm{op}} \longrightarrow$ **InfSL** *is also existential with full comprehensions; in that case the base category* **C** *has a stable proper factorization system. Recall from [Freyd and Kelly, 1972] that a pair* (**E**, **M**) *of families of morphisms in* **C** *is a factorization system if they satisfy*

– ***factorization:*** *every map in* **C** *factors as* $\begin{array}{ccc} A & \xrightarrow{\quad f \quad} & B \\ & {\scriptstyle e}\searrow \; I \; \nearrow{\scriptstyle m} & \end{array}$ *where e is in* **E** *and m is in* **M***;*

– ***orthogonality:*** *for every commutative diagram* $\begin{array}{ccc} & J & \\ {\scriptstyle f}\nearrow & & \searrow{\scriptstyle m} \\ X & & Y \\ {\scriptstyle e}\searrow & & \nearrow{\scriptstyle g} \\ & K & \end{array}$ *such that the morphism e in in* **E** *and the morphism m is in* **M***, then there is a unique morphism* $\begin{array}{ccc} & J & \\ {\scriptstyle f}\nearrow & \uparrow{\scriptstyle h} & \searrow{\scriptstyle m} \\ X & & Y \\ {\scriptstyle e}\searrow & & \nearrow{\scriptstyle g} \\ & K & \end{array}$ *making the two triangles commute.*

The factorization system is stable if the family **E** *is closed under pullbacks—***M** *is automatically closed under pullbacks, thanks to orthogonality—, and it is proper if* **M** *consists of monos and* **E** *consists of epis.*

In the factorization system determined on **C** *by the doctrine P, the family* **M** *consists of all maps isomorphic to a comprehension and the family* **E** *consists of surjective epis. Given a morphism* $f\colon A \to B$ *in* **C***, a suitable factorization is obtained as follows:*

- *The mono is* $\{b: B \mid \exists x: A.f(x) =_B b\}: I \to B$*, the comprehension of the "image in B of the graph of f".*
- *Then the first factor is obtained by the universal property of a comprehension since*

$$a: A \mid \top \vdash \exists x: A.f(x) =_B f(a)$$

A more general connection between fibrations and not necessarily proper factorization systems than that sketched here is in [Hughes and Jacobs, 2002].

3 Doctrines with effective quotients

We saw how in an indexed poset, a comprehension allows to determine a "subset" of definition in the base category for any predicate represented by an object in the fibre. Similarly an equivalence relation in a fibre can determine a quotient in the base category. For that reason, we recall from [Maietti and Rosolini, 2013b,a] the notion of equivalence relation related to a doctrine and the associated notion of effective quotient.

Definition 3.1. *Given an elementary doctrine* $P: \mathbf{C}^{op} \longrightarrow \mathbf{InfSL}$*, an object A in* **C** *and an object* ρ *in* $P(A \times A)$*, we say that* ρ *is a **P-equivalence relation on** A if it satisfies*
- ***reflexivity:*** $a: A, a': A \mid a =_A a' \vdash \rho(a, a')$;
- ***symmetry:*** $a: A, a': A \mid \rho(a, a') \vdash \rho(a', a)$;
- ***transitivity:*** $a: A, a': A, a'': A \mid \rho(a, a'), \rho(a', a'') \vdash \rho(a, a'')$.

In elementary doctrines as those presented in 2.3, P-equivalence relations concide with the usual notion for those of the form 2.3.1. or 2.3.2.; more interestingly, in cases like 2.3.3. a Ψ-equivalence relation is a pseudo-equivalence relation in **S** in the sense of [Carboni and Celia Magno, 1982].

For $P: \mathbf{C}^{op} \longrightarrow \mathbf{InfSL}$ an elementary doctrine, the object $(a: A, a': A \mid a =_A a')$ is a P-equivalence relation on A. And for a morphism $f: A \to B$ in **C**, the functor $P_{f \times f}: P(B \times B) \to P(A \times A)$ takes a P-equivalence relation σ on B to a P-equivalence relation on A. Hence, the ***P-kernel equivalence of*** $f: A \to B$, the object $(a: A, a': A \mid f(a) =_B f(a'))$ of $P_{A \times A}$ is a P-equivalence relation on A. In such a case, one speaks of $(a: A, a': A \mid f(a) =_B f(a'))$ as an ***effective*** P-equivalence relation.

Definition 3.2. *Let* $P: \mathbf{C}^{op} \longrightarrow \mathbf{InfSL}$ *be an elementary doctrine. Let* ρ *be a P-equivalence relation on A. A **P-quotient of*** ρ*, or simply a **quotient** when the doctrine is clear from the context, is a morphism* $q: A \to A/\rho$ *in* **C** *such that*

$$a; A, a': A \mid \rho(a, a') \vdash q(a) =_{A/\rho} q(a')$$

and, for every morphism $g: A \to Z$ such that

$$a; A, a': A \mid \rho(a, a') \vdash g(a) =_Z g(a'),$$

there is a unique morphism $\overline{g}: A/\rho \to Z$ such that $g = \overline{g} \circ q$.

We say that such a P-quotient is ***stable*** *if, for every pullback diagram*

$$\begin{array}{ccc} B & \xrightarrow{q'} & C \\ {\scriptstyle f'}\downarrow & & \downarrow{\scriptstyle f} \\ A & \xrightarrow[q]{} & A/\rho \end{array}$$

in **C***, the morphism q' is a P-quotient.*

Definition 3.3. *Given an elementary doctrine P:* $\mathbf{C}^{\mathrm{op}} \longrightarrow$ **InfSL** *and a P-equivalence relation ρ on an object A in* **C***, the poset of descent data* $\mathbf{Des}_\rho$ *is the sub-poset of $P(A)$ on those α such that*

$$a: A, b: A \mid \alpha(a), \rho(a, b) \vdash \alpha(b).$$

Remark 3.4. *Given an elementary doctrine P:* $\mathbf{C}^{\mathrm{op}} \longrightarrow$ **InfSL***, for $f: A \to B$ in* **C***, let χ be the P-kernel $P_{f\times f}(\delta_B)$. The functor $P_f: P(B) \to P(A)$ applies $P(B)$ into* $\mathbf{Des}_\chi$.

Definition 3.5. *Given an elementary doctrine P:* $\mathbf{C}^{\mathrm{op}} \longrightarrow$ **InfSL** *and a morphism $f: A \to B$ in* **C***, let χ be the P-kernel $P_{f\times f}(\delta_B)$. The morphism f is* ***descent*** *if the functor $P_f: P(B) \to$* $\mathbf{Des}_\chi$ *is full (it is trivially faithful). The morphism f is* ***effective descent*** *if the functor $P_f: P(B) \to$* $\mathbf{Des}_\chi$ *is an equivalence (hence an isomorphisms of orders).*

Consider the 2-full 2-subcategory **QED** of **ED** whose objects are elementary doctrines P: $\mathbf{C}^{\mathrm{op}} \longrightarrow$ **InfSL** with descent quotients of P-equivalence relations.

The 1-morphisms are those pairs $(F, \mathfrak{b})$ in **ED**

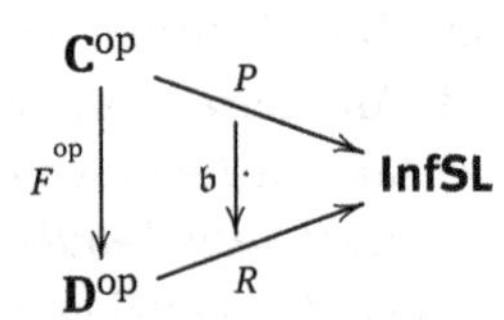

such that F preserves quotients in the sense that, if the morphism $q: A \to C$ in **C** is a quotient of a P-equivalence relation ρ on A, then the morphism $Fq: FA \to FC$ in **D** is a quotient of the R-equivalence relation $R_{\langle F(\mathrm{pr}_1), F(\mathrm{pr}_2)\rangle}(\mathfrak{b}_{A\times A}(\rho))$ on FA.

In [Maietti and Rosolini, 2013a] we proved that it is possible to add freely quotients of equivalence relations to an elementary doctrine. For an elementary doctrine P: $\mathbf{C}^{\mathrm{op}} \longrightarrow$ **InfSL**, the *quotient completion* of P consists of a category $\mathbf{Q}_P$ of "quotients in P", defined as follows:

- **an object of** $\mathbf{Q}_P$ is a pair (A, ρ) such that ρ is a P-equivalence relation on A;
- **a morphism** $f\colon (A, \rho) \to (B, \sigma)$ is a morphism $f\colon A \to B$ in $\mathbf{C}$ such that

$$a\colon A, b\colon A \mid \rho(a, b) \vdash \sigma(f(a), f(b)).$$

Composition is given by that of $\mathbf{C}$, identities are represented by identities of $\mathbf{C}$ and morphism equality is that of $\mathbf{C}$.

The indexed inf-semilattice $\widehat{P}\colon \mathbf{Q}_P^{\mathrm{op}} \longrightarrow \mathbf{InfSL}$ on $\mathbf{Q}_P$ is the categories of descent data: on an object (A, ρ) it is defined as

$$\widehat{P}(A, \rho) := \mathbf{Des}_\rho$$

where the order of descent data $\mathbf{Des}_\rho$ is the sub-order of $P(A)$ on those α such that

$$a\colon A, b\colon A \mid \alpha(a), \rho(a, b) \vdash \alpha(b).$$

There is a 1-morphism $(J, j)\colon P \to \widehat{P}$ in **ED** where $J\colon \mathbf{C}^{\mathrm{op}} \longrightarrow \mathbf{R}_P$ sends an object A in $\mathbf{C}$ to (A, δ_A) and a morphism $f\colon A \to B$ to $f\colon (A, \delta_A) \to (B, \delta_B)$ since $\delta_A \leq_{A\times A} P_{f\times f}(\delta_B)$, and $j_A\colon P(A) \to \widehat{P}(A, \delta_A)$ is the identity since

$$\widehat{P}(A, \delta_A) = \mathbf{Des}_{\delta_A} = P(A).$$

It is immediate to see that J is full and faithful and that (J, j) is a change of base.

Theorem 3.6. *The indexed inf-semilattice* $\widehat{P}\colon \mathbf{Q}_P^{op} \longrightarrow \mathbf{InfSL}$ *is an elementary doctrine with effective quotients, and it is the free such on P in the sense that, for every elementary doctrine* $P\colon \mathbf{C}^{op} \longrightarrow \mathbf{InfSL}$, *pre-composition with the 1-morphism*

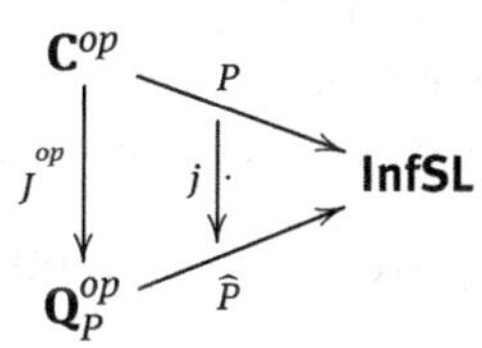

in **ED** *induces an essential equivalence of categories*

$$- \circ (J, j)\colon \mathbf{QED}(\widehat{P}, Z) \equiv \mathbf{ED}(P, Z)$$

for every Z in **QED**.

As noticed in [Maietti and Rosolini, 2013a] the completion with quotients commute with other fibrational structure.

Proposition 3.7. *Let* $P\colon \mathbf{C}^{op} \longrightarrow \mathbf{InfSL}$ *be an elementary doctrine and consider its quotient completion* $\widehat{P}\colon \mathbf{Q}_P^{op} \longrightarrow \mathbf{InfSL}$. *Then the following properties hold.*

1. *If P has comprehensions, then $\widehat{P}$ has comprehensions.*
2. *If P has full comprehensions, then $\widehat{P}$ has full comprehensions and effective quotients are stable.*
3. *If P is existential, then $\widehat{P}$ is existential.*

Note that comprehension is not preserved by the embedding morphism $(J, j): P \to \widehat{P}$ so swiftly (see [Maietti and Rosolini, 2013b,a]).

Proposition 3.8. *Let* $P: \mathbf{C}^{op} \longrightarrow \mathbf{InfSL}$ *be an elementary doctrine with comprehensive diagonals. If P has full comprehensions, then* (J, j) *preserves them.*

Moreover, the property of comprehensive diagonals is not necessarily inherited by the free completion $\widehat{P}$.

Therefore, if one wants to construct from an elementary doctrine $P: \mathbf{C}^{op} \longrightarrow \mathbf{InfSL}$, in a free way, an indexed poset with effective quotients which looks as much as possible like the indexed poset of subobjects of the base category—*i.e.* the indexed poset has full comprehension and comprehensive diagonals—, one must go through a few steps:
1. make sure that P has full comprehensions and comprehensive diagonals;
2. add effective quotients;
3. force comprehensive diagonals.

As for step 1., in case P fails to have comprehensions or comprehensive diagonals, these can be added freely, see *e.g.* [Maietti and Rosolini, 2013a].

Step 2. is just an application of 3.6.

Step 3. is accomplished by the construction after 2.9 once one checks that $\widehat{P}_{\mathrm{x}}$ has effective quotients. Moreover, full comprehensions are inherited by $\widehat{P}_{\mathrm{x}}$, as well as existential quantifiers, if P has them.

In the following sections, we shall analyze a situation which arises when

$$\widehat{P}_{\mathrm{x}}: ((\mathbf{Q}_P)_{\widehat{P}})^{op} \longrightarrow \mathbf{InfSL}$$

is the fibration of subobjects of its base category. We refer to such a doctrine $\widehat{P}_{\mathrm{x}}$ as the ***elementary quotient completion*** of P.

4 The rule of unique choice

In this section we are going to show when an elementary quotient completion of an elementary doctrine happens to be an exact category by making use of a choice principle.

As a first step, it is useful to recall the definition of the exact completion of an existential elementary doctrine introduced in [Maietti and Rosolini, 2015]; we refer the reader to *loc.cit.* for the details, pointing out that there are many sources in the literature which discuss various different presentations of the completion, *e.g.* [Kelly, 1992, Carboni, 1982, Carboni and Walters, 1987, Freyd and Scedrov, 1991, Jacobs, 1999, Hughes and Jacobs, 2002].

For the purpose of the following presentation, it is convenient to introduce a preliminary concept which requires precisely the notion of existential elementary doctrine.

Definition 4.1. *Let $P\colon \mathbf{C}^{op} \longrightarrow$ **InfSL** be an elementary existential doctrine. Let ϕ be an object in $P(A \times B)$. One says that ϕ is **functional from** A **to** B if the following conditions hold:*
- ***ϕ is single-valued:*** $a\colon A, b\colon B, b'\colon B \mid \phi(a,b), \phi(a,b') \vdash b =_B b'$;
- ***ϕ is total:*** $a\colon A \mid \top \vdash \exists b\colon B.\, \phi(a,b)$.

Definition 4.2. *An elementary existential doctrine $P\colon \mathbf{C}^{op} \longrightarrow$ **InfSL** **satisfies the rule of unique choice** if, for pair of objects A and B, and every functional object ϕ from A to B there is a morphism $f\colon A \to B$ in $\mathbf{C}$ such that*

$$a\colon A \mid \top \vdash \phi(a, f(a)).$$

Example 4.3. *The existential elementary doctrine $S\colon \mathbf{Set}^{op} \longrightarrow$ **InfSL** of subsets satisfies the rule of unique choice.*

The reader may appreciate a very close relationship between the concept introduced in 4.2 and that of a relation which is a function. Indeed, like for that one, one can introduce a notion of ***functional composition*** as follows: for ϕ functional from A to B and ψ functional from B to C, it is

$$\exists b\colon B.[\phi(a,b) \wedge \psi(b,c)]$$

with $a\colon A$ and $c\colon C$.

It should now appear clear that, for a given existential elementary doctrine $P\colon \mathbf{C}^{op} \longrightarrow$ **InfSL**, there is a category $\mathbf{F}_P$ of functional morphisms on it whose objects are the objects of $\mathbf{C}$ and where a morphism $\phi\colon A \to B$ is a functional object from A to B. Composition is functional composition; the identity morphism on A is $(a\colon A, a'\colon A \mid a =_A a')$ which is clearly functional from A to A, see [Freyd and Scedrov, 1991].

There is a natural functor $K\colon \mathbf{C} \longrightarrow \mathbf{F}_P$ which is the identity on the objects and maps $f\colon A \to B$ in $\mathbf{C}$ to the functional object $(a\colon A, b\colon B \mid f(a) =_B b)$ in $P(A \times B)$. The functor K is faithful exactly when P has comprehensive diagonals.

As K suggests, the data for $\mathbf{F}_P$ appear as superimposed on $\mathbf{C}$. So, in order to avoid confusion, from now on we shall mark morphisms in $\mathbf{F}_P$ with an asterisk as follows $\phi: A \to_* B$.

Since the initial datum is an existential elementary doctrine, one expects to determine a similar structure on $\mathbf{F}_P$. Indeed, the similarity with the set-theoretic situation continues as one defines the indexed poset P^+ as $P^+(A) := P(A)$ and, for a morphism $\phi: A \to_* B$ in $\mathbf{F}_P$, a homomorphism of inf-semilattices

$$P^+(\phi) := \left[(\, y{:}\, B \mid \alpha(y)\,) \mapsto (\, x{:}\, A \mid \exists y{:}\, B.\, (\phi(x,y) \wedge \alpha(y))\,)\right].$$

The following result follows directly from the definition of the rule of unique choice.

Proposition 4.4. *Let* $P: \mathbf{C}^{op} \longrightarrow$ **InfSL** *be an existential elementary doctrine. P satisfies the rule of unique choice if and only if its extensional reflection* $P_x: \mathbf{C}_P^{op} \longrightarrow$ **InfSL** *does.*

Proof: It is obvious since the fibres of P_x are the same as those of P and there is simply an existence condition on the morphism f in the rule of unique choice. □

Proposition 4.5. *The indexed inf-semilattice* $P^+: \mathbf{F}_P^{op} \longrightarrow$ **InfSL** *is an existential elementary doctrine with comprehensive diagonals which satisfies the rule of unique choice. Moreover*

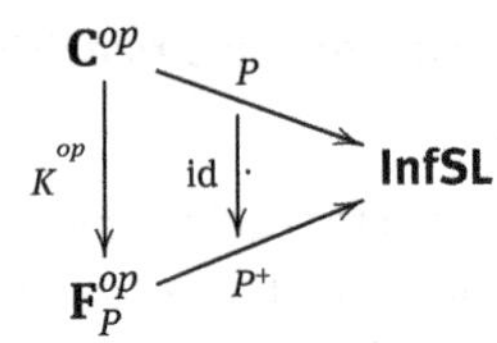

is a 1-morphism in **EED** .

Proof: It is a direct computation after one notices that binary products are induced by K, so the product of A and B is

$$A \xleftarrow{K(\mathrm{pr}_1)} A \times B \xrightarrow{K(\mathrm{pr}_2)} B.$$

□

Remark 4.6. *In fact,* (K, id) *is a 2-reflection of* **EED** *into its full 2-subcategory on the existential elementary doctrines with comprehensive diagonals which satisfies the rule of unique choice. So* (K, id) *is an equivalence exactly when* $P: \mathbf{C}^{op} \longrightarrow$ **InfSL** *has comprehensive diagonals and satisfies the rule of unique choice.*

In fact, K is full if and only if P satisfies the rule of unique choice, and K is faithful if and only if P has comprehensive diagonals.

Note that one obtains the ***exact completion*** $\mathbf{E}_P$ of an elementary existential doctrine P in [Maietti and Rosolini, 2015] as the doctrine $\mathbf{F}_{\widehat{P}}$.

Remark 4.7. *It is easy to recognize that examples of exact completion for the definition given here is the exact completion* $\mathbf{X}_{ex/reg}$ *of a regular category* $\mathbf{X}$*, see [Freyd and Scedrov, 1991, Carboni, 1995, Carboni and Vitale, 1998], by taking the subobject fibration of* $\mathbf{X}$ *as* P.

Other examples come from theories apt to formalize constructive mathematics: the category of total setoids à la Bishop and functional relations based on the intensional type theory of the Minimalist Foundation in [Maietti, 2009], or the category of total setoids à la Bishop and functional relations based on the Calculus of Constructions [Coquand, 1990], which forms a topos as mentioned in [Barthe et al., 2003].

Remark 4.8. *Note that an existential elementary doctrine* $P\colon \mathbf{C}^{op} \longrightarrow$ **InfSL** *with comprehensions has unique choice in the sense of [Jacobs, 1999] if and only if it satisfies the rule of unique choice. So, it follows from 4.4.4 and 4.9.4 of loc.cit. that the following are equivalent.*
1. $\mathbf{C}$ *is a regular category and* P *is the doctrine of its subobject.*
2. P *has full comprehensions, comprehensive diagonals and satisfies the rule of unique choice.*

Theorem 4.9. *Let* $P\colon \mathbf{C}^{op} \longrightarrow$ **InfSL** *be an existential elementary doctrine with stable effective quotients. The following are equivalent.*
1. $\mathbf{C}$ *is an exact category and* P *is the doctrine of its subobjects.*
2. P *has full comprehensions, comprehensive diagonals and satisfies the rule of unique choice.*

Proof: It follows from 4.8 because, for a subobject fibration $S\colon \mathbf{C}^{op} \longrightarrow$ **InfSL**, an S-equivalence relation is precisely an equivalence relation in $\mathbf{C}$ and an exact category is a regular category with effective coequalizers of equivalence relations. □

Recall from [Maietti and Rosolini, 2015] the following rule of choice.

Definition 4.10. *Given an existential elementary doctrine* $P\colon \mathbf{C}^{op} \longrightarrow$ **InfSL** *we say that* P ***satisfies the rule of choice*** *if, for every* $\phi \in P(A \times B)$ *such that*

$$a\colon A \mid \top \vdash \exists b\colon B.\phi(a, b),$$

there is a morphism $f\colon A \to B$ *in* $\mathbf{C}$ *such that*

$$a\colon A \mid \top \vdash \phi(a, f(a)).$$

Clearly the rule of choice is stronger than the rule of unique choice since the relations employed in the rule of unique choice are particular instances of those which appear in the rule of choice.

Proposition 4.11. *Let $P\colon \mathbf{C}^{op} \longrightarrow$ **InfSL** be an existential elementary doctrine with comprehensions. Then P satisfies the rule of choice if and only if its quotient completion $\widehat{P}\colon \mathbf{Q}_P^{op} \longrightarrow$ **InfSL** satisfies the rule of unique choice.*

Proof: ($\Rightarrow$) Immediate.

($\Leftarrow$) Suppose $\phi \in P(A \times B)$ is such that

$$a\colon A \mid \top \vdash \exists b\colon B.\phi(a, b).$$

Consider the comprehension $\ell := \{a\colon A, b\colon B \mid \phi(a, b)\}\colon X \to A \times B$ and consider the object

$$(X, (x\colon X, x'\colon X \mid \mathrm{pr}_1(\ell(x)) =_A \mathrm{pr}_1(\ell(x'))))$$

in $\mathbf{Q}_P$. It is immediate to check that $(a\colon A, x\colon X \mid a =_A \mathrm{pr}_1(\ell(x)))$ is in

$$\widehat{P}((A, (a\colon A, a'\colon A \mid a =_A a')) \times (X, (x\colon X, x'\colon X \mid \mathrm{pr}_1(\ell(x)) =_A \mathrm{pr}_1(\ell(x')))))$$

since it is just

$$a\colon A, a'\colon A, x\colon X, x'\colon X \mid a =_A a', \mathrm{pr}_1(\ell(x)) =_A \mathrm{pr}_1(\ell(x')), a =_A \mathrm{pr}_1(\ell(x)) \vdash a' =_A \mathrm{pr}_1(\ell(x')).$$

It is also easy to prove that $(a\colon A, x\colon X \mid a =_A \mathrm{pr}_1(\ell(x)))$ is functional from $(A, (a\colon A, a'\colon A \mid a =_A a'))$ to $(X, (x\colon X, x'\colon X \mid \mathrm{pr}_1(\ell(x)) =_A \mathrm{pr}_1(\ell(x'))))$ since the existential quantifiers of $\widehat{P}$ are those of P as shown in [Maietti and Rosolini, 2013b].

Thanks to the assumption that $\widehat{P}$ satisfies the rule of unique choice, there is $f\colon A \to X$ such that

$$a\colon A \mid \top \vdash a =_A \mathrm{pr}_1(\ell(f(a))).$$

Hence the morphism $\mathrm{pr}_2 \circ \ell \circ f\colon A \to B$ is such that $a\colon A \mid \top \vdash \phi(a, \mathrm{pr}_2(\ell(f(a))))$. □

5 The rule of choice and exact completions

Lemma 5.1. *Let $P\colon \mathbf{C}^{op} \longrightarrow$ **InfSL** be an existential elementary doctrine, and consider the 1-morphism*

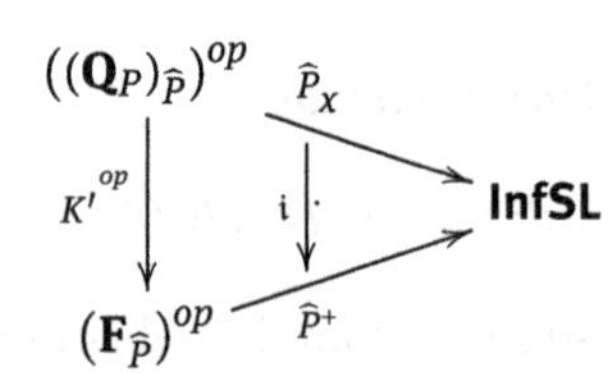

induced by the 1-morphism $(K, \mathrm{id})\colon \widehat{P} \to \mathbf{F}_{\widehat{P}}$. The functor K' is full (hence iso) if and only if P satisfies the rule of choice.

Proof: It follows from 4.4 and 4.11. □

Theorem 5.2. *Let $P\colon \mathbf{C}^{op} \longrightarrow$ **InfSL** be an existential elementary doctrine with comprehensive diagonals, full comprehensions and let $\widehat{P}_x\colon ((\mathbf{Q}_P)_{\widehat{P}})^{op} \longrightarrow$ **InfSL** be its elementary quotient completion. The following are equivalent.*

1. *the base category $(\mathbf{Q}_P)_{\widehat{P}}$ of the elementary quotient completion of P is an exact category and $\widehat{P}_x$ is its subobject doctrine.*
2. *P satisfies the rule of choice.*
3. *the base category $(\mathbf{Q}_P)_{\widehat{P}}$ of the elementary quotient completion of P is equivalent to the exact completion $\mathbf{E}_P \equiv \mathbf{F}_{\widehat{P}}$ of P and $\widehat{P}_x$ is a subobject doctrine.*

Proof: It follows from 5.1 and 4.9. □

Remark 5.1. *Note that proposition 4.11 and theorem 5.2 hold also if P has just weak full comprehensions.*

Examples 5.3. *The setoid model over Martin-Löf's type theory in [Maietti and Rosolini, 2013b] and the elementary quotient completion of the doctrine of a lex category (with a choice of its structure) provide examples of elementary quotient completions which are exact since the rule of unique choice holds in the logic of the quotient model.*

6 The axiom of unique choice in an elementary quotient completion

Proposition 4.11 continues to hold if we replace the rule of choice with the axiom of choice and the rule of unique choice with the axiom of unique choice respectively, provided that the doctrine P has implications and universal quantifications and its base category is closed under (weak) function spaces. We recall the basic notions to state that result.

Definition 6.1. *An elementary doctrine $P\colon \mathbf{C}^{op} \longrightarrow$ **InfSL** is **implicational** if, for every object A in $\mathbf{C}$, every α in $P(A)$, the functor $\alpha \wedge -\colon P(A) \to P(A)$ has a right adjoint $\alpha \Rightarrow -\colon P(A) \to P(A)$.*

Definition 6.2. *An elementary doctrine $P\colon \mathbf{C}^{op} \longrightarrow$ **InfSL** is **universal** if, for A_1 and A_2 in $\mathbf{C}$ and a(ny) projection $\mathrm{pr}_i\colon A_1 \times A_2 \to A_i$, $i = 1, 2$, the functor $P_{\mathrm{pr}_i}\colon P(A_i) \to P(A_1 \times A_2)$ has a right adjoint $\forall_{\mathrm{pr}_i}$, and these satisfy the Beck-Chevalley condition:*

for any pullback diagram

$$\begin{array}{ccc} X' & \xrightarrow{\text{pr}'} & A' \\ {\scriptstyle f'}\downarrow & & \downarrow{\scriptstyle f} \\ X & \xrightarrow{\text{pr}} & A \end{array}$$

with pr *a projection (hence also* pr′ *a projection), for any* β *in* $P(X)$*, the canonical arrow* $P_f \forall_{\text{pr}}(\beta) \leq \forall_{\text{pr}'} P_{f'}(\beta)$ *in* $P(A')$ *is iso.*

Definition 6.3. *An elementary, existential, implicational and universal doctrine* $P\colon \mathbf{C}^{op} \longrightarrow$ **InfSL** *with a cartesian closed base* ***satisfies the axiom of unique choice,*** *(AUC) for short, if*

$$(\forall x\colon A.\exists! y\colon B.\phi(x,y)) \to \big(\exists f\colon B^A.\forall x\colon A.\phi(x,f(x))\big)$$

holds in P.

As already noticed in [Maietti and Rosolini, 2013b] the completion with quotients commute also with the fibrational structure described above.

Proposition 6.4. *Let* $P\colon \mathbf{C}^{op} \longrightarrow$ **InfSL** *be an elementary doctrine with full comprehensions and let* $\widehat{P}\colon \mathbf{Q}_P^{op} \longrightarrow$ **InfSL** *be its quotient completion. Then the following properties hold.*

1. *If* $\mathbf{C}$ *is cartesian closed then* $\mathbf{Q}_P$ *is cartesian closed.*
2. *If P is implicational, then* $\widehat{P}$ *is implicational.*
3. *If P is universal, then* $\widehat{P}$ *is universal.*

Moreover, the 1-arrow $(J,j)\colon P \to \widehat{P}$ *preserves the implicational and universal structure of P.*

Finally, we have:

Proposition 6.5. *Let* $P\colon \mathbf{C}^{op} \longrightarrow$ **InfSL** *be an elementary existential implicational and universal doctrine with a cartesian closed base and full comprehensions. Then P satisfies the axiom of choice if and only if its quotient completion* $\widehat{P}\colon \mathbf{Q}_P^{op} \longrightarrow$ **InfSL** *satisfies the axiom of unique choice.*

Proof: The proof is similar to that of proposition 4.11.

($\Rightarrow$) Immediate by the definition of right adjoints and implications in $\widehat{P}$.

($\Leftarrow$) Suppose $\phi \in P(A \times B)$ is P-relation. Consider the comprehension $\ell := \{a\colon A, b\colon B \mid \phi(a,b)\}\colon X \to A \times B$ and the relation

$$(a\colon A, x\colon X \mid a =_A \text{pr}_1(\ell(x)))$$

which is in

$$\widehat{P}((A,(a\colon A, a'\colon A \mid a =_A a')) \times (X,(x\colon X, x'\colon X \mid \text{pr}_1(\ell(x)) =_A \text{pr}_1(\ell(x')))))$$

Then, (AUC) holds in $\widehat{P}$ for that $\widehat{P}$-relation and that yields that (AC) holds for ϕ in P since the existential quantifiers, the universal quantifiers and implications of $\widehat{P}$ are defined in terms of those of P as shown in [Maietti and Rosolini, 2013b]. □

Remark 6.6. *Observe that, as noticed in [Maietti, 2005], the subobject doctrine of a locally cartesian closed regular category is implicational, universal and validates the axiom of unique choice.*

Remark 6.7. *Note that proposition 6.4 and theorem 6.5 hold also if P has just weak full comprehensions and its base is only weakly cartesian closed.*

An example of this structure is the setoid model over Martin-Löf's type theory of example 5.3. A discussion on the validity of choice principles in such a structure can be found in [Martin-Löf, 2006]. Proposition 6.5 gives an abstract account of what happens in this example.

References

G. Barthe, V. Capretta, and O. Pons. Setoids in type theory. *J. Funct. Programming*, 13(2):261–293, 2003.

A. Carboni. Analysis non-standard e topos. *Rend. Istit. Mat. Univ. Trieste*, 14(1-2):1–16, 1982. ISSN 0049-4704.

A. Carboni. Some free constructions in realizability and proof theory. *J. Pure Appl. Algebra*, 103:117–148, 1995.

A. Carboni and R. Celia Magno. The free exact category on a left exact one. *J. Aust. Math. Soc.*, 33(A):295–301, 1982.

A. Carboni and E.M. Vitale. Regular and exact completions. *J. Pure Appl. Algebra*, 125:79–117, 1998.

A. Carboni and R.F.C. Walters. Cartesian bicategories, I. *J. Pure Appl. Algebra*, 49:11–32, 1987.

T. Coquand. Metamathematical investigation of a calculus of constructions. In P. Odifreddi, editor, *Logic in Computer Science*, pages 91–122. Academic Press, 1990.

P.J. Freyd and G.M. Kelly. Categories of continuous functors I. *J. Pure Appl. Algebra*, 2:169–191, 1972.

P.J. Freyd and A. Scedrov. *Categories Allegories*. North Holland Publishing Company, 1991.

M. Grandis. Weak subobjects and the epi-monic completion of a category. *J. Pure Appl. Algebra*, 154(1-3):193–212, 2000.

J. Hughes and B. Jacobs. Factorization systems and fibrations: Toward a fibred Birkhoff variety theorem. *Electron. Notes Theor. Comput. Sci.*, 11:#, 2002. URL http://www.elsevier.nl/locate/entcs/volume6.html.

J.M.E. Hyland. The effective topos. In A.S. Troelstra and D. van Dalen, editors, *The L.E.J. Brouwer Centenary Symposium*, pages 165–216. North Holland Publishing Company, 1982.

J.M.E. Hyland, P.T. Johnstone, and A.M. Pitts. Tripos Theory. *Math. Proc. Camb. Phil. Soc.*, 88: 205–232, 1980.

B. Jacobs. *Categorical Logic and Type Theory*. North Holland Publishing Company, 1999.

G.M. Kelly. A note on relations relative to a factorization system. In A. Carboni, M.C. Pedicchio, and G. Rosolini, editors, *Category Theory '90*, volume 1488 of *Lecture Notes in Math.*, pages 249–261. Springer-Verlag, Como, 1992.

J. Lambek and P.J. Scott. *Introduction to Higher Order Categorical Logic*. Cambridge University Press, 1986.

F. W. Lawvere. Adjointness in foundations. *Dialectica*, 23:281–296, 1969a.

F.W. Lawvere. Diagonal arguments and cartesian closed categories. In *Category Theory, Homology Theory and their Applications, II (Battelle Institute Conference, Seattle, Wash., 1968, Vol. Two)*, pages 134–145. Springer, 1969b.

F.W. Lawvere. Equality in hyperdoctrines and comprehension schema as an adjoint functor. In A. Heller, editor, *Proc. New York Symposium on Application of Categorical Algebra*, pages 1–14. Amer.Math.Soc., 1970.

F.W. Lawvere and R. Rosebrugh. *Sets for Mathematics*. Cambridge University Press, 2003.

Maria Emilia Maietti. Joyal's arithmetic universe as list-arithmetic pretopos. *Theory Appl. Categ.*, 3:No. 24, 39–83, 2010.

M.E. Maietti. Modular correspondence between dependent type theories and categorical universes including pretopoi and topoi. *Math. Structures Comput. Sci.*, 15(6):1089–1149, 2005.

M.E. Maietti. A minimalist two-level foundation for constructive mathematics. *Ann. Pure Appl. Logic*, 160(3):319–354, 2009.

M.E. Maietti and G. Rosolini. Elementary quotient completion. *Theory Appl. Categ.*, 27:445–463, 2013a.

M.E. Maietti and G. Rosolini. Quotient completion for the foundation of constructive mathematics. *Log. Univers.*, pages 371–402, 2013b.

M.E. Maietti and G. Rosolini. Unifying exact completions. *Appl. Categ. Structures*, 23:43–52, 2015.

M. Makkai and G. Reyes. *First Order Categorical Logic*, volume 611 of *Lecture Notes in Math.* Springer-Verlag, 1977.

P. Martin-Löf. 100 years of Zermelo's axiom of choice: what was the problem with it? *The Computer Journal*, 49(3):10–37, 2006.

P. Taylor. *Practical Foundations of Mathematics*. Cambridge University Press, 1999. URL ftp://ftp.dcs.qmw.ac.uk/pub/lfp/pt/book-contents.

J. van Oosten. *Realizability: An Introduction to its Categorical Side*, volume 152. North Holland Publishing Company, 2008.

Roman Murawski

Some Historical, Philosophical and Methodological Remarks on Proof in Mathematics

Abstract: The paper is devoted to historical and philosophical as well as methodological considerations on the role and meaning of proof in mathematics. The role of informal proofs in mathematical research practice is discussed. Next the concept of formal proof is considered. The problem of the distinction: (formal) provability vs. truth is presented. Relations between informal and formal proofs are discussed.

Keywords: Formal proof, Informal proof, Truth, Mathematical knowledge, Axiomatic method

Mathematics Subject Classification 2010: 00A30, 03A05, 00A35

1 Introduction

Proofs play an important role in mathematics and its methodology (in the context of justification). They form the main method of justifying mathematical statements. Only statements that have been proved can be treated as belonging to the corpus of mathematical knowledge. Proofs are used to convince the readers of the truth of presented theorems. But what is in fact a proof? In mathematical research practice proof is a sequence of arguments that should show the truth of the claim. Of course the particular arguments used in a proof depend on the situation, on the audience, on the type of a claim, etc. Hence a concept of a proof has in fact a cultural, psychological and historical character. In practice mathematicians generally agree whether a given argumentation is or is not a proof. More difficult is the task to define a proof as such. Beside proofs used in the research practice there is a concept of a formal proof developed by logic. What are the relations between them? What roles do they play in mathematics?

Problems of that type will be considered in the paper. We start (Section 1) by some historical remarks showing in what circumstance the idea of a proof (informal and formal) appeared. Next (Section 2) the features and role played by informal proofs will be considered. The subject of Section 3 will be formal proofs and

Roman Murawski: Adam Mickiewicz University, Faculty of Mathematics and Computer Science, ul. Umultowska 87, 61–614 Poznań, Poland, e-mail: rmur@amu.edu.pl

their relation to the concept of truth. In the closing Section 4 some conclusions will be made and a thesis (similar to Church-Turing Thesis in the computation theory) formulated.

2 Historical remarks

The model of mathematics as a science, its paradigm functioning in fact till today was formulated and developed in the ancient Greece about the 4th century B.C. Earlier, for example in ancient Egypt or Babylon mathematics consisted of practical procedures that should help to solve everyday problems such as measuring surface area or the amount of cereal in a granary or oil in a jug. In those both pre-Greek mathematics – though they were advanced and sophisticated (especially the Babylonian mathematics) – there was no need to prove statements. In fact there were no general statements and no attempts were undertaken to deduce the results or to explain their validity. One was satisfied by instructions what should be done in order to receive the result or to perform the required task. In fact mathematics there was a collection of separate algorithms (as one would say today) and resembled in certain sense informatics (though without sophisticated technical equipment). Similar was the situation in China – Chinese mathematics was also a collection of procedures (transferred from generation to generation).

Proofs as deduction from explicitly stated postulates was conceived by the Greeks. It was connected with the axiomatic method. Since Plato, Aristotle and Euclid the axiomatic method was considered as the best method to justify and to organize mathematical knowledge. The first mature and most representative example of its usage in mathematics were the *Elements* of Euclid. They established a pattern of a scientific theory and in particular a paradigm in mathematics. Since Euclid till the end of the 19th century mathematics was developed as an axiomatic – in fact rather a quasi-axiomatic – theory based on axioms, postulates and definitions. Axioms were principles common to all sciences, postulates – specific principles taken for granted by a mathematician engaged in the demonstration of theorems in a particular domain. Definitions should provide meaning to new notions – in practice definitions were rather explanations of notions than proper definitions in the strict sense, moreover, they were explanations in the unprecise everyday colloquial language. Note that the language of a theory was not separated from the natural language. Proofs of theorems contained several gaps – in fact the lists of axioms and postulates were not complete, one freely used in proofs various “obvious” truths or referred to the intuition. Consequently proofs were only partially based on axioms and postulates. In fact proofs were informal and intuitive, they were rather demonstrations and the very concept of a proof was of a psychological and sociological (and not of a logical) nature.

Note that the language of theories was simply the unprecise colloquial language. Till the end of the 19th century mathematicians were convinced that axioms and postulates should be true statements, hence sentences describing the real state of affairs (in the mathematical reality).[1] It seems to be connected with Aristotle's view that a proposition is demonstrated (proved to be true) by showing that it is a logical consequence of propositions already known to be true. Demonstration was conceived here of as a deduction whose premises are known to be true and a deduction was conceived of as a chaining of immediate inferences.

It should be noted that the Euclid's approach (connected with Platonic idealism) to the problem of the development of mathematics and the justification of its statements (which found its fulfilment in the Euclidean paradigm), i.e. justification by deduction (by proofs) from explicitly stated axioms and postulates, was not the only approach and method which was used in the ancient Greek (and later). The other one (call it heuristic) was connected with Democritus' materialism. It was applied for example by Archimedes who used in his mathematical works not only deduction but any methods, such as intuition or even experiments (not only mental ones) to solve problems. One can see this, for example, in his considerations concerning the calculation of the volume of a sphere using cylinder with two excavated cones or in his quadrature of the parable.

Though the Euclidean approach won and dominated in the history, one should note that it formed rather an ideal and not the real scientific practice of mathematicians. In fact rigorous, deductive mathematics was rather a rare phenomenon. On the contrary, intuition and heuristic reasoning were the animating forces of mathematical research practice. The vigorous but rarely rigorous mathematical activity produced "crises" (for example the Pythagoreans' discovery of the incommensurability of the diagonal and the side of a square, Leibniz's and Newton's problems with the explanation of the nature of infinitesimals, Fourier's "proof" that any function is representable in a Fourier series, antinomies connected with Cantor's imprecise and intuitive notion of a set).

New elements appeared in the 19th century with the trend whose aim was the clarification of basic mathematical concepts, especially those of analysis (cf. works by Cauchy, Weierstrass, Bolzano, Dedekind). Still another factor was the discovery of antinomies, in particular in set theory (C. Burali-Forte, G. Cantor,

---

1 In fact the discussion about the truth of mathematical statements started already with the discovery of non-Euclidean geometries but the conviction of the truth of mathematical theses dominated.

B. Russell) and of the semantical antinomies (G.D. Berry, K. Grelling). They forced the revision of some basic ideas of mathematics. Among formulated proposals was the foundational program of David Hilbert and his *Beweistheorie*. Note that "this program was never intended as a comprehensive philosophy of mathematics; its purpose was instead to legitimate the entire corpus of mathematical knowledge" [Rowe, 1989, p. 200].

The main role in Hilbert's program was played by formal (or formalized) proofs. This device was introduced on the basis of (and thanks to) the mathematical logic developed in the 19th century, in particular on the basis of the work of Gottlob Frege who constructed the first formalized system – it was the system of propositional calculus based on two connectives: negation and implication. Investigations carried out in the framework of Hilbert's program established the new scientific discipline, i.e., the metamathematics.

Recall that the concept of a formal proof must be related to a given formal theory. To express such a theory one should at the beginning fix its language. The rules of forming formulas in it should have strictly formal and syntactic character – only the shape of symbols can be taken into account and one should entirely abstract from their possible meaning or interpretation. Next one fixes axioms (logical, non-logical and usually identity axioms) and rules of inference. The latter must have entirely syntactic and formal character. A formal proof of a statement (formula) φ is now a finite sequence of formulas in the given language $\varphi_1, \varphi_2, \ldots, \varphi_n$ such that the last member of the sequence is the formula φ and all members of it either belong to the set of presumed axioms or are consequences of previous members of the sequence according to one of the accepted rules of inference. Observe that this concept of a formal proof has a syntactic character and does not refer to any semantical notions such as meaning or interpretation.

As a result of the described development one has to deal nowadays in mathematics with at least two concepts of a proof: the formal one, used mainly by logicians and specialists in metamathematics in the foundational studies as well as by computer scientists (cf. modern theorem provers) and on the other hand the "normal", "usual" concept of a proof used in mathematical research practice. What are the relations between them? Is the metamathematical concept a precise explication of the "everyday" concept? Do they play similar roles in mathematics or not? To what extent does the concept of a formal proof reflect the main features and the nature of an informal proof used by mathematicians in their researches?

3 Informal proofs and their role

Mathematics was and is developed in an informal way using intuition and heuristic reasonings – it is still developed in fact in the spirit of Euclid (or sometimes of Archimedes) in a quasi-axiomatic way. Moreover, informal reasonings appear not only in the context of discovery but also in the context of justification. Any correct methods are allowed to justify statements. But what does it mean "correct"? In the research practice this was and is decided by the community of mathematicians. Consequently the criteria of being correct have been changing in the history and in the process of developing mathematics. The concept of proof was and is in fact not an absolute notion but it was and is culturally and sociologically dependent and motivated, it had and has a cultural and sociological components. The main aim of a mathematician is always to convince the audience that the given result is justified, correct and true (the latter concept being used in an intuitive and vague way) and not to answer the question whether it can be deduced from stated axioms. In the research practice a proof is in fact an argumentation that should indicate the correctness of a claimed thesis – in particular its form depends on and is relative to a background knowledge of those to whom the proof is presented. And here appears a psychological moment in the understanding and functioning of proofs.

The ultimate aim of mathematics is "to provide correct proofs of true theorems" [Avigad, 2006, p. 105]. In their research practice mathematicians usually do not distinguish concepts "true" and "provable" and often replace them by each other. Mathematicians used to say that a given theorem holds or that it is true and not that it is provable in such and such theory. In fact they do not distinguish concepts 'true' and 'provable' and often replace them by each other Add that axioms of theories being developed are not always precisely formulated and admissible methods precisely described.

Proofs play various roles in the mathematical research practice. One can distinguish (cf. [De Villiers, 1999, CadwalladerOlsker, 2011]) among others the role of:

(1) verification,
(2) explanation,
(3) systematization,
(4) discovery,
(5) intellectual challenge,
(6) communication,
(7) justification of definitions.

The most familiar to research mathematicians is the role of verification. A statement can be treated as belonging to the body of mathematics only when it has been *verified*. The proof should not only show that a given sentence is true and holds but should also *explain* why it is true and holds. This role explains why mathematicians are often looking for new proofs of known theorems – new proofs should have more explanatory power. The role of systematization was exemplified already by Euclid's *Elements*. In this work many theorems known to Greeks have been collected and organized in such a way that they followed from axioms, postulates, definitions and previously proved theorems. It was shown in this way that the accepted axioms, postulates and definitions form a sufficient base on which the whole edifice of mathematics can be developed. Note that the role of discovery may be – *prima facie* – rather seldom associated with proofs but it is not excluded. In fact, e.g., non-Euclidean geometries were arrived at through purely deductive means. Recall that since Euclid one asked the question whether the fifth postulate on parallels formulated in *Elements* is independent of other axioms and postulates or can be deduced from them. After several attempts undertaken through the centuries it has been shown in the 19th century that it is reasonable to consider systems of geometry in which the negation of the fifth postulate is assumed instead of the fifth postulate itself and it is possible to show that such systems are consistent.

Finding proofs is the intellectual challenge for mathematicians: there is a theorem – we want to prove it. Proofs serve in the community of mathematicians as communication means. They communicate not only the reasons why a given statement is a true theorem but introduce also new methods which can be used sometimes in other domains. Proofs can provide also justification of definitions.

The most important roles played by proofs in the research mathematical practice are of course verification and explanation. Note that a proof that verifies a theorem does not have to explain why it holds. One can distinguish between proofs that convince and proofs that explain. The former should show that a statement holds or is true and can be accepted, the latter – why it is so. Of course there are proofs that both convince and explain. The explanatory proof should give an insight in the matter whereas the convincing one should be concise or general.

One can distinguish also between explanation and understanding. Mathematicians often treat simplicity as a characteristic feature of understanding. Observe that, as G.-C. Rota writes in [1997, p. 192]:

> [i]t is an article of faith among mathematicians that after a new theorem is discovered, other, simpler proof of it will be given until a definitive proof is found.

In this context it is worth distinguishing between unveiling proofs and consolidating ones. The former one is "a proof which proves a theorem which was unknown before" and the latter is "a proof of a theorem which is already known to be true" [Kahle, 2015]. Proofs of the second type do not contribute to the truth of a theorem – they consolidate our knowledge of this truth. Such a proof can be quite different from the original one. Aschbacher wrote about this phenomenon in the following way [2005, p. 2403]:

> The first proof of a theorem is usually relatively complicated and unpleasant. But if the result is sufficiently important, new approaches replace and refine the original proof, usually by embedding it in a more sophisticated conceptual context, until the theorem eventually comes to be viewed as an obvious corollary of a larger theoretical construct. Thus proofs are a means for establishing what is real and what is not, but also a vehicle for arriving at a deeper understanding of mathematical reality.

There are many examples from the history of mathematics that confirm this. In this context one can mention Paul Erdős' idea of proofs from The Book in which God maintains the perfect proofs for mathematical theorems – he followed the dictum of G.H. Hardy that there is no permanent place for ugly mathematics. Erdős stressed that one need not believe in God but, as a mathematician, one should believe in The Book.[2]

Note that a proof that convinces can be more (or even quite) formal. Explanatory proofs cannot be strictly formal. Mathematicians set a high value on explanatory proofs. Such a proof is more valued when "it exhibits methods that are powerful and informative" [Avigad, 2006, p. 106]. Hersh says in [1997, p. 60] that "[p]roof is a tool in service of research, not a shackle on the mathematician's imagination".[3]

Sometimes theorems are verified in mathematics by checking all particular cases but this usually does not give an explanation why the theorem holds. The explanation should give a general principle by which the theorem holds. A famous example of a theorem verified by checking cases but not giving reasons is the Four-Color Theorem proved by Appel and Haken and stating that every planar graph is four-colorable, i.e., in another words, that four colors suffice to color every map on the plane in such a way that two regions receive different colors whenever they have a common border.

---

2 In 1998 Springer Verlag published *Proofs From The Book* by M. Aigner and G.M. Ziegler (cf. [Aigner and Ziegler, 1998]). It is treated by the authors as "a first (and very modest) approximation of The Book" (Preface).

3 On the concept of explanation in mathematics see for example [Steiner, 1978] and [Mancosu, 2000] and [Mancosu, 2001].

The example of the Four-Color Theorem indicates other features of proofs in mathematics. Observe that the first purported proof of it given by Kempe in 1879 was accepted for a decade before it was found to be incorrect. This was neither the first nor the last example of such a situation. It means that the community of mathematicians can be fallible.

The Four-Color Theorem opened eyes of philosophers of mathematics to the problem of methods acceptable in a proof or in a verification of cases. The unique known proof – that by Appel and Haken – was obtained by using computer and no traditional proof (without computer) is known so far. Moreover the existing proof cannot be made by a human being because an essential part of it was a computation requiring about 1200 hours of computer time and is beyond the capacities of any mathematician. This initiated a discussion concerning the admissibility of experimental methods in mathematical proofs. Several arguments pro and contra have been formulated – we will not enter here the details of the discussion. Let us say only that the usage of a technical tool (like a computer) seems to refute the commonly accepted thesis that mathematical knowledge is *a priori*. There arises also a question whether a computer-aided proof is (or can be treated as) a mathematical proof and consequently, whether in particular the Four-Color Theorem can be called "theorem" or it is still rather a hypothesis.

In this debate initiated by a paper by Tymoczko [1979] a question was asked what are in fact the characteristic features of a "normal" mathematical proof. Tymoczko says that a proof in mathematics should be:
(1) convincing,
(2) surveyable,
(3) formalizable.

The first feature is – as Tymoczko says – of an "anthropological" character, the other two are treated by him as "deep features". He claims also that "surveyability and formalizability can be seen as two sides of the same coin" [Tymoczko, 1979, p. 61]. Formalizability "idealizes surveyability, analyzes it into finite reiterations of surveyable patters" (*ibidem*). It can be assumed that all surveyable proofs are formalizable. Are also all formalizable proofs surveyable? Tymoczko answers this negatively: "[w]e know that there must exist formal proofs that cannot be surveyed by mathematicians if only because the proofs are too long or involve formulas that are too long" and the phrase "too long" means here "can't be read over by a mathematician in a human life time" (*ibidem*). On the other hand one should observe that "it is not at all obvious that mathematicians could come across formal proofs and recognize them as such without being able to survey them" [Tymoczko, 1979, p. 62].

Considering surveyability one should distinguish local and global surveyability. Bassler in [2006, p. 100] characterizes them in the following way:

> local surveyability requires the surveying of each of the individual steps in a proof in some order, while global surveyability requires the surveying of the entire proof as a comprehensive whole.

Hence a local surveyability does not mean that a proof is practically surveyable. One can say that the proof of the Four-Color Theorem is globally surveyable without being locally surveyable (provided one is willing to countenance a distinction between proof and calculation). On the other hand if one accepts the assumption that global surveyability receives its foundation in local surveyability then this statement is false. Add that one should also distinguish the surveyability of a proof and the fact that it can be (formally) checked on the one hand and the fact that it gives an understanding, that it reveals the deep reasons for the theorem being proved.

The concept of surveyability is not precise enough. In the 20th century there was a trend to link surveyability with the development of formal and complete foundations of mathematics and formalization was treated as a method providing the local surveyability. The works of Frege, Russell and their followers, especially Hilbert, were guided by the desire to find a perspicuous syntactic representation of the relations of semantic content within a proposition.

Computers and methods connected with them were and are used in mathematics not only in the proof of the Four-Color Theorem. They are used in various contexts in mathematics, in particular:

(1) to perform numerical calculations,
(2) to find (usually approximate) solutions of equations and systems of algebraic or differential equations or of integrals,
(3) in automating proofs of theorems,
(4) in checking the correctness of mathematical proofs,
(5) in proving theorems (one says then about computer-aided proofs),
(6) in various experiments with mathematical objects (e.g., in the theory of fractals).

From our point of view the most important applications are (3) and (4) – the application of the type (5) has been discussed above on the example of the Four-Color Theorem.

The automating proving of theorems is connected with the idea of mechanization and automatization of reasoning due to Leibniz (cf. [Marciszewski and

Murawski, 1995]). This idea (as one of the factors) led to the development of formal logic and in consequence to the idea of a formal proof.

4 Formal proofs and their role

Formal proofs were introduced to provide an explication of the informal notion of a proof and to solve some metamathematical problems. They should explain the virtue by which usual proofs used in the research practice are judged to be correct. They should also explain what does it mean that a given statement is a logical or deductive consequence of certain assumptions. As indicated above they were introduced in the atmosphere of a crisis in the foundations of mathematics. For Frege and Russell they were means to an end, a way of precisely isolating the permissible proofs and making sure that all use of axioms was explicit. On the other hand observe that Hilbert was not really interested in actually formalizing proofs and replacing the "normal" research proofs by formalized ones. He treated formalization and formal proofs as a tool to justify mathematics as a science and to establish its consistency. They should serve theoretical purposes – in particular to prove results about mathematics, hence to obtain metamathematical results.

In fact, the development of logic and the concept of a formal proof based only on syntactical properties made possible the development of metamathematics. A lot of interesting results have been obtained here. First of all the old paradigm of mathematics that was functioning since Euclid has been made precise – in fact it has been replaced by a new logico-settheoretical paradigm (cf. [Batóg, 1996]). The main features of this new paradigm can be described as follows:

1. Set theory became the fundamental domain of mathematics, in particular some set-theoretical notions and methods are present in any mathematical theory and set theory is the basis of mathematics in the sense that all mathematical notions can be defined by primitive notions of set theory and all theorems of mathematics can be deduced from axioms of set theory.
2. Languages of mathematical theories are strictly separated from the natural language, they are artificial languages and the meaning of their terms is described exclusively by axioms; some primitive concepts are distinguished and all other notions are defined in terms of them according to precise rules of defining notions.
3. Mathematical theories have been axiomatized.
4. There is a precise and strict distinction between a mathematical theory and its language on the one hand and metatheory and its metalanguage on the other (the distinction was explicitly made by A. Tarski).

Note also that two concepts, crucial for mathematics: the concept of a syntactical consequence (being provable) and the concept of being a semantical consequence have been precisely defined and strictly distinguished. One could also precisely distinguish provability and truth. In a "normal" research mathematical practice– as we indicated above – they are usually identified or at least not distinguished– one says that a theorem holds, i.e., has been proved (in a "normal", informal sense of this notion), or that it is true and treats both as equivalent and synonymous. The very process of distinguishing them was long and not so simple – cf. [Murawski, 2002a] and [Murawski, 2002b]. The crucial role was played here by Gödel's incompleteness theorems.

The incompleteness results of Gödel showed that any theory (based on a recursive set of axioms and finitary rules of inference) including arithmetic of natural numbers is in fact incomplete, hence there exist sentences that are true but are neither provable nor refutable, i.e., they are undecidable in a given theory. Before Gödel it was believed that formal demonstrability is an analysis of the concept of mathematical truth. Gödel wrote in a letter of 7th March 1968 to Hao Wang (cf. [Wang, 1974, p. 10]):

> [...] formalists considered formal demonstrability to be an *analysis* of the concept of mathematical truth and, therefore were of course not in a position to *distinguish* the two.

Indeed, the informal concept of truth was not commonly accepted as a definite mathematical notion at that time.[4] Gödel wrote in a crossed-out passage of a draft of his reply to a letter of the student Yossef Balas: "[...] a concept of objective mathematical truth as opposed to demonstrability was viewed with greatest suspicion and widely rejected as meaningless" (cf. [Wang, 1987, pp. 84–85]). It is worth comparing this with a remark of R. Carnap. He writes in his diary that when he invited A. Tarski to speak on the concept of truth at the September 1935 International Congress for Scientific Philosophy, "Tarski was very sceptical. He thought that most philosophers, even those working in modern logic, would be not only indifferent, but hostile to the explication of the concept of truth". And indeed at the Congress "[...] there was vehement opposition even on the side of our philosophical friends" (cf. [Carnap, 1963, pp. 61–62]).

All these explains in some sense why Hilbert preferred to deal in his metamathematics solely with forms of formulas, using only finitary reasonings which were considered to be save – contrary to semantical reasonings which were non-finitary and consequently not save. Non-finitary reasonings in mathematics were

4 Note that there was at that time no precise definition of truth — this was given only in 1933 by A. Tarski (cf. [Tarski, 1933]).

considered to be meaningful only to the extent to which they could be interpreted or justified in terms of finitary metamathematics.[5]

On the other hand there was no clear distinction between syntax and semantics at that time. Recall for example that the axiom systems came by Hilbert often with a built-in interpretation. Add also that the very notions necessary to formulate properly the difference syntax–semantics were not available to Hilbert though he was aware of the complex of problems – cf. his statement of the question in Hilbert and Ackermann [1928] which led to Gödel's completeness theorem.

Gödel proved that truth can not be adequately achieved and expressed by provability, that the whole of mathematics (or even parts of it) cannot be included in a formalized system. This indicated certain weakness of the concept of a formal proof. Gödel's results showed also that one should not limit or bound the creative invention of mathematicians. In the framework of formalized theories one can extend them by adding new axioms or by admitting new inference rules. The second possibility means that infinitary rules are admitted – but this changes the whole picture and the whole paradigm! Note that Hilbert had in fact no problem with such a change, to the opposite – in view of Gödel's result – he encouraged it. On the other hand one can ask whether the process of adding new axioms, though necessary to solve problems undecidable in a given theory, is sufficient? Will it suffice to express the creativity of a mathematical mind, the creativity of mathematicians?

The incompleteness theorems of Gödel belong to the so called limitation results. They are results stating that certain properties important and desired from a metamathematical point of view (but also from the point of view of a working mathematician) cannot be achieved. Among them is the theorem of Tarski stating the undefinability of the concept of truth and the theorem of Löwenheim and Skolem showing that a mathematical structure cannot be adequately and uniquely described by a formalized theory (a theory having a model has in fact many various models). Tarski wrote (cf. [Tarski, 1969, p. 74]):

5 Cf. Gödel's letter to Hao Wang dated 7th December 1967 — see [Wang, 1974, p. 9]. He wrote there: „I may add that my objectivist conception of mathematics and metamathematics in general, and of transfinite reasoning in particular, was fundamental also to my other work in logic. How indeed could one think of expressing metamathematics in the mathematical systems themselves, if the latter are considered to consist of meaningless symbols which acquire some substitute of meaning only through metamathematics. [...] it should be noted that the heuristic principle of my construction of undecidable number theoretical propositions in the formal systems of mathematics is the highly transfinite concept of 'objective mathematical truth' as opposed to that of 'demonstrability' (cf. M. Davis, The Undecidable, New York 1965, p. 64 where I explain the heuristic argument by which I arrive at the incompleteness results), with which it was generally confused before my own and Tarski's work."

> It was undoubtedly a great achievement of modern logic to have replaced the old psychological notion of proof, which could hardly ever be made clear and precise, by a new simple notion of a purely formal character. But the triumph of the method carried with it the germ of a future setback.

Considerations concerning formal proofs enlightened also the role played by infinity in mathematics, in particular in the process of proving. Gödel's results show that finite/finitistic methods of syntactical formal provability do not exhaust the variety of mathematical truth. In fact, if one wants to obtain a complete theory then some infinite/infinitistic rules (such as, e.g., the ω-rule) are necessary. Recall that the ω-rule is an inference rule with infinitely many premisses, i.e., it is the following rule:

$$\frac{\varphi(0), \varphi(1), \varphi(2), \ldots, \varphi(n), \ldots \quad (n \in \mathbb{N})}{\forall x \varphi(x)}.$$

In mathematical research practice nobody restricts himself/herself to finite methods, on the contrary, any correct methods, among them infinite (in particular set-theoretical and semantical) are applied.

Being not so ideal as it was hoped, formal proofs play an important role in metamathematics, i.e., in the study of mathematical theories or of mathematics as a collection of theories – but not only there. They enable also the automatization and mechanization of proofs in mathematics, hence they make possible the construction of automated proofs and the verification of proofs by computers. Verification of (formal) proofs is possible because the relation "x is a proof of y" is – as was shown in mathematical logic – recursive, hence effective and can be implemented. On the other hand constructing and finding proofs is not an effective (recursive) procedure, there is no universal method of doing this (as results of Turing, Church and Gödel show). In fact it is only recursively enumerable. Hence every formal proof is a result of a creative invention of a human being. One can say that "[f]ormalization is about checking, and not about discovery" (cf. [Wiedijk, 2008, p. 1414]).

Observe that the concept of a formalized proof is one for all theories, it is in a sense a uniform concept. It is independent of subjective, cultural and sociological elements and factors. Moreover, the completeness theorem of Gödel states that the logical means of the first-order logic (first-order predicate calculus) are sufficient.[6] This concept enables us to make the concept of a proof more objective. It also makes possible the precise study of provability in mathematics – under

6 Note that in the case of, e.g., second-order logic the situation is different – one does not have here the completeness phenomenon.

the assumption that the logical concept of a proof reflects all important and essential features of proofs from the research practice of mathematicians. One can prove results stating that a given statement is not a theorem of a given theory, i.e., that there exist no (formal) proof of a given statement or that a given sentence is (formally) undecidable (in a given theory). It enables also the study of important metamathematical properties of mathematical theories such as consistency, completeness, independence of axioms or axiomatizability in a given way, etc.

The concept of a formal proof is also helpful in the philosophy of mathematics. It can be used in attempts to answer the question about the existence and character of mathematical objects as well as in considerations concerning the epistemology of mathematics. On the other hand all philosophies of mathematics reducing mathematics to formalized axiomatic theories (among them logicism and formalism) have a reductionist character and do not take into account the actual research practice of mathematicians. Their aim is to justify mathematics and not to explain the real mathematical practice. It is worth noting that in recent trends in the philosophy of mathematics still more and more attention is paid to the study of the research practice in mathematics – one takes into account various sociological, psychological and cultural factors. Unfortunately it is done only by the analysis of particular single discoveries and achievements, hence by case studies. There are no general conceptions. But is it possible to develop such general conceptions?

In fact formal proofs are connected rather with foundational studies than with research practice. Observe that a formal proof does not give an understanding, it does not explain the deep reasons of a theorem. They are also not suitable for the practice – they are simply too long, they are too tedious and painstaking. In such a proof the underlying intuition may get lost. Formalized mathematics may be also more error-prone than the usual informal one – in fact formal manipulations may become very complicated. As Bourbaki wrote (cf. [Bourbaki, 1968]):

> If formalized mathematics were as simple as the game of chess, then once our chosen formalized language had been described there would remain only the task of writing out proofs in this langauge, [...] But the matter is far from being as simple as that, and no great experience is necessary to perceive that such a project is absolutely unrealizable: the tinest proof at the beginning of the Theory of Sets would already require several hundreds of signs for its complete formalization. [...] formalized mathematics cannot in practice be written down in full, [...] We shall therefore very quickly abandon formalized mathematics, [...]

CadwalladerOlsker writes in [CadwalladerOlsker, 2011, p 42]:

> A purely formal proof [...] cannot be very complex without becoming so lengthy as to be incomprehensible to a human reader. Such a formal proof is rarely able to be explanatory,

> and may only be convincing to the degree that it can be read and understood by the reader or checked by a computer.

Add that a transcription of a single traditional (hence informal) proof into a formal one is a major undertaking.

5 Conclusion

We have shown that one has two concepts of a proof in mathematics: an informal one used by mathematicians in their usual research practice and the concept of a formal or formalized proof used mainly in logic and the foundations of mathematics. The first one is not defined precisely, it is simply practised and any attempts to define it fail. It is – so to speak – a practical notion. It has a psychological, sociological and cultural character. The second one is precisely defined in terms of logical concepts. Hence it is a logical concept having rather theoretical than practical character. The first one has – in a part at least – semantical character, the second is entirely syntactical in nature.

Hence this situation can be compared with the situation concerning Church-Turing thesis. This thesis states the equivalence of two concept: effective computability (in the intuitive sense) and recursivity (or Turing computability or computability in the sense of Markov or any other precisely defined and equivalent sense). As is known this equivalence cannot be proved with the degree of precision usual and required in mathematics. The reason is that one part of this equivalence contains an intuitive vague concept formulated in the everyday language and the other a precise concept defined in the language of mathematics (cf. [Murawski, 2004], [Murawski and Woleński, 2006]).

With such a situation one has to do also in other parts of mathematics – see for example the concepts of function, of truth, of logical validity or of limit (cf. [Mendelson, 1990]). In fact, till the 19th century a function was tied to a rule for calculating it, generally by means of a formula. In 19th and 20th centuries mathematicians started to define a function as a set of ordered pairs satisfying appropriate conditions (hence a function is identified here with its graph). The identification of those notions, i.e., of an intuitive notion and the precise set-theoretical one, can be called "Peano thesis". Similarly "Tarki's thesis" is the thesis identifying the intuitive notion of truth and the precise notion of truth given by Tarski in [1933]. The intuitive notion of a limit widely used in mathematical analysis in the 18th century and then in the 19th century applied by A. Cauchy to define basic notions of the calculus has been given a precise form only by K. Weierstrass in the language of $\varepsilon - \delta$. There are many other such examples: the notion of a measure as an explication of area and volume, the definition of dimension in topology, the definition of velocity as a derivative, etc.

Comparing the two concepts of a proof in mathematics one can formulate a thesis stating that they are equivalent – one can call such a thesis a proof-theoretical thesis.[7] As in the case of Church Thesis, no precise and strict proof of it can be given.[8] One can only formulate arguments in favor or against it. The main argument for this thesis is the conviction – being popular among mathematicians and logicians – that every "normal" mathematical proof can be formalized, i.e., can be written as a formal proof in a suitable axiomatic theory. There are, of course, no general rules describing how this can and should be done. In fact a formalization of an informal proof requires often some original and not so obvious ideas.

In this way we come to the following conclusion: there are two concepts of a proof in mathematics. They play different but complementary roles: formal proofs are used mainly in metamathematical and logical considerations, whereas informal proofs are used in the research practice of mathematicians.

Acknowledgment: The paper is based on my talks at the conferences "Philosophy in Science" (Kraków, Poland, 2012) and "Proof" (International Conference within the frame of Humboldt-Kollegs; Bern, Switzerland, 2013). I would like to thank the referee for helpful comments and suggestions.

References

M. Aigner and G. M. Ziegler. *Proofs From The Book.* Springer Verlag, Berlin/Heidelberg/New York 2001, third edition 2004, second edition, 1998.

M. Aschbacher. Highly complex proofs and implications of such proofs. *Philosophical Transactions of the Royal Society A*, 363:2401–2406, 2005.

J. Avigad. Mathematical method and proof. *Synthese*, 153:105–159, 2006.

7 Add that this formulation is consciously rather vague – for example it is not specified here in which formal theory (or theories) formal proofs should be constructed and what should be the underlying logic. A similar but stronger thesis was formulated by Barwise [1977, p. 41] under the name "Hilbert's Thesis" where he wrote: "... the informal notion of *provable* used in mathematics is made precise by the formal notion *provable in first-order logic.* Following the sug[g]estion of Martin Davis, we refer to this view as *Hilbert's Thesis*". This thesis says that first-order logic is *the* logic of mathematics.

8 Note that in the case of Church Thesis the formal framework is precisely specified: the intuitive notion of computability should be captured by one specific formal model of computation.

J. Barwise. An introduction to first-order logic. In J. Barwise, editor, *Handbook of Mathematical Logic*, pages 5–46. North-Holland, Amsterdam–New York–Oxford, 1977.

G. B. Bassler. The surveyability of mathematical proof: a historical perspective. *Synthese*, 148: 99–133, 2006.

T. Batóg. *Dwa paradygmaty matematyki. Studium z dziejów i filozofii matematyki ()*. Wydawnictwo Naukowe Uniwersytetu im. Adama Mickiewicza, Poznań, 1996. Two Paradigms of Mathematics. A Study in the history and Philosophy of Mathematics.

N. Bourbaki. *Theory of Sets*. Addison-Wesley, Reading, Mass, 1968.

T. CadwalladerOlsker. What do *we* mean by mathematical proof? *Journal of Humanistic Mathematics*, 1:33–60, 2011.

R. Carnap. Intellectual autobiography. In P. A. Schilpp, editor, *The Philosophy of Rudolf Carnap*, pages 3–84. Open Court, La Salle, Ill., 1963.

M. D. De Villiers. *Rethinking Proof with the Geometer's Sketchpad*. Key Curriculum Press, Emeryville, CA, 1999.

R. Hersh. *What Is Mathematics, Really?* Oxford University Press, New York, 1997.

D. Hilbert and W. Ackermann. *Grundzüge der theoretischen Logik*. Springer, Berlin, 1928.

R. Kahle. What is a proof? *Axiomathes*, 25:79–91, 2015. 10.1007/s10516-014-9252-9.

P. Mancosu. On mathematical explanation. In E. Grosholz and H. Breger, editors, *The Growth of Mathematical Knowledge*, pages 103–119. Kluwer Academic, Dordrecht, 2000.

P. Mancosu. Mathematical explanation: problems and prospects. *Topoi*, 20:97–117, 2001.

W. Marciszewski and R. Murawski. *Mechanization of Reasoning in a Historical Perspective*. Editions Rodopi, Amsterdam/Atlanta, GA, 1995.

E. Mendelson. Second thoughts about church's thesis and mathematical proofs. *The Journal of Philosophy*, 87:225–233, 1990.

R. Murawski. Truth vs. provability – philosophical and historical remarks. *Logic and Logical Philosophy*, 10(2002):93–117, 2002a.

R. Murawski. On the distinction proof-truth in mathematics. In P. Gärdenfors et al., editors, *In the Scope of Logic, Methodology and Philosophy of Science*, pages 287–303. Kluwer Academic, Dordrecht–Boston–London, 2002b.

R. Murawski. Church's thesis and its epistemological status. *Annales Universitatis Mariae Curie-Skłodowska, Sectio AI, Informatica*, 2(2004):57–70, 2004.

R. Murawski and J. Woleński. The status of Church's thesis. In J. Woleński A. Olszewski and R. Janusz, editors, *Church's Thesis After 70 Years*, pages 310–330. Ontos, Frankfurt, 2006.

G.-C. Rota. The phenomenology of mathematical proof. *Synthese*, 111:183–196, 1997.

D. E. Rowe. Klein, Hilbert, and the Göttingen mathematical tradition. *Osiris*, 2(5):186–213, 1989.

M. Steiner. Mathematical explanation. *Philosophical Studies*, 34:133–151, 1978.

A. Tarski. *Pojęcie prawdy w językach nauk dedukcyjnych*. Nakładem Towarzystwa Naukowego Warszawskiego, Warszawa, 1933.

A. Tarski. Truth and proof. *Scientific American*, 220(6):63–77, 1969.

T. Tymoczko. The four-color problem and its philosophical significance. *The Journal of Philosophy*, 76:57–83, 1979.

Hao Wang. *From Mathematics to Philosophy*. Routledge and Kegan Paul, London, 1974.

Hao Wang. *Reflections on Kurt Gödel*. M.I.T. Press, Cambridge, Mass, 1987.

F. Wiedijk. Formal proof – getting started. *Notices of the American Mathematical Society*, 55: 1408–1414, 2008.

Sara Negri and Jan von Plato

Cut Elimination in Sequent Calculi with Implicit Contraction, with a Conjecture on the Origin of Gentzen's Altitude Line Construction

Abstract: Sequent calculi are given in which contexts represent finite sets of formulas. Standard cut elimination will not work if the principal formula of a logical rule is already found in a premiss, i.e., if there is an implicit contraction on it. A procedure is given in which cut with the original cut formula is first permuted up, followed by cuts on its immediate subformulas. It is next adapted to sequent calculi with multisets and explicit contraction, by which Gentzen's mix rule trick is avoided, a procedure strikingly similar to the peculiar "altitude line" construction that Gentzen used in his second proof of the consistency of arithmetic in 1938. The conjecture is close at hand that this is indeed the way Gentzen originally proved cut elimination in 1933.

Keywords: Cut elimination, Contraction, Altitude line

Mathematics Subject Classification 2010: 01A60, 03-03, 03F07

1 Introduction

Axiomatic logic began with Frege and was perfected in the Hilbert school in the 1920s, with the aim to apply it in the formalization of mathematical proofs. The crucial *deduction theorem* shows how this goes: Given some axiomatic system with axioms H expressed in the language of logic, it should be possible to treat it, not as a collection of *truths* added to the logical axioms, but as *hypotheses*. Thus, derivations in axiomatic logic were extended to include derivations under hypotheses, as in the book that first defined a complete formal system of predicate logic, [Hilbert and Ackermann, 1928], and very soon the aim was reached: A proof of a proposition P by the mathematical axioms can be converted into a purely logical proof of the implication $H \supset P$. The matter is not as trivial as it might seem: Say, if $A(x)$ is provable without assumptions, $\forall x A(x)$ can be concluded, whereas if $A(x)$ is proved under assumptions, nothing of the kind need follow.

Sara Negri and Jan von Plato: Department of Philosophy, University of Helsinki, Finland, e-mail: jan.vonplato@helsinki.fi, sara.negri@helsinki.fi

Turing in a manuscript of 1944, *The reform of mathematical notation and phraseology*, was keenly aware of how mathematical proof hangs on the deduction theorem. A version of Turing's manuscript is found printed in [Cooper and van Leeuwen, 2013] to which we refer. The first of his two central points was (p. 246): 1. "Free and bound variables should be understood by all and properly respected." He then gives an example of constants and variables and adds: "The difference between the constants and the free variables is somewhat subtle. The constants appear in the formula as if they were free variables, but we cannot substitute for them. In these cases there has always been some assumption made about the variable (or constant) previously."

Turing's second point was (*ibid.*): 2. "The deduction theorem should be taken account of." The deduction theorem is the main way of handling free variables: "This process whereby we pass from P proved under an assumption H to 'If H then P' may be called 'absorption of hypotheses'. The process converts constants or 'restricted variables' into free variables."

His example, slightly rephrased, is: Let the radius a and volume v of a sphere be given. Then $v = \frac{4}{3}\pi a^3$.

> The 'deduction theorem' states that in such a case, where we have obtained a result by means of some assumptions, we can state the result in a form in which the assumptions are included in the result, e.g., 'If a is the radius and v is the volume of the sphere then $v = \frac{4}{3}\pi a^3$. In this statement a and v are no longer constants.

A sufficiently detailed proof of the deduction theorem, as in [Hakli and Negri, 2012] or [von Plato, 2013, section 3.6] gives an algorithm for converting hypothetical proofs that use mathematical axioms as assumptions into purely logical proofs without assumptions, but the algorithm is hopelessly clumsy for any practical use. Axiomatic logic was put to rest as a proof system of logic in an act that rendered the famous deduction theorem a proof-theoretical triviality: it became the rule of implication introduction of natural deduction. Indeed, if a translation from the latter to axiomatic logic is defined in sufficient detail, the deduction theorem will come out simply as the image of implication introduction under that translation (as in [von Plato, 2013, p. 55]). When written in terms of sequent calculus with the sequent arrow $\Gamma \to C$ that indicates the *derivability* of C under the collection of assumptions Γ, the deduction theorem becomes the rule: If $H \to P$, then $\to H \supset P$, with no assumptions left.

In proving the deduction theorem of axiomatic logic, there are two limiting cases: an axiom H was not at all used in a proof, and secondly, it was used repeatedly. The former is easy and the latter routinely skimmed over in logic books. Now we have come to the central problem: How to count the multiplicity of assumptions in proofs? Many books in proof theory are as light-hearted about this problem as are books on axiomatic logic. The latter may include Frege's *contractive axiom* $(A\supset (A\supset B))\supset (A\supset B)$ that can be used for reducing a multiple deduction theorem to a singular one.

The easiest way with assumptions would be if they could be just named and counted once. In a second approach, we could list them in the order in which they are taken, in a third, we could repeat in this list assumptions that were used more than once. Finally, we could ignore the order that seems perhaps not to be intrinsic but at least in part imposed by the linearity of writing proofs as texts, which makes the first and second approach collapse into one that uses *finite sets*. The third uses *lists with multiplicity*, or *lists* for short, and the fourth *finite multisets*.

Logical rules are next to the formulas the central part of the syntax of a logical proof system; thus, the rules act on *expressions*. Lists in contrast to sets seem to be, by their very nature, already formal objects of a syntax. We shall therefore begin with a little theory of expressions for finite sets, then put up a logical sequent calculus and prove the crucial cut elimination theorem directly by the use of sets. In a second part, starting with Sec. 4, we draw conclusions about our proof that point at so far unnoticed passages and constructions in Gentzen's original work connected to his cut elimination procedure: first, his actual use of the "sequents with sets" approach in the proof of decidability of intuitionistic propositional logic in 1933 that can be easily turned into a new kind of cut elimination procedure, detailed out in Sec. 5; whether Gentzen actually saw this possibility is not known at present. Secondly there is his, quite opposite, strict adherence to lists in the "altitude line" construction in the consistency proof of arithmetic of 1938, presented in detail in Sec. 6, that gives an alternative cut elimination procedure known to Gentzen.

2 Sequents with sets

Sequent calculus is the proper setting for a precise discussion of the treatment of assumptions in proofs. The rules of sequent calculus display the collection of *open assumptions* Γ at the left of an arrow and the *consequence* C of those assumptions at right, in a sequent $\Gamma \rightarrow C$. Each logical rule modifies derivability relations, as in rule $R\supset$ that gives the result $\Gamma \rightarrow A\supset B$ from the premiss $A, \Gamma \rightarrow B$, as in

$$\frac{A, \Gamma \rightarrow B}{\Gamma \rightarrow A \supset B} R\supset$$

Here A and B are the *active* formulas of the rule, $A\supset B$ its *principal* formula, and

Γ the *context*. It is customary to speak of derivations even in sequent calculus; no harm, if the double sense of derivability is kept in mind. The word "conclusion" can have a similar double use, either as the consequence C in a sequent $\Gamma \rightarrow C$, or the conclusion of a rule of sequent calculus. Assumptions A are presented in sequent calculus as *initial sequents* $A \rightarrow A$, as a limiting case in which the conclusion A depends on the assumption A.

In Gentzen's doctoral work of 1933, published as [Gentzen, 1934–35], the assumptions Γ were given as a *list* of formulas, thus, the order and multiplicity of occurrences of formulas was counted. Later, Katudi Ono [Ono, 1938] introduced lists in which order is not counted and proved that the resulting calculi are equivalent to Gentzen's, a discovery known to the latter (cf. his review of Ono's work, [Gentzen, 1939]), and reinvented by Kleene, Curry, and others in the 1950s. In sequent calculi as well as natural deduction, formal derivations have the form of a tree, and there is no intrinsic linear order of assumptions. Perhaps in reflection of this latter fact, it has become customary to formulate sequent calculi so that the open assumptions form finite *multisets*, and similarly for the open cases if a symmetric sequent calculus with more than one succedent formula is used. Below we pay only marginal attention to Gentzen's original exchange rules by which the order of formula occurrences in lists of assumptions and cases can be changed.

The idea has been often entertained of dispensing, next to order, also with multiplicity, known as the "sequents with sets" idea. There will be some difficulties in carrying the idea through on the level of formalization: Logical languages are inductively defined classes of formulas, and a system of logical rules gives an inductive definition of the class of derivations. Sets are objects and two sets are equal if they have the same members. However, two different expressions for the same set cannot be just like that substituted one for the other in logical rules, because syntactically incorrect rule instances can be produced: For example, a single left rule of conjunction has the premiss $A, B, \Gamma \rightarrow C$ and the conclusion $A \& B, \Gamma \rightarrow C$. If A and B are identical, the premiss collapses into $A, \Gamma \rightarrow C$ with no match with the premiss of the rule. Therefore a little theory of expressions for sets needs to be put up. This theory will contain rules by which, say , an expression

$$\{A, \ldots, B, B, \ldots, C\}$$

for a set of formulas can be replaced by the expression

$$\{A, \ldots, B, \ldots, C\},$$

thus, a *rule of contraction*. Details such as the above have a bearing on the modularity of sequent calculus proof systems, as expressed by the rule of *cut* that shows how two derivations can be combined, and how it is removed in a process of cut elimination. Formally, we have a derivation of some result A, and another derivation in which A occurs as an assumption, combined as in the scheme:

$$\frac{\Gamma \rightarrow A \quad A, \Delta \rightarrow C}{\Gamma, \Delta \rightarrow C}\ Cut$$

In the cut elimination procedure to be presented we shall, instead of a theory of expressions for sets, use *canonical expressions* for sets, as lists without repetition and with a prescribed lexicographical order, and the principle:

> *At each rule instance, any potential multiplication of formula occurrences in the conclusion is erased when the conclusion is written down.*

Initial sequents have already the property of a single occurrence, and our principle makes conclusions of logical rules inherit that property. The effect is that the rule of contraction, needed when different expressions for the same set are allowed, will become *implicit*.

Cut elimination for sequent calculi with implicit contraction is different from the Gentzen style cut elimination procedure in which, whenever the cut formula is concluded in both premisses of cut by the rules that correspond to its logical form, cuts on shorter formulas replace the given cut. The difference is seen when, for example, the right premiss of cut has been concluded by the left conjunction rule, with an implicit contraction on the cut formula $A \& B$. In this case, the cut on $A \& B$ does not get removed, but is first permuted up, followed by cuts on the two immediate subformulas of $A \& B$. We show first that the cut elimination procedure to be defined works for a single-succedent calculus, then extend it to a calculus with symmetric sequents of the form $\Gamma \rightarrow \Delta$ that can have several cases in the succedent part. These proofs of cut elimination use the standard rule of cut, whereas in Gentzen's original work, cut elimination is done through the "mix rule," or rule of multicut as one often says, in which any number of cut formulas in the two premisses can be deleted:

$$\frac{\Gamma \rightarrow A \quad A^*, \Delta \rightarrow C}{\Gamma, \Delta \rightarrow C}\ Mix$$

Here A^* denotes any number $n \geqslant 1$ of copies of A. [Gentzen, 1934–35] has the comment that the rule is used "to make the proof easier." The problematic case of cut elimination that led Gentzen to the mix rule is that the right premiss has been concluded by a rule of *contraction*.

Here is such a derivation:

$$\dfrac{\overset{\vdots}{\Gamma \to A} \qquad \dfrac{A, A, \Delta \to C}{A, \Delta \to C}\,Ctr}{\Gamma, \Delta \to C}\,Cut$$

The obvious idea of cutting twice with $\Gamma \to A$ gives:

$$\dfrac{\overset{\vdots}{\Gamma \to A} \qquad \dfrac{\overset{\vdots}{\Gamma \to A} \quad A, A, \Delta \to C}{A, \Gamma, \Delta \to C}\,Cut}{\Gamma, \Gamma, \Delta, \Theta \to C}\,Cut$$

The derivation of the right premiss of the lower cut has clearly grown in some sense, with an instance of contraction replaced by an instance of cut, and no simple solution has been found to correct the situation. Intrigued by Gentzen's remark about making the proof easier, the second author produced in [von Plato, 2001] a proof of cut elimination with the standard cut rule, based on an analysis of how the premiss of contraction was derived. Contraction was then reduced by *inversion lemmas* to contractions on shorter formulas. These lemmas require subtle changes in the sequent calculi used, unknown to Gentzen.

2.1 Cut elimination for sequents with sets

The cut elimination procedure to be defined is rather straightforward for a single-succedent sequent calculus. Logical rules with two premisses have contexts that are added up in the conclusion, with possible multiplications of formulas erased. When a rule is written, no premiss has duplications of formulas. A "bra–ket" notation $\langle \Gamma \rangle$, adopted from a manuscript of Gentzen's of 1944, indicates a context Γ in the conclusion of a rule such that no duplications in the antecedent remain. Such duplications can arise either because the principal formula of a rule was already found in Γ or because the antecedents of two premisses of a rule overlapped with some formulas. Rules $R\&$ and $L\vee$ are written in the manner of rule $L\supset$, with contexts added up in the conclusion.

Table 1. Initial sequents of LIS

Initial sequents of the calculus LIS
$C \to C \qquad \bot \to C$ \quad (C an arbitrary formula)

Table 2. Logical rules of LIS

Logical rules of the calculus LIS		
$\dfrac{A, \Gamma \to C}{\langle A \& B, \Gamma\rangle \to C} L\&$	$\dfrac{B, \Gamma \to C}{\langle A \& B, \Gamma\rangle \to C} L\&$	$\dfrac{\Gamma \to A \quad \Delta \to B}{\langle \Gamma, \Delta\rangle \to A \& B} R\&$
$\dfrac{A, \Gamma \to C \quad B, \Delta \to C}{\langle A \vee B, \Gamma, \Delta\rangle \to C} L\vee$	$\dfrac{\Gamma \to A}{\Gamma \to A \vee B} R\vee$	$\dfrac{\Gamma \to B}{\Gamma \to A \vee B} R\vee$
$\dfrac{\Gamma \to A \quad B, \Delta \to C}{\langle A \supset B, \Gamma, \Delta\rangle \to C} L\supset$	$\dfrac{A, \Gamma \to B}{\Gamma \to A \supset B} R\supset$	
$\dfrac{A(t), \Gamma \to C}{\langle \forall x A(x), \Gamma\rangle \to C} L\forall$	$\dfrac{\Gamma \to A(y)}{\Gamma \to \forall x A(x)} R\forall$	
$\dfrac{A(y), \Gamma \to C}{\langle \exists x A(x), \Gamma\rangle \to C} L\exists$	$\dfrac{\Gamma \to A(t)}{\Gamma \to \exists x A(x)} R\exists$	

For the structural rules, note that if the weakening formula is found in the context, the conclusion is identical to the premiss.

Table 3. Structural rules of LIS

Structural rules of the calculus LIS	
$\dfrac{\Gamma \to C}{\langle A, \Gamma\rangle \to C} Wk$	$\dfrac{\Gamma \to A \quad A, \Delta \to C}{\langle \Gamma, \Delta\rangle \to C} Cut$

Note that in rule $L\&$, for example, the principal formula $A \& B$ can very well occur in Γ, whereas if there is no such occurrence, $\langle A \& B, \Gamma\rangle$ could be written as $A \& B, \Gamma$.

We have chosen to have two left rules for conjunction, instead of a single rule with the premiss $A, B, \Gamma \to C$, to avoid the complication that occurs if A and B are identical. As mentioned, this case would make the premiss collapse into $A, \Gamma \to C$ with no match with the conclusion.

Theorem 1 (Cut elimination for LIS). *The rule of cut can be eliminated from derivations in* LIS.

Proof: We show that uppermost occurrences of rule *Cut* can be permuted upwards until they hit initial sequents and get removed. The *height of a cut* is the sum of the heights of the derivations of its premisses, i.e., of the longest branches in each.

In each case, a cut is replaced by cuts that can have the same cut formula but a lesser height or else they have a shorter cut formula. The cases are:

1. One premiss of cut is an initial sequent.
   (a) If it is of the form $C \to C$, the conclusion is identical to the other premiss and the cut deleted.
   (b) The left premiss of cut is an initial sequent of the form $\bot \to C$ and we have:

$$\frac{\bot \to D \quad D, \Gamma \to C}{\langle \bot, \Gamma \rangle \to C}\, Cut$$

   The conclusion is obtained from the initial sequent $\bot \to C$ by repeated weakenings.
   (c) The right premiss of cut is an initial sequent of the form $\bot \to C$ and we have:

$$\frac{\Gamma \to \bot \quad \bot \to C}{\Gamma \to C}\, Cut$$

   The cut formula cannot be principal in the left premiss and cut is permuted up until a sequent of the form $\bot, \Delta \to \bot$ is reached:

$$\frac{\bot, \Delta \to \bot \quad \bot \to C}{\langle \bot, \Delta \rangle \to C}\, Cut$$

   The conclusion of that cut is obtained from the right premiss by weakenings.

2. The left premiss of cut is a conclusion of a left rule. Cut is permuted up until case 1 is met, or the left premiss has been concluded by a right rule:
3. There are five cases of right rules in the left premiss of cut.
   (a) The first is rule $R\&$ with the cut:

$$\frac{\dfrac{\Gamma \to A \quad \Delta \to B}{\langle \Gamma, \Delta \rangle \to A \,\&\, B}\, R\& \qquad A \,\&\, B, \Theta \to C}{\langle \Gamma, \Delta, \Theta \rangle \to C}\, Cut$$

   Cut is permuted up at right and its height reduced until case 1 is met or $A \,\&\, B$ is the principal formula in rule $L\&$. In the latter case we have, say:

$$\frac{\dfrac{\Gamma \to A \quad \Delta \to B}{\langle \Gamma, \Delta \rangle \to A \,\&\, B}\, R\& \qquad \dfrac{A, \Theta \to C}{\langle A \,\&\, B, \Theta \rangle \to C}\, L\&}{\langle \Gamma, \Delta, \Theta \rangle \to C}\, Cut$$

   If $A \,\&\, B$ does not occur in Θ, cut can be permuted as in [Gentzen, 1934–35].

The derivation with the permuted cut is

$$\frac{\Gamma \to A \quad A, \Theta \to C}{\langle \Gamma, \Theta \rangle \to C}\, Cut$$

The conclusion of the original cut is now obtained through weakenings. If $A \,\&\, B$ instead does occur in Θ, rule $L\&$ produces an implicit contraction on the cut formula, and the Gentzen-style transformation leaves one copy of $A \,\&\, B$ in the conclusion: The cut after the transformation can be written with the notation $A \,\&\, B, \Theta' \equiv \Theta$ as:

$$\frac{\Gamma \to A \quad A, A \,\&\, B, \Theta' \to C}{\langle \Gamma, A \,\&\, B, \Theta' \rangle \to C}\, Cut$$

This problematic case is resolved as follows:

$$\frac{\Gamma \to A \qquad \dfrac{\dfrac{\Gamma \to A \quad \Delta \to B}{\langle \Gamma, \Delta \rangle \to A \,\&\, B}\, R\& \qquad A, A \,\&\, B, \Theta' \to C}{\langle A, \Gamma, \Delta, \Theta' \rangle \to C}\, Cut}{\langle \Gamma, \Delta, \Theta' \rangle \to C}\, Cut$$

The height of the upper cut is diminished by one. The new cut is on a shorter formula.

(b) The case of rule $R\vee$ is similar.

(c) With rule $R\supset$ we have as in 3.(a) the worst case:

$$\frac{\dfrac{A, \Gamma_1 \to B}{\Gamma_1 \to A \supset B}\, R\supset \qquad \dfrac{A \supset B, \Gamma_2 \to A \quad B, A \supset B, \Gamma_3 \to C}{\langle A \supset B, \Gamma_2, \Gamma_3 \rangle \to C}\, L\supset}{\langle \Gamma_1, \Gamma_2, \Gamma_3 \rangle \to C}\, Cut$$

The transformed derivation is, with the $\langle\ \rangle$ notation left out to make it fit:

$$\frac{\dfrac{\dfrac{\dfrac{A, \Gamma_1 \to B}{\Gamma_1 \to A \supset B}\, R\supset \quad A \supset B, \Gamma_2 \to A}{\Gamma_1, \Gamma_2 \to A}\, Cut \qquad A, \Gamma_1 \to B}{\Gamma_1, \Gamma_1, \Gamma_2 \to B}\, Cut \qquad \dfrac{\dfrac{A, \Gamma_1 \to B}{\Gamma_1 \to A \supset B}\, R\supset \quad B, A \supset B, \Gamma_3 \to C}{B, \Gamma_1, \Gamma_3 \to C}\, Cut}{\Gamma_1, \Gamma_1, \Gamma_1, \Gamma_2, \Gamma_3 \to C}\, Cut$$

Contractions are now made to reproduce the original conclusion. There are two cuts with the cut formula $A \supset B$ and a lesser cut height, followed by cuts on the shorter formulas A and B.

(d) The cut formula is $\neg A$ and the transformation nearly a special case of the above.

(e) The cut formula is $\forall x A(x)$ with the derivation:

$$\frac{\dfrac{\Gamma_1 \to A(y)}{\Gamma_1 \to \forall x A(x)}\, R\forall \qquad \dfrac{A(t), \forall x A(x), \Gamma_2 \to C}{\forall x A(x), \Gamma_2 \to C}\, L\forall}{\langle \Gamma_1, \Gamma_2 \rangle \to C}\, Cut$$

The transformed derivation is, with the substitution $[t/y]$ in the derivation of the premiss of rule $R\forall$ of the original derivation that gives as a result the derivable sequent $\Gamma_1 \to A(t)$:

$$\dfrac{\Gamma_1 \to A(t) \quad \dfrac{\dfrac{t}{\Gamma_1 \to \forall x A(x)}\,R\forall \; he\Gamma_1 \to A(y) \quad \forall x A(x), A(t), \Gamma_2 \to C}{\langle A(t), \Gamma_1, \Gamma_2 \rangle \to C}\,Cut}{\langle \Gamma_1, \Gamma_2 \rangle \to C}\,Cut$$

The upper cut has a lesser cut height, the lower a shorter cut formula.

(f) The cut formula is $\exists x A(x)$. This case is dual to the previous.

4. One premiss of cut is derived by rule *Wk*. If it is the left premiss, cut can be permuted up, and the same with the right premiss except when the cut formula is principal in weakening. In this case the conclusion is obtained from the premiss of weakening without any cut.

QED.

The above proof of cut elimination goes through also for a calculus with more than one succedent formula, even if the details turn out somewhat intricate. Initial sequents of this classical sequent calculus with sets of formulas LKS have the form $C \to C$, and negation is treated as a primitive connective. The logical and structural rules of LKS are displayed below.

Table 4. Logical rules of LKS

Logical rules of the calculus LKS		
$\dfrac{A, \Gamma \to \Delta}{\langle A \& B, \Gamma \rangle \to \Delta}\,L\&$	$\dfrac{B, \Gamma \to \Delta}{\langle A \& B, \Gamma \rangle \to \Delta}\,L\&$	$\dfrac{\Gamma_1 \to \Delta_1, A \quad \Gamma_2 \to \Delta_2, B}{\langle \Gamma_1, \Gamma_2 \rangle \to \langle \Delta_1, \Delta_2, A \& B \rangle}\,R\&$
$\dfrac{A, \Gamma_1 \to \Delta_1 \quad B, \Gamma_2 \to \Delta_2}{\langle A \vee B, \Gamma_1, \Gamma_2 \rangle \to \langle \Delta_1, \Delta_2 \rangle}\,L\vee$	$\dfrac{\Gamma \to \Delta, A}{\Gamma \to \langle \Delta, A \vee B \rangle}\,R\vee$	$\dfrac{\Gamma \to \Delta, B}{\Gamma \to \langle \Delta, A \vee B \rangle}\,R\vee$
$\dfrac{\Gamma_1 \to \Delta_1, A \quad B, \Gamma_2 \to \Delta_2}{\langle A \supset B, \Gamma_1, \Gamma_2 \rangle \to \langle \Delta_1, \Delta_2 \rangle}\,L\supset$	$\dfrac{A, \Gamma \to \Delta, B}{\Gamma \to \langle \Delta, A \supset B \rangle}\,R\supset$	
$\dfrac{\Gamma \to \Delta, A}{\langle \neg A, \Gamma \rangle \to \Delta}\,L\neg$	$\dfrac{A, \Gamma \to \Delta}{\Gamma \to \langle \Delta, \neg A \rangle}\,R\neg$	
$\dfrac{A(t), \Gamma \to \Delta}{\langle \forall x A(x), \Gamma \rangle \to \Delta}\,L\forall$	$\dfrac{\Gamma \to \Delta, A(y)}{\Gamma \to \langle \Delta, \forall x A(x) \rangle}\,R\forall$	

For the structural rules, note that if the weakening formula is found in the context, the conclusion is identical to the premiss.

Table 5. Structural rules of LKS

Structural rules of the calculus LKS		
$\dfrac{\Gamma \rightarrow \Delta}{\langle A, \Gamma\rangle \rightarrow \Delta} LW$	$\dfrac{\Gamma \rightarrow \Delta}{\Gamma \rightarrow \langle \Delta, A\rangle} RW$	$\dfrac{\Gamma_1 \rightarrow \Delta_1, A \quad A, \Gamma_2 \rightarrow \Delta_2}{\langle \Gamma_1, \Gamma_2\rangle \rightarrow \langle \Delta_1, \Delta_2\rangle} Cut$

Theorem 2 (Cut elimination for LKS). *The rule of cut can be eliminated from derivations in* LKS.

Proof: The proof is similar to that of theorem 1 in which uppermost instances of cut are eliminated. The cases are:

1. One premiss of cut is an initial sequent. The conclusion is identical to the other premiss and the cut is deleted.
2. Cut is permuted up in the left and right premisses until case 1 is met.
3. Cut is permuted up in the left and right premisses until the cut formula is principal in both premisses of cut. There are six cases:
   (a) The cut formula is $A \& B$, and we may assume that the principal formula in rules $L\&$, $R\&$ occurs in each premiss. Should it lack from some, either a simpler cut elimination procedure is made, but with more case distinctions, or, for a uniform procedure, possible missing principal formulas in premisses of $L\&$, $R\&$ are added through weakenings. Further, to fit the derivations on a page, we leave out the $\langle\,\rangle$ -notation:

$$\dfrac{\dfrac{\Gamma_1 \rightarrow \Delta_1, A \& B, A \quad \Gamma_2 \rightarrow \Delta_2, A \& B, B}{\Gamma_1, \Gamma_2 \rightarrow \Delta_1, \Delta_2, A \& B} R\& \quad \dfrac{A, A \& B, \Gamma_3 \rightarrow \Delta_3}{A \& B, \Gamma_3 \rightarrow \Delta_3} L\&}{\Gamma_1, \Gamma_2, \Gamma_3 \rightarrow \Delta_1, \Delta_2, \Delta_3} Cut$$

   The transformed derivation is shown in Figure 1 below. The two upper cuts have a lesser height of cut, the lower cut a shorter cut formula. The transformation is similar with the second form of rule $L\&$.

   (b) The cut formula is $A \vee B$. This case is dual to cut formula $A \& B$.

   (c) The cut formula is $A \supset B$ with the derivation:

$$\dfrac{\dfrac{A, \Gamma_1 \rightarrow \Delta_1, A \supset B, B}{\Gamma_1 \rightarrow \Delta_1, A \supset B} R\supset \quad \dfrac{A \supset B, \Gamma_2 \rightarrow \Delta_2, A \quad B, A \supset B, \Gamma_3 \rightarrow \Delta_3}{A \supset B, \Gamma_2, \Gamma_3 \rightarrow \Delta_2, \Delta_3} L\supset}{\Gamma_1, \Gamma_2, \Gamma_3 \rightarrow \Delta_1, \Delta_2, \Delta_3} Cut$$

   The transformed derivation is too broad to be displayed as such. We shall show the active and principal formulas and indicate the contexts by numbers on top of the sequent arrows so that, for example, $A \xrightarrow{1} A \supset B, B$ stands for the sequent $A, \Gamma_1 \rightarrow \Delta_1, A \supset B, B$:

Here, the active and principal formulas of the transformed derivation.

$$\dfrac{\dfrac{\dfrac{\dfrac{A \xrightarrow{1} A \supset B, B}{\xrightarrow{1} A \supset B}\,R\supset \qquad A \supset B \xrightarrow{2} A}{\xrightarrow{1,2} A}\,\mathit{Cut} \qquad \dfrac{A \xrightarrow{1} B, A \supset B \qquad \dfrac{A \supset B \xrightarrow{2} A \quad B, A \supset B \xrightarrow{3}}{A \supset B \xrightarrow{2,3}}\,L\supset}{A \xrightarrow{1,2,3} B}\,\mathit{Cut}}{\xrightarrow{1,2,3} B}\,\mathit{Cut} \qquad \dfrac{\dfrac{A \xrightarrow{1} A \supset B, B}{\xrightarrow{1} A \supset B}\,R\supset \qquad B, A \supset B \xrightarrow{3}}{B \xrightarrow{1,3}}\,\mathit{Cut}}{\xrightarrow{1,2,3}}\,\mathit{Cut}$$

There are three cuts with the cut formula $A \supset B$ and a lesser cut height, followed by cuts on the shorter formulas A and B.

(d) The cut formula is $\neg A$ and the transformation nearly a special case of the above.

(e) The cut formula is $\forall x A(x)$ with the derivation:

$$\dfrac{\dfrac{\Gamma_1 \to \Delta_1, \forall x A(x), A(y)}{\Gamma_1 \to \Delta_1, \forall x A(x)}\,R\forall \qquad \dfrac{A(t), \forall x A(x), \Gamma_2 \to \Delta_2}{\forall x A(x), \Gamma_2 \to \Delta_2}\,L\forall}{\Gamma_1, \Gamma_1, \Gamma_2, \Gamma_2 \to \Delta_1, \Delta_1, \Delta_2, \Delta_2}\,\mathit{Cut}$$

The transformed derivation is, with the substitution $[t/y]$ in the derivation of the premiss of rule $R\forall$ of the original derivation that gives as a result the derivable sequent $\Gamma_1 \to \Delta_1, \forall x A(x), A(t)$.

$$\dfrac{\dfrac{\Gamma_1 \to \Delta_1, \forall x A(x), A(t) \qquad \dfrac{A(t), \forall x A(x), \Gamma_2 \to \Delta_2}{\forall x A(x), \Gamma_2 \to \Delta_2}\,L\forall}{\Gamma_1, \Gamma_2 \to \Delta_1, \Delta_2, A(t)}\,\mathit{Cut} \qquad \dfrac{\dfrac{\Gamma_1 \to \Delta_1, \forall x A(x), A(y)}{\Gamma_1 \to \Delta_1, \forall x A(x)}\,R\forall \qquad \forall x A(x), A(t), \Gamma_2 \to \Delta_2}{A(t), \Gamma_1, \Gamma_2 \to \Delta_1, \Delta_2}\,\mathit{Cut}}{\Gamma_1, \Gamma_2 \to \Delta_1, \Delta_2}\,\mathit{Cut}$$

The two upper cuts have a lesser cut height, the last cut a shorter cut formula.

(f) The cut formula is $\exists x A(x)$. This case is dual to the previous.

4. One premiss of cut has been derived by weakening. If the weakening formula is the cut formula, the weakening and cut are removed. Else cut is permuted above the weakening.

QED.

$$\dfrac{\dfrac{\Gamma_1 \to \Delta_1, A \& B, A \qquad \dfrac{A, A \& B, \Gamma_3 \to \Delta_3}{A \& B, \Gamma_3 \to \Delta_3}\,L\&}{\Gamma_1, \Gamma_3 \to \Delta_1, \Delta_3, A}\,\mathit{Cut} \qquad \dfrac{\dfrac{\Gamma_1 \to \Delta_1, A \& B, A \quad \Gamma_2 \to \Delta_2, A \& B, B}{\Gamma_1, \Gamma_2 \to \Delta_1, \Delta_2, A \& B}\,R\& \qquad A, A \& B, \Gamma_3 \to \Delta_3}{A, \Gamma_1, \Gamma_2, \Gamma_3 \to \Delta_1, \Delta_2, \Delta_3}\,\mathit{Cut2}}{\Gamma_1, \Gamma_2, \Gamma_3 \to \Delta_1, \Delta_2, \Delta_3}\,\mathit{Cut}$$

Fig. 1. Transformed derivation of case 3.(a)

3 Proof of decidability of intuitionistic propositional logic

The first published application of sequent calculus was a proof of the decidability of intuitionistic propositional logic. Gentzen's argument here is purely verbal, without a single formula [Gentzen, 1934–35, IV Sec. 1]. A sequent is defined to be *reduced* if it has at most three copies of the same formula on either side. Given a derivation of a reduced sequent, all contexts are rewritten so that they have no multiple occurrences of formulas. As Gentzen notes, sequents that appear as conclusions of a rule and as premisses of a successive rule need after this modification not conform to the form of the rules as defined earlier in the setting up of the calculus. He adds, however, that a syntactically correct derivation can always be produced, by the structural rules of weakening, contraction, and exchange.

Gentzen's derivations reduced to single occurrences of each formula are precisely derivations in a sequent calculus with sets.

His main observation is that the occurrences of formulas on either side of a sequent in a derivation in his calculi LI and LK, the rule of cut included, can be limited to at most three, provided this holds for the sequent to be derived. It is easy to see where this number comes from: In Gentzen's rule $L\supset$, we can have one occurrence of $A\supset B$ in both premisses, with the condition of a single occurrence in each context respected. With $\Gamma \equiv A\supset B, \Gamma'$ and $\Delta \equiv A\supset B, \Delta'$, respectively, we have:

$$\frac{A\supset B, \Gamma' \rightarrow A \quad B, A\supset B, \Delta' \rightarrow C}{A\supset B, A\supset B, A\supset B, \Gamma', \Delta' \rightarrow C}\, L\supset$$

Next the conclusion is rewritten as $A\supset B, \Gamma', \Delta' \rightarrow C$, a sequent that can be equally well obtained from the conclusion of $L\supset$ by two contractions on $A\supset B$. Thus, the result is that all derivations of a reduced sequent can be transformed into a reduced form, with at most three copies of a formula in each antecedent or succedent of a sequent and, as we may add, with at most two successive contractions on the same formula. Gentzen's other two-premiss rules $R\&$ and $L\vee$ have shared contexts and behave like one-premiss rules in this respect, with at most two copies of the same formula in the conclusion and at most one contraction. Cut has no principal formula and can produce at most duplications.

4 Cut elimination without the mix rule

The observations at the end of the previous Section can be turned into a cut elimination procedure for Gentzen's calculi LI and LK, with no need for the mix rule trick. Sequents will have multisets as antecedents and succedents, with logical rules as in the above tables but without the bra-ket notation, and the insubstantial difference that we have independent contexts in rules $R\&$ and $L\lor$. Gentzen has shared contexts in these rules, which takes away one step in cut elimination. His rule $L\supset$ is as above, with independent contexts for the reason that that is the only way to arrive at a single-succedent instance of the rule; the proof of cut elimination was designed so that a proof for the intuitionistic calculus LI came out as a single-succedent special case.

To the structural rules are added left and right contraction:

$$\frac{A, A, \Gamma \to \Delta}{A, \Gamma \to \Delta}\,LC \qquad \frac{\Gamma \to \Delta, A, A}{\Gamma \to \Delta, A}\,RC$$

Theorem 3 (Cut elimination for LI). *The rule of cut can be eliminated from derivations in* LI.

Proof: By the above, a given derivation with cuts can be transformed into one in which there are at most two successive contractions on the principal formula with rule $L\supset$ and one with the rest.

All cases except when the right premiss of cut has been derived by contraction have been covered in the proof of theorem 1. We show the most involved case of rule $L\supset$. If between the rule and the cut there are other rules than the two contractions, they can be permuted, so we have the derivation:

$$\dfrac{\Gamma_1 \to A\supset B \qquad \dfrac{\dfrac{\dfrac{A\supset B, \Gamma_2 \to A \quad B, A\supset B, \Gamma_3 \to C}{A\supset B, A\supset B, A\supset B, \Gamma_2, \Gamma_3 \to C}\,L\supset}{A\supset B, A\supset B, \Gamma_2, \Gamma_3 \to C}\,LC}{A\supset B, \Gamma_2, \Gamma_3 \to C}\,LC}{\Gamma_1, \Gamma_2, \Gamma_3 \to C}\,Cut$$

Cut is permuted up at left until the cut formula is principal, concluded by rule $R\supset$, and the transformation of case 3.3 in theorem 1 applies, and similarly for all the other cases. QED.

As Gentzen notes, the notion of a reduced sequent applies as well to the symmetric classical calculus LK. Therefore theorem 2 and its proof turns equally well into a proof for LK:

Theorem 4 (Cut elimination for LK). *The rule of cut can be eliminated from derivations in* LK.

Proof: Similar to the previous. For reasons of comparison in Sec. 7, we show the case of cut formula $\forall x A(x)$ and assume the worst case, with the principal formula found already in the premisses of rules $R\forall$, $L\forall$. After suitable permutations, we have single contractions right after these logical rules, followed by a cut:

$$\dfrac{\dfrac{\dfrac{\Gamma_1 \to \Theta_1, \forall x A(x), A(y)}{\Gamma_1 \to \Theta_1, \forall x A(x), \forall x A(x)}\,R\forall}{\Gamma_1 \to \Theta_1, \forall x A(x)}\,RC \qquad \dfrac{\dfrac{A(t), \forall x A(x), \Gamma_2 \to \Theta_2}{\forall x A(x), \forall x A(x), \Gamma_2 \to \Theta_2}\,L\forall}{\forall x A(x), \Gamma_2 \to \Theta_2}\,LC}{\Gamma_1, \Gamma_2 \to \Theta_1, \Theta_2}\,Cut$$

The premiss of $R\forall$ has the eigenvariable y that can be changed into the term t throughout the derivation of the premiss, after which the transformation is into:

$$\dfrac{\dfrac{\Gamma_1 \to \Theta_1, \forall x A(x), A(t) \qquad \dfrac{\dfrac{A(t), \forall x A(x), \Gamma_2 \to \Theta_2}{\forall x A(x), \forall x A(x), \Gamma_2 \to \Theta_2}\,L\forall}{\forall x A(x), \Gamma_2 \to \Theta_2}\,LC}{\Gamma_1, \Gamma_2 \to \Theta_1, \Theta_2, A(t)}\,Cut \qquad \dfrac{\dfrac{\dfrac{\Gamma_1 \to \Theta_1, \forall x A(x), A(t)}{\Gamma_1 \to \Theta_1, \forall x A(x), \forall x A(x)}\,R\forall}{\Gamma_1 \to \Theta_1, \forall x A(x)}\,RC \qquad A(t), \forall x A(x), \Gamma_2 \to \Theta_2}{A(t), \Gamma_1, \Gamma_2 \to \Theta_1, \Theta_2}\,Cut}{\Gamma_1, \Gamma_1, \Gamma_2, \Gamma_2 \to \Theta_1, \Theta_1, \Theta_2, \Theta_2}\,Cut$$

A sufficient number of contractions gives the original conclusion of cut. QED.

Another case in the proof is covered in the discussion in Sec. 7. It might seem at a first sight that nothing is gained by the above transformation, for the contractions are still there. However, in the left branch, the left premiss of the upper cut is not derived by right contraction but by some other rule, and similarly for the right branch. The elimination of these upper cuts proceeds now by an analysis of how the respective left and right premisses were derived.

4.1 The notion of an altitude line

The method of cut elimination for LKS above, in theorem 2, is quite similar to one connected with the intricate "altitude line" construction (*Höhenlinie*) found in Gentzen's 1938 proof of the consistency of Peano arithmetic [Gentzen, 1938], the origin of which has been wondered by many. It is quite plausible, in the light of how cut elimination is adapted to a calculus with an explicit rule of contraction, theorem 4, that Gentzen had done his original cut elimination theorem of 1933 along the lines given here, then changed for the simpler mix rule proof for expository reasons. As we shall point out, the mix rule cannot be used in the proof theory of arithmetic, so five years later, the original methods of cut elimination resurfaced in his consistency proof.

After the new consistency proof of 1938 that used the classical calculus LK, Gentzen worked hard with the proof theory of intuitionistic arithmetic and analysis. In the Summer of 1944, he prepared a summary of calculi and reductions in consistency proofs, the extant shorthand series WKRd that is explained in some detail in [von Plato, 2012, pp. 356–358] and in detail in the introduction to the first volume of Gentzen's shorthand notes, [Gentzen, 2016]. The central problem was always the multiplication of formulas in steps of reduction, and one suggested remedy in WKRd was a "contractive cut" (*Zusammenziehungsschnitt*). None of Gentzen's attempts at keeping contraction a business separate from the logical rules worked, though. Above we used already Gentzen's notation, here given directly from the edited shorthand manuscript WKRd [Gentzen, 1944, p. 5]:

$$\frac{\Gamma \to \mathfrak{D} \quad \mathfrak{D}\Delta \to \mathfrak{C}}{\langle \Gamma\Delta \rangle \to \mathfrak{C}}$$

Gentzen explains the $\langle\ \rangle$ notation by: "Meaning: the formulas $\Gamma\Delta$ contracted throughout." On p. 4, he uses a similar notation $[\Gamma\Delta]$ for lists without order, with the explanation: "[ ]: arbitrary order of the formulas in question."

The altitude line construction of 1938: Gentzen's 1938 'New formulation of the consistency proof for pure number theory' is in terms of the classical sequent calculus LK. In a letter to Bernays dated 12 May 1938, included in [Gentzen, 2016], he writes that the proof of consistency is organized similarly to the way in which he found the original proof of cut elimination in 1933. Inconsistency is expressed as the derivability of the empty sequent $\to$. Logical rules import a formula on one or the other side of a sequent. Therefore, if the empty sequent is derivable, there is in the derivation an *endpiece* in which only structural rules and steps of inductive inference are found. The uppermost sequents of the endpiece are conclusions of logical rules. Instances of the rule of induction in the endpiece become reduced into repeated cuts, after which contractions and cuts are used to arrive at the empty endsequent. A crucial lemma 3.4.3 in the proof states that there is at least one cut formula in the endpiece such that the cut formula is principal in a left, resp. a right logical rule that delimits the endpiece. Such a cut is permuted so that it applies right after the logical rules, by which the cut formula becomes the principal formula in both premisses of cut. Thus, there is a precise analogy to cut elimination for pure logic, but with a price:

In the proof that the reduction of derivations of the empty sequent through cut elimination terminates, Gentzen uses a strange proof transformation and the related notion of "altitude line."

Here is his example of an altitude line, with the given part of a derivation [Gentzen, 1938, p. 34]:

$$\dfrac{\Gamma_1 \to \Theta_1, F(a)}{\Gamma_1 \to \Theta_1, \forall x F(x)}\,R\forall \qquad \dfrac{F(n), \Gamma_2 \to \Theta_2}{\forall x F(x), \Gamma_2 \to \Theta_2}\,L\forall$$

$$\vdots \qquad\qquad\qquad\qquad \vdots$$

$$\dfrac{\Gamma \to \Theta, \forall x F(x) \qquad\qquad \forall x F(x), \Delta \to \Lambda}{\Gamma, \Delta \to \Theta, \Lambda}\,\textit{Cut}$$

$$\vdots$$

$$\dfrac{\quad}{\Gamma_3 \to \Theta_3}\,\textit{Altitude line}$$

$$\vdots$$

$$\dfrac{\quad}{\to}$$

An *altitude line* in a derivation is any inference line of a cut with a formula such that all cut formulas from the line to the endsequent are shorter. Gentzen uses the word *Höhe* (altitude) even for the length of a cut formula that determines an altitude line, a notion not to be confused with other uses of English equivalents of *Höhe*, such as in the height of a cut in Section 3 or the height of a derivation.

The cut formula $\forall x F(x)$ of the example has a length ϱ that is also assumed to be an altitude; thus, the line of cut is an altitude line. At the indicated *Altitude line* lower down, this altitude diminishes from ϱ to some σ through another cut, so we have $\sigma < \varrho$.

The transformed derivation is, disregarding rules of exchange in the original [Gentzen, 1938, p. 35]:

$$\vdots \qquad\qquad\qquad \vdots \qquad\qquad\qquad \vdots \qquad\qquad\qquad \vdots$$

$$\dfrac{\Gamma_1 \to \Theta_1, F(n)}{\Gamma_1 \to \Theta_1, F(n), \forall x F(x)}\,RW \quad \dfrac{F(n), \Gamma_2 \to \Theta_2}{\forall x F(x), \Gamma_2 \to \Theta_2}\,L\forall \quad \dfrac{\Gamma_1 \to \Theta_1, F(a)}{\Gamma_1 \to \Theta_1, \forall x F(x)}\,R\forall \quad \dfrac{F(n), \Gamma_2 \to \Theta_2}{\forall x F(x), F(n), \Gamma_2 \to \Theta_2}\,LW$$

$$\vdots \qquad\qquad\qquad \vdots \qquad\qquad\qquad \vdots \qquad\qquad\qquad \vdots$$

$$\dfrac{\Gamma \to \Theta, F(n), \forall x F(x) \qquad \forall x F(x), \Delta \to \Lambda}{\Gamma, \Delta \to \Theta, \Lambda, F(n)} \qquad \dfrac{\Gamma \to \Theta, \forall x F(x) \qquad \forall x F(x), F(n), \Delta \to \Lambda}{F(n), \Gamma, \Delta \to \Theta, \Lambda}$$

$$\vdots \qquad\qquad\qquad\qquad\qquad\qquad \vdots$$

$$\dfrac{\dfrac{\quad}{\Gamma_3 \to \Theta_3, F(n)} \qquad\qquad \dfrac{\quad}{F(n), \Gamma_3 \to \Theta_3}}{\dfrac{\Gamma_3, \Gamma_3 \to \Theta_3, \Theta_3}{\Gamma_3 \to \Theta_3}\,\textit{Contractions}}\,\textit{New Cut}$$

$$\vdots$$

$$\dfrac{\quad}{\to}$$

Gentzen indicates only in the text but not in the figure that altitude lines with a lowered altitude $< \varrho$ appear higher up in the transformed derivation, above the sequents $\Gamma_3 \to \Theta_3, F(n)$ and $F(n), \Gamma_3 \to \Theta_3$. On the whole, the idea is that altitude lines are being pushed up in the endpiece. The empty endsequent has altitude 0, and the transformations force this to be the case for the whole endpiece, by which there is no derivation of the empty sequent.

When explaining the transformation, Gentzen writes (p. 34):

> The inference figures represent an introduction and an elimination of ∀ in $\forall x F(x)$. Following the original basic idea, both should be put aside and $\forall x F(x)$ should be substituted by $F(n)$– its grade [length] is one less; in place of the cut with the cut formula $\forall x F(x)$ there would occur a cut with the cut formula $F(n)$. There appears, however, the difficulty already mentioned, namely that the formula $\forall x F(x)$ can have been used and even introduced in several places.

To resolve the difficulty, the mix rule cannot be used, as will be made clear in our closing paragraph. Instead, cuts on the formula $\forall x F(x)$ are maintained, but Gentzen notes that the essential point of the transformation is that above the two cuts on the old cut formula, the inferences have been simplified because one step of logical inference has been removed in each. This, precisely, is what happened in the transformations in our proofs of cut elimination for LIS and LKS, theorems 1 and 2 of Section 3. As to why the steps of weakening are added in which the formula $\forall x F(x)$ reappears, Gentzen notes that "this is a matter of convenience, for one would have to count anyway with its appearance further down, so in this way the new form of the derivation is taken over from the old one in the most convenient way."

5 Contraction and altitude lines

As Gentzen notes, copies of the formula $\forall x F(x)$ can appear through branches above the step of the original cut other than the branch shown. (In fact, Gentzen's figure has three-pronged upward-pointing "forks" that we failed to reproduce.) These formula occurrences would have to be contracted before the cut. Assume now that instead of a weakening, the extra copy of the cut formula was already there in the premisses of the logical rules of the original derivation to be transformed, and vanished right after the step in which it became principal, in a step of contraction. As noted above in Section 4, with a one-premiss rule, there need be at most one such contraction present:

$$\dfrac{\dfrac{\dfrac{\dfrac{\Gamma_1 \to \Theta_1, \forall x F(x), F(a)}{\Gamma_1 \to \Theta_1, \forall x F(x), \forall x F(x)}\, R\forall}{\Gamma_1 \to \Theta_1, \forall x F(x)}\, RC}{\begin{array}{c}\vdots\\ \Gamma \to \Theta, \forall x F(x)\end{array}} \qquad \dfrac{\dfrac{\dfrac{F(n), \forall x F(x), \Gamma_2 \to \Theta_2}{\forall x F(x), \forall x F(x), \Gamma_2 \to \Theta_2}\, L\forall}{\forall x F(x), \Gamma_2 \to \Theta_2}\, LC}{\begin{array}{c}\vdots\\ \forall x F(x), \Delta \to \Lambda\end{array}}}{\begin{array}{c}\Gamma, \Delta \to \Theta, \Lambda\\ \vdots\end{array}}\, Cut$$

$$
\begin{array}{c}
\vdots \\
\Gamma, \Delta \rightarrow \Theta, \Lambda \\
\vdots \\
\hline
\Gamma_3 \rightarrow \Theta_3 \quad \textit{Altitude line} \\
\vdots \\
\hline
\rightarrow
\end{array}
$$

Permuting *Cut* up, the essential case to consider is:

$$
\dfrac{\dfrac{\dfrac{\Gamma_1 \rightarrow \Theta_1, \forall x F(x), F(a)}{\Gamma_1 \rightarrow \Theta_1, \forall x F(x), \forall x F(x)}\,R\forall}{\Gamma_1 \rightarrow \Theta_1, \forall x F(x)}\,RC \qquad \dfrac{\dfrac{F(n), \forall x F(x), \Gamma_2 \rightarrow \Theta_2}{\forall x F(x), \forall x F(x), \Gamma_2 \rightarrow \Theta_2}\,L\forall}{\forall x F(x), \Gamma_2 \rightarrow \Theta_2}\,LC}{\Gamma_1, \Gamma_2 \rightarrow \Theta_1, \Theta_2}\,Cut
$$

$$
\begin{array}{c}
\vdots \\
\Gamma, \Gamma_2 \rightarrow \Theta, \Theta_2 \\
\vdots \\
\Gamma, \Delta \rightarrow \Theta, \Lambda \\
\vdots \\
\hline
\Gamma_3 \rightarrow \Theta_3 \\
\vdots \\
\hline
\rightarrow
\end{array}
$$

After the cut, there follow the steps that led originally from $\Gamma_1 \rightarrow \Theta_1, \forall x F(x)$ to $\Gamma \rightarrow \Theta, \forall x F(x)$, with $\forall x F(x)$ kept intact, then the steps that led originally from $\forall x F(x), \Gamma_2 \rightarrow \Theta_2$ to $\forall x F(x), \Delta \rightarrow \Lambda$. This derivation is transformed following strictly Gentzen's general procedure for the specific case of contractions. We have the substitution $[n/a]$ in the derivation of the sequent $\Gamma_1 \rightarrow \Theta_1, \forall x F(x), F(a)$ and just one step of logical inference instead of two above each of the cuts on $\forall x F(x)$:

$$
\dfrac{\dfrac{\Gamma_1 \rightarrow \Theta_1, \forall x F(x), F(n) \qquad \dfrac{\dfrac{F(n), \forall x F(x), \Gamma_2 \rightarrow \Theta_2}{\forall x F(x), \forall x F(x), \Gamma_2 \rightarrow \Theta_2}\,L\forall}{\forall x F(x), \Gamma_2 \rightarrow \Theta_2}\,LC}{\Gamma_1, \Gamma_2 \rightarrow \Theta_1, \Theta_2, F(n)}\,Cut \qquad \dfrac{\dfrac{\dfrac{\Gamma_1 \rightarrow \Theta_1, \forall x F(x), F(a)}{\Gamma_1 \rightarrow \Theta_1, \forall x F(x), \forall x F(x)}\,R\forall}{\Gamma_1 \rightarrow \Theta_1, \forall x F(x)}\,RC \qquad F(n), \forall x F(x), \Gamma_2 \rightarrow \Theta_2}{F(n), \Gamma_1, \Gamma_2 \rightarrow \Theta_1, \Theta_2}\,Cut}{\Gamma_1, \Gamma_2 \rightarrow \Theta_1, \Theta_2}\,Cut
$$

$$
\begin{array}{c}
\vdots \\
\Gamma, \Gamma_2 \rightarrow \Theta, \Theta_2 \\
\vdots \\
\Gamma, \Delta \rightarrow \Theta, \Lambda \\
\vdots \\
\hline
\Gamma_3 \rightarrow \Theta_3 \\
\vdots \\
\hline
\rightarrow
\end{array}
$$

As in the proof of theorem 4, the two contractions in the derivation of the premisses of the upper cuts remain, but the other premiss has in both cases a reduced derivation.

In the proof of theorem 2, cut elimination for LKS, we assumed the principal formula to occur in each premiss. Should it be lacking from some, either a simpler cut elimination procedure is made, but with more case distinctions as in the proof of theorem 1, or, for a uniform procedure as in our proof of theorem 2, possible missing principal formulas in premisses are added through weakenings, just as in Gentzen's 1938 proof.

If contraction is an explicit rule, we have now precisely the problematic case of cut elimination in the calculus LK, namely the one in which the cut formula has been contracted and that led Gentzen to introduce the mix rule idea [Gentzen, 1934–35]:. For a second example, consider a cut on $A\supset B$ with the "worst case" of $A\supset B$ repeated in each of the three premisses, as in the proof of theorem 2 (case 3.3), and with contractions before the instance of *Cut*. Even this case would be one in a detailed proof of theorem 4:

$$\dfrac{\dfrac{\dfrac{A,\Gamma_1\rightarrow\Delta_1,A\supset B,B}{\Gamma_1\rightarrow\Delta_1,A\supset B,A\supset B}\,R\supset}{\Gamma_1\rightarrow\Delta_1,A\supset B}\,RC \qquad \dfrac{\dfrac{\dfrac{A\supset B,\Gamma_2\rightarrow\Delta_2,A \quad B,A\supset B,\Gamma_3\rightarrow\Delta_3}{A\supset B,A\supset B,A\supset B,\Gamma_2,\Gamma_3\rightarrow\Delta_2,\Delta_3}\,L\supset}{A\supset B,A\supset B,\Gamma_2,\Gamma_3\rightarrow\Delta_2,\Delta_3}\,LC}{A\supset B,\Gamma_2,\Gamma_3\rightarrow\Delta_2,\Delta_3}\,LC}{\Gamma_1,\Gamma_2,\Gamma_3\rightarrow\Delta_1,\Delta_2,\Delta_3}\,Cut$$

The contracted premisses of *Cut* cannot be resolved by simply cutting several times, say twice in the premiss of *RC*, as in (with duplications of the contexts omitted):

$$\dfrac{\dfrac{\dfrac{A,\Gamma_1\rightarrow\Delta_1,A\supset B,B}{\Gamma_1\rightarrow\Delta_1,A\supset B,A\supset B}\,R\supset \quad \begin{matrix}\vdots\\ A\supset B,\Gamma_2,\Gamma_3\rightarrow\Delta_2,\Delta_3\end{matrix}}{\Gamma_1,\Gamma_2,\Gamma_3\rightarrow\Delta_1,\Delta_2,\Delta_3,A\supset B}\,Cut \qquad \begin{matrix}\vdots\\ A\supset B,\Gamma_2,\Gamma_3\rightarrow\Delta_2,\Delta_3\end{matrix}}{\Gamma_1,\Gamma_2,\Gamma_3\rightarrow\Delta_1,\Delta_2,\Delta_3}\,Cut$$

The "vertical dots" stand for a derivation of the right premiss of the original cut, and it is seen that the lower cut is not reduced, because the derivation of the left premiss has grown. Therefore the derivation with contracted premisses of cut has to be transformed analogously to case 3.3 of our theorem 2. There will be altogether five cuts in the new derivation, three of them on $A\supset B$ with the height of cut reduced in each, followed by two on the shorter formulas A and B. The transformation of the derivation is displayed for typographical reasons in parts.

$$\dfrac{\dfrac{\dfrac{A,\Gamma_1\rightarrow\Delta_1,A\supset B,B}{\Gamma_1\rightarrow\Delta_1,A\supset B,A\supset B}\,R\supset}{\Gamma_1\rightarrow\Delta_1,A\supset B}\,RC \qquad A\supset B,\Gamma_2\rightarrow\Delta_2,A}{\Gamma_1,\Gamma_2\rightarrow\Delta_1,\Delta_2,A}\,Cut$$

The second part is:

$$\dfrac{A, \Gamma_1 \to \Delta_1, A \supset B, B \qquad \dfrac{\dfrac{\dfrac{A \supset B, \Gamma_2 \to \Delta_2, A \qquad B, A \supset B, \Gamma_3 \to \Delta_3}{A \supset B, A \supset B, A \supset B, \Gamma_2, \Gamma_3 \to \Delta_2, \Delta_3}\, L\supset}{A \supset B, A \supset B, \Gamma_2, \Gamma_3 \to \Delta_2, \Delta_3}\, LC}{A \supset B, \Gamma_2, \Gamma_3 \to \Delta_2, \Delta_3}\, LC}{A, \Gamma_1, \Gamma_2, \Gamma_3 \to \Delta_1, \Delta_2, \Delta_3, B}\, Cut$$

The third part is:

$$\dfrac{\dfrac{\dfrac{A, \Gamma_1 \to \Delta_1, A \supset B, B}{\Gamma_1 \to \Delta_1, A \supset B, A \supset B}\, R\supset}{\Gamma_1 \to \Delta_1, A \supset B}\, RC \qquad B, A \supset B, \Gamma_3 \to \Delta_3}{B, \Gamma_1, \Gamma_3 \to \Delta_1, \Delta_3}\, Cut$$

These parts are combined by two cuts as follows, with the duplications of the contexts omitted:

$$\dfrac{\dfrac{\begin{array}{c}\vdots\\ \Gamma_1, \Gamma_2 \to \Delta_1, \Delta_2, A\end{array} \qquad \begin{array}{c}\vdots\\ A, \Gamma_1, \Gamma_2, \Gamma_3 \to \Delta_1, \Delta_2, \Delta_3, B\end{array}}{\Gamma_1, \Gamma_2, \Gamma_3 \to \Delta_1, \Delta_2, \Delta_3, B}\, Cut \qquad \begin{array}{c}\vdots\\ B, \Gamma_1, \Gamma_3 \to \Delta_1, \Delta_3\end{array}}{\Gamma_1, \Gamma_2, \Gamma_3 \to \Delta_1, \Delta_2, \Delta_3}\, Cut$$

As can be seen from this transformation, from the one for the cut formula $\forall x F(x)$, as well as from the transformations displayed in the proof of theorem 2, the procedure is somewhat involved, even if in principle clear.

The proof transformations we have found for sequent calculi with implicit contraction are not altogether unknown: In the contraction-free intuitionistic sequent calculus **G3i**, the principal formula $A \supset B$ of rule $L\supset$ is repeated in the antecedent of the left premiss, with the effect of an implicit contraction at that point. In the proof transformation, there is first a cut on $A \supset B$ with a reduced height of cut, followed by cuts on A and B (as in [Negri and von Plato, 2001, p. 40]).

We cited Gentzen's reason for the use of the mix rule in the published proof, namely to make the proof easier. What he writes clearly indicates that he had some previous proof at hand that was changed. The most likely candidate for such a proof is the one presented here. Gentzen had, as is now well known, written down a detailed proof of normalization for intuitionistic natural deduction in 1933 [Gentzen, 2008]. Somehow the level of complexity escaped out of hands with cut elimination for LK, in comparison to natural deduction, so for reasons of exposition, Gentzen took into use the mix rule.

Finally, one may wonder why Gentzen went to the trouble of introducing the cut rule and the altitude line construction in 1938, instead of the well-behaving mix rule of 1933. It might seem a routine matter to treat the proof theory of arithmetic with the sequents-as-sets idea. Instead, what happens is that the cut elimination procedures of Section 3 are useless, because two cuts cannot necessarily be permuted with each other when sets are used, contrary to a calculus with multisets or lists and explicit contractions. The same is true of Gentzen's mix rule, so here is the true reason for the intricacies of the 1938 paper: The essential step in the 1938 proof of consistency, namely the permutation of a "suitable cut" as given by Gentzen's lemma 3.4.3 to the upper limit of the endpiece, would get blocked. Were it possible to do the consistency proof with sets and thus no contractions in the endpiece, an ordinal assignment could be given that beats Gödel's theorem! This particular detail about "sequents with sets" was found out in 2005 and remained a puzzle until Michael Rathjen pointed out that two instances of the mix rule do not necessarily permute.

References

B. Cooper and J. van Leeuwen. *Alan Turing: His Work and Impact.* Elsevier, 2013.

G. Gentzen. Untersuchungen über das logische Schliessen. *Mathematische Zeitschrift*, 39: 176–210, 1934–35.

G. Gentzen. Neue Fassung des Widerspruchsfreiheitsbeweises für die reine Zahlentheorie. *Forschungen zur Logik und zur Grundlegung der exakten Wissenschaften*, 4:19–44, 1938.

G. Gentzen. Review of Ono (1938). *Zentralblatt für Mathematik und ihre Grenzgebiete*, 19:242, 1939.

G. Gentzen. **WKRd** (Widerspruchsfreiheitsbeweis: Kalküle und Reduktionen). Shorthand notes dated VIII.44, 8 pages. Available at the Philosophisches Archiv der Universität Konstanz. 1944.

G. Gentzen. The normalization of derivations. *The Bulletin of Symbolic Logic*, 14:245–257, 2008.

G. Gentzen. *Saved from the Cellar: Gerhard Gentzen's Shorthand Notes on Logic and Foundations of Mathematics.* With an Introduction and English Translation by Jan von Plato, Springer, 2016.

R. Hakli and S. Negri. Does the deduction theorem fail for modal logic? *Synthese*, 187:849–867, 2012.

D. Hilbert and W. Ackermann. *Grundzüge der theoretischen Logik.* Springer, 1928.

S. Negri and J. von Plato. *Structural Proof Theory.* Cambridge University Press, 2001.

K. Ono. Logische Untersuchungen über die Grundlagen der Mathematik. *Journal of the Faculty of Science, Imperial University of Tokyo*, 3:329–389, 1938.

J. von Plato. A proof of Gentzen's *hauptsatz* without multicut. *Archive for Mathematical Logic*, 40:9–18, 2001.

J. von Plato. Gentzen's proof systems: byproducts in a work of genius. *The Bulletin of Symbolic Logic*, 18:313–367, 2012.

J. von Plato. *Elements of Logical Reasoning*. Cambridge University Press, 2013.

A. Turing. *The reform of mathematical notation and phraseology.* Manuscript in the Turing Archives, 1944.

Wolfram Pohlers

Hilbert's Programme and Ordinal Analysis

Abstract: We claim that there are two aspects of Hilbert's programme. The quest for finitistic consistency proofs and the elimination of ideal objects. Though rigid finitistic consistency proofs turned out to be impossible by Gödel's second incompleteness theorem we argue that ordinal analysis not only yields an elimination of ideal elements but is actually based on elimination of ideal methods.

Keywords: Hilbert's programme, Ordinal analysis, Ordinal notations, Semiformal calculi, Cut elimination, Normalization, Provably recursive functions

Mathematics Subject Classification 2010: 03-02, 03A99, 03F05, 03F15, 03F35

1 Introduction

In his famous talk on unsolved mathematical problems presented at the international congress of mathematicians in Paris 1900 Hilbert named the "Consistency of the Arithmetical Axioms" as the second grand mathematical problem, directly after the continuum hypothesis. Towards a solution of this problem Hilbert himself launched a programme—Hilbert's programme—to solve this problem. He published his programme as a "first notification (Erste Mitteilung)" 1922 in [Hilbert, 1922]. The aim of Hilbert's programme was to establish the consistency of Analysis by intangible means, finitistic means as he called them.

That this was an illusion became clear when Kurt Gödel in 1931 published his incompleteness results ([Gödel, 1931]). According to Gödel's results a rigid consistency proof for sufficiently rich axiom systems by intangible finitistic means is impossible. In this sense one can state that Hilbert's programme failed.

Nevertheless, unperturbed by Gödel's results[1] Gentzen published 1936 in [Gentzen, 1936] the first version of his famous consistency proof for Peano arithmetic and a bit later in 1938 the improved version. Together with his 1943 paper [Gentzen, 1943] these papers gave birth to a branch of proof theory which we nowadays call "ordinal analysis".

1 In page 560 in [Gentzen, 1936] he addresses Gödel's results and remarks that his proof does not conflict with Gödel's theorem.

Wolfram Pohlers: Institut für mathematische Logik und Grundlagenforschung, Westfälische Wilhelms-Universität, Münster, Germany, e-mail: pohlers@math.uni-muenster.de

From a malicious point of view one could state that Gentzen's proofs did not really contribute to Hilbert's programme since his methods did not—and by Gödel's results even could not—meet the strict finitistic ambitions of the programme.

In this paper we want to argue that ordinal analysis, launched by Gentzen's work, not only pinpoints the infinitary content of formal theories but also helps us to a deep understanding of another aspect of Hilbert's programme which is independent of the aspect of finitistic consistency proofs: *the elimination of ideal elements.* We will even indicate that the ordinal analysis of a formal theory rests on the elimination of ideal objects.

I want to thank the anonymous referee for helpful comments.

2 Hilbert's programme revisited

In his 1927 talk given in Hamburg (published in [Hilbert, 1928]) Hilbert mentions the importance of the addition of "ideal propositions" to the "real propositions" in a mathematical theory and claims that only *"the introduction of ideal propositions makes mathematical reasoning possible."*[2] To secure that the addition of ideal propositions does not lead to inconsistencies we have to check that *"after the elimination of the ideal entities the old entities keep their original validity"*.[3] So we again have the problem of a consistency proof but in the milder form that we only have to secure that the "old entities" keep their validity.

Admittedly the whole section in page 9 of [Hilbert, 1928] appears rather vague to me. Although there are examples for "ideal entities" it becomes—at least for my reading—not completely clear which entities have to be regarded as "ideal". He apparently subsumes under "ideal" any form of abstract reasoning. I will come back to this later.

On the other hand there is a passage in page 15 of [Hilbert, 1928] which I think makes pretty clear what Hilbert understood by "real entities". There Hilbert compares mathematical reasoning with the work of a physicist. In my translation the passage says *"The physicist requires for a theory that its theorems can be formally derived from the laws of nature and its hypotheses alone without referring to outside perceptions. Only certain combinations and conclusions of physical laws are check-*

2 "[...], so ist durch Einführung der idealen Aussagen erst wissenschaftliche Mathematik möglich". Page 9 in [Hilbert, 1928].

3 "[...], wenn also die Beziehungen, die sich bei der Elimination der idealen Gebilde für die alten Gebilde herausstellen, stets im alten Bereich gültig sind." [ibd]

able by experiments—this is also true for my proof theory in which only real statements are verifiable."[4] According to this passage "real statements" are statements that are verifiable corresponding to the physical statements that are experimentally checkable. We cannot perform experiments in mathematics. But the mathematical activity which is closest to an experiment is checking a statement by a formal computation, i.e., a computation that in principle could be performed by a machine.

Clearly, we cannot expect to check a universally quantified sentence, i.e., an arithmetical Π^0_1-sentence, by a computation but we can check any arbitrary numerical instances of it by a computation. This again resembles the situation in physics where we also can only experimentally check *instances* of a predicted physical phenomenon.[5] That this interpretation of Hilbert's intention is not completely absurd is substantiated by the passage in page 194 of [Hilbert, 1935] (a passage taken from [Hilbert, 1931]) in which he introduces an ω-rule for Π^0_1-sentences which he even regards as finitistic.[6]

Based on this interpretation we may subsume Π^0_1-statements under "real statements". Extending this interpretation we may even subsume Π^0_2-statements, i.e. sentences of the form $(\forall x)(\exists y)F(x, y, \vec{n})$ with a primitive recursive matrix $F(x, y, \vec{n})$, under "real statements" provided that we can find a test-function, i.e., a computable function f which for any numeral n provides us with a value $f(n, \vec{n})$ such that $(\exists y)F(n, y, \vec{n})$ becomes checkable by testing all instances $k \leq f(n, \vec{n})$ for the variable y. In such a situation we can check every instance of $(\forall x)(\exists y)F(x, y, \vec{n})$ by an "experiment", i.e., by finitely many computations. For a true Π^0_2-sentence $(\forall x)(\exists y)F(x, y, \vec{n})$ the function $f_F\colon x \mapsto \min\{y \mid F(x, y, \vec{n})\}$ is computable and thus a test-function. Therefore any Π^0_2-sentence that is provable in a sound theory T possesses a test-function.

Given an axiomatized theory T we of course cannot expect to obtain a global test-function which simultaneously tests all the provable Π^0_2-sentences of T. Let f be any arithmetical function and $m := f(0) + 1$. Then f cannot be a testfunction

4 Der Physiker verlangt gerade von einer Theorie, dass ohne die Heranziehung von anderweitigen Bedingungen aus den Naturgesetzen oder Hypothesen die besonderen Sätze allein durch Schlüsse, also auf Grund eines reinen Formelspiels abgeleitet werden. Nur gewisse Kombinationen und Folgerungen der physikalischen Gesetze können durch Experimente kontrolliert werden—so wie in meiner Beweistheorie nur die realen Aussagen unmittelbar einer Verifikation fähig sind. (Cited from [Hilbert, 1928] page 15)

5 Which is in accordance to Popper's criterion that physical theories can at most be falsifiable.

6 Perhaps this is one of the reasons why Hilbert never directly reacted to Gödel's results which rely on an ω-defect. Though $\neg Bew(n, \ulcorner 0 = 1 \urcorner)$ is provable for any numeral n its universal quantification becomes unprovable. This type of ω-rule was apparently a triviality for Hilbert.

for the sentence $(\forall x)(\exists y)[y = m]$.[7] So all we can expect is a function Φ_T which eventually majorizes all the test-functions for the Π^0_2-sentences with parameters that are provable in T. These considerations motivate the following definition.

Definition 2.1. *A theory T whose language comprises the language of arithmetic is experimentally checkable if there is a computable function Φ_T such that for any Π^0_2-sentence with parameters that is provable in T a test-function is eventually majorized by Φ_T. More precisely this means that for any Π^0_2-sentence $F(\vec{n})$ that is provable in T there is a number k such that $\lambda x\,.\,\Phi_T(k + x, \vec{n})$ is a test-function for $F(\vec{n})$. We then say that Φ_T designs experiments for the theory T.*
In a Π^0_2 analysis of an axiom system T we try to construct a computable function that designs experiments for T.

Note 2.2. *As an aside we want to emphasize the role of the parameters in the above Π^0_2-sentences. For different parameters e in a T-provable Π^0_2-sentence $(\forall x)(\exists y)F(x, y, e)$ there might be different proofs depending on e. As we will indicate later the test-function for $(\forall x)(\exists y)F(x, y, e)$ will depend on its T-proof. So it makes a big difference if T uniformly proves $(\forall e)(\forall x)(\exists y)F(x, y, e)$ or just $(\forall x)(\exists y)F(x, y, e)$ for certain number parameter e.*

This corroborates the importance of the index e in the characterization of the provably recursive functions $\{e\}(x) \simeq U(\mu y.\mathrm{T}(e, x, y))$ of a sound theory T, i.e., the functions for which T proves $(\forall x)(\exists y)\mathrm{T}(e, x, y)$ where T denotes the Kleene predicate.

According to this interpretation of "real entities" all statements above, i.e., more complex than Π^0_2 have to be regarded as "ideal". Loosely speaking real statements are statements which represent a computational content while ideal statements are abstract in that sense that their instances cannot be computationally checked. We will argue that this can be pinpointed by the sentence *all methods that take ω for an "actual infinitary quantity" should be considered as "ideal"*.

Given a formal theory T Hilbert's programme of "elimination of ideal entities" then requires *the elimination of all ideal statements in a T proof of a Π^0_2-statement.*

We will report that an elimination of ideal statements can be obtained by refinements of Gentzen's ordinal analysis and thus exhibits an additional aspect how ordinal analysis contributes to Hilbert's programme. We will moreover indicate that this simultaneously allows for the construction of a test-function for the Π^0_2-statement and moreover a computable function that designs experiments for T.

7 This simple argument is due to the referee.

In course of this paper we will see that it is useful to distinguish different levels of ideality. The strict one, which comprises all statements above Π_2^0—and thus already regards ω as an ideal ordinal—and the more abstract ones which comprise the introduction of ideal notions on the more abstract level, e.g., sets obtained by comprehensions, cardinals, ordinals obtained by reflections and similar "abstract" processes. Before we go into the details of this distinction we briefly resume the main features of ordinal analysis.

3 Ordinal analysis

By an ordinal analysis of a mathematical theory T we commonly understand the computation of its *proof theoretic ordinal*, i.e., the supremum of the order-types of in T elementarily definable well-orderings whose well-foundedness is provable from the axioms of T. Though I do not want to become too technical I will sketch the main lines of an ordinal analysis. To do this we have to distinguish two forms of proof theory. *Predicative proof theory* which analyzes predicative theories and centers around subsystems of second order number theory and and *impredicative proof theory* which is needed in the analysis of impredicative theories and today centers primarily around subsystems of set theory.

3.1 Predicative proof theory

Predicative proof theory originated from Gentzen's ordinal analysis of Peano arithmetic. A central point in predicative proof theory is the fact that in the language of arithmetic well-foundedness is a genuine Π_1^1-notion.

The biggest progress since Gentzen is the use of infinitary logic. The key notion here is the well-known ω-rule which allows us to derive a formula $(\forall x)F(x)$ from the infinitely many premises $F(n)$ for all numerals $n \in \mathbb{N}$. A theorem by Henkin and Orey ([Henkin, 1954], [Orey, 1956]) shows that ω-logic is complete for the Π_1^1-sentences that are true in the standard model $\mathbb{N}$. This can be refined in so far that even cut-free ω-logic is complete for Π_1^1-sentences in that sense that a Π_1^1-sentence $(\forall X)F(X)$ is true in $\mathbb{N}$ iff there is a countable (even recursive) ordinal α such that $\frac{\alpha}{0} F(X)$. Proof figures in cut-free ω-logic are well-founded ω-branching infinitary trees whose lengths are canonically measured by ordinals. The notion $\frac{\alpha}{0} F$ then stands for the fact that there is a proof tree for F in cut-free ω-logic whose length is bounded by the ordinal α.

Using the cut-free ω-calculus we can define the truth-complexity of a Π^1_1-sentence by

$$\mathrm{tc}((\forall X)F(X)) := \begin{cases} \min\{\alpha \mid \vdash^{\alpha}_{0} F(X)\} & \text{if this exist} \\ \omega_1 & \text{otherwise.} \end{cases}$$

If we express the well-foundedness of an arithmetically definable order-relation $\prec$ (with field $\mathbb{N}$, for simplicity) by

$$Wf(\prec) \quad \Leftrightarrow \quad (\forall X)[(\forall x)[(\forall y{\prec}x) y \in X \to x \in X] \to (\forall x)[x \in X]]$$

we obtain the connection between the order-type of $\prec$ and the truth-complexity of $Wf(\prec)$ by the *boundedness property*

$$\mathrm{otyp}(\prec) = \mathrm{tc}(Wf(\prec)) \tag{1}$$

which is at least exactly true if $\prec$ has a limit order-type. In general we only have $\leq$, but this is due to silly syntactical details of the cut-free ω-calculus.

To compute upper bounds for the proof theoretic ordinal of a theory it thus suffices to compute upper bounds for the truth-complexities of the provable (pseudo)[8] Π^1_1-sentences that are provable in the theory.

The classical example for an ordinal analysis is Gentzen's result for Peano arithmetic. Via truth-complexities the computation of an upper bound for the proof theoretic ordinal of Peano arithmetic is obtained by unraveling a formal proof in Peano arithmetic into an infinitary proof in ω-logic and then eliminating the cuts. The length of the such obtained cut free derivation then yields an upper bound for the truth-complexity of the proved formula.

This pattern can be extended to "predicative" subtheories of second order number theory. Second order number theory is Peano arithmetic formulated in a second order language augmented by the comprehension scheme

$$(\exists X)[(\forall y)[y \in X \leftrightarrow F(y)]]. \tag{CA}$$

Here $F(y)$ is a formula in the language of second order number theory which must not contain the second order variable X freely.

Since the formula $F(y)$ in the comprehension scheme may contain second order quantifiers which are supposed to range over sets of natural numbers it incorporates "impredicative definitions". Informally we call an object "impredicatively

8 By a pseudo Π^1_1-sentence we understand an arithmetical formula that must not contain free number variables but may contain free second order variables.

defined" if its definition is circular, i.e., if its definition refers to an entity to which the object itself belongs.

A remedy to avoid circularities is *ramified analysis*. We obtain a ramification of second order number theory by

- saying that all quantifiers ranging over natural numbers are quantifiers of stage 0,
- defining the stage of a formula as the maximum of the stages of all quantifiers occurring in it,
- introducing comprehension terms $\{y \mid F(y)\}$ of stage $\alpha + 1$ if $F(y)$ is a formula of stage α and
- stipulating that quantifiers of stage $\sigma > 0$ are supposed to range over comprehension terms of stages less than σ.

In addition to the ω-rule there is also a canonical infinitary rule for the language of ramified analysis which says that you can derive a formula $(\forall X^{\sigma})F(X)$ from the infinitely many premises $F(T)$ where T ranges over all comprehension terms of stages less than σ. Derivations in ramified analysis are thus infinite well-founded trees whose lengths can be measured by countable ordinals. To make this formally work we have to fix some kind of calculus for ramified analysis. We will not– and need not–introduce such a calculus in all details but, since we want to refer to that also later, at least briefly indicate such a calculus.

We use a one-sided sequent calculus a lâ Tait which derives finite sets of formulas which are to be read as finite disjunctions.[9] We denote by $\vdash^{\alpha}_{\rho} \Delta$ that there is a derivation of Δ of length $\leq \alpha$ whose cut-formulas have all a complexity strictly less than ρ. To give two examples of the rules we display the cut rule

$$\vdash^{\alpha}_{\rho} \Delta, F\,,\quad \vdash^{\alpha}_{\rho} \Delta, \neg F\,,\quad \alpha < \beta \text{ and } \mathrm{rnk}(F) < \rho \quad\Rightarrow\quad \vdash^{\beta}_{\rho} \Delta$$

and the infinitary universal quantification rule

$$\begin{array}{l}\vdash^{\alpha_T}_{\rho} \Delta, F(T) \text{ and } \alpha_T < \beta \text{ for all} \\ \text{comprehension terms of stages} < \tau\end{array} \quad\Rightarrow\quad \vdash^{\beta}_{\rho} \Delta, (\forall X^{\tau})F(X)\,.$$

The main meta-theorem for the calculus of ramified analysis is the following.

Theorem 3.1. *Any derivation* $\vdash^{\alpha}_{\beta+\omega^{\rho}} \Delta$ *can be converted into a derivation* $\vdash^{\varphi_{\rho}(\alpha)}_{\beta} \Delta\,.$

Here φ denotes the binary Veblen function which is defined such that $\varphi_0(\alpha) = \omega^{\alpha}$ and for $\rho > 0$ the function φ_{ρ} enumerates the common fixed-points of the

9 Details about such calculi can e.g. be found in [Pohlers, 2009].

functions φ_ξ for $\xi < \rho$. An ordinal that is closed under the binary Veblen function is called *strongly critical.*

The technique to obtain upper bounds for subsystems of second order number theory is in principle the same as for Peano arithmetic. The first step is to unravel formal proofs into infinitary proofs in ramified analysis and then to use Theorem 3.1 and the boundedness property (1) to obtain an upper bound for the proof theoretic ordinal.

Since there are no circularities in ramified analysis any theory that can be embedded into it is apparently predicative. According to Theorem 3.1 any subtheory of second order number theory that can be embedded into ramified analysis has a proof theoretic ordinal that is less than or equal to the first strongly critical ordinal Γ_0. It is a famous result by S. Feferman and K. Schütte that conversely every ordinal less than Γ_0 can be predicatively justified. An ordinal α is predicatively justified if it can be represented by a well-ordering $\prec$ on the natural numbers for which not only the definition of $x \prec y$ is non-circular but also the proof of its well-foundedness can be embedded into a system RA_α which is ramified analysis in which all formulas and comprehension terms have stages less than α and only derivations of lengths less than α are permitted. In this technical sense the Schütte-Feferman ordinal Γ_0 is regarded as the bounding ordinal for predicativity.

There are series of subsystems of second order number theory that are embeddable into ramified analysis reaching from arithmetical comprehension, which is conservative over Peano arithmetic over Δ^1_1-comprehension, Σ^1_2-choice to the system $(\mathsf{ATR})_0$ of autonomously iterated arithmetical comprehension that reaches Γ_0.

3.2 A remark on semi-formal systems

We call deduction calculi with infinitary rules *semi-formal.* Before we turn to impredicative systems we want to make some general remarks on semi-formal systems. For more information about semi-formal systems in general cf. [Pohlers, 2015].

Let $\mathfrak{S}$ be any countable structure. A semi-formal system for $\mathfrak{S}$ is commonly built around a verification calculus for $\mathfrak{S}$. The verification calculus mimics the truth definition for the elementary language $\mathcal{L}_\mathfrak{S}$ of $\mathfrak{S}$. As to the intended extension to a semi-formal system it makes sense to define the verification for finite sets of $\mathfrak{S}$-sentences which again should be read as finite disjunctions. It is technically convenient to formulate the language $\mathcal{L}_\mathfrak{S}$ in Tait style, i.e., to assume that there is no negation symbol but to define negation via deMorgan's laws. This requires that for any relation in $\mathfrak{S}$ there is also a symbol for its complement. We may then divide the language $\mathcal{L}_\mathfrak{S}$ into sentences in $\bigvee$–type and sentences in $\bigwedge$–type and decorate every sentence F with a characteristic sequence $\mathrm{CS}(F)$ of sentences such that

$$\mathfrak{S} \models G \text{ for some } G \in \mathrm{CS}(F) \text{ implies } \mathfrak{S} \models F \tag{2}$$

for sentences F in $\bigvee$-type, and

$$\mathfrak{S} \models G \text{ for all } G \in \mathrm{CS}(F) \text{ implies } \mathfrak{S} \models F \tag{3}$$

for sentences F in $\bigwedge$-type.

Let us illustrate types and characteristic sentences by examples. To satisfy (3) all sentences in the diagram of $\mathfrak{S}$, i.e. all true atomic sentences of $\mathfrak{S}$, should be put into into $\bigwedge$-type with empty characteristic sequence. Similarly all sentences $(\exists x)F(x)$ starting with a an existential quantifier ranging over the domain of $\mathfrak{S}$ should be put into $\bigvee$-type with $\mathrm{CS}((\exists x)F(x)) = \langle F(\underline{s}) \mid s \in \mathfrak{S}\rangle$. Similarly $(\forall x)F(x)$ should be put into $\bigwedge$-type with characteristic sequence $\langle F(\underline{s} \mid s \in \mathfrak{S}\rangle$ etc.

The verification relation $\models^{\alpha} \Delta$ is then given by two rules.

- ($\bigvee$) From $\models^{\alpha_G} \Delta, G$ for some $G \in \mathrm{CS}(F)$ infer $\models^{\alpha} \Delta, F$ for all $\alpha > \alpha_G$.
- ($\bigwedge$) From $\models^{\alpha_G} \Delta, G$ and $\alpha_G < \alpha$ for all $G \in \mathrm{CS}(F)$ infer $\models^{\alpha} \Delta, F$.

The verification calculus is clearly sound for $\mathfrak{S}$. Therefore we always have

$$\models^{\alpha} \Delta \textit{ implies } \mathfrak{S} \models \bigvee \Delta. \tag{4}$$

The ordinal α in $\models^{\alpha} \Delta$ should be regarded as (an upper bound for) a measure of the complexity of a verification of the sentence $\bigvee\Delta$. If the decoration is given in such a way, that every sentence in the characteristic sequence of a sentence F has a lower complexity than F and that also the opposite directions hold in (2) and (3)—which holds true in case that $\mathfrak{S}$ is the standard model for arithmetic—a simple induction on the complexity of F even shows

$$\mathfrak{S} \models F \textit{ iff } \models^{\alpha} F, \tag{5}$$

where α denotes the complexity of F.

In general the verification calculus does not suffice for an ordinal analysis. We have to extend it to a semi-formal system $\vdash^{\alpha}_{\rho} \Delta$ by replacing $\models^{\alpha} \Delta$ by $\vdash^{\alpha}_{\rho} \Delta$ in the ($\bigvee$) and ($\bigwedge$)-rules and adding additional rules, among them the cut rule as displayed below.

- (cut): from $\vdash^{\alpha}_{\rho} \Delta, F$ and $\vdash^{\alpha}_{\rho} \Delta, \neg F$ for an $\mathcal{L}_{\mathfrak{S}}$-sentence F of rank $< \rho$ infer $\vdash^{\beta}_{\rho} \Delta$ for all $\beta > \alpha$.

Since well-foundedness is Π^1_1-complete in the language of arithmetic we need free second order variables in the definition of the proof theoretic ordinals of arithmetical theories. To extend the verification calculus for $\mathbb{N}$ to a semi-formal system that is suited for an ordinal analysis we thus have to add a "rule" for second order variables, i.e., the rule

- (X-rule): if s and t are arithmetical terms such that $s^{\mathbb{N}} = t^{\mathbb{N}}$ then

  $\vdash^{\alpha}_{\rho} \Delta, s \notin X, t \in X$ holds for all ordinals α and ρ.

We then obtain that $\models^{\alpha} \Delta$ implies $\vdash^{\alpha}_{0} \Delta$, while the opposite direction only holds for arithmetical sentences.[10] In contrast to the completeness of $\models^{\alpha} F$, which is a triviality for arithmetical sentences, the completeness of $\vdash^{\alpha}_{0} F$ is a non-trivial theorem.[11] The truth-complexities for arithmetical sentences are finite while it follows from the boundedness property (1) that for pseudo Π^1_1-sentences truth-complexities up to ω_1^{CK} are possible. This indicates that the presence of free second order variables is indispensable—at least in the ordinal analysis of Peano arithmetic and subsystems of second order number theory.

The presence or absence of second order variables in semi-formal systems makes, however, a big difference. Theorem 3.1 shows that the elimination of cuts in semi-formal systems with second order variables raises the derivation length and gauging this raise is essential for the ordinal analysis. For semi-formal systems that only derive sentences the situation is different. Here cut-elimination is nearly a triviality.

Theorem 3.2. *(Semantical cut elimination) Let $\vdash^{\alpha}_{\rho} \Delta$ be the proof relation for a semi-formal system that is sound for a structure $\mathfrak{S}$ and only talks about sentences. Then $\vdash^{\alpha}_{\rho} \Delta$ already entails $\models^{\alpha} \Delta$.*

Proof: Let

$$\vdash^{\alpha_\iota}_{\rho} \Delta, G_\iota \text{ for } \iota \in I \;\Rightarrow\; \vdash^{\alpha}_{\rho} \Delta, F$$

be a rule of the semi-formal system. Since $\vdash^{\alpha}_{\rho} \Delta$ is sound for $\mathfrak{S}$ we obtain from $\mathfrak{S} \not\models F$ that $\mathfrak{S} \not\models G_\iota$ holds for at least one $\iota \in I$. Using this fact we obtain by an easy induction on α:

$$\text{if } \vdash^{\alpha}_{\rho} \Delta, \Gamma \text{ holds true for a finite set } \Gamma \text{ and } \mathfrak{S} \not\models G \text{ for all } G \in \Gamma \text{ then } \models^{\alpha} \Delta.$$

The lemma then follows for $\Gamma = \emptyset$. □

10 In former publications I defined $\models^{\alpha} \Delta$ such that $\vdash^{\alpha}_{0} \Delta$ also implies $\models^{\alpha} \Delta$. In view of the treatment of impredicative systems it would have been wiser to distinguish these notions as I do it here.

11 Cf.,e. g. [Pohlers, 2009] Theorem 5.4.9.

3.3 Impredicative proof theory

Recall the proven schema to obtain upper bounds for the proof theoretical ordinals of predicative theories which is:

- Unravel a formal proof into a semi-formal proof
- Eliminate the cuts
- Use the boundedness property to obtain upper bounds for the proof theoretical ordinal.

The essential (metamathematical) step in this procedure is apparently cut elimination (Theorem 3.1) although for axiom systems above Peano arithmetic the step of unraveling may also require sophisticated arguments.

If we imagine a situation in which we can define the proof theoretical ordinal of a formal theory T by a *sentence* in the language of its standard model we could replace cut elimination by semantical cut elimination and thus apparently trivialize the procedure.

On the example of subsystems of set theory we are going to show that for impredicative axiom systems there is in fact a possibility to define their proof theoretical ordinals by properties expressible as *sentences* in the language of their "standard models". We will, however, also indicate that in these cases the unraveling procedure can only be obtained by introducing "abstract ideal ordinals" which makes semantical cut elimination useless for ordinal analyses.

3.3.1 Axiom systems for set theory

When dealing with axiom systems in the language of set theory the first problem is that there is no commonly accepted standard model for set theory for which we could design the semi-formal system. However, since we are still struggling with axiom systems far below ZFC we may regard the constructible universe L as the standard model. So in some sense L plays the role that $\mathbb{N}$ plays in the ordinal analysis of subsystems of second order number theory.

Our basis axiom system BST for basis set theory comprises the axioms of extensionality, pair, union and the schemes of absolute separation, i.e., Δ_0-separation, and foundation. Adding also the scheme of Σ_1-collection we obtain the system KP of Kripke-Platek set theory (cf. [Barwise, 1975]). KPω is Kripke-Platek set theory plus the axiom of infinity.

Using the constructible hierarchy L we are able to define characteristic ordinals for axiom systems of set theory.

Definition 3.3. *Let T be an axiom system in the language of set theory that possesses a model in the constructible hierarchy. Put*

$$\|T\| := \min\{\alpha \mid (\forall F \in \Pi_2)[T \vdash F \Rightarrow \mathsf{L}_\alpha \models F]\}$$

and for admissible ordinals $\pi > \omega$

$$\|T\|_\pi := \min\{\alpha \mid (\forall F \in \Pi_2)[T \vdash F^{\mathsf{L}_\pi} \Rightarrow \mathsf{L}_\alpha \models F]\}.$$

Clearly the second definition requires that T can talk about L_π.

From the proof theoretical point of view we are of course only interested in $\|T\|_\pi$ for countable admissible ordinals π. Of special interest are the ordinals $\|T\|_{\omega_1^{CK}}$ and later the ordinal $\|T\|_\omega$, whose definition we have to postpone.

For axiom systems which comprise $\mathsf{KP}\omega$ the ordinal $\|T\|_{\omega_1^{CK}}$ corresponds[12] to the proof theoretic ordinal $|T|$. We briefly sketch why.

For a theory T in the language of arithmetic the proof theoretic ordinal $|T|$ is the supremum of the order-types of arithmetically definable well-orderings $\prec$ which T proves to be well-founded. Now let T be a theory in the language of set theory comprising $\mathsf{KP}\omega$. Then the language of arithmetic can be regarded as sublanguage of T in such a way that all arithmetical formulas become Δ_0-formulas. All arithmetically definable relations on $\mathbb{N}$ are thus sets by Δ_0-separation. The canonical generalization of the proof theoretic ordinal $|T|$ for a theory T in the language of set theory is therefore the supremum of the order-types of order-relations that are sets in T and T proves their well-foundedness.

So let $\prec$ be a well-ordering which is a set. The function $\mathrm{otyp}_\prec(x)$ which assigns to every element x in the field of $\prec$ its order-type is Σ-definable. If we assume that T is strong enough to prove the basic properties of $\mathsf{L}_{\omega_1^{CK}}$ we thus may assume

$$T \vdash (\forall x \in \mathit{field}(\prec))(\exists \alpha \in \mathsf{L}_{\omega_1^{CK}})[\mathrm{otyp}_\prec(x) = \alpha]$$

where we tacitly presuppose that lower case Greek letters denote ordinals. Since this is a $\Pi_2^{\mathsf{L}_{\omega_1^{CK}}}$-sentence there is an ordinal below $\|T\|_{\omega_1^{CK}}$ that bounds the order-type of the relation $\prec$. Hence $|T| \leq \|T\|_{\omega_1^{CK}}$.

For the opposite direction we refer to the fact that the least model for the provable $\Pi_2^{\mathsf{L}_{\omega_1^{CK}}}$-sentences in the constructible hierarchy coincides with the least model for the provable $\Sigma_1^{\mathsf{L}_{\omega_1^{CK}}}$-sentences.[13] Assume $\beta := |T| < \|\mathrm{T}\|_{\omega_1^{CK}}$. Again we will assume that T is strong enough to prove Σ_1-reflection on $\mathsf{L}_{\omega_1^{CK}}$, i.e.

$$G^{\mathsf{L}_{\omega_1^{CK}}} \rightarrow (\exists x \in \mathsf{L}_{\omega_1^{CK}})(\exists \xi \in \mathsf{L}_{\omega_1^{CK}})[x = \mathsf{L}_\xi \wedge G^x] \tag{6}$$

12 If T is $\mathsf{KP}\omega$ the ordinal $\|T\|_{\omega_1^{CK}}$ is not defined. In this case replace it by $\|\mathsf{KP}\omega\|$.

13 Cf.,e.g. [Pohlers, 1998] Lemma 1.2.3.

for Σ_1-sentences G. Moreover we assume that T allows for the definition of the Σ-function $\xi \mapsto \mathsf{L}_\xi$ and the definition of the canonical Δ_1-relation $<_\mathsf{L}$ that well-orders the constructible sets. The assumption $\beta < \|T\|_{\omega_1^{CK}}$ entails that there is a Σ_1-sentence G such that $T \vdash G^{\mathsf{L}_{\omega_1^{CK}}}$ but $\mathsf{L}_\beta \not\models G$. Then we obtain by (6)

$$T \vdash (\exists u \in \mathsf{L}_{\omega_1^{CK}})(\exists \xi \in \mathsf{L}_{\omega_1^{CK}})[u = \mathsf{L}_\xi \wedge G^u \wedge (\forall x \in u)(\exists y)[x <_\mathsf{L} y]]$$

where $<_\mathsf{L}$ is the cononical well-ordering of L. Using Σ-reflection this entails

$$T \vdash (\exists z \in \mathsf{L}_{\omega_1^{CK}})(\exists u \in \mathsf{L}_{\omega_1^{CK}})(\exists \xi \in \mathsf{L}_{\omega_1^{CK}})[u = \mathsf{L}_\xi \wedge G^u \wedge (\forall x \in u)(\exists y \in z)[x <_\mathsf{L} y]].$$

Since this is a Π_2-sentence relativized to $\mathsf{L}_{\omega_1^{CK}}$ we get $\xi < \|T\|_{\omega_1^{CK}}$. Because of $\mathsf{L}_\beta \not\models G$ it follows $\beta < \xi$. Since $\mathrm{otyp}(<_\mathsf{L} \cap (\mathsf{L}_\xi \times z)) \geq \xi > \beta$ and $<_\mathsf{L} \cap (\mathsf{L}_\xi \times z)$ is a set this contradicts the definition of β. Hence $\|T\|_{\omega_1^{CK}} \leq |T|$.

Because of the correspondence of $\|T\|_{\omega_1^{CK}}$ to $|T|$ we call the computation of $\|T\|_{\omega_1^{CK}}$ still a Π_1^1-analysis of T.

To define Π_2^0-analysis also for theories in the language of set theory we have to modify Definition 2.1.

The role of Π_2^0-sentences is played by $\Pi_2^{\mathsf{L}_\omega}$-sentences, i.e. sentences of the form $(\forall x \in \mathsf{L}_\omega)(\exists y \in \mathsf{L}_\omega)F(x, y, \vec{a})$ where $F(x, y, \vec{x})$ is a Δ_0-formula and $\vec{a} = a_1, \ldots, a_n$ a tuple of elements in L_ω.

A function $\Phi\colon \mathsf{L}_\omega \longrightarrow \omega$ is a test-function for $(\forall x)(\exists y)F(x, y, \vec{a})^{\mathsf{L}_\omega}$ if for every $a \in \mathsf{L}_\omega$ there is a $t \in \mathsf{L}_m$ such that $\mathsf{L}_\omega \models F(a, t, \vec{a})$ and $m \leq \Phi(\{a\} \cup \{\vec{a}\})$.

A function $\Phi\colon \mathsf{L}_\omega \longrightarrow \omega$ is *eventually majorized* by a function $\Psi\colon \mathsf{L}_\omega \longrightarrow \omega$ if there is a $k \in \omega$ such that $\Phi(a) < \Psi(a)$ holds true for all $a \in \mathsf{L}_\omega \setminus \mathsf{L}_k$.

Definition 3.4. *A function $\Phi\colon \mathsf{L}_\omega \longrightarrow \omega$ designs experiments for a theory T in the language of set theory, if for any $\Pi_2^{\mathsf{L}_\omega}$ sentence with parameters from L_ω which is provable in T there is a test-function that is eventually majorized by Φ.*

In a Π_2^0-analysis of a theory T we try to construct a function that designs experiments for T.

We will see that a function that designs experiments for a theory T can be obtained via a subrecursive hierarchy indexed by ordinals. We may thus define $\|T\|_\omega$ as the least ordinal at which the designing function appears. This definition is, however, less robust than the ordinal $\|T\|_{\omega_1^{CK}}$ since it depends on an effectively definable norm on the ordinals for which, in general, there are different choices.[14]

14 Actually the resulting subrecursive hierarchy is pretty immune to different choices.

3.3.2 Ramified set theory

To fix the language for a verification calculus for L we define the language $\mathcal{L}_{RS}$ of *ramified set theory*. The terms and sentences of $\mathcal{L}_{RS}$ are given by the following clauses:

- For every ordinal α the term L_α is an atomic term of stage α and
- if $a_1, \dots, a_n$ are terms of stages less than α and $F(x, x_1, \dots, x_n)$ is a formula in the language of set theory then $\{x \in \mathsf{L}_\alpha \,|\, F(x, a_1, \dots, a_n)^{\mathsf{L}_\alpha}\}$ is a composed term of stage α.
- If $F(x_1, \dots, x_n)$ is a Δ_0-formula in the language of set theory whose free variables occur all in the list $x_1, \dots, x_n$ and $a_1, \dots, a_n$ are terms of ramified set theory then $F(a_1, \dots, a_n)$ is a sentence of ramified set theory.

By $\mathsf{L}_{<\alpha}$ we denote the sequence of all $\mathcal{L}_{RS}$-terms of stages less than α ordered by the relation $<_{\mathsf{L}}$.

The $\bigvee$–type of $\mathcal{L}_{RS}$ comprises all sentences of the form $a \in b$, $F \vee G$ and $(\exists x \in a)F(x)$ and dually the $\bigwedge$–type comprises the sentences $a \notin b$, $F \wedge G$ and $(\forall x \in a)F(x)$.

For the sentences in $\bigvee$–type the decoration is given by

- $\mathrm{CS}(a \in \mathsf{L}_\alpha) := \langle a = s \,|\, s \in \mathsf{L}_{<\alpha}\rangle$,
- $\mathrm{CS}(a \in \{x \in \mathsf{L}_\alpha \,|\, F(x)\}) := \langle a = s \wedge F(s) \,|\, s \in \mathsf{L}_{<\alpha}\rangle$,
- $\mathrm{CS}(F \vee G) := \langle F, G\rangle$,
- $\mathrm{CS}((\exists x \in \mathsf{L}_\alpha)F(x)) := \langle F(s) \,|\, s \in \mathsf{L}_{<\alpha}\rangle$,
- $\mathrm{CS}((\exists x \in a)F(x)) := \langle F(s) \wedge G(s) \,|\, s \in \mathsf{L}_{<\alpha}\rangle$ if a is a composed term $\{x \in \mathsf{L}_\alpha \,|\, G(x)\}$.

Dually we define a decoration $\mathrm{CS}(F) = \langle \neg G \,|\, G \in \mathrm{CS}(\neg F)\rangle$ for the sentences in $\bigwedge$–type.

Having defined a decoration for $\mathcal{L}_{RS}$ we obtain a verification calculus $\models^{\alpha} \Delta$ for finite sets of $\mathcal{L}_{RS}$-sentences.

The semantics for ramified set theory is obvious. Each atomic term L_α is interpreted by the stage L_α in the constructible hierarchy and every composed term is interpreted by the corresponding set in the constructible hierarchy. We did not even make the effort to distinguish the $\mathcal{L}_{RS}$-terms and the corresponding sets in the constructible hierarchy notationally.

Observe that we only talk about Δ_0-sentences in $\mathcal{L}_{RS}$. It is now easy to see that the verification calculus is sound and complete for L in the form that an $\mathcal{L}_{RS}$-sentence is true in L_β iff there is an ordinal α such that $\models^{\alpha} F^{\mathsf{L}_\beta}$.

There is also a counterpart to the boundedness property (1) which roughly says that $\models^{\alpha} F^{\mathsf{L}_\beta}$ for a Σ-sentence F containing only parameters from $\mathsf{L}_{<\alpha}$ entails $\mathsf{L}_\alpha \models F$.

To unravel formal proofs of Peano arithmetic (and other predicative theories) into semi-formal derivations it suffices to augment the verification calculus by the (X-rule) for second order variables and the cut rule. This turns out not to be sufficient for impredicative systems. While there is no problem to embed BST into a system augmented by a cut-rule there is no chance to embed the scheme of Σ-collection, i.e., to get an embedding of KPω let alone of even stronger systems.

To be able to embed stronger theories we need "abstract ideal" objects, e.g., for KPω an abstract ideal ordinal Ω which is axiomatized by a reflection rule

$$\models^{\alpha}_{\rho} \Delta, F^{\mathsf{L}_\Omega} \quad \Rightarrow \quad \models^{\beta}_{\rho} \Delta, (\exists a \in \mathsf{L}_\Omega)[a \neq \emptyset \wedge Tran(a) \wedge F^a] \tag{7}$$

for $\beta > \alpha$ where F is supposed to be a Π_2-sentence.

If we interpret Ω by ω_1^{CK} the thus obtained semi formal system is sound and therefore enjoys semantical cut elimination. Interpreting Ω by ω_1^{CK} implies that we need at least ω_1^{CK} many steps to embed the foundation scheme. The embedding therefore produces semi-formal derivations with lengths above ω_1^{CK} in case that the foundation scheme is among the axioms of our formal theory.[15] Semantical cut elimination together with the boundedness property will therefore not help us in the computation of reasonable upper bounds for $\|T\|_{\omega_1^{CK}}$ which is always an ordinal below ω_1^{CK}.

On the other hand it is clear by semantical reasons that for a tuple $\vec{a} \in \mathsf{L}_{<\omega_1^{CK}}$ and a Σ_1-formula $F(\vec{a})$ for which we have $\mathsf{L}_{\omega_1^{CK}} \models F(\vec{a})$ there is an ordinal $\alpha < \omega_1^{CK}$ such that $\models^{\alpha} F(\vec{a})^{\mathsf{L}_{\omega_1^{CK}}}$. So there should exist a collapsing procedure that allows us to collapse derivations of $\Pi_2^{\mathsf{L}_\Omega}$ sentences into derivations of lengths below ω_1^{CK}. Here we encounter the necessity for a collapsing procedure which is apparently inevitable in any ordinal analysis of impredicative systems.

4 Elimination of ideal objects

In this section we want to discuss how ordinal analysis contributes to Hilbert's programme of elimination of ideal objects. We start with the elimination of "ab-

15 One might ask what happens if the foundation scheme is lacking or modified. This leads to a series of theories which are either predicative (cf. [Jäger, 1986], [Pohlers, 2009]) or *metapredicative* as Jäger and his school call them. Cf. [Jäger, 2005] for further information.

stract ideal objects" which is affiliated to the ordinal analysis of impredicative axiom systems. The elimination of "all ideal objects" is more involved and will be treated later.

4.1 Impredicative proof theory and the elimination of "abstract ideal objects"

In the previous section we emphasized that the ordinal analysis of impredicative axiom systems needs in general a collapsing procedure for semi-formal derivations. This clearly implies that we also need a collapsing function on the ordinals that measure the lengths of semi-formal derivations. Since ordinals are transitive and thus incollapsible this means that the lengths of the semi-formal derivations that arise in the unravelling procedure must only be measured by ordinals that are taken from a subset of the ordinals containing gaps which are large enough to allow for the necessary collapsing. Finding the suitable sets of ordinals is actually the most difficult problem in impredicative ordinals analysis.

4.1.1 Iterated Skolem hull operators

A proven technique to obtain suitable sets of ordinals are iterations of Skolem hull operators. A technique whose foundations have been laid by (mostly unpublished) work of S. Feferman, P. Aczel, J. Bridge ([Bridge, 1975]) and W. Buchholz ([Buchholz, 1974], [Buchholz, 1975]).

Let $\mathfrak{F}$ be a set of functions that assign ordinals to tuples of ordinals. The Skolem hull operator $\mathcal{H}^{\mathfrak{F}}\colon \mathrm{Pow}(On) \longrightarrow \mathrm{Pow}(On)$ generated by $\mathfrak{F}$ is inductively defined by the clauses

- $\mathcal{H}_0^{\mathfrak{F}}(X) = X$,
- $\mathcal{H}_{n+1}^{\mathfrak{F}}(X) = \mathcal{H}_n^{\mathfrak{F}}(X) \cup \{f(\xi_1, \ldots, \xi_k) \mid \xi_1, \ldots, \xi_k \in \mathcal{H}_n^{\mathfrak{F}}(X) \land f \in \mathfrak{F}\}$.
- $\mathcal{H}^{\mathfrak{F}}(X) = \bigcup_n \mathcal{H}_n^{\mathfrak{F}}(X)$.

A Skolem hull operator is *Cantorian closed* if ordinal addition and exponentiation to the basis ω are among its generating functions.

Generated Skolem hull operator are apparently inflationary, monotone and idempotent and thus not iterable. To obtain iterations of a Skolem hull operator we therefore have to increase the set of generating functions in each iteration step. An obvious way to do this is to add ordinals which are inaccessible by the so far iterated Skolem hull operator. An ordinal ξ is $\mathfrak{F}$-inaccessible if $\xi \notin \mathcal{H}^{\mathfrak{F}}(\xi)$, where ξ is viewed as the set of its predecessors. So we define simultaneously iterations

$\mathcal{H}^{\mathfrak{F},\alpha}$ and functions $\psi^{\mathfrak{F}}$ such that

$$\psi^{\mathfrak{F}}(\alpha) := \min\{\xi \mid \xi \notin \mathcal{H}^{\mathfrak{F},\alpha}(\xi)\} \tag{8}$$

and augment for $\alpha > 0$ the generators of $\mathcal{H}^{\mathfrak{F},\alpha}$ by $\psi^{\mathfrak{F}}\restriction\alpha$. As an example let us assume that $\mathfrak{F} = \{+\}$. Then $\mathcal{H}^{\mathfrak{F}}(\emptyset) = \emptyset$, hence $\psi^{\mathfrak{F}}(0) = 0$; $\mathcal{H}^{\mathfrak{F},1}(1) = 1$, hence $\psi^{\mathfrak{F}}(1) = 1$; $\mathcal{H}^{\mathfrak{F},2}(\omega) = \omega$, hence $\psi^{\mathfrak{F}}(2) = \omega$; $\psi^{\mathfrak{F}}(n+1) = \omega^n$; $\psi^{\mathfrak{F}}(\omega) = \omega^\omega$ and the least ordinal σ such that $\mathcal{H}^{\mathfrak{F},\sigma}(\sigma) = \sigma$, i.e., $\psi^{\mathfrak{F}}(\sigma) = \sigma$ is ε_0, the proof theoretic ordinal of the axiom system PA. So ε_0 is the largest transitive segment of the ordinals which we can access by iterations of the Skolem hull operator generated by $\{+\}$. Increasing the set $\mathfrak{F}$ of generators to $\{+, \lambda\xi.\omega^\xi\}$ would give $\psi^{\mathfrak{F}}(2) = \varepsilon_0$. But even if we augment the set of generating functions by the Veblen function φ the eventually attained segment is still too small as to capture the proof theoretic ordinals of "small" impredicative axiom systems such as ID_1, the axiomatization of non-iterated arithmetically definable positive inductive definitions, or $\mathsf{KP}\omega$.

The operators and their iterations in these examples are transitive (or replete as Feferman calls them in [Feferman, 1968]) in the sense that applied to a transitive set of ordinals they return a transitive set of ordinals. For these operators we have

$$\psi^{\mathfrak{F}}(\alpha) = \min\{\xi \mid \mathcal{H}^{\mathfrak{F},\alpha}(\xi) = \xi\}.$$

As pointed out at the begin of this section we need a set of ordinals with gaps that are large enough for the needed collapsing procedure. This can be obtained by augmenting the set of generating functions by functions (or ordinals) which we regard as "abstract ideal". The simplest example is the addition of an ideal ordinal Ω which lays safely outside the eventually obtained initial segment. Since the eventually obtained segment is countable a possible interpretation for Ω could be the first uncountable cardinal. The such obtained subset of the ordinals suffices for an ordinal analysis of ID_1, $\mathsf{KP}\omega$ or $\Pi_2 - \mathsf{Ref}$ which is Kripke–Platek set theory with the scheme of Π_2-reflections. Adding the single ideal ordinal Ω to the functions $\{+, \lambda\xi.\omega^\xi\}$[16] may thus be regarded as "the first step into impredicativity".

An obvious extension of the "first step" is to add the function $\alpha \mapsto \alpha^+$ to the generators of $\mathfrak{F}$ where α^+ denotes the first cardinal that is larger than α.

Augmenting the generating functions of an operator by ideal ordinals or functions will in general spoil the transitivity of the generated operator. However, what is kept in our examples is transitivity within number classes, i.e., $\mathcal{H}^{\mathfrak{F}'}(\kappa) \cap \kappa^+$ is a transitive set for all cardinals κ. To obtain iterations of the so generated operator

16 Adding also the Veblen function is possible—and common—but does not increase the eventually obtained transitive segment.

we simultaneously define the iterations $\mathcal{H}^\alpha$ (omitting now the superscript $\mathfrak{F}'$ and suppressing technicalities) and functions ψ_{κ^+} by

- $\psi_{\kappa^+}(\alpha) := \min\{\xi \mid \mathcal{H}^\alpha(\xi) \cap \kappa^+ = \xi\}$,
- and add for $\alpha > 0$ all functions $\psi_{\lambda^+} \restriction \alpha$ to the generators of $\mathcal{H}^\alpha$.

Due to this definition we get $\mathcal{H}^\alpha(\xi) = \mathcal{H}^{\alpha+1}(\xi)$ in case that $\alpha \notin \mathcal{H}^\alpha(\xi)$. So $\psi_{\lambda^+}(\alpha) = \psi_{\lambda^+}(\alpha+1)$ and the functions are not one-to-one. To obtain one-to-one functions we may build in the normal-form condition $\alpha \in \mathcal{H}^\alpha(\xi)$, define

$$\Psi^\alpha_{\kappa^+} := \min\{\xi \mid \alpha \in \mathcal{H}^\alpha(\xi) \wedge \mathcal{H}^\alpha(\xi) \cap \kappa^+ = \xi\}$$

and augment the generators in the iteration step by all functions $\Psi_{\lambda^+} \restriction \alpha$.

The price to pay is that the new functions Ψ_{λ^+} are partial but it is not too hard to prove that all the iterations $\mathcal{H}^\alpha$ are again transitive on number classes such that

$$\begin{aligned}\mathcal{H}^\alpha(\kappa) \cap \kappa^+ &= \min\{\xi \mid \kappa \le \xi \notin \mathcal{H}^\alpha(\xi) \wedge \alpha \in \mathcal{H}^\alpha(\xi)\}\\ &= \min\{\xi \mid \mathcal{H}^\alpha(\xi) \cap \kappa^+ = \xi \wedge \alpha \in \mathcal{H}^\alpha(\xi)\} = \Psi^\alpha_{\kappa^+}.\end{aligned}$$

Since $\Psi^\alpha_{\kappa^+} < \kappa^+$ holds true for all ordinals α the functions Ψ_{κ^+} are collapsing.[17]

Adding Ω_ω, the first cardinal that is closed under cardinal successors, the so obtained set of ordinals suffices for the ordinal analysis of KPl, an axiom system based on Kripke-Platek set theory with a limit axiom that axiomatizes that every set is contained in an admissible set.

Stronger systems can be obtained by adding further abstract ordinals, e.g., the first weakly inaccessible cardinal. The thus obtained ordinal set suffices for the analysis of KPi, an axiom system which axiomatizes an admissible universe which is also the union of admissible universes (cf. [Jäger and Pohlers, 1983]).

Adding a Mahlo cardinal (together with the necessary modification of the generating functions, cf. [Rathjen, 1990]) leads to a set of ordinals that suffices for the ordinal analysis of KPM, a theory which axiomatizes a recursively Mahlo universe.

Yet stronger systems require a refinement in the addition of abstract ideal ordinals. The iteration of the Skolem hull operator has to be combined with iterations of thinning procedures on the cardinals, i.e., on the abstract ideal ordinals. Thinning operations are a familiar tool in the development of ordinal notations systems. A well-known example is the derivative of a class of ordinals which is the class of fixed-points of its enumerating function. So we obtain the Veblen functions as enumerating functions of classes that are obtained as transfinitely iterated derivatives of the class of additively indecomposable ordinals (cf., e.g. [Pohlers, 2009] Section 3.4).

17 Cf.[Buchholz, 1986] for more information.

An important thinning operation on the cardinals is the Mahlo-operation

$$\mathfrak{M}(X) := \{\pi \in X \mid X \text{ is stationary in } \pi\}.$$

Starting with a cardinal κ we get an iteration of the Mahlo-operation by defining $\mathfrak{M}_\kappa^0 = \kappa$, $\mathfrak{M}_\kappa^{\alpha+1} = \mathfrak{M}(\mathfrak{M}_\kappa^\alpha)$ and taking the diagonal intersection

$$\mathfrak{M}_\kappa^\lambda = \{\xi \in \lambda \mid (\forall \eta < \xi)[\xi \in \mathfrak{M}(\mathfrak{M}_\kappa^\eta)]\}$$

at limit levels λ. To combine this with the iterations of a Skolem hull operator $\mathcal{H}$ we then define the sets

$$\mathfrak{M}_\kappa^\alpha := \{\xi \in \kappa \mid (\forall \eta \in \mathcal{H}^\alpha(\xi) \cap \alpha)[\xi \in \mathfrak{M}(\mathfrak{M}_\kappa^\eta)] \land \mathcal{H}^\alpha(\xi) \cap \kappa = \xi \land \alpha \in \mathcal{H}^\alpha(\xi)\}$$

for regular cardinals κ, put $\Psi_\kappa^\alpha := \min \mathfrak{M}_\kappa^\alpha$ and let $\mathcal{H}^\alpha$ be the iterations of the Skolem hull operator $\mathcal{H}$ whose generators are augmented by all functions $\Psi_\kappa \restriction \alpha$ in the iteration step. Michael Rathjen has shown in [Rathjen, 1992] that, starting with the first weakly compact cardinal, (a refinement of) this procedure the technical details do not matter in our discussion—suffices to obtain a set of ordinals that allows for an analysis of Kripke-Platek set theory augmented by the scheme of Π_3-reflection.

For stronger theories more sophisticated thinning operations become necessary which make the technical details *much* more complicated. (cf. e.g. [Rathjen, 2005a], [Rathjen, 2005b], [Pohlers and Stegert, 2012]). To explain the principle we do not need these details. However, we should mention that for successor cardinals we can put $\mathfrak{M}(\mathfrak{M}_{\kappa^+}^\alpha) = \emptyset$. All cardinals that are relevant in the iteration of the thinning procedure are regular limit cardinals. Observe that this implies

$$\Psi_{\kappa^+}^\alpha = \min\{\xi \mid \mathcal{H}^\alpha(\xi) \cap \kappa^+ = \xi \land \alpha \in \mathcal{H}^\alpha(\xi)\}$$

which shows that these functions are still collapsing and continue the previously defined Ψ_{κ^+} functions without thinning operations.

The cardinals that are needed in the ordinal analysis of Π_{n+3}-reflection, i.e., basis set theory **BST** plus the reflection scheme for Π_{n+3}-formulas are Π_{n+1}^1-indescribable cardinals. Recall that the first weakly compact cardinal is the first Π_1^1-indescribable cardinal.

For the theory **Ref** which is **BST** plus full reflection we need a cardinal Ξ that is Π_n^1-indescribable for all $n \in \omega$. (cf. [Pohlers and Stegert, 2012]).

The theory of stability, i.e., Kripke-Platek set theory augmented by the weak stability axiom which requires for every ordinal α and every ordinal ξ the existence of an ordinal $\kappa \geq \alpha$ such that L_κ is a Σ_1-elementary substructure of $\mathsf{L}_{\kappa+\xi}$, needs cardinals that are ν-indescribable. For an ordinal $\nu =_{NF} \omega \cdot \lambda + m$ a cardinal κ is ν-indescribable if for any Π_m^1-formula $F(X_1, \dots, X_n)$ with variables

$X_1, \ldots, X_n$ and sets $P_i \subseteq V_\kappa$ such that $(V_{\kappa+\lambda}, P_1, \ldots, P_n) \models F(X_1, \ldots, X_n)$ there is an ordinal $\pi < \kappa$ such that $(V_{\pi+\lambda}, P_1 \cap V_\pi, \ldots, P_n \cap V_\pi) \models F(X_1, \ldots, X_n)$.[18]. Below a subtle cardinal there is a least cardinal Υ such that for all $\nu < \Upsilon$ there is a $\kappa < \Upsilon$ such that κ is ν-indescribable and for every ν-indescribable cardinal $\kappa < \Upsilon$ we already have $\nu < \Upsilon$. Adding this cardinal together with the associated thinning operations (which actually are pretty involved) the eventually obtained ordinal set suffices for an ordinal analysis of stability. Details will appear in [Pohlers, 201xb].

Even lager cardinals have to be added to obtain a subset of the ordinals which allows for an ordinal analysis of the theory STAB, which is Kripke-Platek set theory augmented by the axiom of full stability which says that for every ordinal α there is a cardinal $\kappa \geq \alpha$ such that L_κ is a Σ_1-elementary substructure of L. Here we need cardinals that are gap-indescribable, similar to the ν-indescribable cardinals, but also move the length of the reflected gap. The theory STAB is equivalent to Kripke-Platek set theory with the scheme of Σ_1-separation which in turn is equivalent to second order number theory with the axiom of parameter free Π^1_2-comprehension. These theories represent the present summit of theories for which an ordinal analysis is obtainable.

Observe that every ordinal in $\mathcal{H}^\alpha(X)$ can be represented by an *ordinal-term*, i.e. a finite string of symbols that is built up from symbols for the generators, the symbol Ψ and the parameters in X. It thus provides a notation system for the ordinals in the initial transitive segment.

An essential point to mention is that already the construction of the ordinal sets via iterated Skolem hull operators has the flavor of *elimination of abstract ideal objects*. Although constructed with the aid of large cardinals the eventually obtained set of ordinals is primitive-recursively definable as a set of ordinal-terms and the "less than"-relation restricted to ordinal-terms is primitive-recursively decidable.

This entails that the eventually obtained transitive segment of the ordinal set is an ordinal less than ω_1^{CK} which means that the function Ψ_{ω_1} is not only collapsing below ω_1 but actually collapsing below ω_1^{CK}. A fact that is essential for the ordinal analysis since we are looking for an upper bound below ω_1^{CK}.

Another consequence is that the eventually obtained set of ordinals may be constructed without reference to large cardinals as a primitive-recursively defined set of *ordinal-terms* with a primitive-recursive order relation on them. This adheres to an elimination of the ideal abstract object "large cardinal".

18 This is a variation of the definition in [Drake, 1974] CH. 9, §4.

In doing so we clearly loose the information that this ordering is well-founded. The well-foundedness of the ordering then has to be secured by a *well-foundedness proof* which (at least in all known examples) exceeds the means of the analyzed theory "by an ω-defect". Here we say that a sentence $(\forall x)F(x)$ *exceeds the means of a theory T by an ω-defect* if T proves $F(n)$ for every standard number n but is unable to prove $(\forall x)F(x)$.

4.1.2 Axiomatization of ideal ordinals

We have seen in (7) that the abstract ideal ordinal Ω is axiomatized by a rule that is affiliated to the complexity class of Π_2-formulas. Similar rules are needed for other abstract ideal objects, e.g., ordinals with stronger reflection properties. Again these abstract ideal ordinals are in general affiliated to certain complexity classes of formulas. Since we want only explain the principles we will not specify them in full detail.

Nevertheless there is one feature which is worth while mentioning. While we use *indescribable cardinals* in iterating the Skolem hull operators we axiomatize them in ramified set theory by rules for *reflecting ordinals*. Recall that an ordinal κ is Π_m-reflecting if for every Π_m-formula $F(a_1, \ldots, a_n)$ with parameters $a_1, \ldots, a_n \in \mathsf{L}_\kappa$ such that $\mathsf{L}_\kappa \models F(a_1, \ldots, a_n)$ there is an ordinal $\pi < \kappa$ such that $a_1, \ldots, a_n \in \mathsf{L}_\pi$ and $\mathsf{L}_\pi \models F(a_1, \ldots, a_n)$. Generalizing that to gap-reflection we say that an ordinal κ is Π_m-λ-gap-reflecting if for every Π_m-sentence $F(a_1, \ldots, a_n)$ with parameters $a_1, \ldots, a_n \in \mathsf{L}_{\kappa+\lambda}$ such that $\mathsf{L}_{\kappa+\lambda} \models F(a_1, \ldots, a_n)$ there is an ordinal $\pi < \kappa$ such that $\mathsf{L}_{\pi+\lambda} \models F({a_1}^{\kappa\downarrow\pi}, \ldots, {a_n}^{\kappa\downarrow\pi})$ where $a^{\kappa\downarrow\pi}$ is a pruning procedure in which all occurrences of L_κ are replaced by L_π.[19]. For $\nu =_{NF} \omega \cdot \lambda + m$ we call an ordinal ν-reflecting if it is Π_m-λ-gap-reflecting.

The correspondence between indescribable cardinals and reflecting ordinals is such that for $m < \omega$ an $m + 1$-indescribable cardinal is axiomatized by a Π_{m+3}-reflection rule while for $\omega < \nu =_{NF} \omega \cdot \lambda + m$ a ν-indescribable cardinal is axiomatized by a rule for a ν-reflecting ordinal. This is in accordance with the studies of Peter Aczel and Wayne Richter who showed that Π_{m+3}-reflecting ordinals are the "recursive counterparts" of Π^1_{m+1}-indescribable cardinals in [Aczel and Richter, 1972] for the case $m = 0$. The generalization to arbitrary finite m is more or less obvious. A result that is corroborated by the fact that the axiomatization of Π^1_{m+1}-indescribable cardinals by Π_{m+3}-reflecting ordinals gives the correct proof theoretic ordinals.

19 Details will appear in [Pohlers, 201xb].

Here we encounter yet another level of “abstract idealizations”. Reflecting ordinals as a “generalized-constructive” idealization and the correponding large cardinals as their “absolute” abstractions. Since the effort to develop the iterations of Skolem hull operators on the basis of ν-reflecting ordinals instead of ν-indescribable cardinals turns out to be extreme this situation may serve as another example for Hilbert’s dictum *“that only the introduction of ideal propositions makes mathematical reasoning possible”*.

The fact that we eventually can eliminate the ideal object “large cardinal” to such an extend that we end up with an “experimentally checkable Π^0_2-statement”—i.e., a statement that is obtainable without any reference to ideal objects, let alone “large cardinals”—may be taken as an affirmation even if not for the *existence* then for the *usefulness* of large cardinals.

4.1.3 Operator controlled derivations

Given a semi-formal system we call the ordinals which have to be axiomatized by rules the “ideal ordinals” of the system. E.g. all regular cardinals are ideal.

To exploit the subsets of the ordinals obtained by iterations of Skolem hull operators proof-theoretically we introduce operator controlled infinitary derivations. A method invented by Wilfried Buchholz in [Buchholz, 1992] as an improvement of the method of local predicativity (cf. [Pohlers, 1981]).

In a first step we define the ordinal parameters of a sentence in the semi-formal system. If F is a sentence of the language $\mathcal{L}_{RS}$ its parameters are the stages of all the terms occurring in it. Let $\mathrm{par}(F)$ denote the set of parameters occurring in F and equally $\mathrm{par}(\Delta)$ denote the set of all parameters occurring in the sentences of the finite set Δ.

Definition 4.1. *We modify the definition of a semi-formal derivation $\vdash^{\alpha}_{\rho} \Delta$ to a semi-formal derivation controlled by an operator $\mathcal{H}$ (denoted by $\mathcal{H} \vdash^{\alpha}_{\rho} \Delta$) by requiring that $\alpha \in \mathcal{H}(\mathrm{par}(\Delta))$ and also $\mathrm{par}(\Delta_\iota) \subseteq \mathcal{H}(\mathrm{par}(\Delta))$ for every inference*

$$\vdash^{\alpha_\iota}_{\rho} \Delta_\iota \;\Rightarrow\; \vdash^{\alpha}_{\rho} \Delta$$

which is not an inference according to $(\bigwedge)$.

The restriction to $\alpha \in \mathcal{H}(\mathrm{par}(\Delta))$ in the ordinal assignment of an operator controlled derivation ensures that only collapsible ordinals are assigned.

The cut elimination theorem for predicative proof theory (Theorem 3.1) remains true for operator controlled derivations provided that there are no ideal ordinals strictly between β and $\beta + \omega^{\rho}$.

The main theorem for operator controlled derivations, formulated for a semi-formal system for ramified set theory $\mathcal{L}_{RS}$, then has (very roughly[20]) the following form.

Theorem 4.2. *Let F be a sentence belonging to the complexity class of the ideal ordinal π. Then*

$$\mathcal{H} \left|\frac{\alpha}{\kappa+1}\right. F^{\mathsf{L}_\pi} \text{ for an ideal ordinal } \kappa \text{ implies } \mathcal{H}^{\omega^{\alpha\#\kappa}\#\mu+1} \left|\frac{\Psi_\pi^{\omega^{\alpha\#\kappa}\#\mu+\omega}}{\Psi_\pi^{\omega^{\alpha\#\kappa}\#\mu}}\right. F^{\mathsf{L}_\mu} \text{ for all } \mu \in \mathfrak{M}_\pi^{\omega^{\alpha\#\kappa}}.$$

Since $\Psi_\pi^{\omega^{\alpha\#\kappa}\#\mu+\omega}$ is always an ordinal below π no rules for ideal ordinals $\geq \pi$ occur in the derivation that is reduced according to Theorem 4.2. The reduction thus incorporates an *elimination of all ideal ordinals $\geq \pi$ in semi-formal proofs of sentences belonging to the complexity class of π.*

Definition 4.3. *We say that a formal theory T (in the language of set theory) has a model in ramified set theory at level π if there are a transitive Cantorian closed Skolem hull operator $\mathcal{H}$ and for every theorem F of T ordinals $\alpha < \varepsilon_{\pi+1}$ and $n < \omega$ such that $\mathcal{H} \left|\frac{\alpha}{\pi+n}\right. F^{\mathsf{L}_\pi}$.*

For a theory T that has a model in ramified set theory at an abstract ideal ordinal π we obtain an ordinal analysis in the following way. Let F be a Π_2-sentence such that $F^{\mathsf{L}_{\omega_1^{CK}}}$ is provable in T. We first get $\mathcal{H} \left|\frac{\alpha}{\pi+1}\right. F^{\mathsf{L}_\Omega}$ for an ordinal $\alpha < \varepsilon_{\pi+1}$ by the fact that T has a model in ramified set theory at π and predicative cut elimination. Since Π_2 is the complexity class which belongs to Ω, and we further have that $\Omega > \Psi_\Omega^{\varepsilon_{\pi+1}} > \gamma := \Psi_\Omega^{\omega^{\alpha\#\pi}\#\Omega} \in \mathfrak{M}_\Omega^{\omega^{\alpha\#\pi}}$ we obtain by Theorem 4.2

$$\mathcal{H}^{\omega^{\alpha\#\pi}\#\gamma+1} \left|\frac{\Psi_\Omega^{\omega^{\alpha\#\pi}\#\gamma+\omega}}{\Psi_\Omega^{\omega^{\alpha\#\pi}\#\gamma}}\right. F^{\mathsf{L}_\gamma}$$

and thus $\|T\|_{\omega_1^{CK}} \leq \Psi_\Omega^{\varepsilon_{\pi+1}}$. More precisely we know from

$$\Psi_\Omega^{\omega^{\alpha\#\pi}\#\gamma} < \Psi_\Omega^{\omega^{\alpha\#\pi}\#\gamma+\omega} < \Omega$$

and $\gamma < \Omega$ that the semi-formal derivation of F^{L_γ} as shown above cannot contain rules for abstract ideal ordinals. *Therefore we obtained an ordinal analysis of T*

20 For the sake of simplicity we suppress all technical details. See [Pohlers and Stegert, 2012] and [Pohlers, 201xb] for details. Theorem 4.2 below can be proved from Theorem 5.9 in [Pohlers and Stegert, 2012] similar to Theorem 5.10 in [Pohlers and Stegert, 2012]. The notion $\alpha \# \beta$ denotes the natural (symmetrical) sum of ordinals.

by eliminating all "abstract" ideal objects in a formal proof of a $\Pi_2^{\mathsf{L}_{\omega_1^{CK}}}$*-sentence for the costs of transforming the formal derivation into an infinitary derivation. However, as mentioned at the end of Section 4.1.1, the "infinity" of this derivation can be measured by the order-type of a relation on the natural numbers which is primitive-recursively definable without the use of abstract ideal objects.*

4.2 Subrecursive hierarchies and the elimination of all idealizations

The elimination of *abstract* ideal objects as demonstrated in the previous section becomes only visible in impredicative theories. It may even be viewed as characteristic for impredicative theories that their ordinal analyses need the introduction of abstract ideal ordinals which (in combination with a foundation scheme) reflect the impredicative content of the analyzed theory.

It is much harder to see how to eliminate *all* ideal objects in the strict sense, i.e., all objects whose descriptions have complexities beyond Π_2^0. Since the computability of a recursive function with index e can be expressed by the Π_2^0-sentence $(\forall y)(\exists y)\mathsf{T}(e, x, y)$ the complexity class Π_2^0 represents the bounding of algorithmic computatiblity. The "ideal objects", i.e. the objects of larger complexity than Π_2^0, may thus be regarded as those which reflect a non-computational content.

On the ordinal side the elimination of all strictly ideal objects means the elimination of all actually infinite ordinals including ω. The actual closure properties of ordinals will be "potentially" approximated by sufficiently increasing functions.

Already in the embedding of formal derivations into the semi-formal system we have to treat ω as an ideal ordinal. While no additional axioms to fix ω as an ideal ordinal are necessary in subsystems of second order number theory—there all first order quantifiers are supposed to range over ω and ω is sufficiently axiomatized by the first order axioms of Peano arithmetic and the scheme of mathematical induction—we need defining axioms for L_ω in the framework of set theory. To define L_ω as the set of hereditarily finite sets turned out to be a good choice.

We are going to indicate that a Π_2^0-analysis of a theory T is obtained by eliminating all objects that are ideal in the strict sense. Recall that a Π_2^0-analysis of a theory T means the computation of a function that designs experiments for T.

For Peano arithmetic such functions have been known for long (cf. [Schwichtenberg and Wainer, 2012] Chapter 4 for more information). Here we want to indicate on the example of subsystems of set theory that such functions can be obtained via a refined ordinal analysis in a very general way.[21] A way the emphasis of which lays on the elimination of all ideal objects. Clearly we can only outline the main features suppressing all technical details. Details are in [Pohlers, 2009] (for Peano Arithmetic) and [Pohlers and Stegert, 2012] (for a subsystem of set theory) and will appear for systems of iterated inductive definitions in [Pohlers, 201xa]. In contrast to the "abstract ideal ordinals" we do not need a defining rule for ω in ramified set theory. It is possible to exploit the fact that L_ω coincides with the hereditarily finite sets. Choosing an adequate starting function Φ there is a fragmented controlled derivation—as defined below—in ramified set theory of the fact that every set in L_ω is hereditarily finite.

4.2.1 Subrecursive hierarchies

To outline the strategy we first cite a definition which is due to Andreas Weiermann (cf. [Buchholz et al., 1994]) . Given a starting function $\Phi\colon \mathbb{N} \longrightarrow \mathbb{N}$ and a set O of ordinals such that every ordinal $\alpha \in O$ possesses a norm $N(\alpha) < \omega$ satisfying some natural prerequisites we can define a function $\psi\colon O \longrightarrow \omega$ by

$$\psi(\alpha) := \sup\{\psi(\beta) + 1 \,|\, \beta < \alpha \wedge N(\beta) < \Phi(N(\alpha))\} \cup \{0\}.$$

Defining $\Phi_\alpha(x) := \psi(\omega \cdot \alpha + x)$ we obtain $\Phi_n(x) \approx \Phi^n(x)$ where the latter means the familiar iteration of the function Φ. Therefore Φ_α is closely connected to the more familiar fast growing subrecursive hierarchy Φ^α which is defined by

$$\Phi^0(x) = x, \quad \Phi^{\alpha+1}(x) = \Phi(\Phi^\alpha(x)) \text{ and } \Phi^\lambda(x) = \Phi^{\lambda[x]}(x)$$

for limit ordinals λ where $\lambda[x]$ denotes the xth member of a fixed fundamental sequence for λ. Starting with a primitive-recursive function Φ satisfying some mild conditions we obtain that Φ_{ε_0} designs experiments for Peano arithmetic.[22]

4.2.2 Elimination of all ideal objects

We will indicate that Weiermann's definition of subrecursive hierarchies combined with (iterations of) Skolem hull operators provides a Π^0_2-analysis of a theory, i.e., a function Φ_T that designs experiments for T. The pioneering work in this direction has been done by Weiermann and Benjamin Blankertz ([Blankertz and Weiermann, 1996], [Blankertz, 1997]).

21 This can be easily transfered to systems of iterated inductive definitions (cf. [Pohlers, 201xa]).

22 Cf. [Pohlers, 2009] Chapter 10.

Recall that the ordinals in $\mathcal{H}(X)$ can be notated by finite strings built up from the generators of $\mathcal{H}$ and the parameters in X. This can be made unique such that every ordinal $\alpha \in \mathcal{H}(X)$ can be equipped with a norm function $N(\alpha)$ which is the length of its notation. Let $\Phi\colon \mathbb{N} \longrightarrow \mathbb{N}$ be a primitive recursive function.

Definition 4.4. *Let X be a set of ordinals. Then we we define recursively*

$$\psi_{\Phi}^{\mathcal{H}(X)}(\alpha) := \sup \{\psi_{\Phi}^{\mathcal{H}(X)}(\beta) + 1 \mid \beta \in \mathcal{H}(X) \cap \alpha \wedge N(\beta) < \Phi(N(\alpha))\} \cup \{\emptyset\}.$$

Since there are only finitely many $\beta < \alpha$ such that $N(\beta) < \Phi(N(\alpha))$ the values of $\psi_{\Phi}^{\mathcal{H}(X)}$ are finite. The function $\psi_{\Phi}^{\mathcal{H}(X)}$ therefore collapses α below ω and we define for regular cardinals κ the function $\Phi_{\kappa}^{\mathcal{H}(X)}\colon \mathbb{N} \longrightarrow \mathbb{N}$ by

$$\Phi_{\kappa,\alpha}^{\mathcal{H}(X)}(x) := \psi_{\Phi}^{\mathcal{H}^{\alpha}(X)}(\Psi_{\kappa}^{\alpha} + x).$$

By the remark in the end of Section 4.1.1 the ordinals Ψ_{Ω}^{α} are definable without use of abstract ideal ordinals. Therefore the functions $\Phi_{\Omega,\alpha}^{\mathcal{H}(X)}$ can be regarded as continuations of their predicative versions Φ_{α} mentioned above.[23]

For a Π_2^0-analysis also the "finite content" of parameters matters. Therefore we define for a finite set X of ordinals

$$|X| := \max(\{N(\alpha) + 1 \mid \alpha \in X\} \cup \{\overline{\overline{X}} + 1\}),$$

where $\overline{\overline{X}}$ stands for the cardinality of the set X, and define the *fragmented* iteration $f_{\Phi}\mathcal{H}_{\kappa}^{\alpha}$ of a Skolem-hull operator $\mathcal{H}$ by[24]

$$f_{\Phi}\mathcal{H}_{\kappa}^{\alpha}(X) := \{\xi \in \mathcal{H}^{\alpha}(X) \mid N(\xi) < \Phi_{\kappa,\alpha}^{\mathcal{H}^{\alpha}(X)}(|X|)\}.$$

Then we define *fragmented controlled derivations*[25] in complete analogy to operator controlled derivations in ramified set theory.

Definition 4.5. *We modify the notion of an operator controlled derivation to a fragmented operator controlled derivation $f_{\Phi}\mathcal{H}_{\kappa}^{\gamma} \overset{\alpha}{\underset{\rho}{\vdash}} \Delta$ by requiring $\alpha \in f_{\Phi}\mathcal{H}_{\kappa}^{\gamma}(\mathrm{par}(\Delta))$ and also $\mathrm{par}(\Delta_{\iota}) \cup \{|\mathrm{par}(\Delta_{\iota})|\} \subseteq f_{\Phi}\mathcal{H}_{\kappa}^{\gamma}(\mathrm{par}(\Delta))$ for every inference*

$$\overset{\alpha_{\iota}}{\underset{\rho}{\vdash}} \Delta_{\iota} \;\Rightarrow\; \overset{\alpha}{\underset{\rho}{\vdash}} \Delta$$

not according to $(\bigwedge)$.

23 Since we defined Ψ_{κ}^{α} only for regular κ, this subrecursive hierarchy is much coarser than the previous one. In fact it skips the whole predicative part. To retrieve also the predicative part one could define $\Psi_0^{\alpha} = \alpha$ and $\Phi_{\kappa,\alpha}^{\mathcal{H}(X)}(x) := \psi_{\Phi}^{\mathcal{H}^{\alpha}(X)}(\omega \cdot \Psi_{\kappa}^{\alpha} + x)$. But this is inessential for our purpose.

24 This notation deviates from [Pohlers and Stegert, 2012] where we denoted it by $\mathcal{F}_{\kappa,\alpha}^{\mathcal{H}}$ suppressing to mention the fixed starting function Φ

25 This technique is due to Jan Carl Steger ([Stegert, 2011]).

Predicative cut-elimination (Theorem 3.1) remains true for fragmented controlled semi-formal derivations in the modification

$$f_\Phi \mathcal{H}_\kappa^\gamma \,\frac{\alpha}{\beta+\omega^\rho}\, \Delta \Rightarrow f_\Phi \mathcal{H}_\kappa^{\gamma \# \varphi_\rho(\alpha)+1} \,\frac{\varphi_\rho(\alpha)}{\beta}\, \Delta$$

and the main theorem (Theorem 4.2) modifies to

Theorem 4.6. *Let F be a formula belonging to the complexity class of the ideal ordinal π. Then*

$$f_\Phi \mathcal{H}_\pi \,\frac{\alpha}{\kappa+1}\, F^{\mathsf{L}_\pi} \text{ implies } f_\Phi \mathcal{H}_\pi^{\omega^{\alpha+\kappa}+\mu+1} \,\frac{\Psi_\pi^{\omega^{\alpha+\kappa}+\mu+\omega}}{\Psi_\pi^{\omega^{\alpha+\kappa}+\mu}}\, F^{\mathsf{L}_\mu} \text{ for all } \mu \in \mathfrak{M}_\pi^{\omega^{\alpha+\kappa}}.$$

We already mentioned that we do not need a defining rule for ω in ramified set theory. The ordinal ω can be characterized by the fact that all sets in L_ω are hereditarily finite. To gauge the exponential growth of the stages of the hereditarily finite sets in a fragmented controlled semi-formal derivation we have to start with an exponentially growing function Φ. Therefore we define $\phi(0) := 1$, $\phi(n+1) := 2^{\phi(n)}$ and $\Phi := \phi^2$. Though there is no rule and thus no complexity class affiliated to a rule for ω there is a theorem which connects ω to the class of Π_2 sentences that are valid in L_ω. To save indices let $\overline{\Phi}_\kappa^\alpha(X) := \Phi_{\kappa,\alpha}^{\mathcal{H}^\alpha(X)}(|X|)$.

Theorem 4.7. *(Witnessing Theorem) Assume $f_\Phi \mathcal{H}_\kappa^\gamma \,\frac{\alpha}{0}\, (\exists x)F(a_1,\dots,a_n,x)^{\mathsf{L}_\omega}$ holds true for a Σ_1-formula $(\exists x)F(x_1,\dots,x_n,x)$ and a tuple $a_1,\dots,a_n$ of $\mathcal{L}_{RS}$-terms of stages less than ω. Then there is a $t \in \mathsf{L}_m$ such that $\mathsf{L}_\omega \models F(a_1,\dots,a_n,t)$ and $m < \overline{\Phi}_\kappa^{\gamma+\omega^\alpha+1}(\{a_1,\dots,a_n\})$.*

From the witnessing theorem we obtain for a Π_2-formula $F \equiv (\forall x)(\exists y)G(x,y,\vec{a})$ and $a, \vec{a} \in \mathsf{L}_\omega$ such that $f_\Phi \mathcal{H}_\kappa^\gamma \,\frac{\alpha}{0}\, F^{\mathsf{L}_\omega}$ the function $\overline{\Phi}_\kappa^{\gamma+\omega^\alpha+1}$ as a test-function for F. Since testing $\mathsf{L}_\omega \models G(a,t,\vec{a})$ for all $t \in \mathsf{L}_m$ with $m < \overline{\Phi}_\kappa^{\gamma+\omega^\alpha+1}(\{a_1,\dots,a_n,a\})$ is just a computation that needs not to refer to ideal objects it constitutes an elimination of all ideal objects.

Observe that even for the ordinal notation system we do not need ω as an "actual infinity" since it can be represented by a primitive-recursive ordering on $\mathbb{N}$ for which we need $\mathbb{N}$, i.e., ω, only as "potential infinity". So we may in fact talk about an elimination of ω as completed infinity.

We sharpen Definition 4.3 by saying that T has a *strict model in ramified set theory at level π* if there are a primitive-recursive function Φ, a Cantorian closed Skolem hull operator $\mathcal{H}$ and for all theorems F of T ordinals $\alpha < \varepsilon_{\pi+1}$ and $n < \omega$ such that $f_\Phi \mathcal{H}_\Omega \,\frac{\alpha}{\pi+n}\, F^{\mathsf{L}_\pi}$.

If a formal theory T has a strict model at level π in ramified set theory and $T \vdash F^{\mathsf{L}_\omega}$ for a Π_2 sentence F we obtain from predicative elimination (Theorem 4.6) and the Witnessing Theorem an $\alpha < \Psi_\Omega^{\varepsilon_{\pi+1}}$ such that $\overline{\Phi}_\Omega^\alpha$ is a test-function for F. However, $\alpha < \beta$ does not yet entail $\overline{\Phi}_\Omega^\alpha(X) < \overline{\Phi}_\Omega^\beta(X)$, since we not necessarily have $N(\Psi_\Omega^\alpha + |X|) < \overline{\Phi}_\Omega^\beta(N(\Psi_\Omega^\beta + |X|))$. This, however, becomes true for large enough $|X|$. Therefore $\overline{\Phi}_\Omega^{\Psi_\Omega^{\varepsilon_{\pi+1}}}$ eventually majorizes a test-function for any $\Pi_2^{\mathsf{L}_\omega}$-sentence F provable in T and thus designs experiments for T. So we obtain $\|T\|_\omega = \Psi_\Omega^{\varepsilon_{\pi+1}} = \|T\|_{\omega_1^{CK}}$.

5 Conclusion

The aim of the paper was to discuss an aspect of the relation of ordinal analyses to Hilbert's programme. Though consistency is a byproduct of ordinal analysis it can hardly be argued that it fully meets Hilbert's demand for a finitistic consistency proof. Of course one could debate in how far the reduction of the consistency of an abstract axiom system to the well-foundedness of a primitive-recursively definable ordering on the natural numbers constitutes a step towards an "effective" consistency proof. But this was not the concern of this paper.

Here we wanted contrariwise argue that ordinal analysis itself rests on another aspect of Hilbert's programme, the elimination of ideal elements. We had to distinguish three levels of ideal objects which can be symbolized by the existence of certain ordinals. Ideal objects in the strict sense, symbolized by the existence of ω. Ideal objects of the first level of abstraction, symbolized by the existence of certain admissible ordinals above ω and ideal objects of the second level of abstraction, symbolized by the existence of indescribable cardinals.

Elimination of abstract ideal objects is the main feature of impredicative proof theory. Ideal objects of the first level of abstraction, formalized by rules for admissible ordinals above ω, are needed to unravel formal proofs into semi-formal proofs. Abstract ideal objects of the second level of abstraction are needed in the construction of the sets of ordinals that contain the necessary gaps for collapsing.

While indescribable cardinals are already eliminated during the construction of the ordinal sets the elimination of rules for admissible ordinals ($> \omega$) takes place in the Π_1^1-analysis. Π_2^0-analysis in addition needs the elimination of all complexites above Π_2^0.

Predicative systems do not need abstract ideal objects. The only ideal object needed there is ω, axiomatized formally by the scheme of mathematical induction and by the ω-rule in the semi-formal systems. Similar to Section 4.2 this rule is eliminated—in a controlled way—in a Π_2^0-analysis of predicative systems. Details for the case of Peano arithmetic are in [Pohlers, 2009] Chapter 10 and for iterated inductive definitions in [Pohlers, 201xa].

References

Peter Aczel and Wayne H. Richter. Inductive definitions and analogues of large cardinals. In Wilfrid Hodges, editor, *Conference on Mathematical Logic – London 1970*, number 255 in Lecture Notes in Mathematics, pages 1–9, Heidelberg/New York, August 1972. Springer.

Jon Barwise. *Admissible Sets and Structures*. Perspectives in Mathematical Logic. Springer, Berlin/Heidelberg/New York, 1975.

Benjamin Blankertz. *Beweistheoretischen Techniken zur Bestimmung von Π^0_2–Skolem Funktionen*. Dissertation, Westfälische Wilhelms-Universität, Münster, 1997.

Benjamin Blankertz and Andreas Weiermann. *How to characterize provably total functions by the Buchholz operator method*. Number 6 in Lecture Notes in Logic. Springer, Heidelberg/New York, 1996.

Jane Bridge. A simplification of the Bachmann method for generating large countable ordinals. *Journal of Symbolic Logic*, 40:171–185, 1975.

Wilfried Buchholz. *Rekursive Bezeichnungssysteme für Ordinalzahlen auf der Grundlage der Feferman–Aczelschen Normalfunktionen Θ_α*. Dissertation, Ludwig Maximilians–Universität, München, 1974.

Wilfried Buchholz. Normalfunktionen und konstruktive Systeme von Ordinalzahlen. In Justus Diller and Gert H. Müller, editors, $\models$ *ISILC Proof Theory Symposium*, number 500 in Lecture Notes in Mathematics, pages 4–25, Heidelberg/New York, July 1975. Springer.

Wilfried Buchholz. A new system of proof-theoretic ordinal functions. *Annals of Pure and Applied Logic*, 32:195–207, 1986.

Wilfried Buchholz. A simplified version of local predicativity. In Peter Aczel, Harold Simmons, and Stanley S. Wainer, editors, *Proof Theory*, pages 115–147, Cambridge, July 1992. Cambridge University Press.

Wilfried Buchholz, E. Adam Cichon, and Andreas Weiermann. A uniform approach to fundamental sequences and hierarchies. *Mathematical Logic Quarterly*, 40:273–286, 1994.

Frank R. Drake. *Set Theory: An introduction to large cardinals*. Number 76 in Studies in Logic and the Foundations of Mathematics. North-Holland, Amsterdam, 1974.

Solomon Feferman. Systems of predicative analysis II: Representation of ordinals. *Journal of Symbolic Logic*, 33:193–220, 1968.

Kentaro Fujimoto. Notes on some second order systems of inductive definitions and Π^1_1–comprehensions and relevant subsystems of set theory. *Annals of Pure and Applied Logic*, 166:409–463, 2014.

Gerhard Gentzen. Die Widerspruchsfreiheit der reinen Zahlentheorie. *Mathematische Annalen*, 112:493–565, 1936.

Gerhard Gentzen. Beweisbarkeit und Unbeweisbarkeit von Anfangsfällen der transfiniten Induktion in der reinen Zahlentheorie. *Mathematische Annalen*, 119:140–161, 1943.

Kurt Gödel. Eine Interpretation des intuitionistischen Aussagenkalküls. *Ergebnisse eines Mathematischen Kolloquiums*, 4:39–40, 1931.

Leon Henkin. A generalization of the notion of ω-consistency. *Journal of Symbolic Logic*, 19: 183–196, 1954.

David Hilbert. Neubegründung der Mathematik. *Abhandlungen aus dem Math. Seminar d. Hamb. Univ.*, I:157–122, 1922.

David Hilbert. Die Grundlagen der Mathematik. Vortrag gehalten auf Einladung des Mathematischen Seminars im Juli 1927 in Hamburg. *Hamburger Mathematische Einzelschriften*, 5. Heft:1–21, 1928.

David Hilbert. Die Grundlegung der elementaren Zahlenlehre. *Mathematische Annalen*, 104: 485–494, 1931.

David Hilbert, editor. *Gesammelte Abhandlungen*, volume III. Springer, Berlin, 1935.

Gerhard Jäger. *Theories for Admissible Sets. A Unifying Approach to Proof Theory.* Number 2 in Studies in Proof Theory, Lecture Notes. Bibliopolis, 1986.

Gerhard Jäger. Metapredicative and explicit mahlo: a proof-theoretic perspective. In Rene Cori, Alexander Razborov, Stevo Todorcevic, and Carol Wood, editors, *Logic Colloquium '00*, volume 19 of *Lecture Notes in Logic*, pages 272–293, Wellesley, MA, 2005. AK Peters.

Gerhard Jäger and Wolfram Pohlers. Eine beweistheoretische Untersuchung von (Δ^1_2-CA) + (BI) und verwandter Systeme. *Bayerische Akademie der Wissenschaften, Sitzungsberichte 1982*, pages 1–28, 1983.

S. Orey. On ω-consistency and related properties. *Journal of Symbolic Logic*, 21:246–252, 1956.

Wolfram Pohlers. Proof-theoretical analysis of ID_ν by the method of local predicativity. In Wilfried Buchholz, Solomon Feferman, Wolfram Pohlers, and Wilfried Sieg, editors, *Iterated Inductive Definitions and Subsystems of Analysis: Recent Proof-Theoretical Studies*, number 897 in Lecture Notes in Mathematics, pages 261–357. Springer, Heidelberg/New York, 1981.

Wolfram Pohlers. Subsystems of set theory and second order number theory. In Samuel R. Buss, editor, *Handbook of Proof Theory*, Studies in Logic and the Foundations of Mathematics, pages 209–335. North-Holland, 1998.

Wolfram Pohlers. *Proof Theory. The first step into impredicativity.* Universitext. Springer, Berlin/Heidelberg/New York, 2009.

Wolfram Pohlers. Semi-formal calculi and their applications. In Reinhard Kahle and Michael Rathjen, editors, *Gentzen's Centenary: The quest for consistency*, pages 195–232. Springer, 2015.

Wolfram Pohlers. Iterated inductive definitions revisited. *submitted*, page 28, 201xa.

Wolfram Pohlers. *Stability. A proof theoretic study.* In preparation, 201xb.

Wolfram Pohlers and Jan Carl Stegert. Provably recursive functions of reflections. In Ulrich Berger, Hannes Diener, Peter Schuster, and Seisberger Monika, editors, *Logic, Construction, Computation*, volume 3 of *Ontos Mathematical Logic*, pages 381–474. Ontos, Frankfurt, Paris, Lancaster, New Brunswick, 2012.

Michael Rathjen. Ordinal notations based on a weakly Mahlo cardinal. *Archive for Mathematical Logic*, 29:249–263, 1990.

Michael Rathjen. *Eine Ordinalzahlanalyse der Π_3-Reflexion.* Habilitationsschrift, Westfälische Wilhelms-Universität, Münster, 1992.

Michael Rathjen. An ordinal analysis of stability. *Archive for Mathematical Logic*, 48/2:1–62, 2005a.

Michael Rathjen. An ordinal analysis of parameter free Π^1_2-comprehension. *Archive for Mathematical Logic*, 48/3:263–362, 2005b.

Helmut Schwichtenberg and Stanley Wainer. *Proofs and Computations*. Perspectives in Logic. Cambridge University Press, Cambridge, UK, 2012.

Jan-Carl Stegert. *Ordinal proof theory of Kripke–Platek set theory augmented by strong reflection principles*. PhD thesis, Westfälische Wilhelms-Universität, Münster, 2011.

Jan von Plato

Aristotle's Deductive Logic: a Proof-Theoretical Study

Abstract: Aristotle's deductive logic, as presented in his book *Prior Analytics*, is a system of rules of proof. The structure of derivations by these rules is analyzed. It is shown that derivations can be so transformed that steps of indirect proof are applied at most once as a last rule, the only way in which Aristotle used indirect proof.

Keywords: Aristotle's logic, Syllogism, Tree derivation

Mathematics Subject Classification 2010: 01A20, 03-03, 03F07

Introduction

Aristotle's system of deductive logic, also known as the "theory of syllogisms," has been interpreted in various ways in the long time since it was conceived. The situation is not different from the reading of other chapters of the formal sciences of antiquity, such as Euclid's geometry and works of Archimedes. With the advent of predicate logic, it became common to read the Aristotelian forms of propositions, such as *Every A is B*, as $\forall x(A(x) \supset B(x))$. Here we have a universal quantification over some domain, and the predicates A and B, and the sentence is *For every x, A(x) implies B(x)*. Aristotelian inferences, such as the conclusion of *Every A is C* from the premisses *Every A is B* and *Every B is C*, were similarly interpreted in predicate logic. Some, [Lukasiewicz, 1957] in the first place, even claimed that Aristotle intended his logic in the sense given to it by predicate logic, but failed to make explicit the underlying logical principles.

In the early 1970's, John Corcoran and Timothy Smiley took decisively distance from the predicate-logical interpretation, and read the *Prior Analytics* for what it is: There are certain basic forms of propositions, analyzed in the Aristotelian way into a subject, a copula, and a predicate. Thus, in *Every A is B*, A is the subject, *Every ... is ...* the copula, and B the predicate. Both authors concentrated on the smallest system of rules that suffices for Aristotelian syllogistic derivations. It has two one-premiss rules, two two-premiss rules, and a schematic rule of indirect proof.

Jan von Plato: Department of Philosophy, University of Helsinki, Finland
e-mail: jan.vonplato@helsinki.fi

Aristotle's proofs, i.e., deductions within his system, are given as texts, one sentence following another. They are thus linearly arranged, as a logician would say today, instead of having, say, tree form. Even so, they are at ground formal in nature.

An impressive feature of the Aristotelian proof texts is the following: Whenever a rule of inference is applied, the premisses of the rule appear in the text right before the rule, and in a standard order that conforms to the pattern of the rule. Such an arrangement is otherwise found in the axiomatic systems of logic of Frege and Bernays and Hilbert. In a derivation with *modus ponens* as the only rule, when the rule is applied repeatedly, not both premisses of the rule can be instances of the logical axioms. In a linear arrangement of formulas, a number or other marker is given to each formula in a derivation so that its use as a premiss in a rule lower down in the derivation can be indicated. Aristotle simply repeats a premiss earlier assumed or derived and adds words such as "But it was assumed that ...," to have an exact formal match of premisses and conclusion in each rule instance. This "rule of repetition," as it is called in [Corcoran, 1974], is nothing but part of the inductive definition of a formal derivation: Whenever there is a derivation of some formula P and a rule instance in which that same P acts as a premiss, the rule can be applied to yield a derivation of its conclusion. More generally, if there is a derivation of P and another derivation in which P appears as an assumption, it should be possible to **compose** the two derivations into one that derives the conclusion of the second derivation from the assumptions of the first and second derivation together, P excluded.

Aristotle uses a symbolic notation only for subjects and predicates and natural language for the rest, but the times are long past when in logical research it was thought that an exclusive use of symbols instead of words is the hallmark of formality. All steps of inference, what the premisses are and what the conclusion, are clearly indicated in Aristotle's texts; nothing is left to guesswork in this respect.

[Corcoran, 1974], and later [Smith, 1989] following the former found that Aristotle's main interest with his system of deductive logic was to study its general properties, not to use it for the actual construction of derivations. Thus, in their view, Aristotle gave his system of deductive logic as an example of the organization of such systems in general, and not as a closed system of logic to which nothing need be added.

A serious limitation of Aristotle's formal analyses of the proofs should be noted: With a few exceptions, the derivations Aristotle considers contain just one syllogistic rule. What are called conversions are not considered proper deductive steps. We shall go through those places in which Aristotle refers to the combination of several syllogistic rules. The overall conclusion seems to be that he did not succeed, or even try, to extend the meticulous analysis of one-syllogism derivations into the general case.

Both of the last-mentioned authors continued to present Aristotle's formal proofs as linear derivations. We shall give them as derivation trees, which gives a much improved control over their structure. In this representation, all assumptions are topformulas in derivations, and it will be easy to show, for example, that derivations that contain indirect steps of proof can be so transformed that there is at most one such step as a last rule. The result conforms to Aristotle's practice of always using indirect proof in that way.

Most of the *Prior Analytics* is devoted to syllogisms in which the premisses and conclusion are modal formulas. We shall not consider such modal syllogisms here.

This article is in three parts. Section 1 goes through the first chapters of the *Prior Analytics* from the point of view of proof systems. Section 2 gives a proof-theoretical analysis of Aristotle's system of deductive logic. This part can be read independently, when the notation from table 2 is gathered, as well as the rule of indirect proof of table 3, the two first rules of table 5, and the two rules of table 8. Section 3 discusses the issue of Aristotle's "underlying logic."

1 Justification of the rules of Aristotle's deductive logic

1.1 The forms of propositions

There are five forms of propositions in the logic of the *Prior Analytics*:

1. The Aristotelian forms of propositions	
Universal affirmative:	*Every A is B*
Universal negative:	*No A is B*
Particular affirmative:	*Some A is B*
Particular negative:	*Some A is not-B*
Indefinite:	*A is B*

The last one is not usually present in the rules of inference, even though it can so be (*cf.* 26a28).

Subjects and predicates together are **terms**. The indefinite form *A is B* has various other readings: *Subject A has the predicate B, predicate B belongs to subject*

A, B belongs to A, etc. The last one is preferred by Aristotle, and he writes the other forms similarly:

B belongs to every A, B belongs to no A, B belongs to some A, B does not belong to some A.

Here the copula is written as one connected expression between the predicate and the subject, which underlines the formal character of the sentence construction.

Another useful way of expressing the Aristotelian propositions is:

Every A is B, No A is B, Some A is B, Not every A is B.

Now the indefinite form *A is B* is a constant part of the propositions and the varying quantifier structure is singled out. It is seen clearly that the first and last are opposites, and that the second and third are likewise opposites.

The main principle in the formation of propositions is that subjects and predicates are treated symmetrically: Whenever *Every A is B* is a proposition, also *Every B is A* is one, and similarly for the universal negative and the particular forms. A formal structure is imposed that is not a natural feature of natural language, as in *Some man is wise*, the **converse** of which, *Some wise is man*, would not be a natural expression, but would have to be paraphrased, as in *Some wise being is a man.*

The symmetry of the subject and predicate in a proposition can break down with indefinite propositions. As Aristotle says, *individuals cannot be predicated of other things, but instead other things are predicated of them* (43a40). An example such as *Socrates is wise* confirms this: It does not convert, because there is no sense in this context in *Wise is Socrates*, or, *The predicate Socrates belongs to the subject wise.*

1.2 Semantical explanation of assertions

Logical rules seem to act on propositions, but the impression is wrong. When the notion of a logical derivation is defined, it is stipulated how a derivation can begin. In axiomatic systems, it can begin with instances of axioms. These are logical truths, not plain propositions. Similarly, in proof systems that have just rules and no axioms, as in Aristotle, derivations begin with **assumptions**. These are propositions assumed to hold, not plain propositions.

By the above, the conclusion of a logical rule is then either a logical truth, as in axiomatic systems of logic, or a claim that holds if the assumptions hold, as in systems that have just rules. For the latter, we can equally call the assumptions and conclusions from them **assertions**, a notion that is more general than truth because assertions can be hypothetical, but truth in axiomatic systems of logic is categorical.

We shall use the following notation for the propositions of Aristotle's logic:

2. Notation for the Aristotelian propositions

$\Pi^+(A, B)$	for	*Every A is B*
$\Pi^-(A, B)$	for	*No A is B*
$\Sigma^+(A, B)$	for	*Some A is B*
$\Sigma^-(A, B)$	for	*Some A is not-B*

In Aristotle's times, there were no lower-case letters in Greek. Today it is customary in logic to denote individuals by lower case and predicates by upper case letters. We shall write $(c : A)B(c)$ for *some A, say c, is B*, a construction Aristotle uses at places (e.g., 25a16). It is not used below, except for the semantical explanations of the four forms of expression of tables 1 and 2. From today's point of view, such an arbitrary instance of an existential assertion is an **eigenvariable** used to eliminate the existential quantifier. The reading of $(c : A)B(c)$ is *c in A such that B(c).*

In Aristotle's terminology, the terms in the first four forms of propositions are **universal** and **particular**, respectively, and those in the fifth **indefinite.** The four kinds of propositions contain terms that designate, in the subject position, **types** of objects, whereas the fifth form can have in subject position **instances.** This feature is captured by the lower case letters in our notation for the indefinite form of proposition.

The **universal quantifier** of the Aristotelian form $\Pi^+(A, B)$ is explained as follows in the *Prior Analytics* (24b28):

> *A thing is predicated of all of another when there is nothing to be taken of which the other could not be said.*

This passage has been translated in various ways, even as ... *whenever no instance of the subject can be found of which the other term cannot be asserted* (in the Oxford edition of Aristotle's works, 1928).

Aristotle's explanation of the universal quantifier is not unlike some later explanations found in logic that, together with the one of Aristotle, can be termed **no-counterexample** interpretations: The formula $\forall x B(x)$ is true in a given discourse (i.e., in a given domain $\mathcal{A}$ and interpretation of B) if there is no object c in $\mathcal{A}$ such that $B(c)$ fails. Put the type A in place of $\mathcal{A}$ and you have the Aristotelian formulation (24b29–30): *There is nothing to be taken* [no $c : A$] *of which the other* [$B(c)$] *could not be said.* In correspondence with the explanation, the syntactic form of the universal quantifier makes it **bounded** to a domain of quantification that varies from one quantifier occurrence to another.

Instances of types of objects are used sometimes, as when Aristotle shows unprovabilities through counterexamples in chapter 4. The notation is just a juxtaposition of two terms, as in *B Γ, B belongs to some Γ* (*cf.* 26b29). Say, if the premisses are $\Sigma^{-}(B, A)$ and $\Pi^{+}(C, B)$, then $\Sigma^{-}(A, C)$ does not necessarily follow.

- **The universal negative quantifier.** *There is nothing in A of which B could be said.*

  With the above notation, we have: *There is no c of type A such that B(c).* Next, the explanation of $\Sigma^{+}(A, B)$ builds on that of $\Pi^{-}(A, B)$ (25a21–23):
- **The existential quantifier.** *It is impossible that no A is B.*

  The explanation is quite natural, on the face of Aristotle's treatment of $\Pi^{-}(A, B)$ and $\Sigma^{+}(A, B)$ as a **contradictory pair**. Similarly with $\Sigma^{-}(A, B)$; Its contradictory is $\Pi^{+}(A, B)$ and the explanation becomes:
- **The existential negative quantifier.** *It is impossible that an A could not be taken such that not-B could not be said.*

  By Aristotle's principle of indirect proof, direct and indirect existence claims are equated, as we shall point out in detail later, so that $\Sigma^{+}(A, B)$ becomes explained as *A c of type A can be taken such that B(c)*, and $\Sigma^{-}(A, B)$ as *A c of type A can be taken such that not-B(c).*

1.3 Indirect proof

The pairs $\Sigma^{+}(A, B)$, $\Pi^{-}(A, B)$ and $\Pi^{+}(A, B)$, $\Sigma^{-}(A, B)$ are **contradictory opposites.** Furthermore, because from $\Pi^{-}(A, B)$ the weaker $\Sigma^{-}(A, B)$ follows, also $\Pi^{+}(A, B)$ and $\Pi^{-}(A, B)$ together lead to a contradictory pair. We shall indicate the contradictory opposite of a formula P by the orthogonality symbol, $P^{\perp}$. In general, if an assumption P has led to contradictory consequences Q and $Q^{\perp}$, $P^{\perp}$ can be concluded and the assumption P **closed.** The rule of indirect proof thus takes on the following schematic form:

3. The scheme of indirect proof

$$\dfrac{\begin{matrix} \overset{1}{P^{m}} \\ \vdots \\ Q \end{matrix} \qquad \begin{matrix} \overset{1}{P^{n}} \\ \vdots \\ Q^{\perp} \end{matrix}}{P^{\perp}}\, Raa{,}1$$

This schematic proof figure is to be understood as follows: The assumption may appear among those that were used in the derivations of Q and $Q^{\perp}$, respectively. Any numbers $m, n \geqslant 0$ of occurrences of P in the two **subderivations** can be closed at the inference. The closed ones are indicated by a suitable label, such as a number, so that each instance of rule *Raa* (for *reductio ad absurdum*) clearly shows which occurrences of P are closed at the inference. It is typical of Aristotle's proofs that an assumption closed in indirect proof occurs just once, i.e., either $m = 1, n = 0$ or $m = 0, n = 1$. We then have one of

4. Aristotelian special cases of indirect proof

$$\dfrac{\begin{matrix}\overset{1}{P}\\ \vdots \\ Q\end{matrix} \quad Q^{\perp}}{P^{\perp}}\,Raa,1 \qquad\qquad \dfrac{Q \quad \begin{matrix}\overset{1}{P}\\ \vdots \\ Q^{\perp}\end{matrix}}{P^{\perp}}\,Raa,1$$

As mentioned, Aristotle's derivations have at most one instance of indirect proof, as a last rule. A rule of indirect proof in which the premisses of *Raa* are $\Pi^{+}(A, B)$ and its **contrary** $\Pi^{-}(A, B)$ can be derived from the rest of the rules.

1.4 The rules of conversion

Aristotle's system of deductive logic begins properly with his **rules of conversion**:

5. The rules of conversion

$$\dfrac{\Pi^{-}(A,B)}{\Pi^{-}(B,A)}\,\Pi^{-}C \qquad \dfrac{\Pi^{+}(A,B)}{\Sigma^{+}(B,A)}\,\Pi^{+}C \qquad \dfrac{\Sigma^{+}(A,B)}{\Sigma^{+}(B,A)}\,\Sigma^{+}C$$

As noted by [Smiley, 1973, p. 142], the third rule of conversion is a **derivable rule**: Its conclusion is derivable from its premiss by the first conversion rule and the rule of indirect proof. Aristotle, in fact, notes the same (24a22) as we shall see in a while.

The semantical justification of the rules of conversion in Aristotle is as follows. For rule $\Pi^{-}C$ assume $\Pi^{-}(B, A)$. To justify the conclusion $\Pi^{-}(A, B)$, assume the contradictory $\Sigma^{+}(A, B)$. Aristotle's argument is (25a15):

> *If B did not belong to some A, say C, it would not be true that A belongs to no B, for C is a B.*

By the semantical explanation, we can conclude from $\Sigma^+(A, B)$ to some instance $(c : A)B(c)$ in the notation of Section 2. The crucial point of Aristotle's conversion argument is that if $(c : A)B(c)$, then $(c : B)A(c)$. Let us call this principle **instance conversion.** In this light, the semantical argument for rule $\Pi^- C$ can be given as an indirect one: If $\Pi^-(A, B)$, then, by the semantical explanation, *there is no instance* $c : A$ *such that* $B(c)$, i.e., $(c : A)B(c)$ is impossible. On the other hand, assuming $\Sigma^+(B, A)$, we have $c : B$ *such that* $A(c)$, or $(c : B)A(c)$, because this is the same as the semantical explanation *no A is B is impossible.* By instance conversion of $(c : B)A(c)$, we have $(c : A)B(c)$, a contradiction. Therefore the contradictory $\Pi^-(B, A)$ of the assumption $\Sigma^+(B, A)$ follows.

The distinction into terms as types and singular terms is central in the above reading of Aristotle: The ground for the semantical explanation of the four Aristotelian quantifiers lies in what we called instance conversion. It, together with the principle of indirect proof, led to the justification of the rule of Π^--conversion that in turn leads to the justification of the other two convertibilities, and to the rest of the rules of deduction. The same division into, let's say, **generic** and **singular** terms, is discussed at length by Aristotle. We saw above an example of the failure of convertibility when a singular term occurred in an indefinite proposition (*Socrates is wise.*) Similarly, singular terms cannot appear as subjects in propositions that are not indefinite: An example such as *Some Socrates is wise* is sufficient to bring home this point.

Behind the principle of instance conversion we find the following: A generic term can be treated both as a type and a property. In $c : A$ *such that* $B(c)$, we can turn the predicate B into a subject, the type of all objects that have the property. Likewise, we can turn the type A into a property A. At this point it is clear that singular terms, even though they appear in notation (upper case letters) to be like any terms, do not have the type-property correspondence.

Next we have the second of the conversion rules, from $\Pi^+(B, A)$ to conclude $\Sigma^+(A, B)$. Aristotle's semantical justification is as follows (25a18):

> If A belongs to every B, then B will belong to some A. For if it belongs to none, neither will A belong to any B. But it was assumed to belong to everyone.

These words correspond exactly to the following formal derivation:

$$\frac{\Pi^+(B,A) \quad \dfrac{\overset{1}{\Pi^-(A,B)}}{\Pi^-(B,A)}\,\Pi^- C}{\Sigma^+(A,B)}\,Raa,1$$

Aristotle uses indirect proof with a formula and its contrary, instead of contradictory. From the point of view of a system of proof, either that has to be done, or the conversion of $\Pi^{+}(B,A)$ into $\Sigma^{+}(A,B)$ taken as a primitive rule.

The second conversion rule has been criticized because its premiss does not presuppose the existence of any A, whereas the conclusion does. The predicate-logical interpretation shows this: The premiss would be $\forall x(A(x) \supset B(x))$, the conclusion $\exists x(A(x)\&B(x))$, and the former holds but the latter fails if A is always false. It is seen from Aristotle's arguments that he assumed every subject to be **inhabited**, which resolves the conflict.

The third conversion rule is, from $\Sigma^{+}(B,A)$ to conclude $\Sigma^{+}(A,B)$. Aristotle's justification is again in terms of rule $\Pi^{-}C$: Assume the contradictory of $\Sigma^{+}(A,B)$. Then (25a22): *If B belongs to no A, then neither A to no B.* In our notation, the justification is the derivation

$$\dfrac{\Sigma^{+}(B,A) \quad \dfrac{\overset{1}{\Pi^{-}(A,B)}}{\Pi^{-}(B,A)}\,\Pi^{-}C}{\Sigma^{+}(A,B)}\,Raa{,}1$$

Only one of the conversion rules is needed and the rest are derivable if indirect proof is used in Aristotle's way. If, instead, the second conversion rule is taken, the contrary premisses $\Pi^{+}(A,B)$ and $\Pi^{-}(A,B)$ lead to a contradiction. The contradictory of the latter is $\Sigma^{+}(A,B)$. It is derived by

$$\dfrac{\dfrac{\Pi^{+}(A,B)}{\Sigma^{+}(B,A)}\,\Pi^{+}C}{\Sigma^{+}(A,B)}\,\Sigma^{+}C$$

Thus, we have as a derived indirect rule the following.

6. The scheme of indirect proof

$$\dfrac{\begin{matrix}\overset{1}{P^{m}} \\ \vdots \\ \Pi^{+}(A,B)\end{matrix} \qquad \begin{matrix}\overset{1}{P^{n}} \\ \vdots \\ \Pi^{-}(A,B)\end{matrix}}{P^{\perp}}\,Raa{,}1$$

We sum up the above results as follows:

Proposition 1. *The system with the conversion rules $\Pi^{-}C$ and $\Pi^{+}C$ and rule Raa is equivalent to the system with rule $\Pi^{-}C$, Raa, and indirect proof with contrary premisses.*

Having proved that two of the conversion rules follow from that for the quantifier Π^- and indirect proof, Aristotle shows that the existential negative quantifier expression does not convert. The example is (25a25): *Man does not belong to all living creatures, but living creature belongs to all men.* It should be noted that two readings are given (in 25a15–18) to what we wrote as $\Sigma^-(A, B)$: *B does not belong to some A* and *B does not belong to all A*, i.e., *some A is not-B* and *not all A are B*. Aristotle's counterexample uses the latter wording.

1.5 The syllogistic rules

So far we have one primitive rule of conversion and two schemes of indirect proof, or the other way around. Two more rules enter into Aristotle's deductive logic, the proper **syllogisms** as this word has been understood for a long time. Its meaning in Aristotle vacillates between a single syllogism and what today is called a deduction. The major part of the *Prior Analytics* deals with derivations that consist of a single syllogism, conversions, and a single step of indirect inference. The two syllogistic rules are (25b38-26a2):

7. Aristotle's formulation of the syllogistic rules

When *A* of every *B* and *B* of every *C*, it is necessary that A is said of every *C*. For we have explained above what we mean by every.
Correspondingly also when *A* of no *B*, *B* instead of every *C*, then *A* will not belong to any *C*.

Aristotle does not elaborate on the justification of the first rule through the explanation of the universal quantifier, but it is straightforward. With our notation, we have the premisses and their semantical explanations:

1. $\Pi^+(B, A)$, no $d : B$ can be taken such that $A(d)$ could not be asserted,
2. $\Pi^+(C, B)$, no $e : C$ can be taken such that $B(e)$ could not be asserted.

Assume the contradictory of the conclusion $\Pi^+(C, A)$:

3. $\Sigma^-(C, A)$, an $f : C$ can be taken such that not-$A(f)$.

By $f : C$ from 3. and by 2., we have $B(f)$. By the correspondence between types and properties, $f : B$, so that by 1., $A(f)$, a contradiction with 3.

The second syllogistic rule is justified similarly:

1. $\Pi^-(B, A)$, no $d : B$ can be taken such that $A(d)$ could be asserted,
2. $\Pi^+(C, B)$, no $e : C$ can be taken such that $B(e)$ could not be asserted.

Assume the contradictory of the conclusion $\Pi^-(C, A)$:

3. $\Sigma^+(C, A)$, an $f : C$ can be taken such that $A(f)$.

By $f : C$ and 2., we have $B(f)$. By the correspondence between types and properties, $f : B$, so by 1., $A(f)$ cannot be asserted, a contradiction with $A(f)$ from 3.

We write the above two rules as:

8. The syllogistic rules

$$\frac{\Pi^+(A,B) \quad \Pi^+(B,C)}{\Pi^+(A,C)}\,\Pi^+S \qquad \frac{\Pi^+(A,B) \quad \Pi^-(B,C)}{\Pi^-(A,C)}\,\Pi^-S$$

The order of the premisses, from left to right, is the reverse of Aristotle's proof texts. At some stage, it became customary to read the propositions with the subject first, so, to have the middle term in the middle, the order of premisses was changed.

1.6 Two example proofs

Some examples will show the advantage of a proof tree form instead of a linear one. The following proof text is from (27a10):

9. An Aristotelian proof text

If *M* belongs to every *N* but to no *X*, then neither will *N* belong to any *X*. For if *M* belongs to no *X*, then neither does *X* belong to any *M*; but *M* belonged to every *N*; therefore, *X* will belong to no *N* (for the first figure has come about). And since the privative converts, neither will *N* belong to any *X*.

[Corcoran, 1974] writes this as a linear succession of formulas, each line numbered and justified. In our notation, the linear derivation is:

10. A linear derivation	
1. $\Pi^+(S,R)$	assumption.
2. $\Sigma^-(S,P)$	assumption.
3. $\Pi^+(R,P)$	assumption contradictory to $\Sigma^-(R,P)$.
4. $\Pi^+(S,R)$	1.
5. $\Pi^+(S,P)$	3,4.
6. $\Sigma^-(S,P)$	2.
7. $\Sigma^-(R,P)$	5,6.

The premisses of each rule instance are written right above the instance, in the proper order. To achieve this, line 1 has to be **repeated** on line 4. The same is seen in Aristotle's proof text.

The next example gives an indirect proof (28b17):

11. An Aristotelian proof text

If *R* belongs to every *S* but *P* does not belong to some, then it is necessary for *P* not to belong to some *R*. For if *P* belongs to every *R* and *R* to every *S*, then *P* will also belong to every *S*; But it did not belong.

The linear indirect derivation of $\Sigma^-(R,P)$ is, as adapted from [Corcoran, 1974, p. 111]:

12. An indirect linear derivation	
1. $\Pi^+(S,R)$	assumption.
2. $\Sigma^-(S,P)$	assumption.
3. $\Pi^+(R,P)$	assumption contradictory to $\Sigma^-(R,P)$.
4. $\Pi^+(S,R)$	1.
5. $\Pi^+(S,P)$	3,4.
6. $\Sigma^-(S,P)$	2.
7. $\Sigma^-(R,P)$	5,6.

The temporary assumption from line 3 has led to the contradictory conclusions 5 and 6. Therefore $\Sigma^-(R,P)$ follows. It has been added as a seventh line. The temporary assumption on line 3 is closed at this step. Note the two repetitions, on lines 4 and 6.

We shall translate linear derivations into a tree form:

13. Translation algorithm from linear to tree form

1. Take the last formula, draw an inference line above it, with the name of the rule that concluded it next to the line. Write the formulas that correspond to the lines of the premisses of the last rule, as indicated by the numbers, in the given order above the inference line.
2. Repeat the procedure until assumptions are arrived at. If an assumption closed by Raa is reached, write the number of its line above it.

The translation produces derivations in which all assumptions are **topformulas** that begin derivation branches.

The result of the application of this translation algorithm to example derivations 10 and 12 is:

14. The example derivations translated into tree form

$$\cfrac{\cfrac{\cfrac{\Pi^-(X,M)}{\Pi^-(M,X)}\,\Pi^-C \qquad \Pi^+(N,M)}{\Pi^-(N,X)}\,\Pi\;S}{\Pi^-(X,N)}\,\Pi^-C \qquad\qquad \cfrac{\cfrac{\overset{3}{\Pi^+(R,P)} \quad \Pi^+(S,R)}{\Pi^+(S,P)}\,\Pi^+S \qquad \Sigma^-(S,P)}{\Sigma^-(R,P)}\,Raa,3$$

The order of the premisses in the two-premiss rules is the reverse of our rule, but this is just a notational convention. As can be seen, the tree form does not mix assumptions and conclusions of rules, but the former are always topformulas and it is seen at once what depends on what. The purely formal character of the translation algorithm is shown by the label 3 produced for the assumption closed in indirect proof. Otherwise, constructing tree derivations directly, one would start numbering labels from 1. No repetitions appear. Had an assumption been used more than once, that would also be seen. Most remarkably, it becomes possible to compose two derivations, and to permute the order of application of the rules.

In the earlier parts of the *Prior Analytics*, deductions are identified with single instances of syllogistic rules, modulo conversions, possibly finished with an indirect proof. We say modulo conversions, because Aristotle doesn't seem to take them as steps of deduction proper (*cf.* 40b36 where he writes that nothing results from a single premiss). In the latter parts of the book, Aristotle considers situations with more than one syllogistic rule. The places are chapter 25 of book A, and chapters 5, 18, and 19 of book B. Aristotle expresses himself in different ways. One is to say that there are *several conclusions* (42a5) or *syllogisms* 42a18), or that they *are multiple* (42a21).[1] He has some difficulty in taking such several syllogisms as one united deductive argument. They can be either related or unrelated to each other (as in 42a21, where multiple syllogistic conclusions *do not touch each other*). Another expression is that several syllogisms can *lie together composed* (68a11). He also writes at one place (42b6) that *the conclusion is reached by means of prior syllogisms.*

If we think of the Aristotelian proof texts, combination of syllogisms would not pose a problem. In the application of a syllogistic or other rule, one would just repeat a premiss that has been deduced earlier by another syllogism. One such case is in (42a1): There are the premisses D and E that give a conclusion A, next the premisses F and G that give a conclusion B. The conclusions A and B can in turn lead to a conclusion C. Aristotle writes that there are *several syllogisms, for the conclusions are many* (42a5).[2] Thus, a whole chain of syllogistic inferences is possible in Aristotle. The structure of these derivations is clear if they are given in tree form, as for the above example:

$$\frac{\dfrac{D \quad E}{A} \qquad \dfrac{F \quad G}{B}}{C}$$

Aristotle does not really come into grips with the composition of several single syllogisms into a deduction. At one place (42a36) he uses the expression *deductive argument.*[3] His analysis of one-syllogism deductions is complete, but he has no general results about deductive arguments with several syllogisms, except for one: Namely, at (42b12–13), he proves that *when the premisses are even, the terms will be odd, and when the terms are even, the premisses will be odd.* The proof is a clear structural induction on the number of steps in a combined deduction.[4]

In general, given a derivation of P from the assumptions $A_1, \dots, A_m$ and another derivation of Q from the assumptions $P, B_1, \dots, B_n$, these two derivations can be **composed** into a derivation of Q from the assumptions $A_1, \dots, A_m$ and $B_1, \dots, B_n$. Composition can be represented schematically as

1 Smith translates *several deductions* when *several syllogisms* would be clearer. Similarly, at (50a5) Aristotle talks about conclusions from a single figure, and these are not just any deductions, but single instances of a syllogistic rule.

2 Again, Smith has *several deductions*.

3 In Smith's translation, from *logos sullogistikos*.

4 A related proof by induction in antiquity has been observed by [Acerbi, 2000].

15. The scheme of composition of derivations

$$\begin{array}{c} A_1,\dots,A_m \\ \vdots \\ P \end{array} \quad \textit{and} \quad \begin{array}{c} P, B_1,\dots,B_n \\ \vdots \\ Q \end{array} \quad \textit{compose into} \quad \begin{array}{c} \begin{array}{c} A_1,\dots,A_m \\ \vdots \\ P \end{array}, B_1,\dots,B_n \\ \vdots \\ Q \end{array}$$

2 Proof-theoretical analysis of Aristotle's deductive logic

2.1 Properties of tree derivations

We shall write in general $P, Q, \dots, R \vdash S$ to indicate that there is a derivation of S from the assumptions $P, Q, \dots, R$. We also count the multiplicity of occurrences of assumption formulas, and gather them into **multisets**, denoted $\Gamma, \Delta, \dots$. In particular, the assumption P is at the same time a degenerate derivation of P from P. In terms of the derivability relation, we have thus $P \vdash P$. The derivability relation also allows us to write the composition of two derivations as a rule:

$$\frac{\Gamma \vdash P \quad P, \Delta \vdash Q}{\Gamma, \Delta \vdash Q}\,\textit{Comp}$$

This is an **admissible rule** in Aristotle's system of logic: Whenever its premisses are derivable by the rules of the system, also the conclusion is derivable. Admissibility is a weaker notion than the derivability of a rule. The latter requires that the conclusion be derivable from the premisses.

Proposition 2 (Composition of derivations). *Given two derivations $\Gamma \vdash P$ and $P, \Delta \vdash Q$, if all labels in instances of Raa in the two derivations are distinct, they can be composed into a derivation $\Gamma, \Delta \vdash Q$.*

Proof: The proof is by induction on the greatest number of successive steps of inference in the derivation $P, \Delta \vdash Q$. If it is an assumption, we have

$$\begin{array}{c} \Gamma \\ \vdots \\ P \end{array} \quad \text{and} \quad P$$

Composition with the degenerate derivation P just gives the original derivation. Assume next that Q has been derived by a proper rule, say a one-premiss rule. We then have the derivations

$$\begin{array}{c} \Gamma \\ \vdots \\ P \end{array} \quad \text{and} \quad \begin{array}{c} P, \Delta \\ \vdots \\ \dfrac{R}{Q}\,\textit{Rule} \end{array}$$

The derivation of R from P, Δ is one step shorter, so we can make by the inductive hypothesis the composition

$$\begin{array}{c} \Gamma \\ \vdots \\ P, \Delta \\ \vdots \\ R \end{array}$$

Now *Rule* is applied to conclude Q. If *Rule* is *Raa*, the assumption closed at the original inference is kept open, the composition with a shorter derivation done, and then *Raa* applied and the assumption closed. QED.

Whenever there is a clash of the numerical labels in *Raa*, these can be renamed.

In general, whenever a formula P appears twice in a derivation branch, the part between the two occurrences can be cut out, to eliminate the **loop**. It can happen that assumptions are used in the part eliminated, so that such assumptions get removed from the list of open assumptions of the overall derivation. If this happens, it just improves the overall result, because the **endformula** of the derivation then follows from fewer assumptions.

The tree form makes it possible to **permute** the order of application of two rules. Say, we have

$$\dfrac{\dfrac{\Pi^{+}(A,B) \quad \Pi^{+}(B,C)}{\Pi^{+}(A,C)}\,\Pi^{+}S \qquad \Pi^{+}(C,D)}{\Pi^{+}(A,D)}\,\Pi^{+}S$$

The middle term B is removed first, then the middle term C. The order of removal can be changed, as in

$$\dfrac{\Pi^{+}(A,B) \qquad \dfrac{\Pi^{+}(B,C) \quad \Pi^{+}(C,D)}{\Pi^{+}(B,D)}\,\Pi^{+}S}{\Pi^{+}(A,D)}\,\Pi^{+}S$$

Any number of successive instances of rule Π^{+} can be so transformed that, say, the left premisses are assumptions.

Finally, we note a remarkable property of the rules of Aristotle's deductive logic:

Each term in a conclusion of a rule other than indirect proof is a term in some premiss.

By this property, any term in a derivation without indirect proof can be traced up to at least one topformula, i.e., an assumption.

2.2 Reflexivity

Aristotle forbids the "self-predication" of a subject, such as in the **reflexive instance** $\Pi^+(A, A)$. This ban on reflexivity is not categorical, though, for in (68a19), self-predication is applied.

We shall show that reflexive instances of the four kinds of Aristotelian quantified formulas produce no harm in derivations. In most cases, they just produce what Aristotle forbids in his definition of a deduction (24b19), namely a conclusion that is identical to a premiss. In other cases, they produce an impossibility.

We notice that the first syllogistic rule produces a loop whenever one premiss is reflexive, as in

$$\frac{\Pi^+(A, A) \quad \Pi^+(A, B)}{\Pi^+(A, B)}\,\Pi^+S$$

For the second syllogistic rule, there is a loop if the universal positive premiss is a reflexivity:

$$\frac{\Pi^+(A, A) \quad \Pi^-(A, B)}{\Pi^-(A, B)}\,\Pi^-S$$

Universal negative conversion produces likewise a loop if its premiss is the negative reflexivity $\Pi^-(A, A)$.

If we grant the positive reflexivity $\Pi^+(A, A)$, then the positive reflexivity $\Sigma^+(A, A)$ follows by conversion. The contradictories of these two, the negative reflexivities $\Sigma^-(A, A)$ and $\Pi^-(A, A)$, respectively, are then impossible.

If the second premiss in rule Π^-S is a reflexivity, we have

$$\frac{\Pi^+(A, B) \quad \Pi^-(B, B)}{\Pi^-(A, B)}\,\Pi^-S$$

The conclusion is the contrary of the first premiss.

Thus, the admission of reflexivities in the Aristotelian forms of propositions causes no essential change in derivability: Their only use is in the conclusion of $\Sigma^+(A, A)$ from $\Pi^+(A, A)$.

2.3 A normal form for tree derivations

The deductions found in the *Prior Analytics* are such that the rule of indirect proof is applied at most once, and always as a last rule. The reason is that Aristotle does not even begin the analysis of combined derivations. We shall show that all derivations by the rules of Aristotle's deductive logic can be brought to a **normal form** such that the stated property for indirect proof holds. We show in particular that two consecutive instances of rule *Raa* contract into one, and that any other instance of *Raa* than one as a last rule can be removed.

Theorem 3 (Normal form for derivations). *All derivations in Aristotle's deductive logic can be so transformed that the rule of indirect proof is applied at most once as a last rule.*

Proof: Consider an uppermost instance of *Raa* in a derivation. If it is followed by another instance of *Raa*, we have a part of derivation such as

$$\dfrac{\dfrac{\begin{matrix}\overset{1}{P} & \\ \vdots & \vdots \\ Q & Q^{\perp}\end{matrix}}{P^{\perp}}\,Raa,1 \qquad \begin{matrix}\overset{2}{R} \\ \vdots \\ P\end{matrix}}{R^{\perp}}\,Raa,2$$

This derivation is transformed into

$$\dfrac{\begin{matrix}\overset{1}{R} & \\ \vdots & \\ P & \vdots \\ \vdots & \vdots \\ Q & Q^{\perp}\end{matrix}}{R^{\perp}}\,Raa,1$$

Admissibility of composition guarantees that the derivation of P from R can be continued by the derivation of Q from P.

The above transformation is repeated until there is just one instance of *Raa*. If the conclusion $R^{\perp}$ is existential, it cannot be a premiss in any rule and the claim of the theorem follows. If the conclusion is universal, we have one of

$$\dfrac{\begin{matrix}\overset{1}{\Sigma^{-}(A,B)} & \\ \vdots & \vdots \\ Q & Q^{\perp}\end{matrix}}{\Pi^{+}(A,B)}\,Raa,1 \qquad\qquad \dfrac{\begin{matrix}\overset{1}{\Sigma^{+}(A,B)} & \\ \vdots & \vdots \\ Q & Q^{\perp}\end{matrix}}{\Pi^{-}(A,B)}\,Raa,1$$

There are by assumption no instances of *Raa* above the ones shown. Therefore the existential formulas that are closed cannot be premisses in any other rules than the instances of *Raa* shown. Then the derivations of the left premisses of *Raa* are degenerate, with $\Sigma^-(A,B) \equiv Q$ and $\Sigma^+(A,B) \equiv Q$, respectively. The derivations are therefore

$$\dfrac{\overset{1}{\Sigma^-(A,B)} \quad \overset{\vdots}{\Pi^+(A,B)}}{\Pi^+(A,B)}\,Raa,1 \qquad \dfrac{\overset{1}{\Sigma^+(A,B)} \quad \overset{\vdots}{\Pi^-(A,B)}}{\Pi^-(A,B)}\,Raa,1$$

A loop is produced in both, and therefore the instances of *Raa* can be removed. QED.

The formulas of Aristotle's deductive logic are atomic formulas in today's terminology, and his rules act only on such atomic formulas. The question of the derivability of an atomic formula from given atomic formulas used as assumptions is known as the **word problem**. The terminology stems from algebra where the word problem concerns the derivability of an equality from given equalities. The solution of this problem in Aristotle's deductive logic, i.e., the decidability of derivability by his rules, follows at once by the above results. It is sufficient to show that the terms in a derivation of P from the assumptions Γ can be restricted to those included in the assumptions and the conclusion. If the proof is direct, this is so by the property of the rules noted at the end of Section 2.1, by which terms in formulas can be traced up to assumptions. Otherwise, the last step is indirect, but the closed assumption is $P^\perp$ so that the terms in a derivation are as in the direct case. With a bounded number of terms, there is a bounded number of distinct formulas. The number of possible consecutive steps of inference in a loop-free derivation, i.e., the height of a branch in a derivation tree, is bounded by the number of distinct formulas and we have:

Theorem 4 (Word problem for Aristotle's deductive logic). *The derivability of a formula P from given formulas Γ used as assumptions is decidable.*

3 The issue of Aristotle's underlying logic

3.1 From rules to axioms

When the study of logic was revived in the mid-nineteenth century, Aristotle's logic was interpreted as a logic of classes. Venn diagrams showed easily what its basic forms of inference amounted to. These diagrams can be equally well used to model monadic predicate logic, and the latter in turn gives one interpretation of Aristotle's logic.

We have followed Corcoran and Smiley and taken Aristotle's logic simply as a system of rules of inference. Such rules stand on their own feet without any underlying logical interpretation. On the other hand, proof systems based on rules can be turned into systems that use axioms and maybe just one or two rules. Today's predicate logic was born, in the hands of Frege and Russell, as precisely such an axiomatic system of logic. The formulation as a system of rules of proof was found by Gentzen in the 1930s.

Frege and Russell followed a trend of the time of axiomatizing logic as a branch of mathematics, similarly to axiomatizations of geometry and arithmetic. Many such axiomatizations can be turned, in perfect analogy to Gentzen's work in pure logic, into systems of rules. In the other direction, systems of rules can be turned into axiomatic systems. An example, close to Aristotle's system of rules, illustrates this correspondence: The theory of equality consists of a two-place relation over a given domain of elements $a, b, c, \ldots$ and the following axioms:

1. $a = a$ (*reflexivity*),
2. $a = b \supset b = a$ (*symmetry*),
3. $a = b \& b = c \supset a = c$ (*transitivity*).

These are "free-variable" axioms, i.e., the terms in the equations are parameters that can be instantiated in any way.

A system of rules that corresponds to the axioms of equality is:

16. A rule system for equality

$$\frac{}{a = a}\,Ref \qquad \frac{a = b}{b = a}\,Sym \qquad \frac{a = b \quad b = c}{a = c}\,Tr$$

Derivations begin with assumptions and instances of the zero-premiss rule of reflexivity. Consider a derivation by the rules such as

$$\cfrac{\cfrac{a = b \qquad \cfrac{c = b}{b = c}\,Sym}{a = c}\,Tr}{c = a}\,Sym$$

The equality $c = a$ is derived by the rules from the equalities $a = b$ and $c = b$ used as assumptions. The corresponding derivation that uses axioms is:

$$\cfrac{a = c \supset c = a \qquad \cfrac{a = b \,\&\, b = c \supset a = c \qquad \cfrac{a = b \qquad \cfrac{c = b \supset b = c \quad c = b}{b = c}\supset E}{a = b \,\&\, b = c}\&I}{a = c}\supset E}{c = a}\supset E$$

The use of logic in the writing of the axioms of equality forces a use of logic in the proof that is not quite natural. The uppermost step has as one premiss an instance of symmetry, followed by rule $\supset E$ that puts to use the symmetry. Rule $\&I$ builds up the minor premiss of another implication elimination that corresponds to the step of transitivity. Finally, there is one more step of $\supset E$ that corresponds to the last rule in the first derivation.

Consider now the Aristotelian derivation

$$\cfrac{\cfrac{\Pi^+(A,B) \qquad \cfrac{\Pi^-(C,B)}{\Pi^-(B,C)}\Pi^-C}{\Pi^-(A,C)}\Pi^-S}{\Pi^-(C,A)}\Pi^-C$$

Aristotle would say: *If every A is B and no C is B then no C is A*. The derivation, however, was not obtained by an application of Aristotle's rules, but directly from the preceding rule-based derivation for equality. The notation for the relations was changed suitably, and the names of the rules, but the geometry of the proof tree that is the difficult part to write out was taken directly from the equality derivation.

It is instructive to see what happens if reflexivity appears as a premiss. The possibilities are

$$\cfrac{\cfrac{}{a = a}Ref}{a = a}Sym \qquad \cfrac{\cfrac{}{a = a}Ref \qquad a = c}{a = c}Tr \qquad \cfrac{a = b \qquad \cfrac{}{b = b}Ref}{a = b}Tr$$

These derivations, similar to those in Section 2.2, show that the only use of rule *Ref* is in the derivation of reflexivities, for in all other cases, a loop is produced.

We shall look at an axiomatic formulation of Aristotle's rules. The axioms are:

17. Aristotle's rules as axioms

$\Pi^-(A,B) \supset \Pi^-(B,A)$	(*conversion*)
$\Pi^+(A,B) \supset \Sigma^+(B,A)$	(*conversion*)
$\Pi^+(A,B)\&\Pi^+(B,C) \supset \Pi^+(A,C)$	(*syllogism*)
$\Pi^+(A,B)\&\Pi^-(B,C) \supset \Pi^-(A,C)$	(*syllogism*)

We could now write, in a manner similar to the case of equality, the axiomatic derivation that corresponds to the above derivation by rules, but it would grow too broad to be printed comfortably.

3.2 Logic in Aristotle's proofs

Aristotle has no explicit connectives for conjunction and implication. These logical particles appear in his proofs, however, as an underlying structure. When two assumptions are used together, there would in logical notation appear a conjunction, as in the axiomatic equality derivation. When a rule is instantiated, there would in logical notation appear an implication, and an implication elimination. Similarly, when Aristotle proves a result of the form *If P, then Q*, he does it by assuming P, then deriving Q. When this has been done, he concludes that the conditional has been proved, which in logical notation is implication introduction. None of this is explicit in Aristotle.

If we state syllogisms with logical notation added, the above syllogistic derivation by Aristotle's rules grows into:[5]

$$\dfrac{\dfrac{\overset{2}{\Pi^{+}(A,B)\,\&\,\Pi^{-}(C,B)} \qquad \dfrac{\dfrac{\overset{1}{\Pi^{+}(A,B)} \quad \dfrac{\overset{1}{\Pi^{-}(C,B)}}{\Pi^{-}(B,C)}\,\Pi C}{\Pi^{-}(A,C)}\,\Pi}{\Pi^{-}(C,A)}\,\Pi C}{\Pi^{-}(C,A)}\,\&E,1}{\Pi^{+}(A,B)\,\&\,\Pi^{-}(C,B) \supset \Pi^{-}(C,A)}\,\supset I,2$$

Now the statement of the syllogism is in a logical notation that uses one conjunction and one implication, and there are correspondingly two logical steps in the derivation. The rest of the derivation, the one that corresponds to Aristotelian rules of proof, is separated from the logical part as a neat subderivation of the minor premiss of rule $\&E$. This separation can be always upheld when the rules of logic are formulated suitably (as explained in the preceding footnote). Were

5 The second-to-last step is a conjunction elimination of a more general form than the one standardly used. With a conjunction $A\&B$ as a major premiss, conjunction elimination would normally allow the conclusion of A and B separately. In the general rule, the conjuncts A and B appear as temporary hypotheses that lead to some conclusion C. When the inference is drawn, C is concluded and the temporary hypotheses A and B closed. With A in place of C (resp. B in place of C), the standard rules follow as special cases. The rest of the connectives and quantifiers can be formulated similarly [von Plato, 2001].

this not the case, the logical rules would not be conservative relative to Aristotle's syllogistic as formulated in theorem 4 above.

Our result is thus that the underlying logic of Aristotle's four rules is that of natural deduction for conjunction and implication, hardly a surprise, because natural deduction arose from an attempt at analyzing the logic of mathematical proofs "as they appear in practice" [Gentzen, 1934–35, p. 176].

One final point can be mentioned: We noticed briefly that Aristotle takes sometimes an arbitrary instance of an existential assertion. The correct way to handle such is to introduce eigenvariables. There has been a lot of discussion as to what the nature of proofs through instantiation (*ekthesis*) is. Proofs in geometry often state things such as: *Let a,b, and c be three points such that A*, where *A* is some property of points. This is undoubtedly an existential assumption. The theorem to be proved, instead, is typically universal in form: *For any three points, if A, then B.* This common practice of any mathematical proof is rendered explicit by the predicate logical equivalence of $\forall x \forall y \forall z(A \supset B)$ and $\exists x \exists y \exists z A \supset B$. To prove the universal $\forall x \forall y \forall z(A \supset B)$, we can assume the existential $\exists x \exists y \exists z A$, then take an instance with the eigenvariables a, b, and c, and prove B.

Acknowledgment: I thank Bianca Boretti and Kati Näätsaari for their help in the reading of the crucial original passages in Aristotle, Per Martin-Löf for pointing me to [Corcoran, 1974], and John Corcoran for making this early work of his readily available.

Primary sources

Aristotles. *Erste Analytik – Translated by Hans Zekl.* Meiner, Hamburg, 1929.

Aristotle. Prior analytics. In Maria C. Pantelia, editor, *Thesaurus Linguae Graecae® Digital Library*. University of California, Irvine, 2015. URL http://www.tlg.uci.edu.

S. J. Jenkinson. *Aristotle: Prior Analytics – Translated by A. J. Jenkinson.* Clarendon Press, Oxford, 1928.

R. Smith. *Aristotle's Prior Analytics.* Hackett, 1989.

References

F. Acerbi. Plato: Parmenides 149a7-c3. A proof by complete induction? *Archive for History of Exact Sciences*, 55:57–76, 2000.

J. Corcoran. Completeness of an ancient logic. *The Journal of Symbolic Logic*, 37:696–702, 1972.

J. Corcoran. Aristotle's natural deduction system. In J. Corcoran, editor, *Ancient Logic and Its Modern Interpretations*, pages 85–131. Reidel, 1974.

G. Gentzen. Untersuchungen über das logische Schliessen. *Mathematische Zeitschrift*, 39: 176–210, 1934–35.

J. Lukasiewicz. *Aristotle's Syllogistic from the Standpoint of Modern Formal Logic, 2nd ed.* Oxford, 1957.

T. Smiley. What is a syllogism? *Journal of Philosophical Logic*, 2:136–154, 1973.

R. Smith. Introduction to the 1989 translation of *Prior Analytics*. 1989.

J. von Plato. Natural deduction with general elimination rules. *Archive for Mathematical Logic*, 40:541–567, 2001.

Michael Rathjen

Remarks on Barr's Theorem: Proofs in Geometric Theories

Abstract: A theorem, usually attributed to Barr, yields that (A) geometric implications deduced in classical $\mathcal{L}_{\infty\omega}$ logic from geometric theories also have intuitionistic proofs. Barr's theorem is of a topos-theoretic nature and its proof is nonconstructive. In the literature one also finds mysterious comments about the capacity of this theorem to remove the axiom of choice from derivations. This article investigates the proof-theoretic side of Barr's theorem and also aims to shed some light on the axiom of choice part. More concretely, a constructive proof of the Hauptsatz for $\mathcal{L}_{\infty\omega}$ is given and is put to use to arrive at a simple proof of (A) that is formalizable in constructive set theory and Martin-Löf type theory.

Keywords: Geometric theory, Topos, Axiom of choice, Cut elimination in $\mathcal{L}_{\infty\omega}$

Mathematics Subject Classification 2010: 18B25, 03F05, 03F50, 03G30

1 Introduction

A *signature* Σ consists of constant symbols, function symbols, and relation symbols together with an assignment of a unique positive integer (arity) to any object of the latter two kinds. A language $\mathcal{L}$ is comprised of a signature Σ and formation rules, i.e., rules for forming formulae over Σ. The familiar Tarskian way of assigning meaning to the symbols of $\mathcal{L}$ proceeds by associating set-theoretic objects to them, notably functions and relations construed set-theoretically, giving rise to the notion of (set-theoretic) *structure for* $\mathcal{L}$ and *model of* T for any theory T in the language of $\mathcal{L}$. There is also a more general notion of structure in a sufficiently rich category $\mathcal{C}$. For example, if $\mathcal{C}$ has finite products, then any equational language (i.e., equality being the sole relation symbol and equations the only formulae) allows for interpretation in $\mathcal{C}$, by viewing terms as morphisms and function symbol application as composition. Another prominent example is the interpretation of the typed λ-calculus in cartesian closed categories (cf. [Lambek and Scott, 1988]). For still richer languages one must impose more conditions on $\mathcal{C}$. If one wants to extend this idea to higher order logic, then $\mathcal{C}$ is required to be a topos (cf. [Johnstone, 2002, D1.2]). This extra level of generality of interpretation, however,

Michael Rathjen: Department of Pure Mathematics, University of Leeds, Leeds LS2 9JT, United Kingdom, e-mail: rathjen@maths.leeds.ac.uk

comes with a penalty to pay in that only intuitionistically valid consequences can be guaranteed to survive the interpretation.

One is often interested in transferring results from the category of sets, **Set**, where classical logic, the axiom of choice and more reign, to an arbitrary topos $\mathcal{E}$. This is possible, for instance, for the following (non first-order) assertion:

All modules over fields are flat.[1]

A result that ensures this transfer is commonly called Barr's Theorem (cf. e.g.[Lane and Moerdijk, 1992, p.515])[2] but it can also be inferred from cut elimination for the infinitary logic $\mathcal{L}_{\omega_1\omega}$ (see later parts of this paper). For this to work, however, the formalization of mathematical notions is important. They have to be chosen carefully, as familiar equivalences are liable to fail in an intuitionistic setting. Moreover, to ensure survival of statements it will be important to develop mathematics within (classical) *geometric theories* and to couch statements as geometric implications. The topos-theoretic result alluded to above is the following.

Theorem: 1.1. *For every Grothendieck topos $\mathcal{E}$ there exists a complete Boolean algebra $\mathbf{B}$ and a surjective geometric morphism* $\mathrm{Sh}(\mathbf{B}) \to \mathcal{E}$. *Here* $\mathrm{Sh}(\mathbf{B})$ *is the topos of sheaves on the Boolean algebra with the usual sup topology. Moreover,* $\mathrm{Sh}(\mathbf{B})$ *is a Boolean topos and satisfies the axiom of choice, in the sense that for any epi $e : Y \twoheadrightarrow X$ there exists $s : Y \to X$ such that $e \circ s = 1_Y$.*

As a consequence[3] one arrives at the following insight.

Corollary: 1.2. *If T is a geometric theory and A is a geometric statement deducible from T with classical logic, then A is also deducible from T with intuitionistic logic, where by logic we mean infinitary $\mathcal{L}_{\infty\omega}$-logic.*

Though this Corollary also follows from a syntactic cut elimination result for $\mathcal{L}_{\infty\omega}$ (see section 7), Barr's theorem is often alleged to achieve more in that it also allows to eliminate uses of the axiom of choice. This is borne out by the following quotes:

1 This is just a simple example. Flatness of a module M, a notion introduced by Serre in 1956, is usually defined by saying that tensoring with M preserves injectivity. An equivalent way of expressing in $\mathcal{L}_{\omega_1\omega}$ that M is a flat R-module for a ring R is the following: For all $m \in \mathbb{N}$, whenever $x_1, \ldots, x_m \in M$ and $r_1, \ldots, r_m \in R$ satisfy $\sum r_i x_i = 0$, then there exist $y_1, \ldots, y_n \in M$ and $a_{ij} \in R$ such that $x_i = \sum a_{ij} y_j$ and $\sum r_i a_{ij} = 0$.

2 Disclaimer applying to the entire paper: This is not a paper on the history of certain pieces of mathematics. The attribution of results to persons is borrowed from standard text books or articles in the area, and therefore may well be historically inaccurate, as is so often the case.

3 There are several steps and further theorems involved; cf. [Johnstone, 2002, 3.1.16].

> METATHEOREM. If a geometric sentence is deducible from a geometric theory in classical logic, with the axiom of choice, then it is also deducible from it intuitionistically." (G.C. Wraith: *Intuitionistic Algebra: Some Recent Developments in Topos Theory.* Proceedings of the International Congress of Mathematicians, Helsinki, 1978 331–337)

> This has the advantage that all such toposes satisfy the Axiom of Choice; so we obtain a further conservativity result ..., asserting that uses of the Axiom of Choice may be eliminated from any *derivation* of a geometric sequent from geometric hypothesis. (P. Johnstone: *Sketches of an elephant*, vol. 2, p. 899)

Judging from conversations with logicians and discussions on internet forums (e.g. MathOverflow), it is probably fair to say that the main appeal of Barr's theorem stems from its mysterious power to utilize AC and then subsequently get rid of it. But can it really perform these wonders? As a backcloth for the discussion it might be useful to recall some famous AC-removal results.

Theorem: 1.3. *Below* GCH *stands for the generalized continuum hypothesis.*

1. *(Gödel 1938–1940) If A is a number-theoretic statement and* $\mathsf{ZFC} + \mathsf{GCH} \vdash A$ *then* $\mathsf{ZF} \vdash A$.
2. *(Shoenfield 1961, Platek, Kripke, Silver 1969) If B is a* Π^1_4*-statement of second order arithmetic and* $\mathsf{ZFC} + \mathsf{GCH} \vdash B$ *then* $\mathsf{ZF} \vdash B$.
3. *(Goodman 1976, 1978) If A is a number-theoretic statement and* $\mathbf{HA}^\omega + \mathsf{AC}_{type} \vdash A$*, then* $\mathbf{HA} \vdash A$*. Here* **HA** *stands for intuitionistic arithmetic also known as Heyting arithmetic.* $\mathbf{HA}^\omega$ *denotes Heyting arithmetic in all finite types with* AC_{type} *standing for the collection of all higher type versions* $\mathsf{AC}_{\sigma\tau}$ *of the axiom of choice with* σ, τ *arbitrary finite types.*

So should Barr's theorem be added to this list of renowned theorems with AC-eliminatory powers? The above quotes by Wraith and Johnstone seem to suggest that the addition of the axiom of choice to a geometric theory does not produce new geometric theorems. But one immediately faces the question of what it means to add AC to a theory T. Here it might be useful to introduce a rough distinction which separates two ways of doing this. The first route, which consists in expressing AC in the same language as T, will be referred to as an *internal* addition of AC. If, on the other hand, AC is expressed in a richer language with a new sort of objects where the choice functions live; we shall term it an *external* addition. If T is a first-order theory, then adding AC internally to T requires the language of T to be sufficiently rich. Moreover, internal AC forces the choice functions to be objects falling under the first order quantifiers of T, and thus, in general, the axioms of T will "interact" with AC in this augmentation. By contrast, an external addition of AC refers to a potentially larger universe, where the choice functions needn't be denizens of the realm that the original theory T speaks about. The foregoing distinction is still very coarse, though. For instance the choice functions might be external but they can certainly act on the original objects of T. Therefore if one also demands principles of T (e.g. induction) to hold for terms that involve these choice functions (like in the Goodman result) it is conceivable that conservativity will be lost (as is the case with the classical version $\mathsf{PA}^\omega + \mathsf{AC}_{type}$ of $\mathbf{HA}^\omega + \mathsf{AC}_{type}$). Notwithstanding that there are multifarious possibilities to add AC, labeling some of them as internal and others as external augmentations provides a useful, if crude, heuristics.

In view of Theorem 1.3 one can also ask if Barr's Theorem can be beefed up to include more than just AC. For instance, how about the continuum hypothesis, $V = L$, $\Diamond$ and other axioms?

2 Geometric and ∞-geometric theories

Below we will work in the extension $\mathcal{L}_{\infty\omega}$ of first order logic ($\mathcal{L}_{\omega\omega}$) which has all the formulae engendering rules of the latter but also allows to form infinitely long conjunctions $\bigwedge \Phi$ and disjunctions $\bigvee \Phi$ from any set Φ of already constructed formulae. A particulary well-behaved fragment of $\mathcal{L}_{\infty\omega}$ is $\mathcal{L}_{\omega_1\omega}$ where the set Φ in $\bigwedge \Phi$ and $\bigvee \Phi$ is always required to be countable. Infinitary logics began to play an important role in logic in the 1950s.[4]

2.1 Geometric theories

Definition: 2.1. *The* geometric formulae *are inductively defined as follows: Every atom is a geometric formula. If A, B, and $C(a)$ are geometric formulae then so are $A \vee B$, $A \wedge B$ and $\exists x\, C(x)$ (where x does not occur in $C(a)$).*

Another way of saying this is that a formula is geometric iff it does not contain any of the particles $\rightarrow, \neg, \forall$.

A formula is called a geometric implication *if it is of either form $\forall \vec{x}\, A$ or $\forall \vec{x}\, \neg A$ or $\forall \vec{x}\, (A \rightarrow B)$ with A and B being geometric formulae. Here $\forall \vec{x}$ may be empty. In particular geometric formulae and their negations are geometric implications.*

A theory is geometric *if all its axioms are geometric implications.*

---

4 "Yet infinitary logic has a long prehistory. Infinitely long formulas were introduced by C.S. Peirce in the 1880s, used by Schröder in the 1890s, developed further by Löwenheim and Lewis in the 1910s, explored by Ramsey and Skolem in the 1920s, extended by Zermelo and Helmer in the 1930s, studied by Carnap, Novikov, and Bochvar (and explicitly rejected by Gödel) in the 1940s, and exploited by A. Robinson (1951)." [Moore, 1997]

Below we shall give several examples of geometric theories.[5]

Examples: 2.2.

1. *Robinson arithmetic. The language has a constant* 0, *a unary successor function* suc *and binary functions* $+$ *and* $\cdot$. *Axioms are the equality axioms and the universal closures of the following.*
   (a) $\neg\mathrm{suc}(a) = 0$.
   (b) $\mathrm{suc}(a) = \mathrm{suc}(b) \to a = b$.
   (c) $a = 0 \lor \exists y\, a = \mathrm{suc}(y)$.
   (d) $a + 0 = a$.
   (e) $a + \mathrm{suc}(b) = \mathrm{suc}(a + b)$.
   (f) $a \cdot 0 = 0$.
   (g) $a \cdot \mathrm{suc}(b) = a \cdot b + a$

   A classically equivalent axiomatization is obtained if (3) is replaced by

   $$\neg a = 0 \to \exists y\, a = \mathrm{suc}(y)$$

   but this is not a geometric implication.
2. *The theories of groups, rings, local rings and division rings have geometric axiomatizations. Local rings are commutative rings with* $0 \neq 1$ *having just one maximal ideal. On the face of it, the latter property appears to be second order but it can be rendered geometrically as follows:*

   $$\forall x\, (\exists y\, x \cdot y = 1 \lor \exists y\, (1 - x) \cdot y = 1).$$

3. *The theories of fields, ordered fields, algebraically closed fields and real closed fields have geometric axiomatizations. To express invertibility of non-zero elements one uses* $\forall x\, (x = 0 \lor \exists y\, x \cdot y = 1)$ *rather than the non-geometric axiom* $\forall x\, (x \neq 0 \to \exists y\, x \cdot y = 1)$.
   To express algebraic closure replace axioms

   $$s \neq 0 \to \exists x\, sx^n + t_1 x^{n-1} + \ldots + t_{n-1} x + t_n = 0$$

   by

   $$s = 0 \lor \exists x\, sx^n + t_1 x^{n-1} + \ldots + t_{n-1} x + t_n = 0$$

   where sx^k *is short for* $s \cdot x \cdot \ldots \cdot x$ *with* k *many* x.

   Also the theory of differential fields *has a geometric axiomatization. This theory*

5 For more detailed descriptions of these theories and also the ones considered in Section 2.2, see e.g. [Barwise, 1977], [Chang and Keisler, 1990, 1.4], [Hodges, 1993, Appendix: Examples].

is written in the language of rings with an additional unary function symbol δ. *The axioms are the field axioms plus the axioms* $\forall x \forall y\, \delta(x+y) = \delta(x)+\delta(y)$ *and* $\forall x \forall y\, \delta(x \cdot y) = x \cdot \delta(y) + y \cdot \delta(x)$.

4. *The theory of projective geometry has a geometric axiomatization.*
5. *The theories of equivalence relations, dense linear orders, infinite sets and graphs also have geometric axiomatizations.*

2.2 The infinite geometric case

Infinitary logics are much more expressive and it is interesting to investigate notions of geometricity in these expanded settings. The infinitary languages we have in mind are such that they accommodate infinite disjunctions $\bigvee \Phi$ and conjunctions $\bigwedge \Phi$, where Φ is set of (infinitary) formulae.[6] This language is customarily denoted by $\mathcal{L}_{\infty\omega}$.

In this richer language a formula is said to be *infinite geometric*, notated ∞-geometric, if in addition to $\vee, \wedge, \exists$ one also allows infinite disjunctions $\bigvee \Phi$, where Φ is already a set of ∞-geometric formulae satisfying the above proviso on the number of variables.

An example of an axiom expressible in this richer language via a ∞-geometric implication is the Archimedean axiom:

$$\forall x\,(x < 1 \vee x < 1+1 \vee \ldots \vee x < 1+\ldots+1 \vee \ldots)$$

or in more compact way with $\mathbb{N}^+ = \{n \in \mathbb{N} \mid n > 0\}$:

$$\forall x \bigvee_{n \in \mathbb{N}^+} x < n.$$

One often only considers the sublanguage $\mathcal{L}_{\omega_1\omega}$ of $\mathcal{L}_{\infty\omega}$ where the formation of $\bigvee \Phi$ and $\bigwedge \Phi$ is only permissible for countable sets of formulae Φ.

Definition: 2.3. *The* ∞-geometric formulae *are inductively defined as follows: Every atom is a ∞-geometric formula. If A and B are ∞-geometric formulae then so are $A \vee B$ and $A \wedge B$. If $C(a)$ is a ∞-geometric formula with all occurrences of a indicated and x is a bound variable that does not occur in $C(a)$ then $\exists x\, C(x)$ is a ∞-geometric formula. If Φ is a set of ∞-geometric formulae having a finite number of free variables then $\bigvee \Phi$ is a ∞-geometric formula.*

6 It will be assumed that the total number of free variables occurring in the formulae of Φ is finite. The reason for this commonly found restriction appears to be that in this language only finitely many variables can be quantified at a time. So if one allowed infinitely many free variables there would be formulae which cannot be closed.

Another way of saying this is that a formula is ∞*-geometric iff it does not contain any of the particles* $\rightarrow, \neg, \forall, \bigwedge$.

The collection of ∞-geometric implications *is generated as follows:*

1. *If* A, B *are* ∞*-geometric formulae then* A, $\neg A$ *and* $A \rightarrow B$ *are* ∞*-geometric implications.*
2. *If* $C(a)$ *is a* ∞*-geometric implication and* a *is a free variable with all occurrences indicated and* x *does not occur in* $C(a)$*, then* $\forall x\, C(x)$ *is a* ∞*-geometric implication.*
3. *If* Ψ *is a set of* ∞*-geometric implications having a finite number of free variables then* $\bigwedge \Psi$ *is a* ∞*-geometric implication.*

A theory is ∞-geometric *if all its axioms are* ∞*-geometric implications.*

Examples: 2.4. *We list some examples of* $\mathcal{L}_{\omega_1\omega}$ *theories.*

1. *The theory of* torsion groups *is characterized by the group axioms plus the axiom*
$$\forall x \bigvee \{\underbrace{x \circ \ldots \circ x}_{n\text{-times}} = e \mid n \geq 1\}.$$
2. *The theory of fields with characteristic a prime is characterized by the field axioms together with the axiom*
$$\bigvee \{\underbrace{1 + \ldots + 1}_{n\text{-times}} = 0 \mid n \geq 2\}.$$
3. *The theory of archimedean ordered fields is characterized by the ordered field axioms together with the axiom*
$$\forall x \bigvee \{x < \underbrace{1 + \ldots + 1}_{n\text{-times}} \mid n \geq 1\}.$$
4. *The class of structures isomorphic to the standard model of Peano arithmetic is characterized by the axioms of* PA *conjoined with the axiom*
$$\forall x \bigvee \{x = 0 + \underbrace{1 + \ldots + 1}_{n\text{-times}} \mid n \geq 0\}. \tag{1}$$
5. *The theory of connected graphs has the usual axioms for graphs and additionally has the axiom*
$$\forall x \forall y [x = y \vee \bigvee \{\exists z_0 \ldots \exists z_n (x = z_0 \wedge y = z_n \wedge E z_0 z_1 \wedge \ldots E z_{n-1} z_n) \mid n \geq 1\}],$$
where E *is a two-place relation such that* $E z_i z_{i+1}$ *expresses that there is an edge going from* z_i *to* z_{i+1}.

The above theories, with the exception of the fourth example, are ∞*-geometric. However, in the fourth example the induction axioms are not really needed as they are* *implied in infinitary logic by the axiom (1) and the axioms of Robinson arithmetic, i.e. the axioms of* PA *pertaining to* 0, suc, +, ·. *They can be expressed by means of geometric formulae as shown in 2.2.*

In the logic $\mathcal{L}_{\infty\omega}$ one has rules for $\bigwedge$ and $\bigvee$ that generalize those for $\wedge$ and $\vee$, respectively. In the sequent calculus version they can be rendered thus.

- $\bigwedge$-Conjunction:

$$\frac{A, \Gamma \Rightarrow \Delta \quad \text{and } A \in \Phi}{\bigwedge \Phi, \Gamma \Rightarrow \Delta} \bigwedge \mathrm{L} \qquad \frac{\Gamma \Rightarrow \Delta, A \quad \text{for all } A \in \Phi}{\Gamma \Rightarrow \Delta, \bigwedge \Phi} \bigwedge \mathrm{R}$$

- $\bigvee$-Disjunction:

$$\frac{\Gamma \Rightarrow \Delta, A \quad \text{and } A \in \Phi}{\Gamma \Rightarrow \Delta, \bigvee \Phi} \bigvee \mathrm{R} \qquad \frac{A, \Gamma \Rightarrow \Delta \quad \text{for all } A \in \Phi}{\bigvee \Phi, \Gamma \Rightarrow \Delta} \bigvee \mathrm{L}$$

A detailed proof system for the logic $\mathcal{L}_{\infty\omega}$ will be provided in section 5. Since the technique of cut elimination will be an essential tool in our investigations, the sequent calculus is most appropriate.

3 Adding the axiom of choice (internally) to geometric theories does not preserve conservativity

This section features two examples of geometric theories where the internal addition of AC does not preserve geometric conservativity. In the subsequent section we will argue that the external addition of AC, in a certain sense, just amounts to arguing in a stronger background theory. It might produce interesting results but perhaps nothing that's not easily obtainable from the Boolean-valued approach[7] to forcing combined with the completeness result for $\mathcal{L}_{\omega_1\omega}$ (both from the 1960s).

---

7 In the Boolean-valued model approach one starts with a complete Boolean algebra $\mathbb{B}$ and builds the $\mathbb{B}$-valued extension $V^{\mathbb{B}} = \bigcup_\alpha V^{\mathbb{B}}_\alpha$ of the universe V by letting $V^{\mathbb{B}}_\alpha$ be the set of all functions taking values in $\mathbb{B}$ whose domain is a subset of $V^{\mathbb{B}}_\xi$ for some $\xi < \alpha$. Each formula of set theory with parameters in $V^{\mathbb{B}}$ gets assigned a 'truth value' in $\mathbb{B}$. So $V^{\mathbb{B}}$ is a model in many-valued logic, with truth values lying in $\mathbb{B}$.

The completeness theorem for $\mathcal{L}_{\omega_1\omega}$ says that a sentence of $\mathcal{L}_{\omega_1\omega}$ is deducible if and only if it holds in all models of the underlying language. Note that while each formula of $\mathcal{L}_{\omega_1\omega}$ is

3.1 First example

The example to be presented is a first-order theory. To define it we draw on a simple method, that is sometimes called *Morleyisation*, by which every theory can be given a geometric axiomatization in a richer language.[8]

Definition: 3.1. *Let $\mathcal{L}$ be a language In this subsection we shall only be concerned with first order formulae. $\forall\vec{x}\,(A_1(\vec{x}\,) \rightleftarrows A_2(\vec{x}\,))$ will stand for two formulae namely $\forall\vec{x}\,(A_1(\vec{x}\,) \to A_2(\vec{x}\,))$ and $\forall\vec{x}\,(A_2(\vec{x}\,) \to A_1(\vec{x}\,))$.*

For each formula $A(u_1, \dots, u_n)$ of $\mathcal{L}$ with all free variables indicated we add two new n-ary relation symbols $P_{A(\vec{u}\,)}$ and $N_{A(\vec{u}\,)}$ to the language, where $\vec{u} = u_1, \dots, u_n$. Call the new language $\mathcal{L}^a$. The first-order theory M^a in the language $\mathcal{L}^a$ has the following axioms:

1. $\forall\vec{x}\,\neg(P_{A(\vec{u}\,)}(\vec{x}) \wedge N_{A(\vec{u}\,)}(\vec{x}))$.
2. $\forall\vec{x}\,(P_{A(\vec{u}\,)}(\vec{x}) \vee N_{A(\vec{u}\,)}(\vec{x}))$.
3. *If $A(\vec{u}\,)$ is atomic add the axioms $\forall\vec{x}\,(P_{A(\vec{u}\,)}(\vec{x}) \rightleftarrows A(\vec{x}\,))$.*
4. *If $A(\vec{u}\,)$ is $B(\vec{u}\,) \wedge C(\vec{u}\,)$ add $\forall\vec{x}\,(P_{A(\vec{u}\,)}(\vec{x}\,) \rightleftarrows P_{B(\vec{u}\,)}(\vec{x}\,) \wedge P_{C(\vec{u}\,)}(\vec{x}\,))$.*
5. *If $A(\vec{u}\,)$ is $B(\vec{u}\,) \vee C(\vec{u}\,)$ add $\forall\vec{x}\,(P_{A(\vec{u}\,)}(\vec{x}\,) \rightleftarrows P_{B(\vec{u}\,)}(\vec{x}\,) \vee P_{C(\vec{u}\,)}(\vec{x}\,))$.*
6. *If $A(\vec{u}\,)$ is $\neg B(\vec{u}\,)$ add $\forall\vec{x}\,(P_{A(\vec{u}\,)}(\vec{x}\,) \rightleftarrows N_{B(\vec{u})}(\vec{x}\,))$.*
7. *If $A(\vec{u}\,)$ is $B(\vec{u}\,) \to C(\vec{u}\,)$ add $\forall\vec{x}\,(P_{A(\vec{u}\,)}(\vec{x}\,) \rightleftarrows N_{B(\vec{u}\,)}(\vec{x}\,) \vee P_{C(\vec{u}\,)}(\vec{x}\,))$.*
8. *If $A(\vec{u}\,)$ is $\exists y B(\vec{u}, y)$ add $\forall\vec{x}\,(P_{A(\vec{u}\,)}(\vec{x}\,) \rightleftarrows \exists y\, P_{B(\vec{u},v)}(\vec{x}, y))$.*
9. *If $A(\vec{u}\,)$ is $\forall y B(\vec{u}, y)$ add $\forall\vec{x}\,(N_{A(\vec{u}\,)}(\vec{x}\,) \rightleftarrows \exists y\, N_{B(\vec{u},v)}(\vec{x}, y))$.*

If T is a first-order theory, we denote by T^a the theory M^a augmented by the axioms

$$\forall\vec{x}\,P_{A(\vec{u})}(\vec{x}\,)$$

for all axioms $\forall\vec{x}\,A(\vec{x}\,)$ of T.

Clearly, M^a and T^a are finite geometric theories.

---

countable, the collection of all formulae is uncountable. The completeness result also holds for nice countable sets of formulae called fragments. For any countable fragment $\mathcal{L}_A$ of $\mathcal{L}_{\omega_1\omega}$ and a sentence φ in $\mathcal{L}_A$, φ is deducible in $\mathcal{L}_A$ (i.e., φ has a proof in $\mathcal{L}_{\omega_1\omega}$ such that every formula of the proof belongs to $\mathcal{L}_A$) if and only if φ holds in all models.

8 One place where one can find this terminology is Sacks' book from 1972 [Sacks, 1972, p. 256]. The technique was used by Skolem in the 1920s and conceivably could have even older roots. Albeit Skolemization would be more appropriate, that name is already used for something else. Keisler in his 1977 paper [Keisler, 1977, Theorem 2.18] refers to this gadget as the introduction of *Skolem relations*. Hodges, in his book [Hodges, 1993, p. 62] from 1993, called this method of gaining a ∀∃ axiomatization and quantifier elimination in a richer language *atomization*. For Morleyization in a topos-theoretic setting see e.g. Johnstone's book [Johnstone, 2002, p. 858] from 2002.

Lemma: 3.2. *Let M^a, T and T^a as above. Let $\vdash^i$ signify intuitionistic deducibility.*

1. *For every formula $A(\vec{u})$ of $\mathcal{L}$ with all free variables indicated,*

$$M^a \vdash^i \forall \vec{x}\,[A(\vec{x}) \leftrightarrow P_{A(\vec{u})}(\vec{x})].$$

2. *As a classical theory, T^a is conservative over T, that is, for every $\mathcal{L}$-sentence B,*

$$T \vdash^c B \text{ iff } T^a \vdash^c B.$$

This is in general not true for T based on intuitionistic logic.

Proof: (1.) is proved by induction on the generation of $A(\vec{u})$, making use of the excluded middle principle for $P_{A(\vec{u})}$ that is encapsulated in the first two axioms of M^a.

(2.) This can be shown syntactically but the model-theoretic proof is shorter. Every $\mathcal{L}$-structure $\mathfrak{A}$ can be expanded in just one way to an $\mathcal{L}^a$-structure $\mathfrak{A}^a$ which is a model of M^a. Hence every model $\mathfrak{M}$ of T can be expanded in just one way to an $\mathcal{L}^a$-structure. Moreover, $\mathfrak{M}^a$ is a model of T^a. Also, by (1.), T^a comprises T as it proves all axioms of T. □

Corollary: 3.3. *Let $\mathcal{L}$ be the language of set theory and ZF^a be the Morleyization of Zermelo-Fraenkel set theory. $\mathsf{ZF}^a + \mathsf{AC}$ is not conservative over ZF^a for geometric implications of $\mathcal{L}^a$.*

Proof: Let AC be the statement $\forall \vec{x}\, B(\vec{x})$. By Lemma 3.2(1.) we have

$$\mathsf{ZF}^a \vdash \mathsf{AC} \leftrightarrow \forall \vec{x}\, P_{B(\vec{u})}(\vec{x}),$$

and hence $\mathsf{ZF}^a + \mathsf{AC} \vdash \forall \vec{x}\, P_{B(\vec{u})}(\vec{x})$. If $\mathsf{ZF}^a + \mathsf{AC}$ were conservative over ZF^a for geometric formulae we could infer that $\mathsf{ZF}^a \vdash \forall \vec{x}\, P_{B(\vec{u})}(\vec{x})$ and hence $\mathsf{ZF}^a \vdash \mathsf{AC}$, which would yield $\mathsf{ZF} \vdash \mathsf{AC}$ by Lemma 3.2(2.). □

3.2 Second example

Here we study an infinitary theory. Let $\mathcal{L}'$ be the language with a set of constants X and infinitely many unary predicates P_n and Q_n for $n \in \mathbb{N}$. Let T' be the $\mathcal{L}'_{\infty\omega}$-theory with the following axioms:

1. $\forall z\, \neg[P_n(z) \wedge Q_n(z)]$ for all $n \in \mathbb{N}$.
2. $\bigvee_{a \in X} P_n(a)$ for all $n \in \mathbb{N}$;
3. $\bigvee_{n \in \mathbb{N}} Q_n(f(n))$ for all $f \in X^{\mathbb{N}}$.

Note that f does not appear as a function symbol in (3.); $f(n)$ is just a constant from X. T' is clearly a ∞-geometric theory.

In $\mathcal{L}'_{\infty\omega}$ we can express an instance A_{cc} of countable choice as follows:

$$\bigwedge_{n\in\mathbb{N}} \bigvee_{a\in X} P_n(a) \rightarrow \bigvee_{f\in X^{\mathbb{N}}} \bigwedge_{n\in\mathbb{N}} P_n(f(n)).$$

Now observe that $T' + A_{cc}$ is a syntactically inconsistent theory, where the latter means that an inconsistency $B \wedge \neg B$ for some formula B can be deduced with the help of the usual logical rules and the infinitary proof rules given at the end of Section 2.2.[9] By contrast, T' is syntactically consistent as long as X has at least two elements; although T' does not have a model in **Set** if we assume countable choice to hold in **Set**. That T is syntactically consistent can be seen as follows. Let $V[G]$ be a forcing extension of the ground model V in which the set $Y := X^{\mathbf{N}}$ of V becomes countable. In $V[G]$ there is an enumeration of all functions $f \in Y$, say $Y = \{f_0, f_1, f_2, \ldots\}$. Let $g : \mathbb{N} \rightarrow X$ be defined in such a way that $g(n) \neq f_n(n)$. This is possible since X has more than one element. Now define a model $\mathfrak{M}$ for T' in $V[G]$ by letting $M = X$ and interpreting $P_n^{\mathfrak{M}}$ as $\{g(n)\}$ and $Q_n^{\mathfrak{M}}$ as $X \setminus \{g(n)\}$. This shows that T' has a model in $V[G]$ and thus T' is syntactically consistent.

4 Adding the axiom of choice externally to geometric theories does preserve conservativity

Let $\mathcal{L}$ be a language and T be a $\mathcal{L}_{\infty\omega}$-theory. We extend $\mathcal{L}$ to $\mathcal{L}'$ by adding two unary predicate symbols S and U and the binary relation symbol $\in$. The idea is to define a set theory with urelements where the axioms of T are supposed to hold for the urelements. Formally this means that every axiom A of T has to be relativized to U, denoted A^{U}, i.e. all quantifier occurrences $\forall x \ldots x \ldots$ and $\exists y \ldots y \ldots$ in A have to be replaced by $\forall x(\mathrm{U}(x) \rightarrow \ldots x \ldots)$ and $\exists x(\mathrm{U}(x) \wedge \ldots x \ldots)$, respectively. T^{U} denotes the theory with language $\mathcal{L}'$ and all axioms A^{U} where A is an axiom of T.

The axioms of set theory then hold for the objects in S. The axiom of extensionality has to be given in the form

$$\forall x, y[\mathrm{S}(x) \wedge \mathrm{S}(y) \wedge \forall z(z \in x \leftrightarrow z \in y) \rightarrow x = y].$$

Further axioms proclaim that everything is either an urelement or a set but not both, that urelements have no elements, and that the urelements form a set: $\forall x\, [\mathrm{U}(x) \vee \mathrm{S}(x)]$, $\forall x\, \neg[\mathrm{U}(x) \wedge \mathrm{S}(x)]$, $\forall x, y[\mathrm{U}(x) \rightarrow y \notin x]$, $\exists y[\mathrm{S}(y) \wedge \forall x[x \in y \leftrightarrow \mathrm{U}(x)]$.

9 Details of the infinitary proof system will be provided in Section 5.

Let $\mathsf{ZF}_n^{\mathrm{U}}$ denote the set theory with language $\mathcal{L}'$ with urelement axioms having the above axioms, the usual axioms of set theory (Pairing, Union, Foundation, Powerset) expressed for objects of sort S, Separation extended to the language $\mathcal{L}'$, but with Replacement restricted to Σ_n-formulae of $\mathcal{L}'$.

Below we refer to *definable global choice* by which we mean that a formula of set theory (usually with extra parameters) defines a well-ordering on the entire universe (see [Levy, 1979, V.3.9] for details). The actual formula will be revealed in the proof of the next theorem. We then have the following conservativity result.

Theorem: 4.1 (ZFC). *Let B be a sentence of $\mathcal{L}_{\infty\omega}$. Then:*

$$T \vdash B \quad \textit{iff} \quad \mathsf{ZF}_n^{\mathrm{U}} + \textit{definable global choice} + \mathsf{GCH} + T^{\mathrm{U}} \vdash B^{\mathrm{U}}.$$

If T is ∞-geometric and A is a geometric implication, then also

$$T \vdash^i B \quad \textit{iff} \quad \mathsf{ZF}_n^{\mathrm{U}} + \textit{definable global choice} + \mathsf{GCH} + T^{\mathrm{U}} \vdash B^{\mathrm{U}}.$$

Proof: We argue in our background universe satisfying ZFC. Suppose $\mathsf{ZF}_n^{\mathrm{U}}$ + global choice+GCH+$T^{\mathrm{U}} \vdash B^{\mathrm{U}}$. We then switch to a forcing extension $V[G]$ in which the language $\mathcal{L}'$, the formula B and its subformulae as well as the axioms of T^{U} together with their subformulae belong a countable transitive set X. Let $f : X \to \mathbf{N}$ be a bijection. Arguing in $V[G]$, we shall work in the relativized constructible hierarchy $L(f)$ which starts with TC(f), the transitive closure of $\{f\}$ (see [Jech, 2003, 13.24]). $L(f)$ has a global definable well-ordering since TC(f) is countable in $V[G]$. It's also a model of GCH. Using the reflection principle of ZF, we can take any model $\mathfrak{M}$ of T in $L(f)$ and expand it into a model of $\mathsf{ZF}_n^{\mathrm{U}}$ + global choice + GCH + T^{U}. Thus $\mathfrak{M}$ will satisfy B. But in $L(f)$, T is a $\mathcal{L}_{\omega_1\omega}$-theory and therefore, by the completeness theorem for this logic, there exists a deduction of B from T in $L(f)$. Consequently, if we work in a sequence calculus, invoking Theorem 7.9 yields that there exists also a cut-free deduction of $\bigwedge T \to B$ in $L(f)$, where $\bigwedge T$ signifies the conjunction of all axioms of T.

Now it's crucial to observe that $\bigwedge T$ and B both belong to the ground model. It remains to show that there is also a deduction of $\bigwedge T \to B$ in the ground model V. To this end we shall prove a more general result:

(∗) If there is cut free deduction $\mathfrak{D}$ of the sequent $\Gamma \Rightarrow \Delta$ in $V[G]$ and $\Gamma \Rightarrow \Delta$ belongs to the ground model V, then there already exists a deduction of $\Gamma \Rightarrow \Delta$ in V.

In (∗) we refer to the sequent calculus for $\mathcal{L}_{\omega_1\omega}$ to be described in section 5. A sequent $\Gamma \Rightarrow \Delta$ consists of two finite sequences of $\mathcal{L}_{\omega_1\omega}$-formulae Γ and Δ. We proceed by induction on the rank of $\mathfrak{D}$. It is crucial that $\mathfrak{D}$ contains no cuts lest the end sequents of the immediate subderivations of $\mathfrak{D}$ contain formulae that are not in the ground model and the inductive proof breaks down. The proof is straightforward except for the cases of a $\bigwedge$R or $\bigvee$L inferences that require a bit more attention. So suppose that the last inference of $\mathcal{D}$ was $\bigwedge$R. Then Δ is of the form $\Delta_0, \bigwedge \Phi$ and we have deductions $\mathcal{D}_A$ of $\Gamma \Rightarrow \Delta_0, A$ for all $A \in \Phi$. With $\Phi \in V$ we also have $A \in V$ for all $A \in \Phi$. Thus inductively for every A in Φ there exists a deduction $\mathcal{D}'_A$ of $\Gamma \Rightarrow \Delta_0, A$. Using collection and the axiom of choice we can then compose a deduction $\mathcal{D}'$ of $\Gamma \Rightarrow \Delta_0, \bigwedge \Phi$ in V. The case of an $\bigvee$L inference is similar.

The final issue to be resolved is how the forcing extension $V[G]$ can be accessed from the ground model V. There are several approaches to this (cf. [Kunen, 1980, Ch. VII.9]). One proper formal way is to resort to the Boolean valued approach (cf. [Bell, 1977]). Also note that in case the language $\mathcal{L}$, the theory T and B are all countable, it is not necessary to take a forcing extension. Then $L(f) \subseteq V$ and the main ingredient for proving the theorem is just the completeness of $\mathcal{L}_{\omega_1\omega}$. □

Remark: 4.2. *The declared background theory for the previous Theorem is* ZFC, *however,* ZF *would be sufficient. The axiom of choice can be dropped, though this requires a more careful definition of the notion of infinitary deduction which does not have the axiom of choice built into its very definition. The problem lies with the infinitary rules $\bigwedge R$ and $\bigvee L$.* AC *is needed when we have to pick exactly one deduction for each of the infinitely many premisses of these inferences. But this can be avoided by allowing non-empty sets of subdeductions of the same end sequent to figure in a deduction. Details will be deferred to section 5.1. It should perhaps be mentioned that jettisoning* AC *when dealing with infinitary deductions is also important for the Barwise completeness theorem (see [Barwise, 1975, III.5]).*

A possible interpretation of those earlier quotes to the effect that adding the axiom of choice to a geometric theory T does not produce new geometric theorems is that AC is simply added to an ambient external type theory $\mathcal{IL}$ which is grafted onto T. Here $\mathcal{IL}$ is the intuitionistic type theory that holds in all toposes also known as the *internal logic* of toposes (see [Lambek and Scott, 1988, II]). The axiom of choice can be expressed in the language for the internal logic (the so-called Mitchell-Bénabou language) in a straightforward way ([Lambek and Scott, 1988, II.6]). If one now assumes that the language of T is incorporated into the Mitchell-Bénabou language via relativization to a specific sort U and one also has an appropriate treatment of the infinite connectives then one gets the following result.

Corollary: 4.3. *Let B be a sentence of $\mathcal{L}_{\infty\omega}$. Then:*

$$\begin{aligned} \textit{Internal Logic} + T^{\mathrm{U}} + \mathsf{AC} \vdash^{c} B^{\mathrm{U}} &\quad \textit{iff} \\ \textit{Internal Logic} + T^{\mathrm{U}} \vdash^{i} B^{\mathrm{U}} &\quad \textit{iff} \\ T \vdash^{i} B & \end{aligned}$$

where $\vdash^c$ and $\vdash^i$ signify classical and intuitionistic derivability, respectively.

Proof: As the internal logic can be interpreted in a small fragment of ZF, this is a consequence of Theorem 4.1. □

5 A sequent calculus for $\mathcal{L}_{\infty\omega}$

In his thesis Gentzen introduced a form of the sequent calculus and his technique of *cut elimination*. The sequent calculus can be generalized to $\mathcal{L}_{\infty\omega}$.

A **sequent** is an expression $\Gamma \Rightarrow \Delta$ where Γ and Δ are finite sequences of $\mathcal{L}_{\infty\omega}$-formulae $A_1, \ldots, A_n$ and $B_1, \ldots, B_m$, respectively. We also allow for the possibility that Γ or Δ (or both) are empty. The empty sequent will be denoted by $\emptyset$. $\Sigma \Rightarrow \Delta$ is read, informally, as Γ yields Δ or, rather, the conjunction of the A_i yields the disjunction of the B_j. In particular, we have:

- If Γ is empty, the sequent asserts the disjunction of the B_j.
- If Δ is empty, it asserts the negation of the conjunction of the A_i.
- if Γ and Δ are both empty, it asserts the impossible, i.e. a contradiction.

We use upper case Greek letters $\Gamma, \Delta, \Lambda, \Theta, \Xi \ldots$ to range over finite sequences of formulae. $\Gamma \subseteq \Delta$ means that every formula of Γ is also a formula of Δ. Γ, A stands for the sequence Γ extended by A.

Next we list the axioms and rules of the first-order sequent calculus.

- *Logical Axioms*

$$\Gamma, A \Rightarrow \Delta, A$$

  where A is any formula. In point of fact, one could limit this axiom to the case of atomic formulae A.

- *Cut Rule*

$$\frac{\Gamma \Rightarrow \Delta, A \qquad A, \Lambda \Rightarrow \Theta}{\Gamma, \Lambda \Rightarrow \Delta, \Theta}\ \text{Cut}$$

  The formula A is called the cut formula of the inference.

- *Structural Rules*

$$\frac{\Gamma \Rightarrow \Delta}{\Gamma' \Rightarrow \Delta'} \qquad \text{if } \Gamma \subseteq \Gamma',\ \Delta \subseteq \Delta'.$$

A special case of the structural rule, known as *contraction*, occurs when the lower sequent has fewer occurrences of a formula than the upper sequent. For instance, $A, \Gamma \Rightarrow \Delta, B$ follows structurally from $A, A, \Gamma \Rightarrow \Delta, B, B$.

- Rules for Logical Operations

Left

Right

$$\frac{\Gamma \Rightarrow \Delta, A}{\neg A, \Gamma \Rightarrow \Delta} \qquad \frac{B, \Gamma \Rightarrow \Delta}{\Gamma \Rightarrow \Delta, \neg B}$$

$$\frac{\Gamma \Rightarrow \Delta, A \qquad B, \Lambda \Rightarrow \Theta}{A \rightarrow B, \Gamma, \Lambda \Rightarrow \Delta, \Theta} \qquad \frac{A, \Gamma \Rightarrow \Delta, B}{\Gamma \Rightarrow \Delta, A \rightarrow B}$$

$$\frac{A, \Gamma \Rightarrow \Delta}{A \wedge B, \Gamma \Rightarrow \Delta} \quad \frac{B, \Gamma \Rightarrow \Delta}{A \wedge B, \Gamma \Rightarrow \Delta} \qquad \frac{\Gamma \Rightarrow \Delta, A \qquad \Gamma \Rightarrow \Delta, B}{\Gamma \Rightarrow \Delta, A \wedge B}$$

$$\frac{A, \Gamma \Rightarrow \Delta \qquad B, \Gamma \Rightarrow \Delta}{A \vee B, \Gamma \Rightarrow \Delta} \qquad \frac{\Gamma \Rightarrow \Delta, A}{\Gamma \Rightarrow \Delta, A \vee B} \quad \frac{\Gamma \Rightarrow \Delta, B}{\Gamma \Rightarrow \Delta, A \vee B}$$

$$\frac{F(t), \Gamma \Rightarrow \Delta}{\forall x F(x), \Gamma \Rightarrow \Delta}\ \forall L \qquad \frac{\Gamma \Rightarrow \Delta, F(a)}{\Gamma \Rightarrow \Delta, \forall x F(x)}\ \forall R$$

$$\frac{F(a), \Gamma \Rightarrow \Delta}{\exists x F(x), \Gamma \Rightarrow \Delta}\ \exists L \qquad \frac{\Gamma \Rightarrow \Delta, F(t)}{\Gamma \Rightarrow \Delta, \exists x F(x)}\ \exists R$$

In $\forall$L and $\exists R$, t is an arbitrary term. The variable a in $\forall$R and $\exists L$ is an *eigenvariable* of the respective inference, i.e. a is not to occur in the lower sequent.

The logic $\mathcal{L}_{\infty\omega}$ in addition has rules for $\bigwedge$ and $\bigvee$ that generalize those for $\wedge$ and $\vee$, respectively.

$$\frac{A, \Gamma \Rightarrow \Delta \ \text{ and } A \in \Phi}{\bigwedge \Phi, \Gamma \Rightarrow \Delta}\ \bigwedge \mathrm{L} \qquad \frac{\Gamma \Rightarrow \Delta, A \ \text{ for all } A \in \Phi}{\Gamma \Rightarrow \Delta, \bigwedge \Phi}\ \bigwedge \mathrm{R}$$

$$\frac{\Gamma \Rightarrow \Delta, A \ \text{ and } A \in \Phi}{\Gamma \Rightarrow \Delta, \bigvee \Phi}\ \bigvee \mathrm{R} \qquad \frac{A, \Gamma \Rightarrow \Delta \ \text{ for all } A \in \Phi}{\bigvee \Phi, \Gamma \Rightarrow \Delta}\ \bigvee \mathrm{L}$$

In the rules for logical operations, the formulae highlighted in the premisses are called the *minor formulae* of that inference, while the formula highlighted in the conclusion is the *principal formula* of that inference. The other formulae of an inference are called *side formulae*.

5.1 What are proofs in $\mathcal{L}_{\infty\omega}$?

Proofs in $\mathcal{L}_{\omega\omega}$ are finite objects and as a result its notion of proof is very robust. For instance, if one knows for a fact that a formula A is provable in a primitive recursive theory T, then one can conclude that it is inferrable in Heyting arithmetic that T proves A. In other words, it is immaterial in which background theory (e.g. ZFC plus large cardinals) we gained the insight that this fact is true. Things are very different when it comes to infinite proofs. An example is provided by the $\mathcal{L}_{\infty\omega}$ intuitionistic (∞-geometric) theory $\mathbf{HA}_\infty$ whose axioms are those of Robinson arithmetic augmented by the axiom $\forall x \bigvee_{n\in\mathbb{N}} x = \bar{n}$, where $\bar{n}$ stands for the n-th numeral. In ZF one can show that there exists an intuitionistic $\mathbf{HA}_\infty$-proof of a particular statement A that cannot be shown to exist in intuitionistic Zermelo-Fraenkel set theory **IZF**.[10]

Even in a classical context it may be relevant to choose a suitable formalization of infinite proof. For instance, for Barwise's completeness theorem for admissible fragments is it important to choose a notion of proof that "does not have the axiom of choice built into its very definition" ([Barwise, 1975], p. 96).

Definition: 5.1 (CZF). *We will assume that the language $\mathcal{L}$ is a set. For the formalization of the $\mathcal{L}$-formulae as set-theoretic objects, we proceed in the same way as Barwise in [Barwise, 1975, III.3]. They form an inductively defined (proper) class of sets. An additional assumption we shall make is that the proper subformulae of a formula A are elements of the transitive closure of A which has the pleasant consequence that the rank of a proper subformula of A is an element of the rank of A; in symbols rank(B) $\in$ rank(A). Here we use the usual rank definition for sets, i.e.,*

$$rank(a) = \bigcup\{rank(x) + 1 \mid x \in a\}.$$

Note that rank(a) is always an ordinal and $x + 1$ stands for $x \cup \{x\}$. Just as in the classical world, an ordinal *is a transitive set whose elements are transitive. However, the crucial difference between the classical and the intuitionistic context is the forfeiture of the right to use the trichotomy law for ordinals in the latter, i.e., the assertion $\alpha \in \beta \vee \beta \in \alpha \vee \alpha = \beta$ can no longer be guaranteed to hold.*

Definition: 5.2. *The class of $\mathcal{L}_{\infty\omega}$-proofs (also called* deductions *or* derivations*) will be defined inductively. It is desirable to ensure that the inferred formula and the*

10 A can be taken of complexity Π^0_3, namely $\forall x\exists y\forall z\,(\mathrm{T}(x,x,y) \vee \neg\mathrm{T}(x,x,z))$, where T is the predicate from Kleene's normal form theorem. Now in the presence of countable choice, A implies the existence of a non-computable function. That the existence of a $\mathbf{HA}_\infty$ proof of A cannot be shown in **IZF** follows from the fact that **IZF** plus countable choice is compatible with the statement that all functions from $\mathbb{N}$ to $\mathbb{N}$ are computable.

last inference together with its principal and minor formulae are straightforwardly retrievable from any proof P. Firstly, sequents $A_1, \ldots, A_r \Rightarrow B_1, \ldots, B_s$ are easily coded set-theoretically as a pair of tuples $\langle\langle A_1, \ldots, A_n\rangle, \langle B_1, \ldots, B_s\rangle\rangle$; we will continue to use the former notation even when we refer to its set-theoretic coding.

We shall not write down all the clauses for the inductive definition of proofs. Rather we will provide two illustrative cases, the finitary $(\wedge R)$ and the infinitary $(\bigvee L)$.

Suppose now we have two proofs $\mathcal{D}_1$ and $\mathcal{D}_2$ of sequents $A_1, \ldots, A_r \Rightarrow B_1, \ldots B_s$ and $A_1, \ldots, A_r \Rightarrow B'_1, \ldots B'_s$ and $1 \le i_0 \le s$ such that $B_i = B'_i$ for all $i \neq i_0$. Then

$$\mathcal{D} := \langle\langle P_1, P_2\rangle, \langle \wedge R, i_0, k_0\rangle, C_1, \ldots, C_p \Rightarrow D_1, \ldots, D_q\rangle$$

is a proof of $C_1, \ldots, C_p \Rightarrow D_1, \ldots, D_q\rangle$ if $1 \le k_0 \le q$ and $B_{i_0} \wedge B'_{i_0} = D_{k_0}$, $A_i \in \{C_1, \ldots, C_p\}$ for all $1 \le i \le r$ and $B_i \in \{D_1, \ldots, D_q\}$ whenever $1 \le i \le s$ and $i \neq i_0$. It is clear that from P we can retrieve the last inference together with its principal and minor formulae.

Next assume Φ is a set of $\mathcal{L}_{\infty\omega}$-formula and there is a function f with domain Φ such that there are finite sequences $\Gamma_1, \Gamma_2, \Delta$ of formulae such that for every $A \in \Phi$, $f(A)$ is an inhabited set of proofs of $\Gamma_1, A, \Gamma_2 \Rightarrow \Delta$. Let i_0 be the position of any such A in Γ_1, A, Γ_2. Then

$$\mathcal{D}' := \langle f, \langle \bigvee L, i_0, k_0\rangle, C_1, \ldots, C_p \Rightarrow D_1, \ldots, D_q\rangle$$

is a proof of $C_1, \ldots, C_p \Rightarrow D_1, \ldots, D_q$ if $1 \le k_0 \le p$, $\bigvee \Phi = C_{k_0}$, $C \in \{C_1, \ldots, C_p\}$ for every $C \in \Gamma_i$ with $i \in \{0, 1\}$, and $D \in \{D_1, \ldots, D_q\}$ for all $D \in \Delta$.

It should by now be obvious how to deal with the other inference rules.

Observe that the above definition allows to combine each inference step with a structural rule. This has the advantage that structural rules needn't be treated as separate rules. There is a lot of leeway as to the details of formalizing infinitary proofs constructively. However, observe that the above definition of proof in the case of $(\bigvee L)$ (and dually $(\bigwedge R)$) contains a crucial part that enables one to construct new proofs from a collection of proofs. It has the advantage that in CZF *one can prove from the assumption that there exists a proof of $\Gamma_1, A, \Gamma_2 \Rightarrow \Delta$ for every $A \in \Phi$, that there also exists a proof of $\Gamma_1, \bigvee \Phi, \Gamma_2 \Rightarrow \Delta$ by invoking Strong Collection (see [Aczel and Rathjen, 2001, Aczel and Rathjen:, 2010]). In general, we wouldn't be able to single out a particular proof of $\Gamma_1, A, \Gamma_2 \Rightarrow \Delta$ for every $A \in \Phi$ without relying on the axiom of choice.*

Let $\vdash \Gamma \Rightarrow \Delta$ signify that there is a proof of $\Gamma \Rightarrow \Delta$. We also define provability with length α and cut-degree ρ,

$$\vdash^{\alpha}_{\rho} \Gamma \Rightarrow \Delta$$

to mean that there is a proof P of $\Gamma \Rightarrow \Delta$ *such that* $rank(P) \in \alpha + 1$ *and for all cut formulae C in P we have* $rank(C) \in \rho$. *In particular* $\vdash^{\alpha}_{0} \Gamma \Rightarrow \Delta$ *then conveys that there is a proof without cuts.*

Naturally, proofs in theories will also be considered. An $\mathcal{L}_{\infty\omega}$*-theory T is a set of* $\mathcal{L}_{\infty\omega}$*-formulae without free variables. The T-proofs are defined as the* $\mathcal{L}_{\infty\omega}$*-proofs except that one adds the additional axioms* $\Gamma \Rightarrow \Delta, A$ *with* $A \in T$ *to the sequent calculus.*

That there is a T-proof of $\Phi \Rightarrow \Theta$ *will be conveyed by* $T \vdash \Phi \Rightarrow \Theta$. *Since the theory T can be expressed in* $\mathcal{L}_{\infty\omega}$ *via a single formula* $\bigwedge T$ *we also have*

$$T \vdash \Phi \Rightarrow \Theta \quad \textit{iff} \quad \vdash \Phi, \bigwedge T \Rightarrow \Theta. \tag{2}$$

6 Turning classical ∞-geometric proofs into intuitionistic ones

Recall that intuitionistic $\mathcal{L}_{\infty\omega}$-proofs are those obeying the simple structural restriction that there can be at most one formula on the right hand side of the sequent symbol $\Rightarrow$. Below we shall indicate intuitionistic proofs in $\mathcal{L}_{\infty\omega}$ by putting I_∞ before the turn style symbol.

The fact that $\mathcal{L}_{\omega\omega}$ geometric proofs can be turned into intuitionistic ones is basically a consequence of Gentzen's Hauptsatz. It could have been proved by Gentzen in 1934. It is not clear to the present author who first made this observation but it can be found in Orevkov's 1968 paper [Orevkov, 1968].[11] As for the $\mathcal{L}_{\infty\omega}$ case it is not clear to him whether there are any syntactic proofs in the published literature (before [Rathjen, 2012]). But the purpose of this part of the article, rather than originality, is to show that there is an easy syntactic proof that can also be formalized in the constructive set theory CZF (see [Aczel and Rathjen, 2001, Aczel and Rathjen:, 2010]). Closer inspection would actually reveal that intuitionistic Kripke-Platek set theory (see [Aczel and Rathjen, 2001, Aczel and Rathjen:, 2010]) suffices. As in the finite case the crucial tool is the Hauptsatz for $\mathcal{L}_{\infty\omega}$.

Theorem: 6.1. *If* $\vdash^{\alpha}_{\rho} \Gamma \Rightarrow \Delta$ *then there exists* α' *such that* $\vdash^{\alpha'}_{0} \Gamma \Rightarrow \Delta$.

11 This is not to say that there are no interesting research questions left. Since cut elimination is costly there are still unsolved problems as to how efficient this procedure can be, in general and for special theories (see e.g. [Mints, 2012]).

The proof of 6.1 in CZF will be deferred to section 7. Without paying attention to constructivity issues, for the countable logic $\mathcal{L}_{\omega_1\omega}$ this was essentially shown by Tait [Tait, 1968].

The main result of this section requires knowledge of some basic facts.

Lemma: 6.2 (Substitution). *Let $\Gamma(a) \Rightarrow \Delta(a)$ be a sequent with all occurrences of the free variable a indicated. Further, let t be an arbitrary term. If $\vdash^{\alpha}_{\rho} \Gamma(a) \Rightarrow \Delta(a)$, then $\vdash^{\alpha}_{\rho} \Gamma(t) \Rightarrow \Delta(t)$.*

Proof: Proceed by induction on α. □

Lemma: 6.3 (Inversion).

1. *If* $\vdash^{\alpha}_{\rho} \Gamma, A \wedge B \Rightarrow \Delta$ *then* $\vdash^{\alpha}_{\rho} \Gamma, A, B \Rightarrow \Delta$.
2. *If* $\vdash^{\alpha}_{\rho} \Gamma \Rightarrow \Delta, A \wedge B$ *then* $\vdash^{\alpha}_{\rho} \Gamma \Rightarrow \Delta, A$ *and* $\vdash^{\alpha}_{\rho} \Gamma \Rightarrow \Delta, B$.
3. *If* $\vdash^{\alpha}_{\rho} \Gamma, A \vee B \Rightarrow \Delta$ *then* $\vdash^{\alpha}_{\rho} \Gamma, A \Rightarrow \Delta$ *and* $\vdash^{\alpha}_{\rho} \Gamma, B \Rightarrow \Delta$.
4. *If* $\vdash^{\alpha}_{\rho} \Gamma \Rightarrow \Delta, A \vee B$ *then* $\vdash^{\alpha}_{\rho} \Gamma \Rightarrow \Delta, A, B$.
5. *If* $\vdash^{\alpha}_{\rho} \Gamma \Rightarrow A \rightarrow B, \Delta$ *then* $\vdash^{\alpha}_{\rho} A, \Gamma \Rightarrow \Delta, B$.
6. *If* $\vdash^{\alpha}_{\rho} \Gamma, A \rightarrow B \Rightarrow \Delta$ *then* $\vdash^{\alpha}_{\rho} \Gamma \Rightarrow \Delta, A$ *and* $\vdash^{\alpha}_{\rho} \Gamma, B \Rightarrow \Delta$.
7. *If* $\vdash^{\alpha}_{\rho} \Gamma \Rightarrow \neg A, \Delta$ *then* $\vdash^{\alpha}_{\rho} \Gamma, A \Rightarrow \Delta$.
8. *If* $\vdash^{\alpha}_{\rho} \Gamma, \neg A \Rightarrow \Delta$ *then* $\vdash^{\alpha}_{\rho} \Gamma \Rightarrow \Delta, A$.
9. *If* $\vdash^{\alpha}_{\rho} \Gamma \Rightarrow \Delta, \forall x\, B(x)$ *then* $\vdash^{\alpha}_{\rho} \Gamma \Rightarrow \Delta, B(s)$ *for any term s.*
10. *If* $\vdash^{\alpha}_{\rho} \Gamma, \exists x\, B(x) \Rightarrow \Delta$ *then* $\vdash^{\alpha}_{\rho} \Gamma, B(s) \Rightarrow \Delta$ *for any term s.*
11. *If* $\vdash^{\alpha}_{\rho} \Gamma, \bigvee \Phi \Rightarrow \Delta$ *then* $\vdash^{\alpha}_{\rho} \Gamma, A \Rightarrow \Delta$ *for every* $A \in \Phi$.
12. *If* $\vdash^{\alpha}_{\rho} \Gamma \Rightarrow \Delta, \bigwedge \Phi$ *then* $\vdash^{\alpha}_{\rho} \Gamma \Rightarrow \Delta, A$ *for every* $A \in \Phi$.
13. *With the exception of 4., 6. and 8. the above inversion properties remain valid for the intuitionistic sequent calculus. One half of 6. also remains valid intutionistically:*

$$\text{if } I_{\infty} \vdash^{\alpha}_{\rho} \Gamma, A \rightarrow B \Rightarrow \Delta \text{ then } I_{\infty} \vdash^{\alpha}_{\rho} \Gamma, B \Rightarrow \Delta.$$

Proof: All can be shown easily by induction on α. □

Below we use $\bigvee(\Phi, A)$ to stand for $\bigvee(\Phi \cup \{A\})$.

Lemma: 6.4.

1. *If* $I_{\infty} \vdash \Gamma \Rightarrow \bigvee(\Phi, F(s))$, *then* $I_{\infty} \vdash \Gamma \Rightarrow \bigvee(\Phi, \exists x\, F(x))$.
2. *If* $I_{\infty} \vdash \Gamma \Rightarrow \bigvee(\Phi, B)$ *and* $I_{\infty} \vdash \Gamma \Rightarrow \bigvee(\Phi, C)$, *then* $I_{\infty} \vdash \Gamma \Rightarrow \bigvee(\Phi, B \wedge C)$.
3. *If* $I_{\infty} \vdash \Gamma \Rightarrow \bigvee(\Phi, A)$, *then* $I_{\infty} \vdash \Gamma, \neg A \Rightarrow \bigvee \Phi$.
4. *If* $I_{\infty} \vdash \Gamma, B \Rightarrow \bigvee \Phi$ *and* $I_{\infty} \vdash \Gamma \Rightarrow \bigvee(\Phi, A)$, *then* $I_{\infty} \vdash \Gamma, A \rightarrow B \Rightarrow \bigvee \Phi$.
5. *If* $I_{\infty} \vdash \Gamma \Rightarrow \bigvee(\Phi, A)$ *and* $A \in \Theta$, *then* $I_{\infty} \vdash \Gamma \Rightarrow \bigvee(\Phi, \bigvee \Theta)$.

Proof: (1.) We have

$$\dfrac{\dfrac{D \Rightarrow D}{D \Rightarrow \bigvee(\Phi, \exists x F(x)) \quad \text{for all } D \in \Phi}(\bigvee R) \qquad \dfrac{\dfrac{\dfrac{F(s) \Rightarrow F(s)}{F(s) \Rightarrow \exists x F(x)}(\exists R)}{F(s) \Rightarrow \bigvee(\Phi, \exists x F(x))}(\bigvee R)}{\bigvee(\Phi, F(s)) \Rightarrow \bigvee(\Phi, \exists x F(x))}(\bigvee L)$$

As $I_\infty \vdash \Gamma \Rightarrow \bigvee(\Phi, F(s))$, a cut with $\bigvee(\Phi, F(s))$ yields $I_\infty \vdash \Gamma \Rightarrow \bigvee(\Phi, \exists x F(x))$.

(2.) We have

$$\dfrac{\dfrac{B, D \Rightarrow D}{B, D \Rightarrow \bigvee(\Phi, B \wedge C) \quad \text{all } D \in \Phi}(\bigvee R) \qquad \dfrac{\dfrac{\dfrac{B, C \Rightarrow B \qquad B, C \Rightarrow C}{B, C \Rightarrow B \wedge C}(\wedge R)}{B, C \Rightarrow \bigvee(\Phi, B \wedge C)}(\bigvee R)}{B, \bigvee(\Phi, C) \Rightarrow \bigvee(\Phi, B \wedge C)}(\bigvee L)$$

and therefore

$$\dfrac{\dfrac{D, \bigvee(\Phi, C) \Rightarrow D}{D, \bigvee(\Phi, C) \Rightarrow \bigvee(\Phi, B \wedge C) \quad \text{all } D \in \Phi}(\bigvee R) \qquad \overline{\overline{B, \bigvee(\Phi, C) \Rightarrow \bigvee(\Phi, B \wedge C)}}}{\bigvee(\Phi, B), \bigvee(\Phi, C) \Rightarrow \bigvee(\Phi, B \wedge C)}(\bigvee L)$$

Cuts with $I_\infty \vdash \Gamma \Rightarrow \bigvee(\Phi, B)$ and $I_\infty \vdash \Gamma \Rightarrow \bigvee(\Phi, C)$ yield the desired outcome

$$I_\infty \vdash \Gamma \Rightarrow \bigvee(\Phi, B \wedge C)\,.$$

(3.) is shown as follows:

$$\dfrac{\dfrac{\dfrac{\dfrac{\Gamma, A \Rightarrow A}{\Gamma, A \Rightarrow \bigvee \Theta}(\bigvee R)}{\Gamma, A \Rightarrow \bigvee(\Phi, \bigvee \Theta)}(\bigvee R) \qquad \dfrac{\Gamma, B \Rightarrow B \quad (B \in \Phi)}{\Gamma, B \Rightarrow \bigvee(\Phi, \bigvee \Theta)}(\bigvee R)}{\Gamma, \bigvee(\Phi, A) \Rightarrow \bigvee(\Phi, \bigvee \Theta)}(\bigvee L) \qquad \overline{\overline{\Gamma \Rightarrow \bigvee(\Phi, A)}}}{\Gamma \Rightarrow \bigvee(\Phi, \bigvee \Theta)}(Cut)$$

(4.) We have

$$\dfrac{\dfrac{\Gamma, A \Rightarrow A \qquad \overline{\overline{\Gamma, B \Rightarrow \bigvee \Phi}}}{A, \Gamma, A \to B \Rightarrow \bigvee \Phi}(\to L) \qquad \dfrac{\Gamma, C, A \to B \Rightarrow C}{\Gamma, C, A \to B \Rightarrow \bigvee \Phi \quad \text{all } C \in \Phi}(\vee R)}{\Gamma, A \to B, \bigvee(\Phi, A) \Rightarrow \bigvee \Phi}(\vee L)$$

Cutting out $\bigvee(\Phi, A)$ with $I_\infty \vdash \Gamma \Rightarrow \bigvee(\Phi, A)$ yields $I_\infty \vdash \Gamma, A \to B \Rightarrow \bigvee \Phi$.

(5.) is shown as follows:

$$\dfrac{\dfrac{\dfrac{\dfrac{\Gamma, A \Rightarrow A}{\Gamma, A \Rightarrow \bigvee \Theta}\,(\bigvee R)}{\Gamma, A \Rightarrow \bigvee(\Phi, \bigvee \Theta)}\,(\bigvee R) \qquad \dfrac{\Gamma, B \Rightarrow B \quad (B \in \Phi)}{\Gamma, B \Rightarrow \bigvee(\Phi, \bigvee \Theta)}\,(\bigvee R)}{\Gamma, \bigvee(\Phi, A) \Rightarrow \bigvee(\Phi, \bigvee \Theta)}\,(\bigvee L) \qquad \overline{\overline{\Gamma \Rightarrow \bigvee(\Phi, A)}}}{\Gamma \Rightarrow \bigvee(\Phi, \bigvee \Theta)}\,(Cut)$$

□

Lemma: 6.5. *Let Δ be a finite set of ∞-geometric formulae and Γ be a finite set of ∞-geometric implications.*

If $\vdash \Gamma \Rightarrow \Delta$ then $I_\infty \vdash \Gamma \Rightarrow \bigvee \Delta$.

Proof: Here we rely on Gentzen's Hauptsatz, Theorem 6.1, for classical $\mathcal{L}_{\infty\omega}$ logic.

Let $\mathcal{D}$ be a cut free deduction of $\Gamma \Rightarrow \Delta$. The proof proceeds by induction on the ordinal height α of $\mathcal{D}$.

If $\Gamma \Rightarrow \Delta$ is an axiom then there exists an atom A such that $A \in \Gamma \cap \Delta$. Thus $I_\infty \vdash \Gamma \Rightarrow A$ and therefore, via $(\bigvee R)$, we get to $I_\infty \vdash \Gamma \Rightarrow \bigvee \Delta$.

Now suppose that $\Gamma \Rightarrow \Delta$ is the result of an inference rule. We inspect the last inference of $\mathcal{D}$. Note that $\forall R$, $\bigwedge R$, $\neg R$ and $\to R$ are ruled out since their principal formulae are not ∞-geometric formulae and would have to occur in the succedent Δ.

If the last inference was of the form $\forall L$, $\bigwedge L$, $\wedge L$, $\exists L$, $\bigvee L$, or $\vee L$ we can simply apply the induction hypothesis to the premisses and re-apply the same inference in the intuitionistic calculus.

If the last inference was $\exists R$ we apply the induction hypothesis to its premiss and subsequently use Lemma 6.4 (1) to get the desired result.

If the last inference was $\wedge R$ we apply the induction hypothesis to its premisses and subsequently use Lemma 6.4 (2).

If the last inference was $\neg L$ then its minor formula must be ∞-geometric. Thus we can apply the induction hypothesis to its premiss and subsequently use Lemma 6.4 (3).

If the last inference was $\to L$ then apply the induction hypothesis to its premisses and subsequently use Lemma 6.4 (4).

If the last inference was $\bigvee R$ then apply the induction hypothesis to its premisses and subsequently use Lemma 6.4 (5).

The case when the last inference was $\vee R$ is similar to the previous one. □

Theorem: 6.6. *Let T be a ∞-geometric theory and suppose that there is a classical proof of a ∞-geometric implication G from T. Then there is an intuitionistic proof of G from the axioms of T.*

Proof: Below we shall write $T \vdash A$ for $T \vdash\Rightarrow A$ and $T \vdash^i A$ if there is an intuitionistic proof of $\Rightarrow A$ from the axioms of T.

We proceed by induction on the buildup of G. First suppose that G is of the form $\forall \vec{x}\, F(\vec{x})$ where $F(\vec{a})$ is a ∞-geometric implication. By (2) we have

$$\vdash \bigwedge T \Rightarrow G$$

where $\bigwedge T$ is the conjunction of all axioms of T. Using the Inversion Lemma 6.3 (9.) we get $\vdash \bigwedge T \Rightarrow F(\vec{a})$ and hence $T \vdash F(\vec{a})$. The induction hypothesis (since $F(\vec{a})$ is a shorter formula than G) thus yields $T \vdash^i F(\vec{a})$ from which $T \vdash^i G$ follows via (several) $\forall R$ inferences.

Now suppose that G is of the form $\bigwedge \Phi$, where Φ is a set of ∞-geometric formulae. By $\bigwedge$-inversion on the right we get

$$\vdash \bigwedge T \Rightarrow H$$

for all $H \in \Phi$, and thus inductively we have

$$I_{\infty} \vdash \bigwedge T \Rightarrow H$$

for all $H \in \Phi$, so that via $(\bigwedge R)$ we arrive at $I_{\infty} \vdash \bigwedge T \Rightarrow G$, thus $T \vdash^i G$.

If G is of the form $\neg G_0$ with G_0 ∞-geometric we apply the Inversion Lemma 6.3 (7.) to get

$$\vdash \bigwedge T, G_0 \Rightarrow .$$

By Lemma 6.5 we infer that $I_{\infty} \vdash \bigwedge T, G_0 \Rightarrow$ and thus, by $\neg R$, we have

$$I_{\infty} \vdash \bigwedge T \Rightarrow \neg G_0 ,$$

thus $T \vdash^i G$.

If F is of the form $F_0 \to F_1$ with F_i geometric formulae we apply the Inversion Lemma 6.3 (5.) to get

$$\vdash \bigwedge T, F_0 \Rightarrow F_1 .$$

By Lemma 6.5 we infer that $I_{\infty} \vdash \bigwedge T, F_0 \Rightarrow F_1$. By employing $\to R$ we get $I_{\infty} \vdash \bigwedge T \Rightarrow F_0 \to F_1$ and hence $T \vdash^i G$.

□

7 Constructive cut elimination for $\mathcal{L}_{\infty\omega}$

The usual cut elimination proof for $\mathcal{L}_{\infty\omega}$ uses the Veblen functions (see [Veblen, 1908]) φ_α in order to measure the "cost" of cut elimination. In a constructive setting, however, one looses the linearity of ordinals as well as the principle that every inhabited set of ordinals has a least element. As a result, the definition of analogs of the φ_α functions has to be carried out in a different way. A central gadget of cut elimination in infinitary systems is the "natural" commutative sum of ordinals $\alpha\#\beta$. Its definition utilizes the Cantor normal form of ordinals to base ω. This normal form is not available in CZF (or **IZF**) and thus a different approach is called for. We shall have use for the following induction and recursion principle on ordinals, henceforth referred to as $\lhd$-*induction* and $\lhd$-*recursion*.

Lemma: 7.1. *Define* $(\alpha, \beta) \lhd (\alpha', \beta')$ *by*

$$\alpha = \alpha' \wedge \beta \in \beta' \text{ or } \alpha \in \alpha' \wedge \beta = \beta' \text{ or } \alpha \in \alpha' \wedge \beta \in \beta'.$$

1. (CZF) $\forall\alpha\forall\beta\,[\forall\gamma\forall\delta((\gamma, \delta) \lhd (\alpha, \beta) \to F(\gamma, \delta)) \to F(\alpha, \beta)] \to \forall\alpha\forall\beta\, F(\alpha, \beta)$.
2. (CZF) *If* G *is a total* $(n + 3$*-ary class function* $G : V^n \times \mathrm{ON} \times \mathrm{ON} \times V \to \mathrm{ON}$ *then there is a (unique) (n+2)-ary class function* $F : V^n \times \mathrm{ON} \times \mathrm{ON} \to \mathrm{ON}$ *such that*

$$F(\vec{x}, \alpha, \beta) = G(\vec{x}, \alpha, \beta, \{\langle\gamma, \delta, F(\vec{x}, \gamma, \delta)\rangle \mid (\gamma, \delta) \lhd (\alpha, \beta)\}).$$

Proof: (1.) Assume

$$\forall\alpha\forall\beta\,[\forall\gamma\forall\delta((\gamma, \delta) \lhd (\alpha, \beta) \to F(\gamma, \delta)) \to F(\alpha, \beta)]. \tag{3}$$

Fix an arbitrary ordinal ρ. We show

$$\forall\xi \in \rho\, F(\alpha, \xi) \tag{4}$$

by induction on $\alpha \in \rho$. So the inductive assumption gives $\forall\alpha_0 \in \alpha\, \forall\xi \in \rho\, F(\alpha, \xi)$. We then use use a further subsidiary induction on $\beta \in \rho$ to show $F(\alpha, \beta)$. By (3) it suffices to show

$$\forall\gamma\forall\delta[(\gamma, \delta) \lhd (\alpha, \beta) \to F(\gamma, \delta)]. \tag{5}$$

So suppose that $(\gamma, \delta) \lhd (\alpha, \beta)$.
Case 1: $\gamma = \alpha$ and $\delta \in \beta$. $F(\gamma, \delta)$ follows by the subsidiary induction hypothesis.
Case 2: $\gamma \in \alpha$ and $\delta = \beta$. $F(\gamma, \delta)$ follows by the main induction hypothesis.
Case 3: $\gamma \in \alpha$ and $\delta \in \beta$. $F(\gamma, \delta)$ also follows by the main induction hypothesis.

Thus we have shown (5). This establishes (4). Since ρ was arbitrary it follows that $F(\alpha, \beta)$ holds for all α, β.[12]

(2.) Noting that $\{(\gamma, \delta) \mid (\gamma, \delta) \lhd (\alpha, \beta)\}$ is a set, (2.) follows from (1.) in the same manner as ordinary $\in$-recursion follows from $\in$–induction. For more details see [Aczel and Rathjen:, 2010]. □

Definition: 7.2. *For a class X, let $X^{\cup} := X \cup \{u \mid \exists y \in X\, u \in y\}$.*

Define $\alpha\#\beta$ by $\lhd$-recursion as follows:

$$\begin{aligned} \alpha\#\beta &= \{\gamma\#\delta \mid (\gamma, \delta) \lhd (\alpha, \beta)\}^{\cup} \\ &= \{\alpha\#\delta \mid \delta \in \beta\}^{\cup} \cup \{\eta\#\beta \mid \eta \in \alpha\}^{\cup} \cup \{\eta\#\delta \mid \eta \in \alpha \wedge \eta \in \beta\}^{\cup}. \end{aligned} \tag{6}$$

Lemma: 7.3.

1. *If X is a set of ordinals then $X^{\cup}$ is an ordinal.*
2. *$\alpha\#\beta$ is an ordinal and $\alpha\#\beta = \beta\#\alpha$.*
3. *If $(\gamma, \delta) \lhd (\alpha, \beta)$, then $\gamma\#\delta \in \alpha\#\beta$.*

Proof: (1.) Let X be a set of ordinals. Then $X^{\cup}$ is also a set of ordinals. It remains to show that $X^{\cup}$ is transitive. Suppose $\alpha \in \beta \in X^{\cup}$. Then $\beta \in X$ or $\beta \in \delta$ for some $\delta \in X$. In the first case we have $\alpha \in \bigcup X \subseteq X^{\cup}$. In the second case we infer that $\alpha \in \delta$ since δ is an ordinal, thus $\alpha \in \bigcup X \subseteq X^{\cup}$.

(2.) follows by $\lhd$-induction (also using (1.)).

(3.) is obvious by definition of $\alpha\#\beta$. □

Definition: 7.4. *Let I be a set and $(f_i)_{i \in I}$ be a definable collection of functions*

$$f_i : \mathrm{ON}^{a(i)} \to \mathrm{ON}$$

with arity $a(i) \in \mathbb{N}$. Let X be a set of ordinals. Then the closure of X under $(f_i)_{i \in I}$, $\mathrm{Cl}(X, (f_i)_{i \in I})$, *is defined as follows:*

$$\begin{aligned} X_0 &= X^{\cup} \cup \{0\} \\ X_{n+1} &= X_n \cup \{f_i(\alpha_1, \ldots, \alpha_{a(i)}) \mid \alpha_1, \ldots \alpha_{a(i)} \in X_n\}^{\cup} \\ \mathrm{Cl}(X, (f_i)_{i \in I}) &= \bigcup_{n \in \mathbb{N}} X_n. \end{aligned}$$

Lemma: 7.5. *Under same assumptions as in the foregoing definition, $\mathrm{Cl}(X, (f_i)_{i \in I})$ is an ordinal which contains 0 and all elements of X. Moreover, $\mathrm{Cl}(X, (f_i)_{i \in I})$ is closed under $(f_i)_{i \in I}$, i.e., if $\vec{\alpha} \in \mathrm{Cl}(X, (f_i)_{i \in I})$ then $f_i(\vec{\alpha}) \in \mathrm{Cl}(X, (f_i)_{i \in I})$ for all $i \in I$.*

---

12 The reason for restricting the quantifier in (4) to ρ is that it shows that $\lhd$-induction with $F(\alpha, \beta)$ follows from $\in$-induction using a formula having no more unbounded quantifiers than F. Of course, this is not essential to the current paper.

Proof: Induction on n shows that $X \subseteq X_0 \subseteq X_1 \subseteq \ldots \subseteq X_n$ and all X_n are ordinals. Hence $\mathrm{Cl}(X, (f_i)_{i \in I})$ is an ordinal. If $\alpha_1, \ldots, \alpha_{a(i)} \in \mathrm{Cl}(X, (f_i)_{i \in I})$, then $\alpha_1, \ldots, \alpha_{a(i)} \in X_n$ for some n since $a(i) \in \mathbb{N}$, and hence $f_i(\alpha_1, \ldots, \alpha_{a(i)}) \in X_{n+1} \subseteq \mathrm{Cl}(X, (f_i)_{i \in I})$. □

Definition: 7.6. *By main recursion on α and subsidiary recursion on β we define the functions $\varphi_\alpha : \mathrm{ON} \to \mathrm{ON}$ by letting $\varphi_\alpha(\beta)$ be the closure of*

$$\{\varphi_\alpha(\xi) \mid \xi \in \beta\}$$

under the functions # and $(\varphi_\eta)_{\eta \in \alpha}$.

Lemma: 7.7. (CZF) *For all α, β, $\varphi_\alpha(\beta)$ exists. Also $0 \in \varphi_\alpha(\beta)$.*

1. *If $\delta, \xi \in \varphi_\alpha(\beta)$ then $\varphi_\delta(\xi) \in \varphi_\alpha(\beta)$.*
2. *If $\delta \in \beta$ then $\varphi_\alpha(\delta) \in \varphi_\alpha(\beta)$.*
3. *If $\delta, \xi \in \varphi_\alpha(\beta)$, then $\delta \# \xi \in \varphi_\alpha(\beta)$.*

Proof: The existence of $\varphi_\alpha(\beta)$ follows by main induction on α and subsidiary induction on β, using Lemma 7.5. (1.), (2.), and (3.) are immediate by the closure properties of $\varphi_\alpha(\beta)$. □

Lemma: 7.8 (Reduction).
Suppose $\rho = rank(C)$. If $\vdash^{\alpha}_{\rho} \Gamma, C \Rightarrow \Delta$ and $\vdash^{\beta}_{\rho} \Xi \Rightarrow \Theta, C$, then

$$\vdash^{\alpha\#\alpha\#\beta\#\beta}_{\rho} \Gamma, \Xi \Rightarrow \Delta, \Theta\,.$$

Proof: The proof is by induction on $\alpha\#\alpha\#\beta\#\beta$. We only look at two cases where C was the principal formula of the last inference in both derivations.
Case 1. The first is when C is of the form $\bigwedge \Phi$. Then we have

$$\vdash^{\alpha_1}_{\rho} \Gamma, C, A_0 \Rightarrow \Delta$$

and

$$\vdash^{\beta_A}_{\rho} \Xi \Rightarrow \Theta, C, A$$

for some $\alpha_1 < \alpha$ and $A_0 \in \Phi$ as well as $\beta_A < \beta$ for all $A \in \Phi$. By the induction hypothesis we obtain

$$\vdash^{\alpha_1\#\alpha_1\#\beta\#\beta}_{\rho} \Gamma, \Xi, A_0 \Rightarrow \Delta, \Theta$$

and

$$\vdash^{\alpha\#\alpha\#\beta_A\#\beta_A}_{\rho} \Gamma, \Xi \Rightarrow \Delta, \Theta, A_0\,.$$

As A_0 is a subformula of C we have $\mathrm{rank}(A_0) \in \rho$. Cutting out A_0 thus gives

$$\vdash^{\alpha\#\alpha\#\beta\#\beta}_{\rho} \Gamma, \Xi \Rightarrow \Delta, \Theta\,.$$

Case 2. The second case is when C is of the form $\forall x\, A(x)$ Then we have

$$\vdash^{\alpha_1}_{\rho} \Gamma, C, A(t) \Rightarrow \Delta$$

and

$$\vdash^{\beta_0}_{\rho} \Xi \Rightarrow \Theta, C, A(a)$$

for some $\alpha_1 < \alpha$, term t as well as $\beta_0 < \beta$ for some eigenvariable a. By Lemma 6.2 we have

$$\vdash^{\beta_0}_{\rho} \Xi \Rightarrow \Theta, C, A(t)\,.$$

By the induction hypothesis we thus get

$$\vdash^{\alpha_1\#\alpha_1\#\beta\#\beta}_{\rho} \Gamma, \Xi, A(t) \Rightarrow \Delta, \Theta$$

and

$$\vdash^{\alpha\#\alpha\#\beta_0\#\beta_0}_{\rho} \Gamma, \Xi \Rightarrow \Delta, \Theta, A(t)\,.$$

Observing that $\mathrm{rank}(A(t)) \in \mathrm{rank}(\forall x\, A(x)) = \rho$, cutting out $A(t)$ gives

$$\vdash^{\alpha\#\alpha\#\beta\#\beta}_{\rho} \Gamma, \Xi \Rightarrow \Delta, \Theta\,.$$

□

Theorem: 7.9 (Cut Elimination Theorem).
If $\vdash^{\alpha}_{\rho} \Gamma \Rightarrow \Delta$ *then* $\vdash^{\varphi_\rho(\alpha)}_{0} \Gamma \Rightarrow \Delta\,.$

Proof: We use induction on ρ with a subsidiary induction on α.

If $\Gamma \Rightarrow \Delta$ is an axiom then we clearly get the desired result. So let's assume that $\Gamma \Rightarrow \Delta$ is not an axiom. Then we have a last inference $(\mathfrak{I})$ with premisses $\Gamma_i \Rightarrow \Delta_i$. Suppose the inference was not a cut. We then have $\vdash^{\alpha_i}_{\rho} \Gamma_i \Rightarrow \Delta_i$ for some $\alpha_i < \alpha$. By the subsidiary induction hypothesis we obtain $\vdash^{\varphi_\rho(\alpha_i)}_{0} \Gamma_i \Rightarrow \Delta_i\,.$ Applying the same inference $(\mathfrak{I})$ yields $\vdash^{\varphi_\nu(\alpha)}_{\rho} \Gamma \Rightarrow \Delta\,.$

Now suppose the last inference was a cut with cut formula C. Then $\mathrm{rank}(C) \in \rho$ and there exist derivations $\vdash^{\alpha_0}_{\rho} \Gamma_1, C \Rightarrow \Delta_1$ and $\vdash^{\alpha_1}_{\rho} \Gamma_2 \Rightarrow \Delta_2, C$ for some $\alpha_0, \alpha_1 \in \alpha$ such that $\Gamma_1, \Gamma_2 \subseteq \Gamma$ and $\Delta_1, \Delta_2 \subseteq \Delta$. By the subsidiary induction hypothesis we conclude that $\vdash^{\varphi_\rho(\alpha_0)}_{0} \Gamma_1, C \Rightarrow \Delta_1$ and $\vdash^{\varphi_\rho(\alpha_1)}_{0} \Gamma_2 \Rightarrow \Delta_2, C\,.$ By the Reduction Lemma 7.8 we can infer

$$\vdash^{\varphi_\rho(\alpha_0)\#\varphi_\rho(\alpha_0)\#\varphi_\rho(\alpha_1)\#\varphi_\rho(\alpha_1)}_{\nu} \Gamma \Rightarrow \Delta$$

where $\nu = \mathrm{rank}(C)$. Since $\nu \in \rho$ we can now employ the main induction hypothesis, yielding

$$\vdash^{\varphi_\nu(\varphi_\rho(\alpha_0)\#\varphi_\rho(\alpha_0)\#\varphi_\rho(\alpha_1)\#\varphi_\rho(\alpha_1))}_{0} \Gamma, C \Rightarrow \Delta\,.$$

Since $\varphi_v(\varphi_\rho(\alpha_0)\#\varphi_\rho(\alpha_0)\#\varphi_\rho(\alpha_1)\#\varphi_\rho(\alpha_1)) \in \varphi_\rho(\alpha)$ we arrive at

$$\left|\frac{\varphi_\rho(\alpha)}{0}\right. \Gamma \Rightarrow \Delta\,.$$

□

Acknowledgment: Part of the material is based upon research supported by the EPSRC of the UK through grant No. EP/K023128/1. This research was also supported by a Leverhulme Research Fellowship and a Marie Curie International Research Staff Exchange Scheme Fellowship within the 7th European Community Framework Programme. This publication was made possible through the support of a grant from the John Templeton Foundation. The opinions expressed in this publication are those of the author and do not necessarily reflect the views of the John Templeton Foundation.

Thanks are owed to the anonymous referee of this paper for valuable comments that helped to improve the paper.

References

P. Aczel and M. Rathjen. Notes on constructive set theory. Technical Report 40, Institut Mittag-Leffler, The Royal Swedish Academy of Sciences, Stockholm, 2001.

P. Aczel and M. Rathjen:. *Constructive set theory.* book draft, 2010.

J. Barwise. *Admissible Sets and Structures.* Springer, Berlin, 1975.

J. Barwise. An introduction to first-order logic. In *Handbook of Mathematical Logic*, pages 5–46. North-Holland, Amsterdam, 1977.

J. R. Bell. *Boolean-valued models and independence proofs in set theory.* Clarendon Press, Oxford, 1977.

C. C. Chang and H. R. Keisler. *Model Theory.* North-Holland, Amsterdam, 3 edition, 1990.

W. Hodges. *Model theory.* Cambridge University Press, 1993.

T. Jech. *Set Theory.* Springer, Berlin, 3 edition, 2003.

P. Johnstone. *Sketches of an elephant: A topos theory compendium*, volume 2. Clarendon Press, Oxford, 2002.

J. Keisler. Fundamentals of model theory. In J. Barwise, editor, *Handbook of Mathematical Logic.* North-Holland, Amsterdam, 1977.

K. Kunen. *Set theory: An introduction to independence proofs.* North-Holland, Amsterdam, 1980.

J. Lambek and P. J. Scott. *Introduction to Higher Order Categorical Logic.* Cambridge University Press, 1988.

S. Mac Lane and I. Moerdijk. *Sheaves in Geometry and Logic.* Springer, Berlin, 1992.

A. Levy. *Basic set theory.* Springer, Berlin, 1979.

G. Mints. Classical and intuitionistic geometric logic. In *Philosophy, Mathematics, Linguistics: Aspects of Interaction 2012.* Talk at Conference, May 2012.

G. H. Moore. The prehistory of infinitary logic: 1885 – 1955. In Mundici Dalla Chiara, Doets and Benthem, editors, *Structures and norms in science*. Kluwer, Dordrecht, 1997.

V. P. Orevkov. Glivenko's sequence classes. In V. P. Orevkov, editor, *Logical-mathematical calculus. Part 1*, pages 131–154. Inst. Steklov, Leningrad, 1968.

M. Rathjen. *Proof Theory*. Lecture notes for a postgraduate course, Leeds, 2012.

G. Sacks. *Saturated model theory*. Benjamin, 1972.

W. W. Tait. *Normal derivability in classical logic*. Springer, Berlin, 1968.

O. Veblen. Continuous increasing functions of finite and transfinite ordinals. *Trans. Amer. Math. Soc.*, 9:280–292, 1908.